Studies in Systems, Decision and Control

Volume 304

The series "Studies in Systems, Decision and Control" (SSDC) covers both new developments and advances, as well as the state of the art, in the various areas of broadly perceived systems, decision making and control–quickly, up to date and with a high quality. The intent is to cover the theory, applications, and perspectives on the state of the art and future developments relevant to systems, decision making, control, complex processes and related areas, as embedded in the fields of engineering, computer science, physics, economics, social and life sciences, as well as the paradigms and methodologies behind them. The series contains monographs, textbooks, lecture notes and edited volumes in systems, decision making and control spanning the areas of Cyber-Physical Systems, Autonomous Systems, Sensor Networks, Control Systems, Energy Systems, Automotive Systems, Biological Systems, Vehicular Networking and Connected Vehicles, Aerospace Systems, Automation, Manufacturing, Smart Grids, Nonlinear Systems, Power Systems, Robotics, Social Systems, Economic Systems and other. Of particular value to both the contributors and the readership are the short publication timeframe and the world-wide distribution and exposure which enable both a wide and rapid dissemination of research output.

** Indexing: The books of this series are submitted to ISI, SCOPUS, DBLP, Ulrichs, MathSciNet, Current Mathematical Publications, Mathematical Reviews, Zentralblatt Math: MetaPress and Springerlink.

More information about this series at http://www.springer.com/series/13304

Oliver Junge · Oliver Schütze ·
Gary Froyland · Sina Ober-Blöbaum ·
Kathrin Padberg-Gehle
Editors

Advances in Dynamics, Optimization and Computation

A volume dedicated to Michael Dellnitz on the occasion of his 60th birthday

Springer

Editors
Oliver Junge
Department of Mathematics
Technical University of Munich
Munich, Germany

Oliver Schütze
Computer Science Department
Cinvestav-IPN
Mexico City, Distrito Federal, Mexico

Gary Froyland
School of Mathematics and Statistics
University of New South Wales
Sydney, NSW, Australia

Sina Ober-Blöbaum
Department of Mathematics
Paderborn University
Paderborn, Germany

Kathrin Padberg-Gehle
Institute of Mathematics and its Didactics
Leuphana University Lüneburg
Lüneburg, Germany

ISSN 2198-4182 ISSN 2198-4190 (electronic)
Studies in Systems, Decision and Control
ISBN 978-3-030-51263-7 ISBN 978-3-030-51264-4 (eBook)
https://doi.org/10.1007/978-3-030-51264-4

This Springer imprint is published by the registered company Springer Nature Switzerland AG
The registered company address is: Gewerbestrasse 11, 6330 Cham, Switzerland

Für Michael

Preface

This book is a collection of recent advances on problems in dynamical systems, optimal control and optimization. In many cases, computational aspects and techniques are central. We dedicate this volume to Michael Dellnitz on the occasion of his 60th birthday. In one way or the other, there is a connection to Michael's research in all of these contributions.

In Part I, we collect chapters related to problems in dynamical systems. We start with a new technique for computing highly degenerate periodic orbits in ordinary differential equations, so-called phase resetting curves which appear, e.g., in models of spiking neurons. The second contribution employs concepts from singularity theory in order to classify and compute homeostasis points, i.e., points in state space of, e.g., biochemical networks, in which some output variable is roughly constant while some input variable is changing. We continue with a review on recent developments on the set-oriented approximation of invariant sets, a technique that has been pioneered by Michael and which will reappear in several other chapters in this volume. In the following paper, this technique is employed to approximate the transfer operator in non-autonomous differential equations, enabling the computation of coherent behavior in otherwise turbulent fluid flows. In high-dimensional systems with high-dimensional invariant sets, other concepts for an approximation of the transfer operator have to be found—and this is the subject of the next chapter where empirical bases are employed to construct a finite-rank approximation of this operator. Eigenfunctions of the (approximate) transfer operator can be used in order to detect, e.g., rare events like transitions between almost–invariant, respectively, metastable subsets in state space—an observation which already appears in one of the earlier works of Michael from the late 1990s. This is built upon in the subsequent chapter where a new, weaker characterization of slowly changing coordinates in noisy dynamical systems is proposed. Another way to address the reliable detection of rare events is based on sampling techniques like importance sampling and this addressed in the next chapter, where several sampling algorithms are compared and validated. We close the first part of this book by a chapter which demonstrates the usefulness of concepts from dynamical systems for solving questions on the computational complexity of certain problems.

Part II is dedicated to optimal control problems. In the first contribution, symmetric optimal control problems are investigated. Lie group symmetries and associated motion primitives of mechanical systems are exploited to develop numerical methods for multi-objective model predictive optimal control problems. In the second contribution, we review the numerical treatment of a mixed-integer optimal control problem governed by linear convection–diffusion equations and binary control variables. Relaxation and sum-up rounding techniques are combined with model order reduction to make the numerical approximation computationally more efficient. In the third contribution, we review set-oriented methods for the construction of globally optimal controllers. Based on a discrete version of Bellman's optimality principle applied to a dynamic game, a discrete feedback is constructed which robustly stabilizes a given nonlinear control system. In the last contribution, we review and highlight some connections between the problem of nonlinear smoothing and optimal control problems involving control of probability densities.

Finally, in Part III, we present three contributions related to optimization. The first contribution deals with the occurrence of "dents" in Pareto fronts of continuous multi-objective optimization problems. This can be helpful to obtain information about the structure of the Pareto front without explicitly computing the entire Pareto set. The second contribution deals with equality constrained bi-level multi-objective optimization problems and proposes a novel set-oriented algorithm that aims for a well-distributed finite-size approximation of the Pareto front of the higher-level problem. The third contribution reviews the gradient subspace approximation which allows one to compute descent directions in a best-fit manner from given neighborhood information. The method works particularly well in combination with set-oriented searchers such as evolutionary algorithms.

April 2020

Oliver Junge
Oliver Schütze
Gary Froyland
Sina Ober-Blöbaum
Kathrin Padberg-Gehle

Acknowledgements

The editors are grateful to the authors for their contributions to this volume recognizing Michael's scientific achievements. All editors have been either graduate students or postdocs with Michael and would like to express their appreciation for Michael's mentoring and their collaborative endeavors, both within and beyond Michael's research group.

O. Schütze acknowledges support from Conacyt Basic Science project no. 285599 and SEP-Cinvestav project no. 231.

Contents

Optimal Control

Optimization

List of Contributors

Fernando Antoneli Escola Paulista de Medicina, Universidade Federal de São Paulo, São Paulo, SP, Brazil

Andreas Bittracher Department of Mathematics and Computer Science, Freie Universität Berlin, Berlin, Germany

Meagan Carney Max Planck Institute for the Physics of Complex Systems, Dresden, Germany

Michael Dellnitz Chair of Applied Mathematics, Paderborn University, Paderborn, Germany

Alessandro Dell'Aere Chair of Applied Mathematics, Institute for Mathematics University of Paderborn, Paderborn, Germany

Kathrin Flaßkamp Systems Modeling and Simulation, Saarland University, Saarbrücken, Germany

Gary Froyland School of Mathematics and Statistics, University of New South Wales, Sydney, NSW, Australia

Raphael Gerlach Paderborn University, Paderborn, Germany

Martin Golubitsky Mathematics Department, The Ohio State University, Columbus, OH, USA

Lars Grüne Mathematical Institute, University of Bayreuth, Bayreuth, Germany

Zhengyuan Huang The Ohio State University, Columbus, OH, USA

Oliver Junge Department of Mathematics, Technical University of Munich, Munich, Germany

Christian Jäkle Department of Mathematics and Statistics, University of Konstanz, Konstanz, Germany

Holger Kantz Max Planck Institute for the Physics of Complex Systems, Dresden, Germany

Jin Won Kim Coordinated Science Laboratory and Department of Mechanical Science and Engineering, University of Illinois at Urbana-Champaign, Urbana, IL, USA

Stefan Klus Department of Mathematics and Computer Science, Freie Universität Berlin, Berlin, Germany

Anna Klünker Institute of Mathematics and its Didactics, Leuphana Universität Lüneburg, Lüneburg, Germany

Bernd Krauskopf Department of Mathematics, The University of Auckland, Auckland, New Zealand

Adriana Lara ESFM del Instituto Politécnico Nacional, Mexico City, Mexico

Peter Langfield IHU Liryc, Electrophysiology and Heart Modeling Institute, Fondation Bordeaux Université, Pessac - Bordeaux, France; Institut de Mathématiques de Bordeaux UMR 5251, Université de Bordeaux, Talence, France

Prashant G. Mehta Coordinated Science Laboratory and Department of Mechanical Science and Engineering, University of Illinois at Urbana-Champaign, Urbana, IL, USA

Mattes Mollenhauer Department of Mathematics and Computer Science, Freie Universität Berlin, Berlin, Germany

Mirko Hessel-von Molo Faculty of Computer Science Fachhochschule Dortmund – University of Applied Sciences and Arts, Dortmund, Germany

Matthew Nicol Department of Mathematics, University of Houston, Houston, TX, USA

Sina Ober-Blöbaum Department of Mathematics, Paderborn University, Paderborn, Germany

Hinke M. Osinga Department of Mathematics, The University of Auckland, Auckland, New Zealand

Kathrin Padberg-Gehle Institute of Mathematics and its Didactics, Leuphana Universität Lüneburg, Lüneburg, Germany

Sebastian Peitz Department of Mathematics, Paderborn University, Paderborn, Germany

Tuhin Sahai Raytheon Technologies Research Center, Berkeley, CA, USA

Christiane Schneide Institute of Mathematics and its Didactics, Leuphana Universität Lüneburg, Lüneburg, Germany

Jörg Schumacher Department of Mechanical Engineering, Technische Universität Ilmenau, Ilmenau, Germany

Ingmar Schuster Zalando Research, Zalando SE, Berlin, Germany

Christof Schütte Department of Mathematics and Computer Science, Freie Universität Berlin, Berlin, Germany; Zuse Institute Berlin, Berlin, Germany

Oliver Schütze Computer Science Department, Cinvestav-IPN, Mexico City, Mexico

Ian Stewart Mathematics Institute, University of Warwick, Coventry, UK

Lourdes Uribe ESFM del Instituto Politécnico Nacional, Mexico City, Mexico

Stefan Volkwein Department of Mathematics and Statistics, University of Konstanz, Konstanz, Germany

Yangyang Wang Department of Mathematics, The University of Iowa, Iowa City, IA, USA

Katrin Witting dSPACE GmbH, Paderborn, Germany

Adrian Ziessler Paderborn University, Paderborn, Germany

Dynamics

A Continuation Approach to Computing Phase Resetting Curves

Peter Langfield[1,2], Bernd Krauskopf[3], and Hinke M. Osinga[3](✉)

[1] IHU Liryc, Electrophysiology and Heart Modeling Institute, Fondation Bordeaux Université, 33600 Pessac - Bordeaux, France
peter.langfield@u-bordeaux.fr

[2] Institut de Mathématiques de Bordeaux UMR 5251, Université de Bordeaux, 33400 Talence, France

[3] Department of Mathematics, The University of Auckland, Private Bag 92019, Auckland 1142, New Zealand
{b.krauskopf,h.m.osinga}@auckland.ac.nz

Abstract. Phase resetting is a common experimental approach to investigating the behaviour of oscillating neurons. Assuming repeated spiking or bursting, a phase reset amounts to a brief perturbation that causes a shift in the phase of this periodic motion. The observed effects not only depend on the strength of the perturbation, but also on the phase at which it is applied. The relationship between the change in phase after the perturbation and the unperturbed old phase, the so-called phase resetting curve, provides information about the type of neuronal behaviour, although not all effects of the nature of the perturbation are well understood. In this chapter, we present a numerical method based on the continuation of a multi-segment boundary value problem that computes phase resetting curves in ODE models. Our method is able to deal effectively with phase sensitivity of a system, meaning that it is able to handle extreme variations in the phase resetting curve, including resets that are seemingly discontinuous. We illustrate the algorithm with two examples of planar systems, where we also demonstrate how qualitative changes of a phase resetting curve can be characterised and understood. A seven-dimensional example emphasises that our method is not restricted to planar systems, and illustrates how we can also deal with non-instantaneous, time-varying perturbations.

1 Introduction

Measuring phase resetting is a common approach for testing neuronal responses in experiments: a brief current injection perturbs the regular spiking behaviour of a neuron, resulting generally in a shifted phase as the neuron returns to its regular oscillating behaviour. This phase shift can be advanced or delayed—meaning that the next spike arrives earlier or later compared with the unperturbed spiking oscillation—and which effect occurs also depends on the moment when the

O. Junge et al. (Eds.): SON 2020, SSDC 304, pp. 3–30, 2020.
https://doi.org/10.1007/978-3-030-51264-4_1

current is applied; see [9] for more details. A plot of the shifted phase ϑ_{new} versus the original phase ϑ_{old} at which the current was applied is known as the *phase transition curve* (PTC). Experimentally, it is often easier to represent the reset in terms of the resulting phase difference $\vartheta_{\text{new}} - \vartheta_{\text{old}}$ as a function of ϑ_{old}, which can be measured as the time to the next spike; such a representation is called a *phase response curve* or *phase resetting curve* (PRC).

The shape of a PTC or PRC of a given system obviously depends on the size of the applied perturbation: already for quite small amplitudes, nonlinear effects can dramatically affect a PTC or PRC. The shape of the PTC or PRC has been used to classify neuronal behaviour [1,7,16], where the underlying assumption is that the size of the applied perturbation is sufficiently small. Hodgkin [17] distinguished between so-called Type-I and Type-II excitable membranes, where neurons with membranes of Type II are not able to fire at arbitrarily low frequencies. Note that transitions from Type-I to Type-II can occur when system parameters are changed [8]. Ermentrout [7] found that the PRC of a Type-I neuron always has the same sign, while that of a Type-II neuron changes sign; this means that the PTC is always entirely above or below the diagonal for Type-I neurons, while it intersects the diagonal for Type-II neurons. In either case, the PTC is invertible for sufficiently small perturbation amplitudes, since it can be viewed as a continuous and smooth deformation of the identity, which is the PTC in the limit of zero amplitude. Invertibility itself has also been used as a distinguishing property of PTCs: noninvertible PTCs are said to be of type 0 (or strong) and invertible PTCs are of type 1 (or weak) [9,37]. If an increasingly stronger perturbation is applied, for example, in the context of synchronisation, it is well known that PTCs can change from type 1 to type 0, that is, become noninvertible [11,37].

A motivation in recent work on phase resetting has been the idea of interpreting the PTC as defining a one-dimensional phase-reduction model that, hopefully, captures the essential dynamics of a possibly high-dimensional oscillating system. The main interest is in coupled systems, formed by two or more (planar) systems with known PRCs; for example, see [31,32] for mathematical as well as experimental perspectives. Unfortunately, the convergence back to the limit cycle after some perturbation can be quite slow for a coupled system, such that only (infinitesimally) small perturbations are accurately described. Furthermore, it makes physiological sense to assume a time-varying input, usually in the form of a short input pulse, rather than the instantaneous perturbation assumed for the theoretical phase reset. Moreover, the perturbation may be repeated at regular intervals. In this context, PTCs and PRCs can be useful for explaining the resetting behaviour, though strictly speaking, the theory is only valid at low firing rates [15,35]. More recently, the idea of a phase-amplitude description has led to a better understanding of the effects resulting from these kinds of repeated time-varying resets [2,3,26,30,34].

From a dynamical systems perspective, the key question of phase resetting is how the perturbed initial conditions relax back to an attracting periodic orbit Γ with period T_Γ of an underlying continuous-time model, which we take here to be a vector field on $\mathbb{R}^n$, that is, a system of n first-order autonomous ordinary differential equations. All points in its basin $\mathcal{B}(\Gamma)$ converge to Γ, and they do so with a given asymptotic phase. The subset of all points in $\mathcal{B}(\Gamma)$ that converge to Γ in phase with the point $\gamma_\vartheta \in \Gamma$, where $\vartheta \in [0, 1)$ by convention, is called the (forward-time) *isochron* of γ_ϑ, which we refer to as $I(\gamma_\vartheta)$. Isochrons were defined and named by Winfree [36]. Guckenheimer [12] showed that $I(\gamma_\vartheta)$ is, in fact, an $(n-1)$-dimensional invariant stable manifold of the attracting fixed point $\gamma_\vartheta \in \Gamma$ under the time-T_Γ map. In particular, it follows that $I(\gamma_\vartheta)$ is tangent to the attracting linear eigenspace of γ_ϑ and, hence, transverse to Γ. Moreover, the ϑ-dependent family of all isochrons $I(\gamma_\vartheta)$ foliates the basin $\mathcal{B}(\Gamma)$. In other words, any point in $\mathcal{B}(\Gamma)$ has a unique asymptotic phase determined by the isochron it lies on.

For a given ϑ_{old}, consider now the perturbed point $\gamma_{\vartheta_{\text{old}}} + A\,\mathbf{d} \in \mathcal{B}(\Gamma)$, obtained from $\gamma_{\vartheta_{\text{old}}} \in \Gamma$ by applying the perturbation of strength A in the given direction $\mathbf{d}$. The asymptotic phase ϑ_{new} is, hence, uniquely determined by the isochron $I(\gamma_{\vartheta_{\text{new}}})$ on which this point lies. This defines a circle map $P : [0, 1) \to [0, 1)$ with $P(\vartheta_{\text{old}}) = \vartheta_{\text{new}}$. Therefore, finding the PTC is equivalent to determining how the perturbed cycle $\Gamma + A\,\mathbf{d} = \{\gamma_{\vartheta_{\text{old}}} + A\,\mathbf{d} \mid \vartheta_{\text{old}} \in [0, 1)\}$ intersects the foliation of $\mathcal{B}(\Gamma)$ by the isochrons $I(\gamma_{\vartheta_{\text{new}}})$ for $\vartheta_{\text{new}} \in [0, 1)$. Notice further that the PTC is the graph of the circle map P on the unit torus $\mathbb{T}^2$, represented by the unit square $[0, 1) \times [0, 1)$.

When considering the amplitude A of the perturbation as a parameter (while keeping the direction $\mathbf{d}$ fixed throughout), one can deduce some important properties of the associated PTC. Suppose that $0 < A_{\max}$ is such that $\Gamma_A := \Gamma + A\,\mathbf{d} \subset \mathcal{B}(\Gamma)$ for all $0 \leq A < A_{\max}$. Then none of these perturbed cycles Γ_A intersects the boundary of the basin $\mathcal{B}(\Gamma)$ and the associated circle map $P = P_A$ is well defined for all $\vartheta_{\text{old}} \in [0, 1)$. The map P_0 for zero perturbation amplitude is the identity on $\mathbb{T}^2$, which means that, as its graph, the PTC is the diagonal on $[0, 1) \times [0, 1)$ and a 1:1 torus knot on $\mathbb{T}^2$; in particular, P_0 is invertible, that is, it is injective and surjective. Because of smooth dependence on the amplitude A and the fact that P_A is a function over $[0, 1)$, the PTC remains a 1:1 torus knot on $\mathbb{T}^2$ and P_A is surjective for all $0 \leq A < A_{\max}$.

Since the isochrons are transverse to Γ, the circle map P_A is C^1-close to the identity, and hence, also injective, for sufficiently small A. As the graph of a near-identity transformation, the PTC is then strictly monotone, invertible, and hence, of type 1 (or weak) in the notation of [9,37]. While surjectivity is preserved, injectivity may be lost before $A = A_{\max}$ is reached. Indeed, the PTC is either invertible for all $0 \leq A < A_{\max}$, or there is a maximal $0 < A_{\text{inv}} < A_{\max}$ such that P_A is invertible only for all $0 < A \leq A_{\text{inv}}$. The loss of injectivity of P_A at $A = A_{\text{inv}}$ happens generically because of the emergence of an inflection point. For $0 \leq A < A_{\max}$ this transition creates a local minimum and a local maximum of the PTC, which is now no longer invertible and so of type 0 (or strong) in the notation of [9,37]. As we will show, an inflection point of P_A corresponds to a cubic tangency between the perturbed cycle Γ_A and an isochron. Indeed, additional inflection points and, hence, local minima and maxima may appear at subsequent cubic isochron tangencies. Since P_A is a circle map, these must come in pairs; hence, counting the number of its local maxima (or minima) would provide a further refinement of the notation of a type 0 (or strong) PTC.

The above discussion shows that, when the applied perturbation A is sufficiently weak, it suffices to consider only the linear approximation to the isochron family, which is given by the ϑ-family of stable eigenspaces of the time-T_Γ map for each ϑ. In practice, nonlinear effects are essential, especially when multiple time scales are present or the phase reset involves relatively strong perturbations. Isochrons are often highly nonlinear objects of possibly very complicated geometry [21,36]. While the geometric idea of isochrons determining the phase resets has been around since the mid 1970s, the practical implementation has proven rather elusive. In practice, it is not at all straightforward to compute the isochrons of a periodic orbit. In planar systems, when such isochrons are curves, three different approaches have been proposed, based on Fourier averages [23,25], a parametrisation formulated in terms of a functional equation [14,18], and continuation of solutions to a suitably posed two-point boundary value problem [21,29]. In principle, all three approaches generalise to higher-dimensional isochrons, but there are only few explicit examples [14,25].

From the knowledge of the isochron foliation of $\mathcal{B}(\Gamma)$, one can immediately deduce geometrically the phase resetting for perturbations of any strength and in any direction. However, already for planar and certainly for higher-dimensional systems, this is effectively too much information when one is after the PTC resulting from a perturbation in a fixed direction and with a specific amplitude. In essence, finding a PTC or PRC remains the one-dimensional problem of finding the asymptotic phase of all points on the perturbed cycle.

In this chapter, we show how this can be achieved with a multi-segment boundary value problem formulation. Specifically, we adapt the approach from [21,22] to set up the calculation of the circle map P_A by continuation, first in A from $A = 0$ for fixed ϑ_{old}, and then in $\vartheta_{\text{old}} \in [0, 1)$ for fixed A. In this way, we obtain accurate numerical approximations of the PTC or PRC as continuous curves, even when the system shows strong phase sensitivity. The set-up is extremely versatile, and the direct computation of a PTC in this way does not require the system to be planar. We demonstrate our method with a constructed example going back to Winfree [37, Chapter 6], where we also show how injectivity is lost in a first cubic tangency of Γ_A with an isochron. The robustness of the method is then illustrated with the computation of a PTC of a perturbed cycle that cuts through a region of extreme phase sensitivity in the (planar) FithHugh–Nagumo system; in spite of very large derivatives due to this phase sensitivity, the PTC is computed accurately as a continuous curve. Our final example of a seven-dimensional system from [20] modelling a type of cardiac pacemaker cell shows that our approach also works in higher dimensions; this system also features phase sensitivity due to the existence of different time scales.

This chapter is organised as follows. In the next section, we provide precise details of the setting and explain the definitions used. Section 3 presents the numerical set-up for computing a resetting curve by continuation of a multi-segment boundary value problem. We then discuss two planar examples in depth, which are both taken from [22]: a variation of Winfree's model in Sect. 4 and the FitzHugh–Nagumo system in Sect. 5. The third and higher-dimensional example from [20] is presented in Sect. 6. A summary of the results is given in Sect. 7, where we also discuss some consequences of our findings and directions of future research.

2 Basic Setting and Definitions

As mentioned in the introduction, we consider a dynamical system with an attracting periodic orbit Γ. For simplicity, we assume that the state space is $\mathbb{R}^n$ and consider the dynamical system

$$\dot{\mathbf{x}} = \mathbf{F}(\mathbf{x}), \tag{1}$$

where $\mathbf{F} : \mathbb{R}^n \to \mathbb{R}^n$ is at least once continuously differentiable. We assume that system (1) has an attracting periodic orbit Γ with period $T_\Gamma > 0$, that is,

$$\Gamma := \{\gamma(t) \in \mathbb{R}^n \mid 0 \le t \le T_\Gamma \text{ with } \gamma(T_\Gamma) = \gamma(0)\},$$

and T_Γ is minimal with this property. We associate a phase $\vartheta \in [0, 1)$ with each point $\gamma_\vartheta \in \Gamma$, defining $\gamma_\vartheta := \gamma(t)$ with $t = \vartheta T_\Gamma$. Here $\gamma_0 := \gamma(0)$ needs to be chosen, which is usually done by fixing it to correspond to a maximum in the first component. The (forward-time) isochron $I(\gamma_\vartheta)$ associated with $\gamma_\vartheta \in \Gamma$ is then defined in terms of initial conditions $\mathbf{x}(0)$ of forward trajectories $\mathbf{x} := \{\mathbf{x}(t) \in \mathbb{R}^n \mid t \in \mathbb{R}\}$ of system (1) that accumulate on Γ, namely, as

$$I(\gamma_\vartheta) := \{\mathbf{x}(0) \in \mathbb{R}^n \mid \lim_{k\to\infty} \mathbf{x}(k\, T_\Gamma) = \gamma_\vartheta \text{ with } k \in \mathbb{N}\}.$$

In other words, the trajectory $\mathbf{x}$ approaches Γ in phase with γ_ϑ. Note that $I(\gamma_\vartheta)$ is the stable manifold of the fixed point γ_ϑ of the time-T_Γ return map; in particular, this means that $I(\gamma_\vartheta)$ is of dimension $n - 1$ and tangent at γ_ϑ to the stable eigenspace $E(\gamma_\vartheta)$, which is part of the stable Floquet bundle of Γ [12,13]; we utilise this property when computing isochrons, and also when computing a PTC or PRC.

We are now ready to give formal definitions of the PTC and PRC; see also [9].

Definition 1 (Phase Transition Curve)
The phase transition curve or PTC associated with a perturbation of amplitude $A \ge 0$ in the direction $\mathbf{d} \in \mathbb{R}^n$ is the graph of the map $P : [0, 1) \to [0, 1)$ defined as follows. For $\vartheta \in [0, 1)$, the image $P(\vartheta)$ is the phase φ associated with the isochron $I(\gamma_\varphi)$ that contains the point $\gamma_\vartheta + A\,\mathbf{d}$ for $\gamma_\vartheta \in \Gamma$.

Definition 2 (Phase Response Curve)
The phase response curve or PRC associated with a perturbation of amplitude $A \ge 0$ in the direction $\mathbf{d} \in \mathbb{R}^n$ is the graph of the phase difference $\Delta(\vartheta) = P(\vartheta) - \vartheta \,(\text{mod } 1)$, where the map P is as above.

The definitions of the PTC and PRC are based on knowledge of the (forward-time) isochron $I(\gamma_\varphi)$ associated with a point $\gamma_\varphi \in \Gamma$. We previously designed an algorithm based on continuation of a two-point boundary value problem (BVP) that computes one-dimensional (forward-time and backward-time) isochrons of a planar system up to arbitrarily large arclengths [21,22,29]. Here, we briefly describe this algorithm in its simplest form, because this is useful for understanding the basic set-up, and for introducing some notation. The description is presented in the style that is used for implementation in the software package AUTO [4,5]. In particular, we consider a time-rescaled version of the vector field (1), which represents an orbit segment $\{\mathbf{x}(t) \mid 0 \leq t \leq T\}$ of (1) as the orbit segment $\{\mathbf{u}(t) \mid 0 \leq t \leq 1\}$ of the vector field

$$\dot{\mathbf{u}} = T\,\mathbf{F}(\mathbf{u}), \tag{2}$$

so that the total integration time T is now a parameter of the system.

We approximate $I(\gamma_0)$ as the set of initial points of orbit segments that end on the linear space $E(\gamma_0)$, the linearised isochron of $I(\gamma_0)$, close to γ_0 after integer multiples of the period T_Γ. These points are formulated as initial points $\mathbf{u}(0)$ of orbit segments $\mathbf{u}$ that end on $E(\gamma_0)$ at a distance η from γ_0; hence, η defines a one-parameter family of orbit segments. Each orbit segment in this family is a solution of system (2) with $T = k\,T_\Gamma$ for $k \in \mathbb{N}$; the corresponding boundary conditions are:

$$[\mathbf{u}(1) - \gamma_0] \cdot \mathbf{v}_0^\perp = 0, \tag{3}$$

$$[\mathbf{u}(1) - \gamma_0] \cdot \mathbf{v}_0 = \eta, \tag{4}$$

where $\mathbf{v}_0$ is the normalised vector that spans $E(\gamma_0)$ and $\mathbf{v}_0^\perp$ is perpendicular to it. Note that Γ itself, when starting from γ_0, is a solution to the two-point BVP (2)–(4) with $T = T_\Gamma$ and $\eta = 0$. This gives us a first solution to start the continuation for computing $I(\gamma_0)$. We fix $T = T_\Gamma$ and continue the orbit segment $\mathbf{u}$ in η up to a maximum prespecified tolerance $\eta = \eta_{\max}$. As the end point $\mathbf{u}(1)$ is pushed away from γ_0 along $E(\gamma_0)$, the initial point $\mathbf{u}(0)$ traces out a portion of $I(\gamma_0)$.

Once we reach $\eta = \eta_{\max}$, we can extend $I(\gamma_0)$ further by considering points that map to $E(\gamma_0)$ after one additional period, that is, after time $T = 2\,T_\Gamma$. We start the continuation with the orbit segment formed by concatenation of the final orbit segment with Γ; here, we rescale time such that this first orbit is again defined for $0 \leq t \leq 1$, we set $T = 2\,T_\Gamma$, and $\eta = 0$. Note that this orbit segment has a discontinuity at $t = \frac{1}{2}$, but it is very small and AUTO will automatically correct and close it as part of the first continuation step. This correction will cause a small shift in η away from 0, but η will still be much smaller than $\eta_{\max}$ (in absolute value). We can keep extending $I(\gamma_0)$ further in this way, by continuation with $T = k\,T_\Gamma$, for integers $k > 2$. See [21,29] for more details on the implementation and, in particular, see [19,29] for details on how to find $E(\gamma_0)$ represented by the first vector $\mathbf{v}_0$ in the stable Floquet bundle of Γ.

The computational set-up forms a well-posed two-point BVP with a one-parameter solution family that can be found by continuation, provided the equality $\mathsf{NDIM} - \mathsf{NBC} + \mathsf{NPAR} = 1$ holds for the dimension NDIM of the problem, the number NBC of boundary conditions, and the number NPAR of free parameters. Indeed, for the computation of $I(\gamma_0)$, we have NDIM $= 2$, because we assumed that the system is planar; NBC $= 2$, namely, one condition to restrict $\mathbf{u}(1)$ to the linearised isochron of $I(\gamma_0)$, and one condition to fix its distance to γ_0; and NPAR $= 1$, because we free the parameter η.

To compute $I(\gamma_\varphi)$ for other $\varphi \in [0, 1)$, this same approach can be used, working with a shifted periodic orbit Γ so that its head point is γ_φ, and determining the associated direction vector $\mathbf{v}_\varphi$ that spans the eigenspace $E(\gamma_\varphi)$ to which $I(\gamma_\varphi)$ is tangent. In [29], approximations of γ_φ and $\mathbf{v}_\varphi$ are obtained by interpolation of the respective mesh discretisations from AUTO. We describe an alternative approach in [21], where we consider $I(\gamma_\varphi)$ as the set of initial points of orbit segments that end in the linear space $E(\gamma_0)$ of $I(\gamma_0)$ sufficiently close to γ_0 after total integration time $T = k\, T_\Gamma + (1 - \varphi)\, T_\Gamma$.

For the computation of a phase resetting curve, we use a combination of these two approaches, but rather than interpolation, we shift the periodic orbit by imposing a separate two-point BVP. More precisely, we set up a multi-segment BVP comprised of several subsystems of two-point BVPs; the set-up for this extended BVP is explained in detail in the next section.

3 Algorithm for Computing a Phase Resetting Curve

Based on the definition of PTC and PRC, one could now calculate a sufficiently large number of isochrons and determine the resetting curve numerically from data. We prefer to compute the PTC or PRC directly also with a BVP set-up and continuation. The major benefit of such a direct approach is that it avoids accuracy restrictions arising from the selection of computed isochrons; in particular, any phase sensitivity of the PTC or PRC will be dealt with automatically as part of the pseudo-arclength continuation with AUTO [4,5].

For ease of presentation, we will formulate and discuss our continuation set-up for the case of a planar system. We remark, however, that it can readily be extended for use in $\mathbb{R}^n$ with $n > 2$, because the dimensionality of the problem is not determined by the dimension $n - 1$ of the isochrons but by the dimension of the PTC or PRC, which is always one; see also the example in Sect. 6.

The essential difference between calculating a resetting curve rather than an isochron is the following: for an isochron $I(\gamma_\vartheta)$, we compute orbit segments with total integration time $T = T_\Gamma$ (or integer multiples), where we move the end point $\mathbf{u}(1)$ along the linear approximation of $I(\gamma_\vartheta)$ to some distance η from Γ, while the initial point $\mathbf{u}(0)$ traces out a new portion of $I(\gamma_\vartheta)$; imagining the same set-up, if we move $\mathbf{u}(0)$ transverse to $I(\gamma_\vartheta)$, the end point $\mathbf{u}(1)$ will move to lie on the linearisation of an isochron $I(\gamma_\varphi)$ with a different phase φ. (Here, one should expect that the distance to Γ also changes, but we assume it is still less than $\eta_{\max}$). The key idea behind our approach is that we find a way to determine the different phase φ, or the phase shift $\varphi - \vartheta$, by allowing Γ and its corresponding stable Floquet bundle to rotate as part of an extended system. We ensure the head point of Γ moves with the phase-shifted point, that is, the first point on Γ will be γ_φ. In this way, we can determine the shifted phase φ along any prescribed arc traced out by $\mathbf{u}(0)$, provided it lies in the basin of attraction of Γ. For the PTC or PRC associated with a perturbation of amplitude $A \geq 0$ in the direction $\mathbf{d} \in \mathbb{R}^n$, this prescribed arc is the perturbed cycle $\Gamma + A\,\mathbf{d}$, that is, $\mathbf{u}(0)$ traces out the closed curve $\{\gamma_\vartheta + A\,\mathbf{d} \mid \vartheta \in [0, 1)\}$.

3.1 Continuation Set-Up for Rotated Representation of Γ

We formulate an extended BVP that represents a rotated version of Γ with a particular phase, meaning that we automatically determine the phase of the head point relative to γ_0. To this end, we assume that the zero-phase point $\gamma_0 \in \Gamma$ and its associated linear vector $\mathbf{v}_0$, or more practical, its perpendicular $\mathbf{v}_0^\perp$, are readily accessible as stored parameters, or constants that do not change. Hence, even when Γ is rotated and its first point is γ_φ for some different $\varphi \in [0, 1)$, we can still access the coordinates of γ_0 and $\mathbf{v}_0^\perp$ from the parameter/constants list.

The extended BVP consists of three components, one to define Γ, one to define the associated (rotated) linear bundle, and one to define the associated phase. We start by representing Γ as a closed orbit segment $\mathbf{g}$ that solves system (2) for $T = T_\Gamma$. Hence, we define

$$\dot{\mathbf{g}} = T_\Gamma\,\mathbf{F}(\mathbf{g}), \tag{5}$$

with periodic boundary condition

$$\mathbf{g}(1) - \mathbf{g}(0) = 0. \tag{6}$$

The stable Floquet bundle of Γ is coupled with the BVP (5)–(6) via the first variational equation. More precisely, we consider a second orbit segment $\mathbf{v}_\mathrm{g}$, such that each point $\mathbf{v}_\mathrm{g}(t)$ represents a vector associated with points $\mathbf{g}(t)$ of the orbit segment that solves (5). The orbit segment $\mathbf{v}_\mathrm{g}$ is a solution to the linearised flow such that $\mathbf{v}_\mathrm{g}(0)$ is mapped to itself after one rotation around Γ. The length of $\mathbf{v}_\mathrm{g}(0)$ is contracted after one rotation by the factor $\exp(T_\Gamma\,\lambda_\mathrm{s})$, which is the stable Floquet multiplier of Γ. We prefer formulating this in logarithmic form, which introduces the stable Floquet exponent λ_s to the first variational equation, rather than affecting the length of $\mathbf{v}_\mathrm{g}(0)$. Therefore, the BVP (5)–(6) is extended with the following system of equations:

$$\dot{\mathbf{v}}_\mathrm{g} = T_\Gamma\,\left[\mathrm{D}_\mathrm{g}\mathbf{F}(\mathbf{g})\,\mathbf{v}_\mathrm{g} - \lambda_\mathrm{s}\,\mathbf{v}_\mathrm{g}\right], \tag{7}$$

$$\mathbf{v}_\mathrm{g}(1) - \mathbf{v}_\mathrm{g}(0) = 0, \tag{8}$$

$$\|\,\mathbf{v}_\mathrm{g}(0)\,\| = 1. \tag{9}$$

In particular, $\mathbf{v}_\mathrm{g}(0) = \mathbf{v}_\mathrm{g}(1)$ is the normalised vector that spans the local linearised isochron associated with $\mathbf{g}(0)$.

We have not specified a phase condition and, indeed, we allow $\mathbf{g}$ to shift and start at any point $\gamma_\vartheta \in \Gamma$. Consequently, the linear bundle $\mathbf{v}_\mathrm{g}$ will also shift such that $\mathbf{v}_\mathrm{g}(0)$ still spans the local linearised isochron associated with $\mathbf{g}(0)$.

Phase shifting the periodic orbit and its linear bundle by continuation in this way has been performed before [8]. However, the implementation in [8] requires accurate knowledge of the coordinates of the point γ_ϑ in order to decide when to stop shifting. Our approach uses another BVP set-up to monitor the phase shift, so that both γ_ϑ and $\mathbf{v}_\vartheta$ are determined up to the accuracy of the computation. To this end, we introduce a third orbit segment $\mathbf{w}$ that lies along Γ, with initial point $\mathbf{w}(0)$ equal to $\mathbf{g}(0)$, and end point $\mathbf{w}(1)$ equal to γ_0. The total integration time associated with this orbit segment $\mathbf{w}$ is the fraction of the period T_Γ that $\mathbf{g}(0)$ lies away from γ_0 along Γ; hence, it is directly related to the phase of $\mathbf{g}(0)$. We extend the BVP (5)–(9) with the following system of equations:

$$\dot{\mathbf{w}} = \nu \, T_\Gamma \, \mathbf{F}(\mathbf{w}), \tag{10}$$

$$\mathbf{w}(0) = \mathbf{g}(0), \tag{11}$$

$$[\mathbf{w}(1) - \gamma_0] \cdot \mathbf{v}_0^\perp = 0. \tag{12}$$

Here, we do not impose $\mathbf{w}(1) = \gamma_0$. Instead, condition (12) allows $\mathbf{w}(1)$ to move in the linearisation of $I(\gamma_0)$ at γ_0; this relaxation is necessary to ensure that the BVP remains well posed and the discretised problem has a solution. In practice, since $\mathbf{w}(0) \in \Gamma$, the difference between $\mathbf{w}(1)$ and γ_0 will be of the same order as the overall accuracy of the computation. Note that it is important to ensure $\nu \geq 0$ in Eq. (10), because $\mathbf{w}(1)$ may diverge from γ_0 along $E(\gamma_0)$ otherwise. We found it convenient to start the calculation with $\nu = 1$, which corresponds to the orbit segment $\mathbf{w} = \mathbf{g}$.

The combined solution $\{\mathbf{g}, \mathbf{v}_\mathbf{g}, \mathbf{w}\}$ to the multi-segment BVP (5)–(12) represents a rotated version of Γ and its stable Floquet bundle so that the head point is γ_φ with phase $\varphi = 1 - \nu \,(\mathrm{mod}\ 1)$. We remark here that this extended set-up can also be used to compute $I(\gamma_\varphi)$, for any phase $0 < \varphi < 1$, with the method for $I(\gamma_0)$ described in Sect. 2; such a computation would approximate each isochron up to the same accuracy, without introducing an additional interpolation error.

3.2 Continuation Set-Up for the Phase Reset

Recall the set-up for computing a phase reset by moving $\mathbf{u}(0)$ transversely to $I(\gamma_\vartheta)$, so that the end point $\mathbf{u}(1)$ will move and lie on the linearisation of an isochron $I(\gamma_\varphi)$ with a different phase φ. Here, the orbit segment $\mathbf{u}$ is a solution of

$$\dot{\mathbf{u}} = k \, T_\Gamma \, \mathbf{F}(\mathbf{u}), \tag{13}$$

for some $k \in \mathbb{N}$. The end point $\mathbf{u}(1)$ should lie close to Γ on the linearisation of $I(\gamma_\varphi)$, for some $\varphi \in [0, 1)$. We stipulate that the rotated version of Γ is shifted such that $\mathbf{u}(1)$ lies close to $\mathbf{g}(0)$ along the direction $\mathbf{v}_\mathbf{g}(0)$. Hence, we require the two boundary conditions

$$[\mathbf{u}(1) - \mathbf{g}(0)] \cdot \mathbf{v}_\mathbf{g}(0) = \eta, \tag{14}$$

$$[\mathbf{u}(1) - \mathbf{g}(0)] \cdot \mathbf{v}_\mathbf{g}(0)^\perp = 0, \tag{15}$$

where $\mathbf{v_g}(0)^\perp$ is the vector perpendicular to $\mathbf{v_g}(0)$. Here, η measures the (signed) distance between $\mathbf{u}(1)$ and $\mathbf{g}(0)$, which is along $\mathbf{v_g}(0)$. Since $\mathbf{u}$ is a solution of (13) and $k \in \mathbb{N}$, the initial point $\mathbf{u}(0)$ has the same phase as the last point $\mathbf{u}(1)$, and the combined multi-segment BVP (5)–(15) ensures that $\mathbf{u}(1)$ has (approximate) phase $1 - \nu \,(\text{mod } 1)$. In practice, we should choose $k \in \mathbb{N}$ large enough such that $\eta < \eta_{\max}$. If $\mathbf{u}(0)$ lies close to Γ, it will be sufficient to set $k = 1$. In order to consider phase resets of large perturbations, for which $\mathbf{u}(0)$ starts relatively far away, we need $k > 1$, to allow for sufficient time to let $\mathbf{u}$ converge and have $\mathbf{u}(1)$ lie close to Γ.

At this stage, the multi-segment BVP (5)–(15) is a system of $\mathsf{NDIM} = 8$ ordinary differential equations (for the case of a planar system), with $\mathsf{NBC} = 10$ boundary conditions, and $\mathsf{NPAR} = 4$ free parameters, namely, T_Γ, λ_s, ν, and η; the period T_Γ and stable Floquet exponent λ_s must remain free parameters to ensure that the discretised problem has a solution, but their variation will be almost zero. Hence, $\mathsf{NDIM} - \mathsf{NBC} + \mathsf{NPAR} = 2 \neq 1$, and one more condition is needed to obtain a one-parameter family of solutions.

The final step in the set-up is to impose an extra condition that specifies how $\mathbf{u}(0)$ moves along an arc or closed curve in the phase plane. Consequently, since $k\,T_\Gamma$ is fixed, the orbit segment $\mathbf{u}$ changes, so that $\mathbf{u}(1)$ will move as well, and $\mathbf{g}(0)$, along with $\mathbf{v_g}(0)$ will shift accordingly. This causes a variation in ν to maintain $\mathbf{w}(0) = \mathbf{g}(0)$, and these ν-values precisely define the new phase in the continuation run as a function of the position along the chosen arc or closed curve.

To compute the PRC, we need to let $\mathbf{u}(0)$ traverse the closed curve $\{\gamma_\vartheta + A\,\mathbf{d} \mid \vartheta \in [0, 1)\}$ obtained by the (instantaneous) perturbation of Γ in the direction $\mathbf{d}$ for distance A. We can impose this relatively complicated path on $\mathbf{u}(0)$ by including another system of equations to the multi-segment BVP, namely, the BVP that defines Γ in terms of another rotated orbit segment $\mathbf{g_u}$. Furthermore, in order to keep track of the phase ϑ along this path, we introduce another segment $\mathbf{w_u}$ that plays the same role as $\mathbf{w}$ in Sect. 3.1; compare with equations (5)–(6) and (10)–(12). In other words, we extend the BVP (5)–(15) by the following system of equations

$$\dot{\mathbf{g}}_{\mathrm{u}} = \widehat{T}_\Gamma \, \mathbf{F}(\mathbf{g}_{\mathrm{u}}), \tag{16}$$

$$\mathbf{g}_{\mathrm{u}}(1) - \mathbf{g}_{\mathrm{u}}(0) = 0. \tag{17}$$

$$\dot{\mathbf{w}}_{\mathrm{u}} = (1 - \vartheta)\,\widehat{T}_\Gamma \, \mathbf{F}(\mathbf{w}_{\mathrm{u}}), \tag{18}$$

$$\mathbf{w}_{\mathrm{u}}(0) = \mathbf{g}_{\mathrm{u}}(0), \tag{19}$$

$$[\mathbf{w}_{\mathrm{u}}(1) - \gamma_0] \cdot \mathbf{v}_0^\perp = 0. \tag{20}$$

Here, we decrease ϑ from 1 to 0, during which $\mathbf{w_u}$ grows and $\mathbf{g_u}$ tracks γ_ϑ. In order for a solution to exist, the periods T_Γ and $\widehat{T}_\Gamma$ must be two different free parameters, although they remain constant (and equal) to within the accuracy of the computation. The phase reset is now obtained by imposing

$$\mathbf{u}(0) = \mathbf{g}_{\mathrm{u}}(0) + A\,\mathbf{d}. \tag{21}$$

The multi-segment BVP (5)–(21) is now a system of dimension NDIM = 12, with NBC = 17 boundary conditions, and NPAR = 6 free parameters, which are T_Γ, λ_s, ν, η, $\widehat{T}_\Gamma$, and either ϑ or A. Since, NDIM − NBC + NPAR = 1, we obtain a one-parameter solution family by continuation. As the first solution in the continuation, we use the known solution $\mathbf{g} = \mathbf{w} = \mathbf{u} = \mathbf{g}_u = \Gamma$, which starts with the head point γ_0, the associated stable linear bundle $\mathbf{v}_0$ that we assumed has been pre-computed, and $\mathbf{w}_u = \gamma_0$; then $T_\Gamma = \widehat{T}_\Gamma$ and λ_s are set to their known computed values, $\eta = A = 0$, and $\nu = \vartheta = 1$. Initially, $k = 1$, and one should monitor η to make sure it does not exceed η_{max}.

To obtain the PTC or PRC we first perform a homotopy step, where we fix $\vartheta = 1$ and vary the amplitude A until the required value is reached. This continuation run produces a one-parameter family of solutions representing the effect of a reset of varying amplitude A from the point γ_0. In the main continuation run, we then fix A and decrease ϑ until $\vartheta = 0$, so that it covers the unit interval; the associated solution family of the multi-segment BVP (5)–(21), hence, provides the resulting phase $\vartheta_{new} := 1 - \nu \,(\mathrm{mod}\ 1)$ as a function of the phase $\vartheta_{old} := \vartheta$ along the perturbed periodic orbit.

4 Illustration of the Method with a Model Example

We illustrate our method for computing a PTC with a constructed example, namely, a parametrised version of the model introduced by Winfree [37, Chapter 6], which we also used in [22]; it is given in polar coordinates as

$$\begin{cases} \dot{r} = (1-r)(r-a)\,r, \\ \dot{\psi} = -1 - \omega\,(1-r). \end{cases}$$

In Euclidean coordinates, the system becomes

$$\begin{cases} \dot{x} = (1-\sqrt{x^2+y^2})\,\left(x\,(\sqrt{x^2+y^2}-a) + \omega y\right) + y, \\ \dot{y} = (1-\sqrt{x^2+y^2})\,\left(y\,(\sqrt{x^2+y^2}-a) - \omega x\right) - x. \end{cases} \tag{22}$$

Note that this system is invariant under any rotation about the origin; moreover, its frequency of rotation only depends on $r = \sqrt{x^2+y^2}$; see [22] for details. We now fix the parameters to $a = 0$ and $\omega = -0.5$, as in [22]. Then the unit circle is an attracting periodic orbit Γ with period $T_\Gamma = 2\pi$ and the origin is an unstable equilibrium $\mathbf{x}^*$.

4.1 Computing the PTC

We choose $\gamma_0 = (1, 0)$ and compute the normalised linear direction associated with its isochron as $\mathbf{v}_0 \approx (-0.83, -0.55)$. As was explained in Sect. 3.2, the computation is performed in two separate continuation runs: first, we apply a perturbation to the point γ_0 in a fixed direction $\mathbf{d}$, where we vary the amplitude A from 0 to 0.75 during the homotopy step. Next, we fix $A = 0.75$ and apply

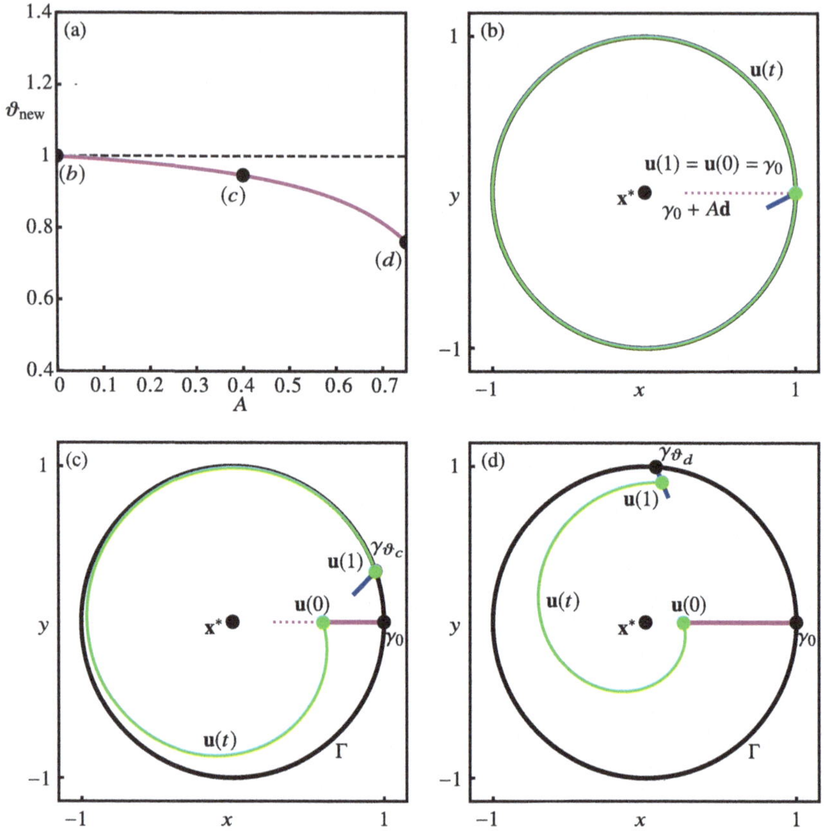

Fig. 1. Phase reset of system (22) at fixed γ_0 in the direction $\mathbf{d} = (-1, 0)$ with amplitude $A \in [0, 0.75]$ (a), and continuation set-up at the three labelled points (b), (c), and (d). Panel (b) shows the initial set-up when $A = 0$ and $\vartheta_{\text{new}} = 1$, in panel (c) the continuation has progressed to $A = 0.4$ and $\vartheta_{\text{new}} = \vartheta_c \approx 0.96$, and in panel (d) $A = 0.75$ has been reached and $\vartheta_{\text{new}} = \vartheta_d \approx 0.76$.

the same perturbation to each point $\gamma_\vartheta \in \Gamma$. For the purpose of visualising the computational set-up, we choose the (somewhat unusual) direction $\mathbf{d} = (-1, 0)$ and set the maximum distance along the linearised isochron to the relatively large value of $\eta_{\max} = 0.2$.

The first continuation run of the multi-segment BVP (5)–(21) is illustrated in Fig. 1. Here, the free amplitude A increases while $\vartheta = 1 = 0 \,(\text{mod } 1)$ is fixed and, hence, the perturbation is always applied at γ_0 and grows in size. Figure 1(a) shows the resulting phase ϑ_{new} as a function of A. Three points are labelled, indicating the three stages during the continuation that are illustrated in panels (b), (c) and (d). In each of these panels we show the periodic orbit Γ in black, and the current orbit segment $\mathbf{u}$ of the continuation run in green. Note that Γ is rotated here and its head point $\mathbf{g}(0)$ lies at the point on Γ with phase ϑ_{new}. A short segment of the associated linearisation of the isochron of $\gamma_{\vartheta_{\rm new}}$ is shown in blue. We do not plot the orbit segment $\mathbf{w}$ that determines the value of $\vartheta_{\rm new}$, but it follows Γ from $\mathbf{g}(0)$ back to $\mathbf{g}(0)$ and then extends (approximately) along Γ to γ_0. Indeed, notice in Fig. 1(a) that $\vartheta_{\rm new}$ is decreasing, which means that $\nu > 1$ is increasing so that $\mathbf{w}$ becomes longer. We also do not show the orbit segments $\mathbf{g_u}$ and $\mathbf{w_u}$ that determine the phase $\vartheta = \vartheta_{\rm old}$ at which the perturbation is applied, because $\vartheta_{\rm old} = 1$ is fixed in this continuation run.

Figure 1(b) shows the initial set-up, with $\mathbf{g} = \mathbf{w} = \mathbf{u} = \mathbf{g_u} = \Gamma$, $\mathbf{w_u} = \gamma_0$, $T_\Gamma = \widehat{T}_\Gamma$ and λ_s set to their known values, and $\nu = 1$, $\eta = A = 0$, with $k = 1$ and $\vartheta = 1$. The dotted line segment in Fig. 1(b) indicates the direction $\mathbf{d}$ of the intended perturbation away from γ_0; its length is the maximal intended amplitude $A = 0.75$. An intermediate continuation step when $A = 0.4$ is shown in Fig. 1(c). The perturbation has pushed $\mathbf{u}(0)$ out along $\mathbf{d}$, such that $\mathbf{u}(1)$ now lies (approximately) on the linearised isochron, parametrised as $\mathbf{g}(0) + \eta\, \mathbf{v_g}(0)$ with $0 < \eta \leq \eta_{\max}$, associated with the rotated head point $\mathbf{g}(0) = \gamma_{\vartheta_c}$, where $\vartheta_c \approx 0.96$. Note that the orbit segment $\mathbf{w}$ (not shown) has now changed from its initialisation to match the solution to subsystem (10)–(12) with $\nu \approx 1.04$. Figure 1(d) illustrates the last step of the first continuation run, when $A = 0.75$. The head point $\mathbf{g}(0) \in \Gamma$ has rotated further to γ_{ϑ_d} with $\vartheta_d = 1 - \nu \approx -0.24 = 0.76 \pmod 1$. Notice that $\mathbf{u}(1)$ lies quite far along the linearised isochron, because we allow a relatively large distance η. The corresponding orbit segment $\mathbf{u}$ is determined for an integration time of only one period, that is, for $k = 1$. We show this case for illustration purposes, but in practice, it would be worth choosing a smaller value for $\eta_{\max}$, so that $\mathbf{u}$ would be extended, and the integer multiple of T_Γ set to $k = 2$, before reaching $A = 0.75$.

The second continuation run uses the fixed perturbation of size $A = 0.75$ along $\mathbf{d} = (-1, 0)$, and varies the phase ϑ at which it is applied. Since ϑ controls the integration time associated with the orbit segment $\mathbf{w_u}$, the multi-segment BVP (16)–(20) with solution $\{\mathbf{g_u}, \mathbf{w_u}\}$ and parameter $\widehat{T}_\Gamma$ now plays an important role. For each ϑ, the head point $\mathbf{g_u}(0)$ of $\mathbf{g_u}$ lies (approximately) at $\gamma_\vartheta \in \Gamma$, and $\mathbf{w_u}$ represents the remaining part of Γ from γ_ϑ to γ_0; hence, the total integration time of $\mathbf{w_u}$ is the fraction $1 - \vartheta$ of $\widehat{T}_\Gamma$, which is equal, up to the computational accuracy, to the period T_Γ of Γ.

Figure 2 illustrates different aspects of this continuation run. As $\vartheta_{\rm old} = \vartheta$ decreases from 1, the multi-segment BVP (5)–(21) determines the orbit segment $\mathbf{u}$ with $\mathbf{u}(0) = \gamma_\vartheta + A\,\mathbf{d}$ and uses the rotated orbit segment $\mathbf{g}$ and $\mathbf{w}$ to establish the resulting phase $\vartheta_{\rm new} = 1 - \nu \pmod 1$ of $\mathbf{u}(1)$. Panel (a) shows the PTC computed for $A = 0.75$. Note that ν takes values in the covering space $\mathbb{R}$; the output is then folded onto the unit torus by taking $\vartheta_{\rm new} = 1 - \nu \pmod 1$, giving the solid curve in Fig. 2. The points labelled (b) and (c) in this panel correspond to $\vartheta_{\rm old} = 0.9$ and $\vartheta_{\rm old} = 0.1$, respectively. The continuation set-up for these two cases is shown in the corresponding panels (b) and (c). As in Fig. 1, the periodic orbit Γ is black and $\mathbf{u}$ is green. The path traced by the initial point $\mathbf{u}(0)$ is the magenta dotted circle, which is Γ shifted by $A = 0.75$ in the direction $\mathbf{d} = (-1, 0)$; hence, for fixed ϑ, the point $\mathbf{u}(0)$ corresponds to the perturbation of the point $\gamma_\vartheta \in \Gamma$

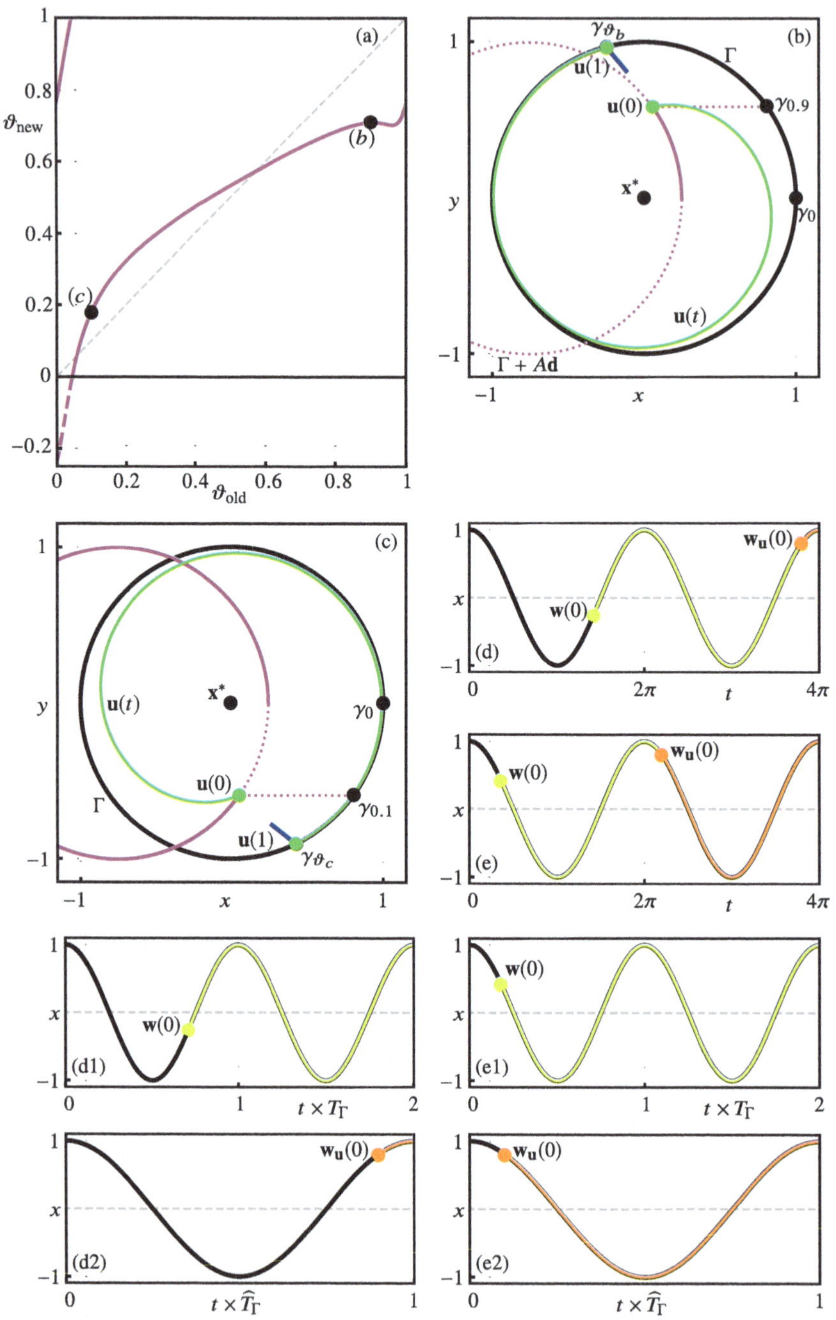

Fig. 2. PTC of Γ in system (22) for $\mathbf{d} = (-1, 0)$ and $A = 0.75$ (a), and continuation set-up at $\vartheta_{\mathrm{old}} = 0.9$ (b) and at $\vartheta_{\mathrm{old}} = 0.1$ (c) with $\mathbf{w}$ and $\mathbf{w_u}$ in (d), (d1), (d2) and (e), (e1), (e2), respectively.

that lies horizontally to the right of $\mathbf{u}(0)$, as indicated by the magenta dotted line segment. The end point $\mathbf{u}(1)$ lies on the linearised isochron, parametrised as $\mathbf{g}(0) + \eta\, \mathbf{v}_\mathbf{g}(0)$ with $0 < \eta \leq \eta_{\max}$, associated with the rotated head point of $\mathbf{g}$, which is determined by subsystem (5)–(9). The phase of this head point is given by $\vartheta_{\text{new}} = 1 - \nu\,(\text{mod } 1)$, where ν is determined from subsystem (10)–(12) that defines the orbit segment $\mathbf{w}$.

Hence, the two orbit segments $\mathbf{w}$ and $\mathbf{w}_\mathrm{u}$ essentially determine the PTC, that is, the map $P : \vartheta_{\text{old}} \mapsto \vartheta_{\text{new}}$. Their x-coordinates are plotted versus time in panel (d) for $\vartheta_{\text{old}} = 0.9$ and in panel (e) for $\vartheta_{\text{old}} = 0.1$, respectively, overlaid on two copies of Γ (black curve), that is, time t runs from 0 to 4π. The further panels (d1) and (d2) for $\vartheta_{\text{old}} = 0.9$ and panels (e1) and (e2) for $\vartheta_{\text{old}} = 0.1$ show $\mathbf{w}$ (yellow curve) and $\mathbf{w}_\mathrm{u}$ (orange curve) individually, relative to the periods T_Γ and $\widehat{T}_\Gamma$, respectively. Note that both $\mathbf{w}$ and $\mathbf{w}_\mathrm{u}$ end at $x = 1$, for $t = 4\pi$ and $t = 2\pi$, respectively, as required, but their initial points differ. As ϑ decreases from 1 to 0 during the continuation, the orbit segment $\mathbf{w}_\mathrm{u}$ lengthens as expected, but note that $\mathbf{w}$ lengthens as well; this is due to the (near-)monotonically increasing nature of the PTC for this example.

4.2 Loss of Invertibility

Recall that any PTC is the identity for $A = 0$, and invertible for sufficiently small amplitude A of the perturbation, because its graph remains a 1:1 torus knot on the torus parametrised by the two periodic variables ϑ_{old} and ϑ_{new}. However, the PTC in Fig. 2(a) for $A = 0.75$ is no longer near the identity: it is not injective and, hence, not invertible.

To show how injectivity of the PTC is lost as A is increased, we consider again model (22), but now with $a = 0.25$; see also [22]. Apart from the attracting unit circle $\Gamma_\mathrm{s} = \Gamma$ with period $T_\Gamma = 2\pi$, there exists then also a repelling circle Γ_u with radius $r = a = 0.25$ and period $2\pi/(1 + \omega\,(1 - a)) = 3.2\,\pi$; note that Γ_u forms the boundary of the basins of attraction of both Γ_s and the equilibrium $\mathbf{x}^*$ at the origin, which is now attracting.

We consider three phase resets of Γ_s of the form $\Gamma_\mathrm{s} + A\,\mathbf{d}$ in the positive direction $\mathbf{d} = (1, 0)$ and with $A = 0.54$, $A = 0.59$, and $A = 0.64$. Figure 3 shows the three corresponding PTCs, the corresponding PRCs, and the perturbed cycles $\Gamma_\mathrm{s} + A\,\mathbf{d}$ in increasingly darker shades of magenta as A increases in panels (a), (b), and (c), respectively. Panel (a) shows that the first PTC for $A = 0.54$ is injective and invertible. As A is increased to approximately $A = 0.59$, the graph has a cubic tangency near $(\vartheta_{\text{old}}, \vartheta_{\text{new}}) = (0.45, 0.24)$, because the associated map P has an inflection point at $\vartheta_{\text{old}} \approx 0.45$. For larger values of A, such as for $A = 0.64$, the PTC has a local maximum followed by a local minimum and is, hence, no longer invertible. Note from Fig. 3(b) that this qualitative change of the PTC does not lead to a corresponding qualitative change of the PRC.

Figure 3(c) and the enlargement near the basin boundary Γ_u in panel (d) show that the loss of injectivity of the PTC is due to a cubic tangency between the perturbed cycle $\Gamma_\mathrm{s} + A\mathbf{d}$ and the foliation of the basin of Γ_s by (forward-time) isochrons; ten isochrons are shown in panel (c) and one hundred in panel (d),

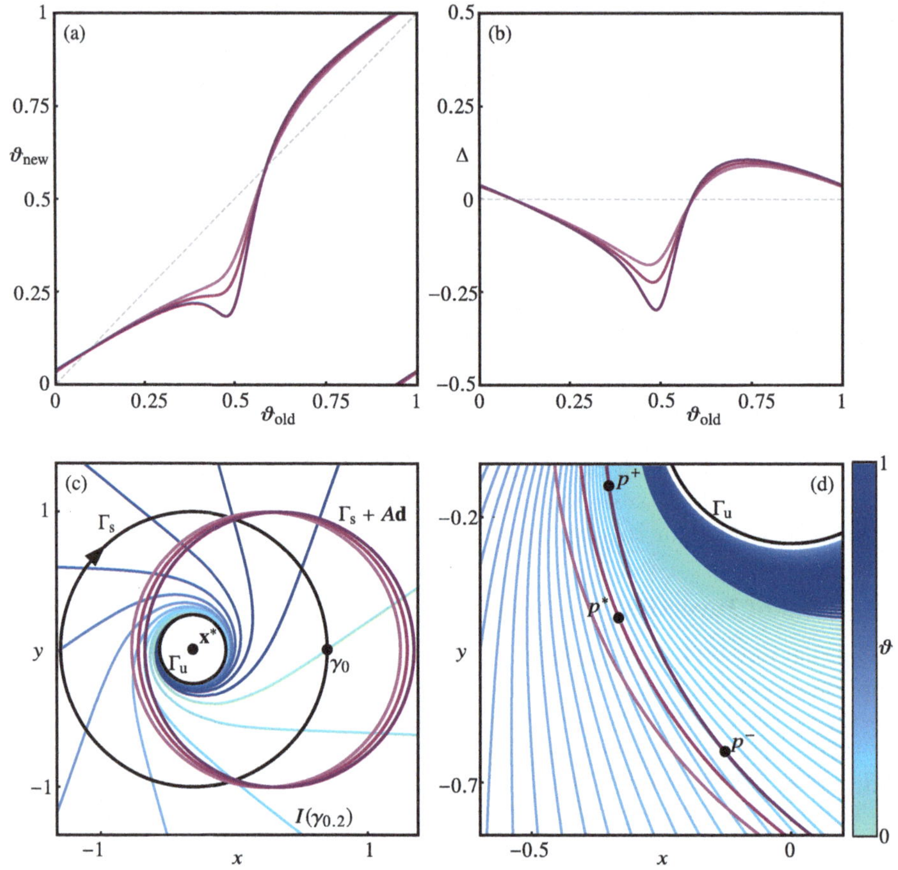

Fig. 3. Phase resets of Γ_s in system (22) with $a = 0.25$ in the direction $\mathbf{d} = (1, 0)$ for amplitudes $A \in \{0.54, 0.59, 0.64\}$ (increasingly darker shades of magenta). The three PTCs are shown in panel (a) and the corresponding PRCs in panel (b). The three perturbed cycles are shown in panel (c) together with Γ_s and ten of its isochrons that are uniformly distributed over one period; the enlargement near Γ_u in panel (d) shows them with 100 uniformly distributed isochrons of Γ_s and points of tangency at p^* and $p^{\pm}$. Isochrons are coloured according to the colour bar.

distributed uniformly in phase and coloured according to the colour bar. The left-most light-magenta cycle for $A = 0.54$ is transverse to all isochrons. The middle magenta cycle for $A = 0.59$, on the other hand, has a single cubic tangency (approximately) with the isochron $I(\gamma_{0.24})$ of phase $\vartheta = 0.24$ at the point $p^* \approx (-0.33, -0.39)$, shown in panel (d). For larger A, as for the right-most dark-magenta cycle for $A = 0.64$, there are now two quadratic tangencies with two different isochrons, namely, (approximately) with $I(\gamma_{0.22})$ and $I(\gamma_{0.19})$ at the points $p^+ \approx (-0.35, -0.14)$ and $p^- \approx (-0.13, -0.64)$, respectively. As a result, all isochrons that intersect the perturbed cycle between p^+ and p^- intersect three times; hence, the map P from ϑ_{old} to ϑ_{new} is no longer invertible. Note that p^+ and p^- correspond to the local maximum and local minimum of the PTC in panel (a), respectively.

5 Phase Resetting in the FitzHugh-Nagumo Model

We now illustrate the capability of our method by computing the phase reset of a periodic orbit in the FitzHugh–Nagumo system [10,27]. This model is an iconic example that motivated early work on isochrons and phase response curves; in particular, it has a very complicated geometry of isochrons with regions of extreme phase sensitivity [21,37]. The FitzHugh–Nagumo system is given by the equations

$$\begin{cases} \dot{x} = c\ \left(y + x - \frac{1}{3}\,x^3 + z\right), \\ \dot{y} = -\dfrac{x - a + by}{c}. \end{cases} \tag{23}$$

We set $a = 0.7$, $b = 0.8$, and $z = -0.4$, as in [37], and fix $c = 2.5$, as was done in [22]. For these parameter values, there exists an attracting periodic orbit Γ with period $T_\Gamma \approx 10.71$ and a repelling equilibrium $\mathbf{x}^* \approx (0.9066, -0.2582)$. The parameter c is a time-scale parameter, the increase of which makes the x-variable faster than the y-variable. It plays an important role in the onset of phase sensitivity due to an accumulation of isochrons in a narrow region close to the slow manifold [21], which is associated with the occurrence of sharp turns in the isochrons of Γ; see also [29]. For the chosen value of $c = 2.5$, one finds both strong phase sensitivity and sharp turns, which makes the computation of any phase resetting curve quite challenging.

Figure 4 illustrates the phase reset for the FitzHugh–Nagumo model (23) after a perturbation in the direction $\mathbf{d} = (1, 0)$ of amplitude $A = 0.25$. Panel (a) and the enlargement near the equilibrium $\mathbf{x}^*$ in panel (b) show how the perturbed cycle $\Gamma + A\,\mathbf{d}$ intersects the isochrons of Γ, of which 100 are shown uniformly distributed in phase and coloured according to the colour bar in Fig. 3. In particular, one notices quite a few instances in panel (b) of quadratic tangencies between the perturbed cycle and different isochrons; one such isochron is the highlighted $I(\gamma_{0.62})$. The green curves O^+ and O^- in panels (a) and (b) are two special trajectories, along which the foliation by forward-time isochrons of Γ has quadratic tangencies with the foliation by backward-time isochrons (not shown) of the focus $\mathbf{x}^*$. Tangencies between these two foliations were introduced in [22], where we argued that such tangencies give rise to sharp turns of isochrons. We remark that the two trajectories O^+ and O^- of quadratic tangencies appear at a specific value $c^* < 2.5$ where one finds a cubic tangency between the two foliations, called a cubic isochron foliation tangency or CIFT for short; see [22] for details. The relevance of the special trajectories O^+ and O^- in the present context is that along them the isochrons of Γ have sharp turns as they approach $\mathbf{x}^*$. This can clearly be seen in Fig. 4(b); as the highlighted isochron $I(\gamma_{0.62})$ illustrates, the turns along O^- are so sharp that $I(\gamma_{0.62})$ appears to retrace itself along certain segments. Since this happens for all isochrons of Γ, one finds extreme

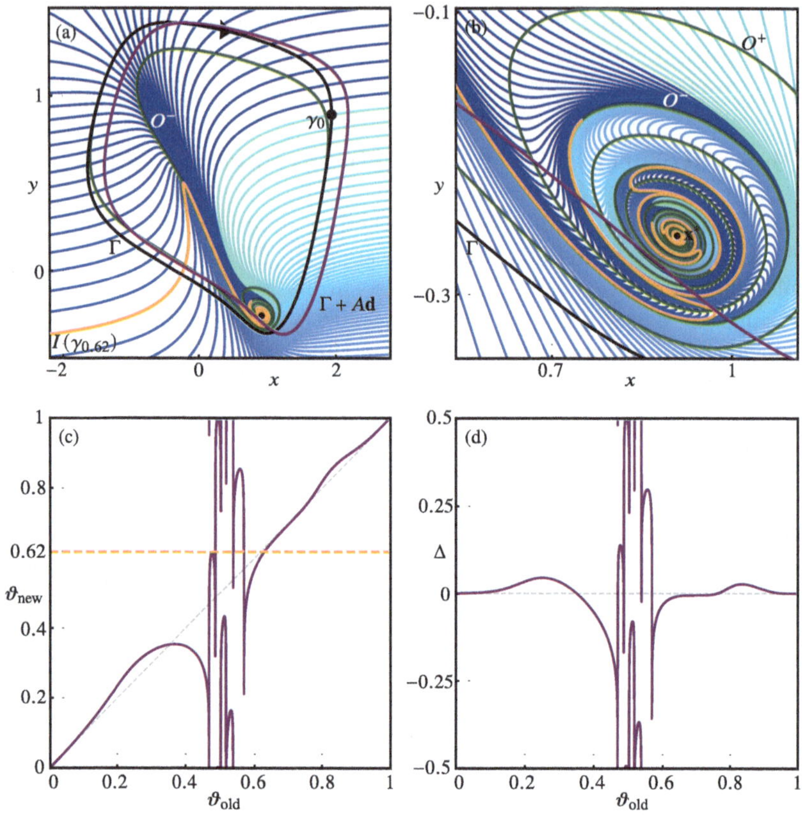

Fig. 4. Phase reset for the FitzHugh–Nagumo model (23). Panel (a) shows the periodic orbit Γ (black), the perturbed cycle $\Gamma + A\mathbf{d}$ (magenta) with $\mathbf{d} = (1, 0)$ and $A = 0.25$, the two trajectories O^+ and O^- (green), and 100 isochrons uniformly distributed in phase; isochrons are coloured according to the colour bar in Fig. 3, and the isochron $I(\gamma_{0.62})$ is highlighted in orange. Panel (b) is an enlargement near the equilibrium $\mathbf{x}^*$, and panels (c) and (d) show the corresponding PTC and PRC, respectively; the dashed orange line in panel (c) indicates the phase of $I(\gamma_{0.62})$.

phase sensitivity near the trajectory O^-. Moreover, quadratic tangencies of the perturbed cycle with isochrons of Γ occur near both O^+ and O^-. Hence, the number of intersection of $\Gamma + A\mathbf{d}$ with O^+ and O^- gives an indication of how many quadratic tangencies the perturbed cycle has with different isochrons.

As we have seen in Sect. 4.2, any such quadratic tangency between $\Gamma + A\,\mathbf{d}$ and an isochron is associated with a local maximum or minimum of the PTC, which is, therefore, not expected to be invertible. Figure 4(c) presents the PTC computed with our method as a continuous curve shown in $(\vartheta_{\text{old}}, \vartheta_{\text{new}})$-coordinates on the unit torus. Clearly, its graph is quite intriguing and features six local maxima and six local minima. Observe that the local maxima correspond to quadratic tangencies near O^+, while the sharper local minima correspond to

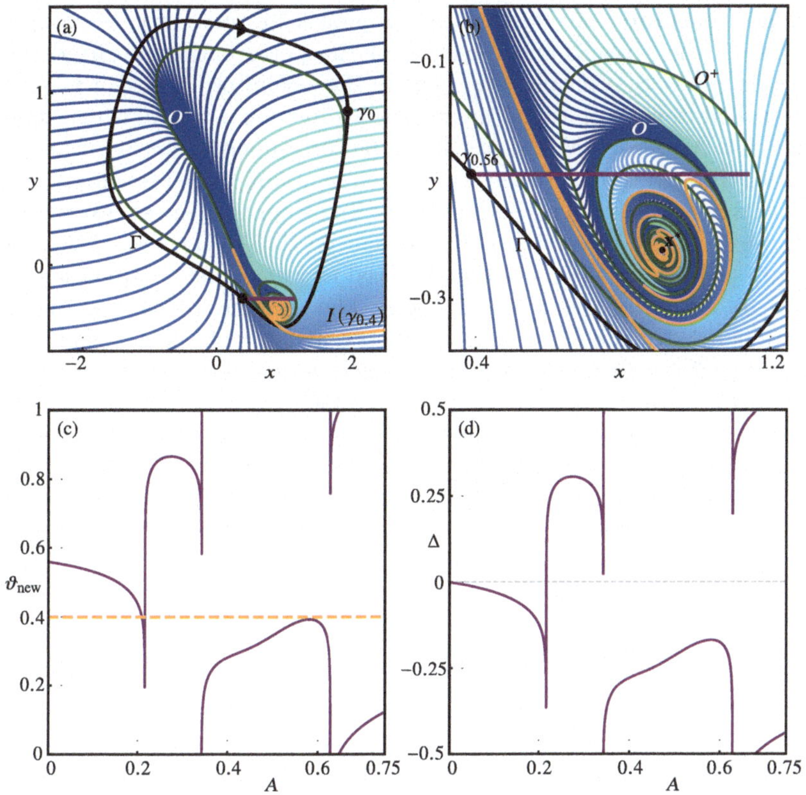

Fig. 5. Phase reset along the line segment $\gamma_{0.56} + A\,\mathbf{d}$ with $\mathbf{d} = (1, 0)$ and $A \in [0, 0.75]$ in the FitzHugh–Nagumo model (23). Panel (a) shows the periodic orbit Γ (black), the line segment of perturbations (magenta) starting at point $\gamma_{0.56} \in \Gamma$, the two trajectories O^+ and O^- (green), and 100 isochrons uniformly distributed in phase; isochrons are coloured according to the colour bar in Fig. 3, and the isochron $I(\gamma_{0.4})$ is highlighted in orange. Panel (b) is an enlargement near the equilibrium $\mathbf{x}^*$, and panels (c) and (d) show the periodic variables ϑ_{new} and $\Delta = \vartheta_{\text{new}} - \vartheta_{\text{old}}$, respectively, as a function of A.

quadratic tangencies near O^-; in particular, the tangency with the highlighted isochron $I(\gamma_{0.62})$ near O^+ in panel (b) gives rise to a local maximum of the PTC in panel (c), where the graph has a quadratic tangency with the dashed orange line at $\vartheta_{\text{new}} = 0.62$. Notice that $I(\gamma_{0.62})$ intersects the perturbed cycle $\Gamma + A\mathbf{d}$ in panel (b), and hence, the PTC in panel (c), five more times. The associated PRC of the change in phase $\Delta = \vartheta_{\text{new}} - \vartheta_{\text{old}}$ is shown in panel (d); it is also quite a complicated curve with local maxima and minima. The PTC and PRC both have six near-vertical segments at $\vartheta_{\text{old}} \approx 0.47$, 0.49, 0.50, 0.52, 0.54, and 0.57; such large gradients arise near the local minima because of the extreme phase sensitivity near O^-.

Figure 5 illustrates our continuation approach for another type of resetting experiment, where phase and direction of the perturbation are fixed but its magnitude varies. Specifically, we calculate the asymptotic phase of points that are perturbed from $\gamma_{0.56} \in \Gamma$ in the positive x-direction $\mathbf{d} = (1, 0)$ with amplitude $A \in [0, 0.75]$. As panels (a) and (b) show, the corresponding line segment $\gamma_{0.56} + A\,\mathbf{d}$ passes through the phase-sensitive region of accumulating isochrons near $\mathbf{x}^*$, where it intersects O^+ and O^- several times. To compute the reset, we first rotate Γ and, consequently, the entire multi-segment BVP (5)–(21), such that the head point $\mathbf{g}(0)$ of Γ lies at $\gamma_{0.56}$. We then proceed as in the first continuation run in Sect. 4.1 to obtain ϑ_{new} as a function of A. The resulting resets ϑ_{new} and Δ are shown in panels (c) and (d), respectively; note that Δ is obtained from ϑ_{new} by a fixed shift of $\vartheta_{\mathrm{old}} = 0.56$. The resulting reset as a function of A also shows near-vertical segments near three local minima, which are again directly associated with the three points where the line segment $\gamma_{0.56} + A\,\mathbf{d}$ intersects the trajectory O^-. Notice that the turns of the isochrons along O^- are so very sharp that one will find a quadratic tangency nearby with respect to the horizontal—or indeed practically any given direction. Along O^+, on the other hand, the turns of the isochrons are more gradual and the local maxima due to intersections of the line segment of perturbations are not associated with strong phase sensitivity. Notice further that the penultimate intersection between $\gamma_{0.56} + A\,\mathbf{d}$ and O^+ does not come with a nearby quadratic tangency and, hence, does not lead to a local maximum of ϑ_{new}.

6 Phase Resetting in a Seven-Dimensional Sinoatrial Node Model

We now illustrate how our computational approach can be applied to systems of dimension higher than two. Indeed, while the multi-segment BVP (5)–(21) now consists of higher-dimensional subsystems that represent the various orbit segments in this higher-dimensional phase space, the necessary input-output information is still given by the two parameters ϑ and ν that determine the relationship $\vartheta_{\mathrm{new}} = P(\vartheta_{\mathrm{old}})$.

We compute the PTC for the seven-dimensional model from [20] of a sinoatrial node of a rabbit, which is a type of cardiac pacemaker cell. The model is described in standard Hodgkin–Huxley formalism: the main variable is voltage V (measured in mV), which depends on five ionic currents that are determined by the dynamic opening and closing of six so-called gating variables, denoted m, h, d, f, p, and q. The five currents (measured in pA) are: a fast inward sodium current I_{Na}, a slow inward current I_{s}, a delayed rectifier potassium current I_{K}, a pacemaker current I_{h}, and time-independent leak current I_{l}. The system of

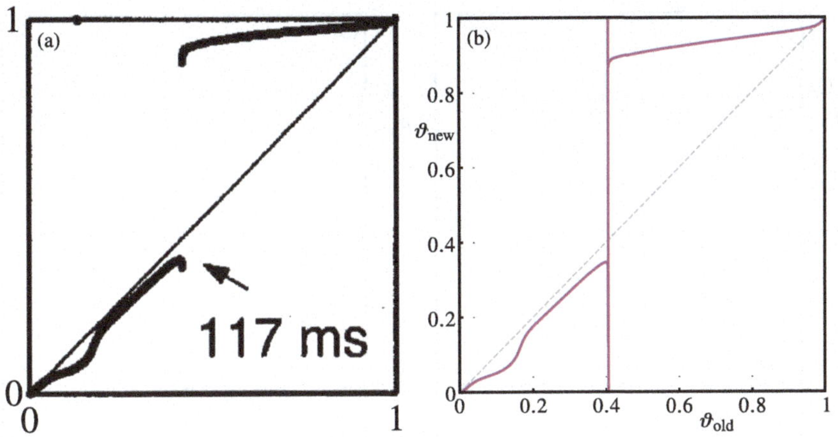

Fig. 6. The PTC of the seven-dimensional model (24), as presented in [20, Fig. 7 (middle)] (a) and as computed with our method (b). Panel (a) is from [Krogh-Madsen, Glass, Doedel and Guevara, Apparent discontinuities in the phase-resetting response of cardiac pacemakers, J. Theor. Biol. **230**(4), 499–519 (2004)] ©Elsevier; reproduced with permission.

seven equations is given by

$$\begin{cases} \dot{V} = -\frac{1}{C_m}\left[I_{Na}(V,m,h) + I_s(V,d,f) + I_K(V,p) + I_h(V,q) + I_l(V)\right], \\ \dot{m} = \alpha_m(V)\,(1-m) - \beta_m(V)\,m, \\ \dot{h} = \alpha_h(V)\,(1-h) - \beta_h(V)\,h, \\ \dot{d} = \alpha_d(V)\,(1-d) - \beta_d(V)\,d, \\ \dot{f} = \alpha_f(V)\,(1-f) - \beta_f(V)\,f, \\ \dot{p} = \alpha_p(V)\,(1-p) - \beta_p(V)\,p, \\ \dot{q} = \alpha_q(V)\,(1-q) - \beta_q(V)\,q, \end{cases} \tag{24}$$

where, $C_m = 0.065\,\mu$F is the capacitance. (Note the minus sign in the right-hand side of the equation for V, which was accidentally omitted in [20].) The precise form of the ionic currents and the various functions $\alpha_{\rm x}$ and $\beta_{\rm x}$ with $\rm x \in \{m, h, d, f, p, q\}$, and associated parameter values, are given in the Appendix; see also [20].

System (24) was presented and studied in [20], because experimental data on similar pacemaker cells suggested that the PTC was discontinuous; see already Fig. 6(a). Without a possibility to compute the PTC directly, the authors of [20] reduced the model to a three-dimensional system and used geometric arguments to explain that the apparent discontinuities were abrupt transitions mediated by the stable manifold of a weakly unstable manifold in the model. Figure 6 shows the relevant PTC image from [20] and the PTC as computed with our method. The comparison confirms that we are able to calculate the PTC directly in the

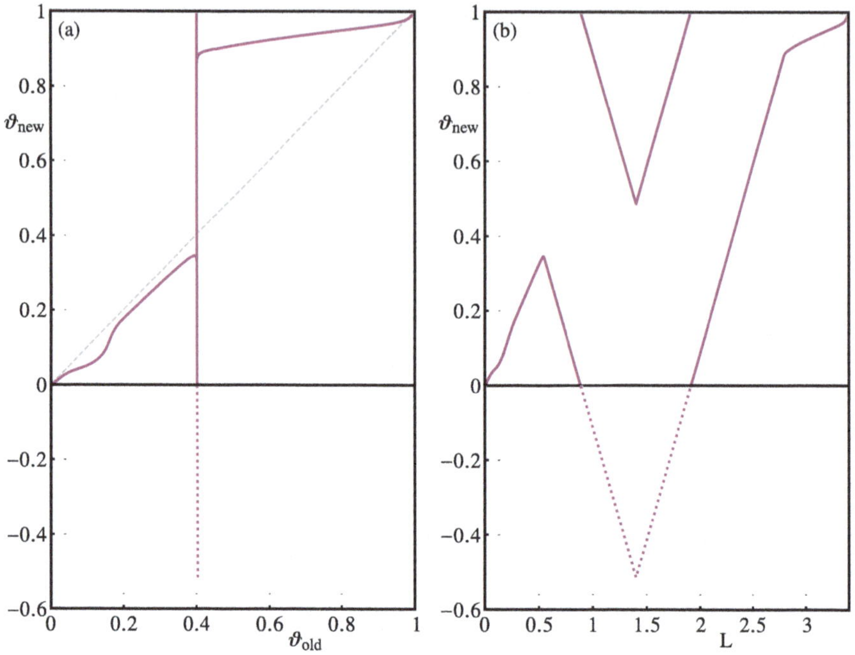

Fig. 7. The computed PTC of the seven-dimensional system (24). Shown is ϑ_{new}, also over the interval $[-0.6, 0]$ (dotted line), as a function of ϑ_{old} in panel (a), and as a function of the arclength L of the PTC in panel (b).

seven-dimensional model as a continuous curve on $\mathbb{T}^2$, even though the PTC has a near-vertical segment at $\vartheta_{old} \approx 0.4$.

Figure 7 illustrates that the PTC for system (24) is indeed continuous. Panel (a) reproduces the PTC from Fig. 6(b), but shows the computed values for ϑ_{new} over the wider range $[-0.6, 1]$ to illustrate that a maximum of ϑ_{new} is quickly followed by a minimum of ϑ_{new} (lowest point of dashed curve). Since it is hard to see that the PTC is indeed continuous, panel (b) shows ϑ_{new} over the same range $[-0.6, 1]$, but now as a function of the arclength L of the PTC in the $(\vartheta_{old}, \vartheta_{new})$-plane from the point $(0, -3.56 \times 10^{-3})$. The near-vertical segment in the $(\vartheta_{old}, \vartheta_{new})$-plane of panel (a) corresponds to the two (almost) linear segments in the (L, ϑ_{new})-plane of panel (b). Hence, this representation resolves the near-vertical part of the PTC in a tiny ϑ_{old}-interval near 0.4. Panel (b) also demonstrates that the PTC is indeed a continuous closed curve on $\mathbb{T}^2$ with exactly one maximum at $\vartheta_{new} \approx 0.35$, followed by one minimum at $\vartheta_{new} \approx 0.49$.

Instead of an instantaneous reset, the reset in [20] is obtained by applying a current with amplitude I_{app} for a fixed duration T_{ON}; the specific case for which a seemingly discontinuous PTC was observed is given by $I_{app} = -150\,\text{pA}$

and $T_{\mathrm{ON}} = 0.02$ s. Mathematically, this amounts to replacing the V-equation in system (24) by

$$\dot{V} = -\tfrac{1}{C_m}\left[I_{Na} + I_s + I_K + I_h + I_l\right] + \tfrac{150}{C_m}, \tag{25}$$

and switching back to the original equation after 0.02 s. In our set-up, this means that we add the perturbation $A\,\mathbf{d}$ to the right-hand side of system (24), where the direction vector $\mathbf{d} = (1, 0, 0, 0, 0, 0, 0)$ is the unit vector pointing purely in the V-direction and the amplitude $A = 150/C_m = 2.31$ (mV/s).

We include this time-varying perturbation in the multi-segment BVP (5)–(21) in much the same way as done in [28], that is, we replace subsystem (13) defining the orbit segment $\mathbf{u}$, with boundary conditions (14) and (15), by two subsystems that define orbit segments $\mathbf{u}_{\mathrm{ON}}$ and $\mathbf{u}_{\mathrm{OFF}}$. Here, $\mathbf{u}_{\mathrm{ON}}$ exists while the applied current is 'on' and $\mathbf{u}_{\mathrm{ON}}(1)$ determines the location of the reset (21) after the first 0.02 s. Hence, $\mathbf{u}_{\mathrm{ON}}$ is a solution to system (24) with equation (25) for V with total integration time $T_{\mathrm{ON}} = 0.02$ s, that is,

$$\dot{\mathbf{u}}_{\mathrm{ON}} = T_{\mathrm{ON}}\,\left[\mathbf{F}(\mathbf{u}_{\mathrm{ON}}) + A\,\mathbf{d}\right].$$

The second orbit segment $\mathbf{u}_{\mathrm{OFF}}$ is a solution to the original system (24), with applied current 'off'. The total integration time over both orbit segments combined should be an integer multiple of the period T_Γ of the periodic orbit (as before for $\mathbf{u}$). Hence, we define

$$\dot{\mathbf{u}}_{\mathrm{OFF}} = (k\,T_\Gamma - T_{\mathrm{ON}})\,\mathbf{F}(\mathbf{u}_{\mathrm{OFF}}).$$

The subsystem for $\mathbf{u}_{\mathrm{ON}}$ can be viewed as an initial value problem, with initial condition

$$\mathbf{u}_{\mathrm{ON}}(0) = \mathbf{g}_{\mathbf{u}}(0).$$

Similarly, the initial point of $\mathbf{u}_{\mathrm{OFF}}$ should start where $\mathbf{u}_{\mathrm{ON}}$ ends, that is,

$$\mathbf{u}_{\mathrm{ON}}(1) = \mathbf{u}_{\mathrm{OFF}}(0).$$

We refer to [28] for further details.

The end point $\mathbf{u}_{\mathrm{OFF}}(1)$ of the second segment $\mathbf{u}_{\mathrm{OFF}}$ plays the same role as $\mathbf{u}(1)$ in the multi-segment BVP (5)–(21). Hence, $\mathbf{u}_{\mathrm{OFF}}(1)$ must satisfy boundary conditions (14) and (15). Unfortunately, this formulation requires knowledge of the Floquet bundle $\mathbf{v}_{\mathbf{g}}$ specified by subsystem (7)–(9), and specifically the vector $\mathbf{v}_{\mathbf{g}}(0)$ to measure the distance of $\mathbf{u}_{\mathrm{OFF}}(1)$ to Γ in boundary condition (14). In the seven-dimensional phase space, this Floquet bundle is no longer unique, because Γ now has six non-trivial Floquet exponents. Note that the perpendicular vectors $\mathbf{v}_0^{\perp}$, used in boundary conditions (12) and (20), and $\mathbf{v}_{\mathbf{g}}(0)^{\perp}$, used in boundary condition (15), are still well defined in a higher-dimensional phase space, because the isochrons are codimension-one manifolds. We get around the issue of non-uniqueness as follows. Firstly, we define subsystem (7)–(9) in terms of the adjoint Floquet bundle $\mathbf{v}_{\mathbf{g}}{}^{\perp}$, that is, the left eigenvector bundle associated with the trivial Floquet exponent 0. In other words, we solve the first variational equation

$$\dot{\mathbf{v}}_{\mathbf{g}}^{\perp} = T_\Gamma\,\mathrm{D}_{\mathbf{g}}\mathbf{F}^{*}(\mathbf{g})\,\mathbf{v}_{\mathbf{g}}{}^{\perp},$$

with the same boundary conditions (8) and (9) for $\mathbf{v_g}^{\perp}$ instead, namely,

$$\begin{cases} \mathbf{v_g}^{\perp}(1) - \mathbf{v_g}^{\perp}(0) = 0, \\ \qquad \| \mathbf{v_g}^{\perp}(0) \| = 1. \end{cases}$$

Here $\mathrm{D_g}\mathbf{F}^*(\mathbf{g})$ is the transpose Jacobian matrix evaluated along the periodic orbit $\mathbf{g}$. We similarly assume that $\mathbf{v}_0^{\perp}$, rather than $\mathbf{v}_0$, is stored as a known vector.

Secondly, we use the Euclidean norm to measure the distance of $\mathbf{u}_{\mathrm{OFF}}$ from $\mathbf{g}(0)$, that is, we stipulate

$$[\mathbf{u}_{\mathrm{OFF}}(1) - \mathbf{g}(0)] \cdot [\mathbf{u}_{\mathrm{OFF}}(1) - \mathbf{g}(0)] = \eta^2, \tag{26}$$

rather than imposing a signed distance. We remark that a formulation in terms of the Euclidean norm does make the continuation numerically less stable, but it still works for our set-up because η is a free parameter that remains positive, and boundary condition (26) effectively plays a monitoring role.

7 Conclusions

We presented an algorithm for the computation of the phase reset for a dynamical system with periodic orbit Γ that is subjected to an (instantaneous or time-varying) perturbation $\Gamma_A := \Gamma + A\,\mathbf{d}$ of a given direction $\mathbf{d}$ and amplitude A. In theory, the phase reset can be determined from the isochron foliation of the basin $\mathcal{B}(\Gamma)$. In particular, for small enough A, it suffices to know only the linear approximation of the isochrons. Our algorithm is designed for the computation of phase resets for larger A in systems that exhibit strong nonlinearities, which is the situation in many practical applications. Our computational set-up has the distinguishing feature that it automatically tracks the necessary nonlinear phase information without computing the isochrons themselves. This approach is efficient, able to deal with extreme phase sensitivity, and suitable for use in higher-dimensional settings.

Our method is formulated in terms of a multi-segment boundary value problem that is solved by continuation and gives the new phase ϑ_{new} as a function of either the perturbation amplitude A or the original phase ϑ_{old} before the reset. The data can readily be used to produce phase transition and phase response curves. We presented the multi-segment BVP set-up in detail for a planar system, but also discussed in Sect. 6 the straightforward adaptation to higher-dimensional systems, and how to implement phase resets arising from time-varying inputs.

Our approach has the advantage that the map $\vartheta_{\mathrm{old}} \mapsto \vartheta_{\mathrm{new}}$ is computed in a single continuation run, even in the presence of extreme phase sensitivity. If the amplitude A is such that $\Gamma_A \subset \mathcal{B}(\Gamma)$, then the associated circle map $P_A : [0,1) \to [0,1)$ is obtained in its entirety, and its graph, the phase transition curve (PTC), is a continuous closed curve on $\mathbb{T}^2$. For A close to 0, the circle map P_A is a near-identity transformation, so that the PTC is a 1:1 torus knot. For large A it is possible that the PTC is a contractible closed curve on the torus, which corresponds to loss of surjectivity of P_A. It is well known that surjectivity is lost as soon as A increases past a value for which $\Gamma_A \not\subset \mathcal{B}(\Gamma)$ [11,37].

There typically exists a maximal amplitude $A_{\max}$ such that $\Gamma_A \subset \mathcal{B}(\Gamma)$ for $0 \leq A < A_{\max}$. Then P_A depends smoothly on A and, hence, the PTC is a 1:1 torus knot for all $0 \leq A < A_{\max}$. Therefore, P_A is surjective for all $0 \leq A < A_{\max}$. We were particularly interested in loss of injectivity of P_A as A increases from 0. We showed that this is typically mediated by a cubic tangency between the PTC and one of the isochrons of Γ. Further tangencies lead to very complicated PTCs, with possibly many local maxima and minima and very sudden phase changes. The associated phase sensitivity is known to occur near the boundary of $\mathcal{B}(\Gamma)$, but our examples illustrate that milder forms of phase sensitivity inside the basin also lead to complicated PTCs.

We remark that P_A is no longer well defined for all $\vartheta_{\text{old}} \in [0, 1)$ when Γ_A intersects the boundary of the basin $\mathcal{B}(\Gamma)$. For example, when Γ_A crosses an equilibrium that forms a single component of the basin boundary in a planar system, there exists exactly one $\vartheta \in [0, 1)$ such that $P_A(\vartheta)$ is not defined, because the perturbed phase point never returns to Γ. Entire intervals of $\vartheta_{\text{old}} \in [0, 1)$ must be excluded, e.g., when Γ_A crosses a repelling periodic orbit of a planar system, such that a closed segment of Γ_A lies outside $\mathcal{B}(\Gamma)$. Changes of the PTC during the transition through different types of boundaries of $\mathcal{B}(\Gamma)$ are beyond the scope of this chapter and will be reported elsewhere.

Phase resets for higher-dimensional systems are expected to exhibit other, more complicated behaviours that lead to possibly different mechanisms of loss of injectivity and/or surjectivity of the circle map associated with the PTC. In particular, the basin $\mathcal{B}(\Gamma)$ can be a lot more complicated, which affects the isochron foliation and, consequently, the PTC [24]. Such higher-dimensional systems are of particular interest when resets are considered in large coupled systems. Even when the coupling is through the mean-field dynamics, such systems can exhibit rich collective dynamics that are reflected in their PTCs [6,33]. We believe that our approach will be useful in this context, in particular, when the perturbation is a time-dependent stimulus.

Acknowledgments. This research of BK and HMO is supported by the Royal Society Te Apārangi Marsden Fund grant #16-UOA-286, and that of PL by a grant from the Fondation Leducq. We thank Leon Glass and Yannis Kevrekidis for their continued interest in planar isochron computations based on our BVP continuation set-up. Their tireless enquiries have led to the results presented in this chapter. We thank Michael Dellnitz for stimulating a friendly competitive environment that encouraged us to develop and apply computational methods for invariant manifolds in new contexts.

Table 1. Parameter values for system (24) as used for the seven-dimensional fast-upstroke model in [20].

$C_m = 6.5 \times 10^{-2}$ (μF),	$g_s = 1950$ (nS),	$g_h = 52$ (nS),
$g_{Na} = 325$ (nS),	$g_K = 354.9$ (nS),	$g_l = 65$ (nS).

Appendix: Details of the Sinoatrial Node Model

System (24) is the seven-dimensional fast-upstroke model from [20]. The five currents in the equation for V are defined as follows:

$$
\begin{aligned}
I_{Na} &= I_{Na}(V, m, h) = g_{Na}\, m^3\, h\, [V - 40.0], \\
I_s &= I_s(V, d, f) = g_s\, d\, f\, \left[e^{(V-40.0)/25.0} - 1.0\right], \\
I_K &= I_K(V, p) = g_K\, p\, \left[e^{(V+90.0)/36.1} - 1.0\right] e^{-(V+40.0)/36.1}, \\
I_h &= I_h(V, q) = g_h\, q\, [V + 25.0], \\
I_l &= I_l(V) \\
&= g_l \left(1.2 \left[1.0 - e^{-(V+60.0)/25.0}\right] + 0.15\, [V - 2.0] \left[1.0 - e^{-(V-2.0)/5.0}\right]^{-1}\right).
\end{aligned}
$$

The parameters g_{Na}, g_s, g_K, g_h, and g_l, are conductances (measured in nS). The capacitance C_m and these five conductances are set to the same values as those of the fast-upstroke model in [20]; see also Table 1. The V-dependent functions for m are defined as

$$
\begin{cases}
\alpha_m(V) = 10^3\, [V + 37.0] \left[1.0 - e^{-(V+37.0)/10.0}\right]^{-1}, \\
\beta_m(V) = 4.0 \times 10^4\, e^{-(V+62.0)/17.9},
\end{cases}
$$

for h, they are defined as

$$
\begin{cases}
\alpha_h(V) = 0.1209\, e^{-(V+30.0)/6.534}, \\
\beta_h(V) = 10^2 \left[e^{-(V+40.0)/10.0} + 0.1\right]^{-1},
\end{cases}
$$

for p, they are

$$
\begin{cases}
\alpha_p(V) = 8.0 \left[1.0 + e^{-(V+4.0)/13.0}\right]^{-1}, \\
\beta_p(V) = 0.17\, [V + 40.0] \left[e^{(V+40.0)/13.3} - 1.0\right]^{-1},
\end{cases}
$$

for d, they are

$$
\begin{cases}
\alpha_d(V) = 1.2 \times 10^3 \left[1.0 + e^{-V/12.0}\right]^{-1}, \\
\beta_d(V) = 2.5 \times 10^2 \left[1.0 + e^{(V+30.0)/8.0}\right]^{-1},
\end{cases}
$$

for f, they are

$$\begin{cases} \alpha_f(V) = 0.7\ [V + 45.0]\ \left[e^{(V+45.0)/9.5} - 1.0\right]^{-1}, \\ \beta_f(V) = 36.0\ \left[1.0 + e^{-(V+21.0)/9.5}\right]^{-1}, \end{cases}$$

and finally, for q, they are defined as

$$\begin{cases} \alpha_q(V) = 0.34\ [V + 100.0]\ \left[e^{(V+100.0)/4.4} - 1.0\right]^{-1} + 0.0495, \\ \beta_q(V) = 0.5\ [V + 40.0]\ \left[1.0 - e^{-(V+40.0)/6.0}\right]^{-1} + 0.0845. \end{cases}$$

References

1. Brown, E., Moehlis, J., Holmes, P.: On the phase reduction and response dynamics of neural oscillator populations. Neural Comput. **16**(4), 673–715 (2004)
2. Castejón, O., Guillamon, A.: Phase-amplitude dynamics in terms of extended response functions: invariant curves and Arnold tongues. Commun. Nonlinear Sci. Numer. Simul. **81**, 105008 (2020)
3. Castejón, O., Guillamon, A., Huguet, G.: Phase-amplitude response functions for transient-state stimuli. J. Math. Neurosci. **3**, 13 (2013)
4. Doedel, E.J.: Auto: a program for the automatic bifurcation analysis of autonomous systems. Congr. Numer. **30**, 265–284 (1981)
5. Doedel, E.J., Oldeman, B.E.: Auto-07P: continuation and bifurcation software for ordinary differential equations. With major contributions from Champneys, A.R., Dercole, F., Fairgrieve, T.F., Kuznetsov, Yu.A., Paffenroth, R.C., Sandstede, B., Wang, X.J., Zhang, C.H. (2007). http://cmvl.cs.concordia.ca/auto/
6. Duchet, B., Weerasinghe, G., Cagnan, H., Brown, P., Bick, C., Bogacz, R.: Phase dependence of response curves to deep brain stimulation and their relationship: from essential tremor patient data to a Wilson-Cowan model. J. Math. Neurosci. **10**(1), 4 (2020)
7. Ermentrout, G.B.: Type I membranes, phase resetting curves, and synchrony. Neural Comput. **8**(5), 979–1001 (1996)
8. Ermentrout, G.B., Glass, L., Oldeman, B.E.: The shape of phase-resetting curves in oscillators with a saddle node on an invariant circle bifurcation. Neural Comput. **24**(12), 3111–3125 (2012)
9. Ermentrout, G.B., Terman, D.H.: Mathematical Foundations of Neuroscience. Springer, New York (2010)
10. FitzHugh, R.: Impulses and physiological states in theoretical models of nerve membrane. Biophys. J. **1**(6), 445–466 (1961)
11. Glass, L., Winfree, A.T.: Discontinuities in phase-resetting experiments. Am. J. Physiol.-Regul. Integr. Comp. Physiol. **246**(2), R251–R258 (1984)
12. Guckenheimer, J.: Isochrons and phaseless sets. J. Math. Biol. **1**(3), 259–273 (1975)
13. Guckenheimer, J., Holmes, P.: Nonlinear Oscillations, Dynamical Systems, and Bifurcations of Vector Fields. Springer, New York (1986)
14. Guillamon, A., Huguet, G.: A computational and geometric approach to phase resetting curves and surfaces. SIAM J. Appl. Dyn. Syst. **8**(3), 1005–1042 (2009)
15. Gutkin, B.S., Ermentrout, G.B., Reyes, A.D.: Phase-response curves give the responses of neurons to transient inputs. J. Neurophys. **94**(2), 1623–1635 (2005)

16. Hansel, D., Mato, G., Meunier, C.: Synchrony in excitatory neural networks. Neural Comput. **7**(2), 307–337 (1995)
17. Hodgkin, A.L.: The local electric changes associated with repetitive action in a non-modulated axon. J. Physiol. **107**(2), 165–181 (1948)
18. Huguet, G., de la Llave, R.: Computation of limit cycles and their isochrons: fast algorithms and their convergence. SIAM J. Appl. Dyn. Syst. **12**(4), 1763–1802 (2013)
19. Krauskopf, B., Rieß, T.: A Lin's method approach to finding and continuing heteroclinic connections involving periodic orbits. Nonlinearity **21**(8), 1655–1690 (2008)
20. Krogh-Madsen, T., Glass, L., Doedel, E.J., Guevara, M.R.: Apparent discontinuities in the phase-resetting response of cardiac pacemakers. J. Theor. Biol. **230**(4), 499–519 (2004)
21. Langfield, P., Krauskopf, B., Osinga, H.M.: Solving Winfree's puzzle: the isochrons in the FitzHugh-Nagumo model. Chaos **24**(1), 013131 (2014)
22. Langfield, P., Krauskopf, B., Osinga, H.M.: Forward-time and backward-time isochrons and their interactions. SIAM J. Appl. Dyn. Syst. **14**(3), 1418–1453 (2015)
23. Mauroy, A., Mezić, I.: On the use of Fourier averages to compute the global isochrons of (quasi) periodic dynamics. Chaos **22**(3), 033112 (2012)
24. Mauroy, A., Mezić, I.: Extreme phase sensitivity in systems with fractal isochrons. Physica D **308**, 40–51 (2015)
25. Mauroy, A., Rhoads, B., Moehlis, J., Mezić, I.: Global isochrons and phase sensitivity of bursting neurons. SIAM J. Appl. Dyn. Syst. **13**(1), 306–338 (2014)
26. Monga, B., Wilson, D., Matchen, T., Moehlis, J.: Phase reduction and phase-based optimal control for biological systems: a tutorial. Biol. Cyber. **113**(1–2), 11–46 (2019)
27. Nagumo, J., Arimoto, S., Yoshizawa, S.: An active pulse transmission line simulating nerve axon. Proc. IRE **50**(10), 2061–2070 (1962)
28. Nowacki, J., Osinga, H.M., Tsaneva-Atanasova, K.T.: Continuation-based numerical detection of after-depolarization and spike-adding thresholds. Neural Comput. **25**(4), 877–900 (2013)
29. Osinga, H.M., Moehlis, J.: Continuation-based computation of global isochrons. SIAM J. Appl. Dyn. Syst. **9**(4), 1201–1228 (2010)
30. Pérez-Cervera, A., Seara, T.M., Huguet, G.: A geometric approach to phase response curves and its numerical computation through the parameterization method. J. Nonlinear Sci. **29**(6), 2877–2910 (2019)
31. Pietras, B., Daffertshofer, A.: Network dynamics of coupled oscillators and phase reduction techniques. Phys. Rep. **819**, 1–150 (2019)
32. Schultheiss, N.W., Prinz, A.A., Butera, R.J. (eds.): Phase Response Curves in Neuroscience: Theory, Experiment, and Analysis. Springer, Cambridge (2012)
33. Ullner, E., Politi, A.: Self-sustained irregular activity in an ensemble of neural oscillators. Phys. Rev. X **6**, 011015 (2016). Erratum: Phys. Rev. X **7**, 029901 (2017)
34. Wedgwood, K.C.A., Lin, K.K., Thul, R., Coombes, S.: Phase-amplitude descriptions of neural oscillator models. J. Math. Neurosci. **3**, 2 (2013)
35. Wilson, D., Ermentrout, G.B.: Augmented phase reduction of (not so) weakly perturbed coupled oscillators. SIAM Rev. **61**(2), 277–315 (2019)
36. Winfree, A.T.: Patterns of phase compromise in biological cycles. J. Math. Biol. **1**(1), 73–93 (1974)
37. Winfree, A.T.: The Geometry of Biological Time, 2nd edn. Springer, New York (2001)

Input-Output Networks, Singularity Theory, and Homeostasis

Martin Golubitsky[1(✉)], Ian Stewart[2], Fernando Antoneli[3], Zhengyuan Huang[4], and Yangyang Wang[5]

[1] Mathematics Department, The Ohio State University, Columbus, OH 43210, USA
golubitsky.4@osu.edu
[2] Mathematics Institute, University of Warwick, Coventry CV4 7AL, UK
i.n.stewart@warwick.ac.uk
[3] Escola Paulista de Medicina, Universidade Federal de São Paulo, São Paulo, SP 04039-032, Brazil
fernando.antoneli@unifesp.br
[4] The Ohio State University, Columbus, OH 43210, USA
huang.3224@buckeyemail.osu.edu
[5] Department of Mathematics, The University of Iowa, Iowa City, IA 52242, USA
yangyang-wang@uiowa.edu

Abstract. Homeostasis is a regulatory mechanism that keeps some specific variable close to a set value as other variables fluctuate, and is of particular interest in biochemical networks. We review and investigate a reformulation of homeostasis in which the system is represented as an input-output network, with two distinguished nodes 'input' and 'output', and the dynamics of the network determines the corresponding input-output function of the system. Interpreting homeostasis as an infinitesimal notion—namely, the derivative of the input-output function is zero at an isolated point—we apply methods from singularity theory to characterise homeostasis points in the input-output function. This approach, coupled to graph-theoretic ideas from combinatorial matrix theory, provides a systematic framework for calculating homeostasis points in models, classifying different types of homeostasis in input-output networks, and describing all small perturbations of the input-output function near a homeostasis point.

1 Introduction

Homeostasis is an important concept, occurring widely in biology, especially biochemical networks, and in many other areas including control engineering. A system exhibits homeostasis if some output variable remains constant, or almost constant, when an input variable or parameter changes by a relatively large amount. In the control theory literature, mathematical models of homeostasis are often constructed by requiring the output to be constant when the input lies in some range. That is, the derivative of the input-output function is identically

O. Junge et al. (Eds.): SON 2020, SSDC 304, pp. 31–65, 2020.
https://doi.org/10.1007/978-3-030-51264-4_2

zero on that interval of input values. Such models have *perfect homeostasis* or *perfect adaptation* [17,41].

An alternative approach is introduced and studied in [22,23,25,37,42], using an 'infinitesimal' notion of homeostasis—namely, the derivative of the input-output function is zero at an isolated point—to introduce singularity theory into the study of homeostasis. From this point of view, perfect homeostasis is an infinite-codimension phenomenon, hence highly non-generic. It is also unlikely to occur *exactly* in a biological system. Nonetheless, perfect homeostasis can be a reasonable modeling assumption for many purposes.

The singularity-theoretic analysis leads to conditions that are very similar to those that occur in bifurcation theory when recognizing and unfolding bifurcations (see [20,24]). These conditions have been used to organize the numerical computation of bifurcations in nonlinear systems, for example in conjunction with continuation methods. See for example Dellnitz [9–11], Dellnitz and Junge [12], Dellnitz *et al.* [13], Jepson and Spence [27], Jepson *et al.* [28], and Moore *et al.* [31]. It might be possible to adapt some of these methods to homeostasis. Donovan [15,16] has used the singularity-theoretic framework to adapt such numerical methods to homeostasis. As well as organizing the numerical calculations, singularity theory and homeostasis matrix techniques may help to simplify them.

Mathematically, homeostasis can be thought of as a network concept. One variable (a network node) is held approximately constant as other variables (other nodes) vary (perhaps wildly). Network systems are distinguished from large systems by the desire to keep track of the output from each node individually. If we are permitted to mix the output from several nodes, then homeostasis is destroyed, since the sum of a constant variable with a wildly varying one is wildly variable. Placing homeostasis in the general context of network dynamics leads naturally to the methods reviewed here.

Summary of Contents

Section 2 opens the discussion with a motivational example of homeostasis: regulation of the output 'body temperature' in an opossum, when the input 'environmental temperature' varies. The graph of body temperature against environmental temperature I is approximately linear, with nonzero slope, when I is either small or large, while in between is a broad flat region, where homeostasis occurs. This general shape is called a 'chair' by Nijhout and Reed [34] (see also [33,35]), and plays a central role in the singularity theory discussion. This example is used in Sect. 4 to motivate a reformulation of homeostasis in terms of the derivative of an output variable with respect to an input being zero at some point, hence approximately constant near that point. We discuss this mathematical reformulation in terms of singularities of input-output functions.

Section 5 introduces input-output networks – networks that have input and output nodes. In such networks the observable is just the value of the output node as a function of the input that is fed into the input node. This simplified form of the observable and the input-output map allows us to use Cramer's rule to simplify the search for infinitesimal homeostasis points. See Lemma 5.2.

As it happens, many nodes and arrows in input-output networks may have no effect on the existence of homeostasis. The end result is that when looking for infinitesimal homeostasis in the original network, we may first reduce that network to a 'core' network. The definition of and reduction to the core are given in Sect. 6. These reductions allow us to discuss three different types of infinitesimal homeostasis in three-node input-output networks. The first is that there are only three core networks in three-node input-output networks (even though there are 78 possible input-output three-node networks) and there are three types of infinitesimal homeostasis (*Haldane*, *null-degradation*, and *structural*) distinguished by the mathematics. The mathematics of three-node input-output networks is presented in Sect. 8, and the relation to the biochemical networks that motivated the mathematics is given in Sect. 3.

Section 9 discusses the relationship between infinitesimal homeostasis and singularity theory—specifically elementary catastrophe theory [19,36,43]. The two simplest singularities are simple homeostasis and the chair. We characterize these singularities, discuss their normal forms (the simplest form into which the singularity can be transformed by suitable coordinate changes), and universal unfoldings, which classify all small perturbations as other system parameters vary. We relate the unfolding of the chair to observational data on two species of opossum and the spiny rat, Fig. 2. Section 9 also provides a brief discussion of how chair points can be calculated analytically by implicit differentiation, and considers a special case with extra structure, common in biochemical applications, where the calculations simplify.

Catastrophe theory enables us to discuss how infinitesimal homeostasis can arise in systems with an extra parameter. In Sect. 10 we see that the simplest such way for homeostasis to evolve is through a chair singularity. This observation gives a mathematical reason for why infinitesimal chairs are important and complements the biological reasons given by Nijhout, Reed, and Best [33,35].

Until this point the paper has dealt with input-output functions having one input variable. This is the most important case; however multiple input systems are also important. We follow [23] and discuss two input systems in Sect. 11. We argue that the hyperbolic umbilic of elementary catastrophe theory plays the role of the chair in systems with two inputs.

The paper ends with a discussion of a possible singularity theory description of housekeeping genes in Sect. 12. Here we emphasize how both the homeostasis network theory and the network singularity theory intertwine. The details of this application are given in Antoneli et al. [1].

2 Thermoregulation: A Motivation for Homeostasis

Homeostasis occurs when some feature of a system remains essentially constant as an input parameter varies over some range of values. For example, in thermoregulation the body temperature of an organism remains roughly constant despite variations in its environment. (See Fig. 1 for such data in the brown opossum where body temperature remains approximately constant over a range of 18 °C in environmental temperature [32,33].) Or in a biochemical network the equilibrium concentration of some important biochemical molecule might not change much when the organism ingests food.

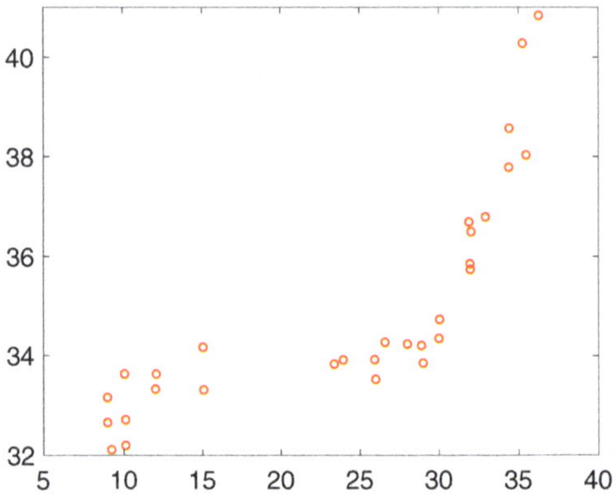

Fig. 1. Experimental data indicating thermoregulatory homeostasis in the brown opossum. The horizontal axis is *environmental temperature* (°C) and the vertical axis is *body temperature* (°C) [32,33].

Homeostasis is almost exactly opposite to bifurcation. At a bifurcation, the state of the system undergoes a change so extensive that some qualitative property (such as number of equilibria, or the topological type of an attractor) changes. In homeostasis, the state concerned not only remains topologically the same: some feature of that state does not even change quantitatively. For example, if a steady state does not bifurcate as a parameter is varied, that state persists, but can change continuously with the parameter. Homeostasis is stronger: the steady state persists, and in addition some feature of that steady state remains almost constant.

Homeostasis is biologically important, because it protects organisms against changes induced by the environment, shortage of some resource, excess of some resource, the effect of ingesting food, and so on. The literature is extensive [44]. However, homeostasis is not merely the existence (and persistence as parameters vary) of a stable equilibrium of the system, for two reasons.

First, homeostasis is a stronger condition than 'the equilibrium varies smoothly with parameters', which just states that there is no bifurcation. In the biological context, approximately *linear* variation of the equilibrium with nonzero slope as parameters change is not normally considered to be homeostasis, unless the slope is very small. For example, in Fig. 1, body temperature appears to be varying linearly when the environmental temperature is either below 10 °C or above 30 °C and is approximately constant in between. Nijhout *et al.* [33] call this kind of variation (linear, constant, linear) a *chair*.

Second, some variable(s) of the system may be homeostatic while others undergo larger changes. Indeed, other variables may have to change dramatically to keep some specific variable roughly constant.

We assume that there is an input-output function, which we consider to be the product of a system black box. Specifically, we assume that for each *input* $\mathcal{I}$ there is an *output* $x_o(\mathcal{I})$. For opossums, $\mathcal{I}$ is the environmental temperature from which the opossum body produces an internal body temperature $x_o(\mathcal{I})$.

Nijhout *et al.* [33] suggest that there is a chair in the body temperature data of opossums [32]. We take a singularity-theoretic point of view and suggest that chairs are better described locally by a homogeneous cubic function (that is, like $x_o(\mathcal{I}) \approx \mathcal{I}^3$) rather than by the previous piecewise linear description. Figure 2(a) shows the least-squares fit of a cubic function to data for the brown opossum, which is a cubic with a maximum and a minimum. In contrast, the least squares fit for the eten opossum, Fig. 2(b), is monotone.

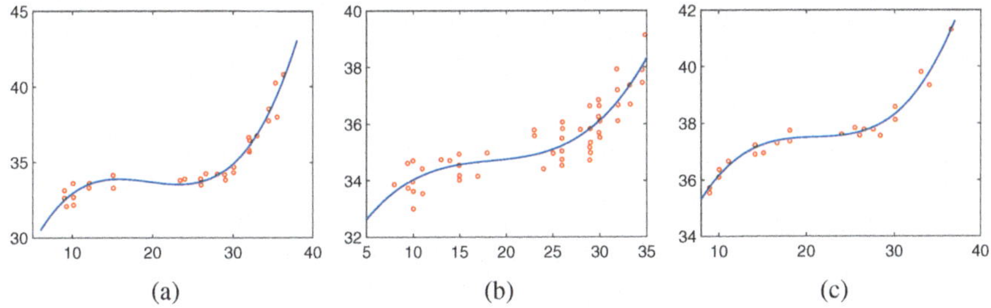

Fig. 2. The horizontal coordinate is *environmental temperature*; the vertical coordinate is *body temperature*. From [32] and [22]: (a) data from the brown opossum; (b) data from the eten opossum; (c) data from the spiny rat. The smooth curves are the least squares best fit of the data to a cubic polynomial.

These results suggest that in 'opossum space' there should be a hypothetical type of opossum that exhibits a chair in the system input-output function of *environmental temperature* to *body temperature*. In singularity-theoretic terms, this higher singularity acts as an organizing center, meaning that the other types of cubic can be obtained by small perturbations of the homogeneous cubic. In fact, data for the spiny rat have a best-fit cubic very close to the homogeneous cubic, Fig. 2(c). We include this example as a motivational metaphor, since we do not consider a specific model for the regulation of opossum body temperature.

This example, especially Fig. 2, motives a formulation of homeostasis in a way that can be analyzed using singularity theory. The first step in any discussion of homeostasis must be the formulation of a model that defines, perhaps only implicitly, the input-output function x_o. Our singularity theory point of view suggests defining *infinitesimal homeostasis* as an input $\mathcal{I}_0$ where the derivative of output x_o with respect to the input vanishes at $\mathcal{I}_0$; that is, $x_o'(\mathcal{I}_0) = 0$.

3 Biochemical Input-Output Networks

We provide context for our results by first introducing some of the biochemical models discussed by Reed in [37]. In doing so we show that input-output networks form a natural category in which homeostasis may be explored.

There are many examples of biochemical networks in the literature. In particular examples, modelers decide which substrates are important and how the various substrates interact. Figure 3 shows a network resulting from the detailed modeling of the production of extracellular dopamine (eDA) by Best *et al.* [3] and Nijhout *et al.* [33]. These authors derive a differential equation model for this biochemical network and use the results to study homeostasis of eDA with respect to variation of the enzyme tyrosine hydroxylase (TH) and the dopamine transporters (DAT).

In another direction, relatively small biochemical network models are often derived to help analyse a particular biochemical phenomenon. We present four examples; three are discussed in Reed et al. [37] and one in Ma et al. [29]. These examples belong to a class that we call biochemical *input-output networks* (Sect. 5) and will help to interpret the mathematical results.

3.1 Feedforward Excitation

The input-output network corresponding to feedforward excitation is in Fig. 4. This motif occurs in a biochemical network when a substrate activates the enzyme that removes a product. The standard biochemical network diagram for this process is shown in Fig. 4a. Here **X**, **Y**, **Z** are the names of chemical substrates and their concentrations are denoted by lower case x, y, z. Each straight arrow represents a flux coming into or going away from a substrate. The differential equations for each substrate simply state that the rate of change of the concentration is the sum of the arrows going towards the substrate minus the arrows going away (conservation of mass). The curved line indicates that substrate is activating an enzyme.

Both diagrams in Fig. 4 represent the same information, but in different ways. The framework employed in this paper for the mathematics focuses on the structure of the model ODEs. Figure 4b uses nodes to represent variables, and arrows to represent couplings. In other areas, conventions can differ, so it is necessary to translate between the two representations. The simplest method is to write down the model ODEs.

In this motif, one path consists of two excitatory couplings: $g_1(x) > 0$ from **X** to **Y** and $g_2(y) > 0$ from **Y** to **Z**. The other path is an excitatory coupling $f(x) > 0$ from **X** to the synthesis or degradation $g_3(z)$ of **Z** and hence is an inhibitory path from **X** to **Z** having a negative sign.

The equations are the first column of:

$$
\begin{aligned}
\dot{x} &= \mathcal{I} - g_1(x) - g_4(x) &\qquad \dot{x}_\iota &= f_\iota(x_\iota, \mathcal{I}) \\
\dot{y} &= g_1(x) - g_2(y) - g_5(y) &\qquad \dot{x}_\rho &= f_\rho(x_\iota, x_\rho) \\
\dot{z} &= g_2(y) - f(x)g_3(z) &\qquad \dot{x}_o &= f_o(x_\iota, x_\rho, x_o)
\end{aligned}
\tag{1}
$$

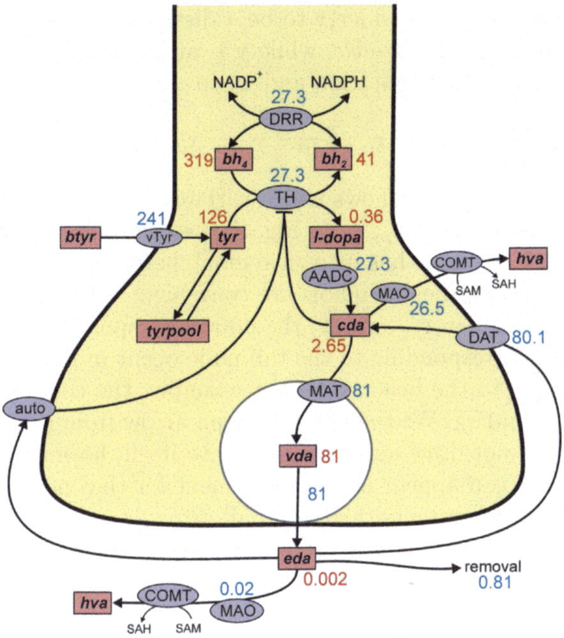

Fig. 3. Biochemical network for dopamine synthesis, release, and reuptake in Nijhout et al. [33] and Golubitsky and Stewart [23].

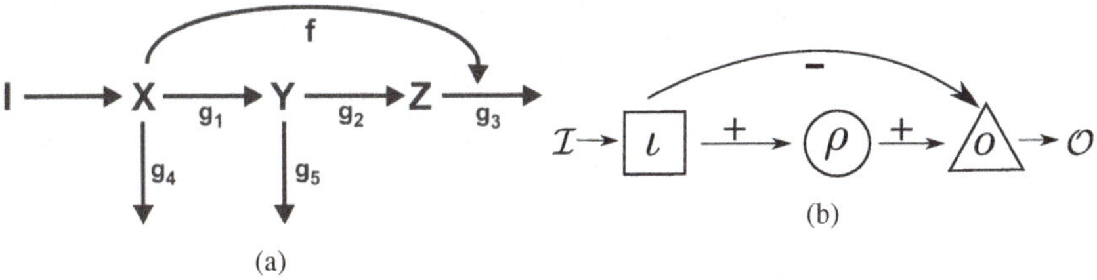

Fig. 4. Feedforward excitation: (a) Motif from [37]; (b) Input-output network with two paths from ι to o corresponding to the motif in (a).

It is shown in [37] (and reproduced using this theory in [25]) that the model system (1) (left) for feedforward excitation leads to infinitesimal homeostasis at X_0 if

$$f_x(x_0) = \frac{g_1'(x_0)g_2'(y_0)}{g_3(z_0)(g_2'(y_0) + g_5'(y_0))}$$

where $X_0 = (x_0, y_0, z_0)$ is a stable equilibrium.

Figure 4b redraws the diagram in Fig. 4a using the math network conventions of this paper, together with some extra features that are crucial to this particular application. We consider x to be a distinguished *input variable*, with z as a distinguished *output variable*, while y is an intermediate *regulatory variable*. Accordingly we change notation and write

$$x_\iota = x \qquad x_\rho = y \qquad x_o = z$$

The second column in (1) shows which variables occur in the components of the model ODE for each of x_ι, x_ρ, x_o. In Fig. 4b these variables are associated with three nodes ι, ρ, o. Each node has its own symbol, here a square for ι, circle for ρ, and triangle for o. Here these symbols are convenient ways to show which type of variable (input, regulatory, output) the node corresponds to. Arrows indicate that the variables corresponding to the tail node occur in the component of the ODE corresponding to the head node. For example, the component for $\dot{x}_o$ is a function of x_ι, x_ρ, and x_o. We therefore draw an arrow from ι to o and an arrow from ρ to o. We do not draw an arrow from o to itself, however: by convention, every node variable can appear in the component for that node. In a sense, the node symbol (circle) represents this 'internal' arrow.

The mathematics described here shows that infinitesimal homeostasis occurs in the system in the second column of (1) if and only if

$$f_{\rho,x_\iota} f_{o,x_\rho} - f_{\rho,x_\rho} f_{o,x_\iota} = 0$$

at the stable equilibrium X_0.

Here Fig. 4b incorporates some additional information. The arrow from $\mathcal{I}$ to node ι shows that $\mathcal{I}$ occurs in the equation for $\dot{x}_\iota$ as a *parameter*. Similarly the arrow from node o to $\mathcal{O}$ shows that node o is the output node. Finally, the $\pm$ signs indicate which arrows are excitatory or inhibitory. This extra information is special to biochemical networks and does not appear as such in the general theory.

3.2 Product Inhibition

Here substrate **X** influences **Y**, which influences **Z**, and **Z** inhibits the flux g_1 from **X** to **Y**. The biochemical network for this process is shown in Fig. 5a.

This time the model equations for Fig. 5a are in the first column of (2)

$$\begin{aligned} \dot{x} &= \mathcal{I} - g_4(x) - f(z)g_1(x) \\ \dot{y} &= f(z)g_1(x) - g_2(y) - g_5(y) \\ \dot{z} &= g_2(y) - g_3(z) \end{aligned} \qquad \begin{aligned} \dot{x}_\iota &= f_\iota(x_\iota, x_o, \mathcal{I}) \\ \dot{x}_\rho &= f_\rho(x_\iota, x_\rho, x_o) \\ \dot{x}_o &= f_o(x_\rho, x_o) \end{aligned} \tag{2}$$

and the input-output equations in the second column of (2) can be read directly from the first column. The input-output network in Fig. 5b then follows.

Reed et al. [37] discuss why the model equations for product inhibition also satisfy

$$f > 0 \quad g_1' < 0 \quad g_2' < 0 \tag{3}$$

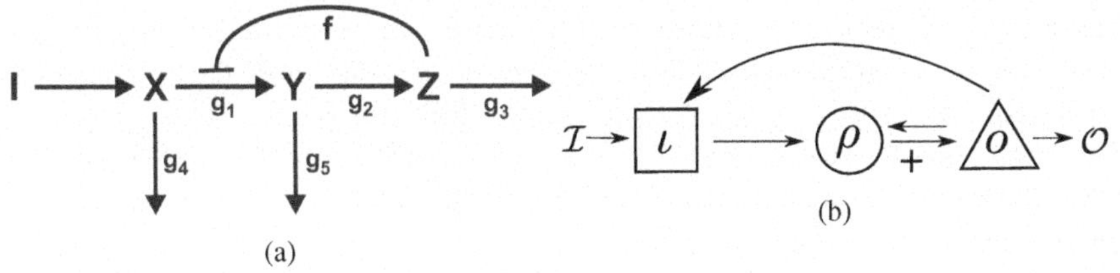

Fig. 5. Product inhibition: (a) Motif from [37]; (b) Input-output network with one path from ι to o corresponding to the motif in (a).

Our general mathematical results show that the system in the second column of (2) exhibits infinitesimal homeostasis at a stable equilibrium X_0 if and only if either

$$f_{\rho,x_\iota} = f(z_0)g_1'(x_0) = 0 \qquad \text{or} \qquad f_{o,x_\rho} = g_2'(y_0) = 0 \tag{4}$$

It follows from (3) and (4) that the model equations cannot satisfy infinitesimal homeostasis. Nevertheless, Reed et al. [37] show that these bichemical network equations do exhibit *homeostasis*; that is, the output z is *almost* constant for a broad range of input values $\mathcal{I}$.

3.3 Substrate Inhibition

The biochemical network model for substrate inhibition is given in Fig. 6a, and the associated model system is given in the first column of (5). This biochemical network and the model system are discussed in Reed et al. [37]. In particular, this paper provides justification for taking $g_1'(x) > 0$ for all relevant x, whereas the coupling (or kinetics term) $g_2'(y)$ can change sign.

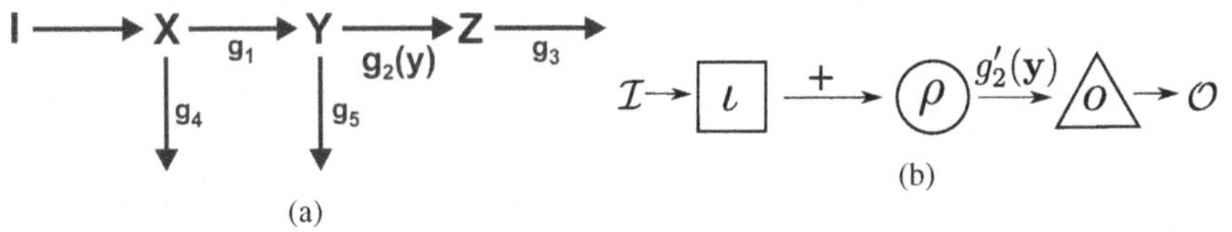

Fig. 6. Substrate inhibition: (a) Motif from [37]; (b) Input-output network corresponding to the motif in (a).

The equations are the first column of:

$$\begin{aligned} \dot{x} &= \mathcal{I} - g_1(x) - g_4(x) \qquad & \dot{x}_\iota &= f_\iota(x_\iota, \mathcal{I}) \\ \dot{y} &= g_1(x) - g_5(y) - g_2(y) \qquad & \dot{x}_\rho &= f_\rho(x_\iota, x_\rho) \\ \dot{z} &= g_2(y) - g_3(z) \qquad & \dot{x}_o &= f_o(x_\rho, x_o) \end{aligned} \tag{5}$$

That model system of ODEs is easily translated to the input-output system in the second column of (5). Our theory shows that the equations for infinitesimal homeostasis are identical to those given in (4) for product inhibition. Given the assumption on g_1' infinitesimal homeostasis is possible only if the coupling is neutral (that is, if $f_{o,x_\rho} = g_2' = 0$ at the equilibrium point). This observation agrees with the observation in [37] that **Z** can exhibit infinitesimal homeostasis in the substrate inhibition motif if the infinitesimal homeostasis is built into the kinetics term g_2 between **Y** and **Z**.

Reed et al. [37] note that neutral coupling can arise from substrate inhibition of enzymes, enzymes that are inhibited by their own substrates. See the discussion in [38]. This inhibition leads to reaction velocity curves that rise to a maximum (the coupling is excitatory) and then descend (the coupling is inhibitory) as the substrate concentration increases. Infinitesimal homeostasis with neutral couplings arising from substrate inhibition often has important biological functions and has been estimated to occur in about 20% of enzymes [38].

3.4 Negative Feedback Loop

The input-output network in Fig. 7b corresponding to the negative feedback loop motif in Fig. 7a has only one simple path $\iota \to o$. Our results imply that infinitesimal homeostasis is possible in the negative feedback loop if and only if the coupling $\iota \to o$ is neutral (Haldane) or the linearized internal dynamics of the regulatory node ρ is zero (null-degradation).

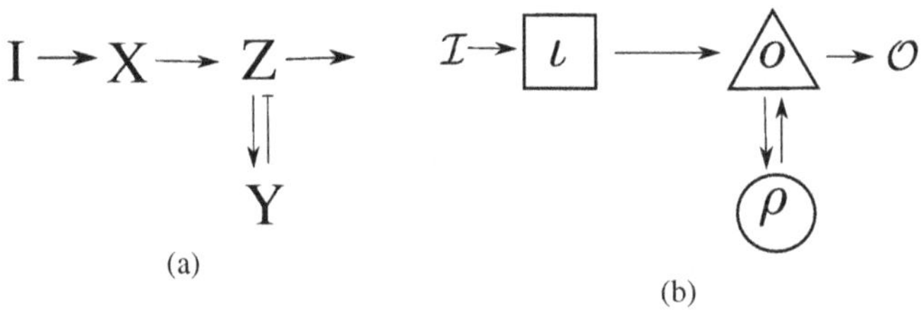

Fig. 7. Negative feedback loop: (a) Motif adapted from [29]. Unlike the arrows in Figs. 4, 6 and 5 that represent mass transfer between substrates, positive or negative arrows between enzymes in this negative feedback motif indicate the activation or inactivation of an enzyme by a different enzyme. (b) Input-output network corresponding to the motif in (a).

The equations are:

$$
\begin{aligned}
\dot{x} &= \mathcal{I}\, k_{\mathcal{I}x} \tfrac{1-x}{(1-x)+K_{\mathcal{I}x}} - F_x k'_{F_x} \tfrac{x}{x+K'_{F_x}} \\
\dot{y} &= z k_{zy} - F_y k'_{F_y} \\
\dot{z} &= x k_{xz} \tfrac{1-z}{(1-z)+K_{xz}} - y k'_{yz} \tfrac{z}{z+K'_{yz}}
\end{aligned}
\qquad
\begin{aligned}
\dot{x}_\iota &= f_\iota(x_\iota, \mathcal{I}) \\
\dot{x}_\rho &= f_\rho(x_\rho, x_o) \\
\dot{x}_o &= f_o(x_\iota, x_\rho, x_o)
\end{aligned}
\tag{6}
$$

where $k_{\mathcal{I}x}, K_{\mathcal{I}x}, F_x, k'_{F_x}, K'_{F_x}, k_{zy}, F_y, k'_{F_y}, k_{xz}, K_{xz}, k'_{yz}, K'_{yz}$ are 12 constants.

Each enzyme **X**, **Y**, **Z** in the feedback loop motif (Fig. 7a) can have active and inactive forms. In the kinetic equations (6, left) the coupling from **X** to

Z is non-neutral according to [29]. Hence, in this model only null-degradation homeostasis is possible. In addition, in the model the $\dot{y}$ equation does not depend on y and homeostasis can only be perfect homeostasis. However, this model is a simplification based on saturation in y [29]. In the original system $\dot{y}$ does depend on y and we expect standard null-degradation homeostasis to be possible in that system.

Stability of the equilibrium in this motif implies negative feedback. The Jacobian of (6, right) is

$$J = \begin{bmatrix} f_{\iota,x_\iota} & 0 & 0 \\ 0 & f_{\rho,x_\rho} & f_{\rho,x_o} \\ f_{o,x_\iota} & f_{o,x_\rho} & f_{o,x_o} \end{bmatrix}$$

At null-degradation homeostasis ($f_{\rho,x_\rho} = 0$) it follows from linear stability that

$$f_{\iota,x_\iota} < 0, \qquad f_{o,x_o} < 0, \qquad f_{\rho,x_o} f_{o,x_\rho} < 0 \tag{7}$$

Conditions (7) imply that both the input node and the output node need to degrade and the couplings $\rho \to o$ and $o \to \rho$ must have opposite signs. This observation agrees with [29] that homeostasis is possible in the network motif Fig. 7a if there is a negative loop between **Y** and **Z** and when the linearized internal dynamics of **Y** is zero.

Another biochemical example of null-degradation homeostasis can be found in [17, Fig. 2].

4 Infinitesimal Homeostasis

In applications, homeostasis is often a property of an observable on a many-variable system of ODEs. Specifically, consider a system of ODEs

$$\dot{X} = F(X, \mathcal{I}) \tag{8}$$

in a vector of variables $X = (x_1, \ldots, x_m) \in \mathbf{R}^m$ that depends on an *input* parameter $\mathcal{I} \in \mathbf{R}$. Although not always valid in applications we assume that F is infinitely differentiable. Suppose that (8) has a linearly stable equilibrium at $(X_0, \mathcal{I}_0)$. By the implicit function theorem there exists a family of linearly stable equilibria $X(\mathcal{I}) = (x_1(\mathcal{I}), \ldots, x_m(\mathcal{I}))$ near $\mathcal{I}_0$ such that $X(\mathcal{I}_0) = X_0$ and

$$F(X(\mathcal{I}), \mathcal{I}) \equiv 0. \tag{9}$$

By assumption, we are interested in homeostasis of a chosen observable $\varphi : \mathbf{R}^m \to \mathbf{R}$. The *input-output* function is

$$x_o(\mathcal{I}) \equiv \varphi(X(\mathcal{I})) \tag{10}$$

This system exhibits homeostasis if the input-output function $x_o(\mathcal{I})$ remains roughly constant as $\mathcal{I}$ is varied.

Often times the observable is just one coordinate of the ODE system; that is, $\varphi(X) = x_j$, which we denote as the *output variable* x_o. This formulation of homeostasis is often a network formulation. The output variable is just a choice of *output node* and the input parameter can be assumed to affect only one node—the *input node* x_ι.

We now introduce a formal mathematical definition of infinitesimal homeostasis, one which opens up a potential singularity-theoretic approach that we discuss later.

Definition 4.1. The equilibrium X_0 is *infinitesimally homeostatic at* $\mathcal{I}_0$ if

$$x_o'(\mathcal{I}_0) = 0$$

where $'$ indicates differentiation with respect to $\mathcal{I}$.

By Taylor's theorem, infinitesimal homeostasis implies homeostasis, but the converse need not be true. See [37] and the discussion of product inhibition in Sect. 3.

5 Input-Output Networks

We now apply the notion of infinitesimal homeostasis to input-output networks—a natural formulation in biochemical networks that we discussed in detail in Sect. 3. We assume that one node ι is the *input node*, a second node o is the *output node*, and the remaining nodes $\rho = (\rho_1, \ldots, \rho_n)$ are the *regulatory nodes.* Our discussion of network infinitesimal homeostasis follows [25]. Input-output networks equations have the form $F = (f_\iota, f_\rho, f_o)$ where each coordinate function f_ℓ depends on the state variables of the nodes coupled to node ℓ in the network graph. We assume that only the input node coordinate function f_ι depends on the external input variable $\mathcal{I}$.

As shown in [25] there are 13 distinct three-node fully inhomogeneous networks and six choices of input and output nodes for each network. Thus, in principle, there are 78 possible ways to find homeostasis in three-node input-output networks. The number of input-output four-node networks increases dramatically: there are 199 fully inhomogeneous networks and more than 2000 four-node input-output networks.

Further motivated by biochemical networks, we assume:

(a) The state space for each node is 1-dimensional and hence the state space for an input-output network system of differential equations is $\mathbf{R}^{n+2}$.
(b) The coordinate functions f_ℓ are usually distinct functions, so the network is assumed to be fully inhomogeneous.
(c) Generically

$$f_{\iota,\mathcal{I}} \neq 0 \tag{11}$$

is valid everywhere, where the notation $f_{\ell,y}$ denotes the partial derivative of the coordinate function f_ℓ with respect to y.

Cramer's Rule and Infinitesimal Homeostasis

The equilibria of an input-output system satisfy the system

$$\begin{aligned} f_\iota(x_\iota, x_\rho, x_o, \mathcal{I}) &= 0 \\ f_\rho(x_\iota, x_\rho, x_o) &= 0 \\ f_o(x_\iota, x_\rho, x_o) &= 0 \end{aligned} \tag{12}$$

The assumption of a stable equilibrium X_0 at $\mathcal{I}_0$ implies that the Jacobian

$$J = \begin{bmatrix} f_{\iota,x_\iota} & f_{\iota,x_\rho} & f_{\iota,x_o} \\ f_{\rho,x_\iota} & f_{\rho,x_\rho} & f_{\rho,x_o} \\ f_{o,x_\iota} & f_{o,x_\rho} & f_{o,x_o} \end{bmatrix} \tag{13}$$

has eigenvalues with negative real part at $(X_0, \mathcal{I}_0)$, so J is invertible.

To state the next result we first need:

Definition 5.1. The *homeostasis matrix* is:

$$H \equiv \begin{bmatrix} f_{\rho,x_\iota} & f_{\rho,x_\rho} \\ f_{o,x_\iota} & f_{o,x_\rho} \end{bmatrix} \tag{14}$$

Lemma 5.2. *The input-output function for the input-output network* (12) *satisfies*

$$x_o' = \pm \frac{f_{\iota,\mathcal{I}}}{\det(J)} \det(H)$$

Infinitesimal homeostasis occurs at a stable equilibrium $X_0 = X(\mathcal{I}_0)$ *if and only if*

$$\det(H)(X_0) = 0 \tag{15}$$

Proof. Implicit differentiation of (12) with respect to $\mathcal{I}$ yields the matrix system

$$J \begin{bmatrix} x_i' \\ x_\rho' \\ x_o' \end{bmatrix} = - \begin{bmatrix} f_{\iota,\mathcal{I}} \\ 0 \\ 0 \end{bmatrix}$$

Cramer's rule implies that

$$x_o' = \frac{1}{\det(J)} \det \begin{bmatrix} f_{\iota,x_\iota} & f_{\iota,x_\rho} & -f_{\iota,\mathcal{I}} \\ f_{\rho,x_\iota} & f_{\rho,x_\rho} & 0 \\ f_{o,x_\iota} & f_{o,x_\rho} & 0 \end{bmatrix}$$

Since $f_{\iota,\mathcal{I}} \neq 0$ by genericity assumption (11), X_0 is a point of infinitesimal homeostasis if and only if $x_o' = 0$, if and only if (15), as claimed. □

6 Core Networks

The results in Sects. 6, 7 and 8 will appear in Wang et al. [42].

Definition 6.1. A node ρ is *downstream* from a node τ if there is a path from τ to ρ and *upstream* if there is a path from ρ to τ. An input-output network is a *core network* if every node is downstream from ι and upstream from o.

A core network $\mathcal{G}_c$ can be associated to any given input-output network $\mathcal{G}$ as follows. The nodes in $\mathcal{G}_c$ are the nodes in $\mathcal{G}$ that lie on a path from ι to o. The arrows in $\mathcal{G}_c$ are the arrows in $\mathcal{G}$ that connect nodes in $\mathcal{G}_c$.

Reduction to the Core
In this section we discuss why every network that exhibits infinitesimal homeostasis can be reduced to a core network in such a way that the core has essentially the same input-output function as the original network. This reduction is performed in two stages.

(a) Homeostasis implies that the output node o is downstream from the input node ι.
(b) Nodes that are not upstream from the output node, and nodes that are not downstream from the input node, may be deleted.

We show that if infinitesimal homeostasis occurs in the original network, then that infinitesimal homeostasis can be computed in the smaller core network.

Lemma 6.2. *In an input-output network, the existence of (generic) infinitesimal homeostasis implies that the output node o is downstream from the input node ι.*

Heuristically, if the input node is not upstream from the output node, then changes in the input node cannot affect the dynamics of the output node. So the input-output map must satistfy $x_o'(\mathcal{I}) \equiv 0$ and the set value $x_o(\mathcal{I})$ is constant (and not generic).

We assume that there is a path from the input node to the output node and show that nodes that are not upstream from o and nodes that are not downstream from ι can be deleted without changing the existence of homeostasis.

Proposition 6.3. *Let $\mathcal{G}$ be a connected input-output network where there is a path from the input node ι to the output node o. Divide the regulatory nodes ρ into three classes $\rho = (u, \sigma, d)$, where*

- *nodes in u are not upstream from o,*
- *nodes in d are not downstream from ι, and*
- *regulatory nodes σ are both upstream from o and downstream from ι.*

Then all nodes u, d and all arrows into nodes in u and out of nodes in d can be deleted to form a core network $\mathcal{G}_c$ without affecting the existence of infinitesimal homeostasis.

Again, heuristically the proof is straightforward. If a node is not upstream from the output node, than its value cannot affect the output node and if a node is not downstream from the input node than its value cannot be affected by the value of the input node. So deleting these nodes should not affect the input-output map.

Core Equivalence

Definition 6.4. Two core networks are *core equivalent* if the determinants of their homeostasis matrices are identical.

The general result concerning core equivalence is given in Theorem 7.2. Here we give an example of arrows that do not affect the homeostasis matrix and therefore the input-output function.

Definition 6.5. A *backward arrow* is an arrow whose head is the input node ι or whose tail is the output node o.

Proposition 6.6. *If two core networks differ from each other by the presence or absence of backward arrows, then the core networks are core equivalent.*

Proof. Backward arrows are not present in the homeostasis matrix (14). □

Therefore, backward arrows can be ignored when computing infinitesimal homeostasis from the homeostasis matrix H. However, backward arrows cannot be completely ignored, since they can be involved in the existence of both equilibria of (12) and their stability.

7 Types of Infinitesimal Homeostasis

Infinitesimal homeostasis is found in an input-output network $\mathcal{G}$ by simultaneously solving two equations: Find a stable equilibrium of an admissible system $\dot{X} = F(X, \mathcal{I})$ and find a zero of the determinant of the homeostasis matrix H. In this section, we discuss the different types of zeros $\det(H)$ can have and (for the most part) ignore the question of finding an equilibrium and its stability.

The homeostasis matrix H of an admissible system has three types of entries: linearized coupling strengths f_{k,x_ℓ} where node ℓ is connected to node k, linearized internal dynamics f_{k,x_k} of node k, and 0. We emphasize that the entries that are forced to be 0 depend specifically on network architecture.

Assume that the input-output network has $n + 2$ nodes: the input ι, the output o, and the n regulatory nodes $\rho = (\rho_1, \ldots, \rho_n)$. It follows that $\det(H)$ is a homogeneous polynomial of degree $n + 1$ in the variables f_{k,x_ℓ}. It is discussed in [42], based on Frobenius-König theory (see [40] for a historical account), that the homeostasis matrix H can be put in block upper triangular form. Specifically, there exist two constant $(n+1)\times(n+1)$ permutation matrices P and Q such that

$$PHQ = \begin{bmatrix} H_1 & * & \cdots & * \\ 0 & H_2 & \cdots & * \\ \vdots & & & \vdots \\ 0 & 0 & \cdots & H_m \end{bmatrix} \tag{16}$$

where the square matrices $H_1, \ldots, H_m$ are unique up to permutation, that is, individually the blocks cannot be brought into the form (16) by permutation of their rows and columns.

Moreover, when $\det(H)$ is viewed as a homogeneous polynomial in the entries of the matrix H there is a factorization

$$\det(H) = \det(H_1) \cdots \det(H_m) \tag{17}$$

into irreducible homogeneous polynomials $\det(H_1), \ldots, \det(H_m)$. That is, the *irreducible* blocks of the decomposition (16) correspond to the irreducible components in the factorization (17) (this follows from Theorem 4.2.6 (pp. 114–115) and Theorem 9.2.4 (p. 296) of [5]). We note that the main nontrivial result that allows us to write Eq. (17)—proved in [5, Theorem 9.2.4 (p. 296)]—is that $\det(H_j)$ is irreducible as a polynomial if and only if the matrix H_j is *irreducible* in the sense that H_j cannot be brought to the form (16) by permutation of H_j's rows and columns.

Low Degree Irreducible Factors of det(H)

Wang et al. [42] show that there can be two types of degree 1 factors (Haldane and null-degradation) and two types of degree 2 factors (structural and appendage). The principal result in [42] is the assertion that these four irreducible factors of $\det(H)$ can be associated with topological characteristics of the network $\mathcal{G}$ that in turn defines a *type of homeostasis*. The connection between the form of a factor $\det(H_j)$ and the topology of the network is given by certain determinant formulas that are reminiscent of the connection between a directed graph and its adjacency matrix and has been rediscovered by many authors [7,8,14,26] (see [6] for a modern account). Before stating the classification we introduce some graph theoretic terminology.

Definition 7.1. Let $\mathcal{G}$ be an input-output network.

(a) A directed path between two nodes is called a *simple path* if it visits each node on the path at most once. An *ιo-simple path* is a simple path connecting the input node ι to the output node o.
(b) A node in an input-output network $\mathcal{G}$ is *simple* if the node is on an ιo-simple path and *appendage* if the node is not simple.
(c) The *appendage subnetwork* $\mathcal{A}_{\mathcal{G}}$ of $\mathcal{G}$ is the subnetwork consisting of appendage nodes and arrows in $\mathcal{G}$ that connect appendage nodes.
(d) The *complementary subnetwork* corresponding to an ιo-simple path S is the network C_S consisting of all nodes not in S and all arrows in $\mathcal{G}$ between nodes in C_S.

Given these definitions we can state necessary and sufficient conditions for core equivalence:

Theorem 7.2. *Two core networks are core equivalent if and only if they have the same set of ιo-simple paths and the Jacobian matrices of the complementary subnetworks to any simple path have the same determinant up to sign.*

We isolate four types of homeostasis.

(A) Haldane homeostasis is associated with the arrow $\ell \to k$, where $k \neq \ell$, if homeostasis is caused by the vanishing of the degree 1 irreducible factor f_{k,x_ℓ} of $\det(H)$.

Theorem 7.3. *Haldane homeostasis associated with an arrow $\ell \to k$ can occur if and only if the arrow $\ell \to k$ is contained in every ιo-simple path.*

(B) Null-degradation homeostasis is associated with a node τ if homeostasis is caused by the vanishing of the degree 1 irreducible factor f_{τ,x_τ} of $\det(H)$.

Theorem 7.4. *Null-degradation homeostasis associated with a node τ can occur if and only if for every ιo-simple path S*
(a) *τ belongs to the complementary subnetwork C_S and*
(b) *τ is not contained in a cycle of C_S.*

(C) Structural homeostasis of degree 2 is caused by the vanishing of a degree 2 irreducible factor of $\det(H)$ that has the form

$$f_{\rho_2,x_{\rho_1}} f_{\rho_3,x_{\rho_2}} - f_{\rho_3,x_{\rho_1}} f_{\rho_2,x_{\rho_2}}$$

that is, the determinant of the homeostasis matrix of a *feedforward loop motif* defined by two ιo-simple path snippets: one snippet is $\rho_1 \to \rho_2 \to \rho_3$ and the other snippet is $\rho_1 \to \rho_3$. A *snippet* of a path is a connected subpath.

Theorem 7.5. *Structural homeostasis of degree 2 can occur if and only if*
(a) *two ιo-simple path snippets form a feedforward loop motif and*
(b) *all ιo-simple paths contain one of the two snippets of the feedforward loop motif.*

Structural homeostasis of degree 2 is exactly the *structural homeostasis* considered in [25] for 3-node core networks; it often arises in biochemical networks associated with the mechanism of *feedforward excitation.*

(D) Appendage homeostasis of degree 2 is caused by the vanishing of a degree 2 irreducible factor of $\det(H)$ that has the form

$$f_{\tau_1,x_{\tau_1}} f_{\tau_2,x_{\tau_2}} - f_{\tau_2,x_{\tau_1}} f_{\tau_1,x_{\tau_2}}$$

where the two node cycle $\mathcal{A} = \{\tau_1 \leftrightarrows \tau_2\}$ consists of appendage nodes.

Theorem 7.6. *Appendage homeostasis of degree 2 associated with a two-node cycle $\mathcal{A} \subset \mathcal{A}_G$ can occur if and only if for every ιo-simple path S*
(a) *$\mathcal{A}$ belongs to the complementary subnetwork C_S and*
(b) *nodes in $\mathcal{A}$ do not form a cycle with other nodes in C_S.*

The four types of infinitesimal homeostasis (A)–(D) correspond to the only possible factors of degree $\leqslant 2$. More precisely:

Theorem 7.7. *Any factor of degree 1 is of type (A) or (B) and any irreducible factor of degree 2 is of type (C) or (D).*

Homeostasis can also occur in blocks of degree 3 or higher. There are three types of such blocks: *structural* (all couplings are between simple nodes), *appendage* (all couplings are between appendage nodes), and *mixed* (both simple and appendage nodes appear in the block). Theorem 7.6 generalizes to higher degree appendage homeostasis. Specifically:

Theorem 7.8. *Let $\mathcal{G}$ be a network with appendage subnetwork $\mathcal{A} \subset \mathcal{A}_{\mathcal{G}}$. Appendage homeostasis associated with $\mathcal{A}$ can occur if and only if for every ιo-simple path S*

(a) $\mathcal{A}$ belongs to the complementary subnetwork C_S and
(b) nodes in $\mathcal{A}$ do not form a cycle with other nodes in C_S.

8 Low Degree Homeostasis Types

The homeostasis matrix H of a three-node input-output network is a 2×2 matrix. It follows that a homeostasis block is either 1×1 or 2×2. If the block is 2×2, it must be structural. For if it were appendage, the network would need to have two appendage nodes and one simple node. If the network had only one simple node, then the input node and the output node would be identical and that is not permitted.

Examples of Haldane, Structural of Degree 2, and Null Degradation The admissible systems of differential equations for the three-node networks in Fig. 8 are:

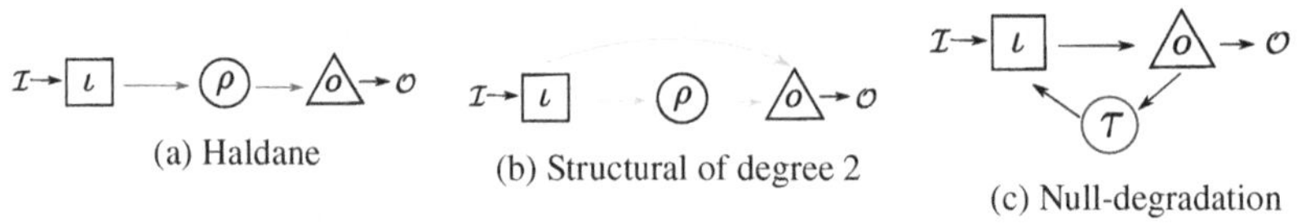

Fig. 8. Homeostasis types in three-node networks.

$$\begin{array}{ccc} \dot{x}_\iota = f_\iota(x_\iota) & \dot{x}_\iota = f_\iota(x_\iota) & \dot{x}_\iota = f_\iota(x_\iota, x_\tau) \\ \dot{x}_\rho = f_\rho(x_\iota, x_\rho) & \dot{x}_\rho = f_\rho(x_\iota, x_\rho) & \dot{x}_\tau = f_\tau(x_\tau, x_o) \\ \dot{x}_o = f_o(x_\rho, x_o) & \dot{x}_o = f_o(x_\iota, x_\rho, x_o) & \dot{x}_o = f_o(x_\iota, x_o) \\ (a) & (b) & (c) \end{array} \tag{18}$$

The determinants of the 2×2 homeostasis matrices are:

$$(a) f_{\rho,x_\iota} f_{o,x_\rho} \qquad (b) f_{\rho,x_\iota} f_{o,x_\rho} - f_{\rho,x_\rho} f_{o,x_\iota} \qquad (c) f_{o,x_\iota} f_{\tau,x_\tau} \tag{19}$$

A vanishing determinant in (19)(a) leads to two possible instances of Haldane homeostasis. A vanishing determinant in (19)(b) leads to balancing of two simple paths and structural homeostasis. Finally, a vanishing determinant in (19)(c) leads to null-degradation or Haldane homeostasis. These types of homeostasis were classified in [25] where it was also noted that Haldane occurs in product inhibition, structural occurs in feedforward excitation, and null-degradation occurs in a negative feedback loop.

Appendage Homeostasis of Degree 2

The admissible systems of differential equations for the four-node network in Fig. 9 have the form:

$$\begin{aligned}\dot{x}_\iota &= f_\iota(x_\iota, x_{\tau_2})\\ \dot{x}_{\tau_1} &= f_{\tau_1}(x_{\tau_1}, x_{\tau_2}, x_o)\\ \dot{x}_{\tau_2} &= f_{\tau_2}(x_{\tau_1}, x_{\tau_2})\\ \dot{x}_o &= f_o(x_\iota, x_o)\end{aligned}$$

The homeostasis matrix is

$$H = \begin{bmatrix} 0 & f_{\tau_1,x_{\tau_1}} & f_{\tau_1,x_{\tau_2}} \\ 0 & f_{\tau_2,x_{\tau_1}} & f_{\tau_2,x_{\tau_2}} \\ f_{o,x_\iota} & 0 & 0 \end{bmatrix}$$

and

$$\det(H) = f_{o,x_\iota}(f_{\tau_1,x_{\tau_1}} f_{\tau_2,x_{\tau_2}} - f_{\tau_1,x_{\tau_2}} f_{\tau_2,x_{\tau_1}})$$

It follows that $\det(H) = 0$ can lead either to Haldane homeostasis or appendage homeostasis of degree 2.

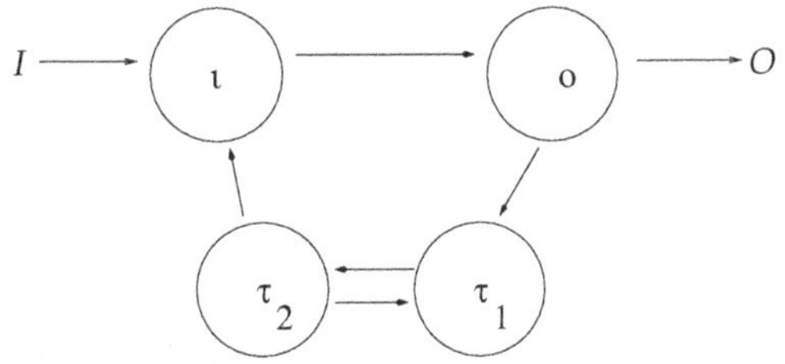

Fig. 9. Appendage homeostasis of degree 2.

9 Singularity Theory of Input-Output Functions

As discussed in Sect. 2, Nijhout et al. [33,35] observe that homeostasis appears in many applications through the notion of a chair. Golubitsky and Stewart [23] observed that a chair can be thought of as a singularity of the input-output function, one where $x_o(\mathcal{I})$ 'looks like' a homogeneous cubic $x_o(\mathcal{I}) \approx \mathcal{I}^3$. More precisely, the mathematics of singularity theory [19,36] replaces 'looks like' by 'up to a change of coordinates.'

Definition 9.1. Two functions $p, q : \mathbf{R} \to \mathbf{R}$ are *right equivalent* on a neighborhood of $\mathcal{I}_0 \in \mathbf{R}$ if

$$q(\mathcal{I}) = p(\Lambda(\mathcal{I})) + K$$

where $\Lambda : \mathbf{R} \to \mathbf{R}$ is an invertible change of coordinates on a neighborhood of $\mathcal{I}_0$ and $K \in \mathbf{R}$ is a constant.

The simplest singularity theory theorem states that $q : \mathbf{R} \to \mathbf{R}$ is right equivalent to $p(\mathcal{I}) = \mathcal{I}^3$ on a neighborhood of the origin if and only if $q'(\mathcal{I}_0) = q''(\mathcal{I}_0) = 0$ and $q'''(\mathcal{I}_0) \neq 0$. Hence we call a point $\mathcal{I}_0$ an *infinitesimal chair* for an input-output function x_o if

$$x_o'(\mathcal{I}_0) = x_o''(\mathcal{I}_0) = 0 \quad \text{and} \quad x_o'''(\mathcal{I}_0) \neq 0 \tag{20}$$

A simple result is:

Lemma 9.2. *An input-output map x_o has an infinitesimal chair at $\mathcal{I}_0$ if and only if*

$$h(\mathcal{I}_0) = h'(\mathcal{I}_0) = 0 \quad \textit{and} \quad h''(\mathcal{I}_o) \neq 0$$

where $h(\mathcal{I}) = \det(H)$.

Proof. Suppose that $x_o'(\mathcal{I}) = k(\mathcal{I})h(\mathcal{I})$ where $k(\mathcal{I})$ is nowhere zero. Then $h(\mathcal{I}_0) = h'(\mathcal{I}_0) = 0$ if and only if $x_0'(\mathcal{I}_0) = x_0''(\mathcal{I}_0) = 0$ because $x_o'' = k'h + kh'$. Moreover, if $h = h' = 0$, then $x_o'' = kh''$. Finally, it follows from the Cramer's rule calculation in Lemma 5.2 that

$$k = \pm \frac{f_{\iota,\mathcal{I}}}{\det(J)}$$

Hence, $k(\mathcal{I})$ is nowhere zero. □

A simpler result states the following. The input-output function defines *simple infinitesimal homeostasis* if

$$x_o' = 0 \quad \text{and} \quad x_o'' \neq 0,$$

which is equivalent to $h = 0$ and $h' \neq 0$. The graph of x_o 'looks like' a parabola near a point of simple infinitesimal homeostasis.

9.1 Chair Points for Blocks of Degree 1 and 2

Lemma 9.2 gives necessary and sufficient conditions for the existence of infinitesimal homeostasis using the function $h = \det(H)$. In general, the homeostasis function can be simplified by recalling from (16) that the homeostasis matrix PHQ is block upper triangular. It follows that if homeostasis stems from block j, then $\det(H)$ is a nonzero multiple of $\det(H_j)$. The results in Sect. 7 imply

$$h_j \equiv \det(H_j) = \begin{cases} f_{k,x_\ell} & \text{Haldane} \\ f_{\ell,x_\ell} & \text{null-degradation} \\ f_{\rho_2,x_{\rho_1}} f_{\rho_3,x_{\rho_2}} - f_{\rho_3,x_{\rho_1}} f_{\rho_2,x_{\rho_2}} & \text{structural of degree 2} \\ f_{\tau_1,x_{\tau_1}} f_{\tau_2,x_{\tau_2}} - f_{\tau_1,x_{\tau_2}} f_{\tau_2,x_{\tau_1}} & \text{appendage of degree 2} \end{cases} \tag{21}$$

Theorem 9.3. *Given an input-output network. Then, the defining conditions for infinitesimal chair homeostasis are given by $h_j = h_j' = 0$ where h_j is defined by* (21).

We now calculate chair equations for the two degree 1 three-node examples.

Lemma 9.4.

(a) If the arrow $\rho \to o$ has Haldane homeostasis in the network $\iota \to \rho \to o$, then

$$h = h' = 0 \quad \Longleftrightarrow \quad f_{o,x_\rho} = f_{o,x_\rho x_\rho} = 0$$

(b) If the node τ has null-degradation homeostasis in the network $\iota \to o$, $o \to \tau$, $\tau \to \iota$, then

$$h = h' = 0 \quad \Longleftrightarrow \quad f_{\tau,x_\tau} = f_{\tau,x_\tau x_\tau} = 0$$

Proof. Suppose $h(\mathcal{I}) = h_j(\mathcal{I})k(\mathcal{I})$, where $k(\mathcal{I}_0)$ is nonzero at I_0, then $h(\mathcal{I}_0) = h'(\mathcal{I}_0) = 0$ if and only if $h_j(\mathcal{I}_0) = h_j'(\mathcal{I}_0) = 0$. The proof proceeds in two parts.

(a) Observe that

$$h_j = f_{o,x_\rho}(x_\rho, x_o) = 0$$

is one equation for a Haldane chair and the second equation is

$$h_j' = f_{o,x_\rho x_\rho} x_\rho' + f_{o,x_\rho x_o} x_o' = 0$$

Since h_j' is evaluated at a point of homeostasis, $x_o' = 0$. It follows that either $f_{o,x_\rho x_\rho} = 0$ or $x_\rho' = 0$. We can use Cramer's rule to solve for x_ρ'; it is a nonzero multiple of $f_{\rho,x_\iota} f_{o,x_o}$. If $f_{\rho,x_\iota} = 0$, then we would have a second Handane in the $\iota \to \rho$ arrow - a codimension 2 homeostasis. So, generically, we can assume $f_{\rho,x_\iota} \neq 0$. By computing the Jacobian at the assumed Haldane point we see that f_{o,x_o} is an eigenvalue and therefore negative by the assumed stabilty.

(b) We use the admissible system equilibrium equations from (18) (c) to see that null-degradation is defined by $h_j = f_{\tau,x_\tau}(x_\tau, x_o) = 0$ and a chair by

$$h_j' = f_{\tau,x_\tau x_\tau} x_\tau' + f_{\tau,x_\tau x_o} x_o' = 0$$

Since $x_o' = 0$ and $x_\tau' \neq 0$ at the generic homeostasis point, it follows that $f_{\tau,x_\tau x_\tau} = 0$ is the chair equation, as claimed. □

9.2 Elementary Catastrophe Theory and Homeostasis

The transformations of the input-output map $x_o(\mathcal{I})$ given in Definition 9.1 are just the standard change of coordinates in elementary catastrophe theory [19,36, 43]. We can therefore use standard results from elementary catastrophe theory to find normal forms and universal unfoldings of $x_o(\mathcal{I})$, as we now explain.

Because $x_o(\mathcal{I})$ is 1-dimensional, we consider singularity types near the origin of a 1-variable function $g(\mathcal{I})$. Such singularities are determined by the first nonvanishing $\mathcal{I}$-derivative $g^{(k)}(0)$ (unless all derivatives vanish, which is an 'infinite codimension' phenomenon that we do not discuss further). Informally, the *codimension* of a singularity is the number of conditions on derivatives that determine it. This is also the minimum number of extra variables required to specify all small perturbations of the singularity, up to changes of coordinates. These perturbations can be organized into a family of maps called the *universal unfolding*, which has that number of extra variables.

Definition 9.5. $G(\mathcal{I}, a)$ is an *unfolding* of $g(\mathcal{I})$ if $G(\mathcal{I}, 0) = g(\mathcal{I})$. G is a *universal unfolding* of g if every unfolding of $H(\mathcal{I}, b)$ *factors through* G. That is,

$$H(\mathcal{I}, b) = G(\Lambda(\mathcal{I}, b), A(b)) + K(b) \tag{22}$$

It follows that *every* small perturbation $H(\cdot, b)$ is equivalent to a perturbation $G(\cdot, A(b))$ of g in the G family.

If such k exists, the normal form is $\pm\mathcal{I}^k$. Simple infintesimal homeostasis occurs when $k = 2$, and an infinitesimal chair when $k = 3$. When $k \geq 3$ the universal unfolding for catastrophe theory equivalence is

$$\pm\mathcal{I}^k + a_{k-2}\mathcal{I}^{k-2} + a_{k-3}\mathcal{I}^{k-3} + \cdots + a_1\mathcal{I}$$

for parameters a_j and when $k = 2$ the universal unfolding is $\pm\mathcal{I}^2$. The codimension in this setting is therefore $k - 2$. See [4] Example 14.9 and Theorem 15.1; [18] chapter IV (4.6) and chapter VI (6.3); and [30] chapter XI Sect. 1.1 and chapter XII Sects. 3.1, 7.2.

To summarize: the normal form of the input-output function for simple infinitesimal homeostasis is

$$x_o(\mathcal{I}) = \pm\mathcal{I}^2 \tag{23}$$

and no unfolding parameter is required. Similarly,

$$x_o(\mathcal{I}) = \pm\mathcal{I}^3 \tag{24}$$

is the normal form of the input-output function for a chair, and

$$x_o(\mathcal{I}; a) = \pm\mathcal{I}^3 + a\mathcal{I} \tag{25}$$

is a universal unfolding.

10 Evolving Towards Homeostasis

Control-theoretic models of homeostasis often build in an explicit ‘target’ value for the output, and construct the equations to ensure that the input-output function is exactly flat over some interval. Such models are common, and provide useful information for many purposes. In singularity theory an exactly flat input-output function has ‘infinite codimension’, so our approach is not appropriate for models of this type.

However, in biology, homeostasis is an emergent property of biochemical networks, not a preset target value, and the input-output function is only approximately flat, for example as in Fig. 2 (left). Many of the more recent models of homeostasis do not assume a preset target value; instead, this emerges from the dynamics of a biochemical network. Here we expect typical singularities to have finite codimension, and our approach is then potentially useful. For example, in [21, Section 8] we proved that for one such model, of feedforward inhibition [33,39], the input-output map has a ‘chair’ singularity, with normal form $x^3 + \lambda x$. Other examples of chair singularities are given in [37].

A key question is: In a mathematical sense, how does a biological system evolve towards homeostasis? Imagine a system of differential equations depending on parameters. Suppose that initially the parameters are set so that the associated input-output function has no regions of homeostasis. Now vary the parameters so that a small region of homeostasis appears in the input-output function. Since this region of homeostasis is small, we can assume that it is spawned by a singularity associated with infinitesimal homeostasis. How can that happen?

Singularities Organizing Evolution Towards Homeostasis
A plausible answer follows from the classification of elementary catastrophes. If there is one input and one output, the assumption of no initial homeostasis implies that the input-output function $x_o : \mathbf{R} \to \mathbf{R}$ is strictly increasing (or strictly decreasing). Generically, evolving towards infinitesimal homeostasis can occur in only one way. As a parameter β is varied, at some point $\mathcal{I}_0$ the function $x_o(\mathcal{I})$ approaches a singularity, so there is a point $\mathcal{I}_0$ where $x_o'(\mathcal{I}_0) = 0$. This process can happen only if $x_o''(\mathcal{I}_0) = 0$ is also satisfied. That is, from a singularity-theoretic point of view, the simplest way that homeostasis can evolve is through an infinitesimal chair.

This process can be explained in the following way. The system can evolve towards infinitesimal homeostasis only if the universal unfolding of the singularity has a parameter region where the associated function is nonsingular. For example, simple homeostasis ($x_o(\mathcal{I}) = \mathcal{I}^2$, which is structurally stable) does not have this property. All small perturbations of $\mathcal{I}^2$ have a Morse singularity. The simplest (lowest codimension) singularity that has nonsingular perturbations is the fold singularity $x_o(\mathcal{I}) = \mathcal{I}^3$; that is, the infinitesimal chair.

At least two assumptions underlie this discussion. First, we have assumed that all perturbations of the input-output function can be realized by perturbations in the system of ODEs. This is true; see Lemma 10.1. Second, we assume that when evolving towards homeostasis the small region of homeostasis that forms is one that could have grown from a point of infinitesimal homeostasis.

When x_o depends on one parameter, generically the infinitesimal chair is the only possible singularity that can underlie the formation of homeostasis.

Lemma 10.1. *Given a system of ODEs $\dot{x} = F(x, \mathcal{I})$ whose zero set is defined by*

$$F(X(\mathcal{I}), \mathcal{I}) \equiv 0$$

and a perturbation $\tilde{X}(\mathcal{I}) = X(\mathcal{I}) + P(\mathcal{I})$ of that zero set. Then $\tilde{X}$ is the zero set of the perturbation

$$\tilde{F}(x, \mathcal{I}) = F(x - P(\mathcal{I}), \mathcal{I})$$

Therefore any perturbation of the input-output function $x_o(\mathcal{I})$ can be realized by perturbation of F.

Proof. Clearly

$$\begin{aligned}\tilde{F}(\tilde{X}(\mathcal{I}),\mathcal{I}) &= \tilde{F}(X(\mathcal{I})+P(\mathcal{I}),\mathcal{I})\\ &= F(X(\mathcal{I})+P(\mathcal{I})-P(\mathcal{I}),\mathcal{I})\\ &= F(X(\mathcal{I}),\mathcal{I})\\ &= 0\end{aligned}$$

If we write $P(\mathcal{I}) = (0, P_o(\mathcal{I}))$ where $P_o(\mathcal{I})$ is a small perturbation of $x_o(\mathcal{I})$, then we can obtain the perturbation $x_o + P_o$ of x_o by the associated perturbation of F. □

Theorem 10.2. *Consider input-output functions with one input and one output. Then the only singularities of codimension ≤ 3 that have perturbations with no infinitesimal homeostasis are the fold (chair) and the swallowtail.*

Proof. It is easy to see that perturbations of $\mathcal{I}^k$ always have a local minimum when k is even. So the only normal forms with perturbations that have no infinitesimal homeostasis occur when k is odd. Those that have codimension at most 3 are the fold ($k = 3$) and the swallowtail ($k = 5$). □

We remark that folds occur in the unfoldings of swallowtails and that the generic non-homeostatic approach to a swallowtail would also give a non-homeostatic approach to a fold (or chair).

11 Input-Output Maps with Two Inputs

Suppose now that the input $\mathcal{I}$ consists of several variables. In general terms, consider a parametrized family of ODEs

$$\dot{X} = F(X,\mathcal{I}) \tag{26}$$

where $X = (x_1,\ldots,x_m) \in \mathbf{R}^m$, $\mathcal{I} \in \mathbf{R}^k$, and F is infinitely differentiable. We assume that (26) stems from an input-output network where one of the nodes (or coordinates of X) is the output node that is denoted, as before, by o. We also assume that (26) has a stable equilibrium at X_0 when $\mathcal{I} = \mathcal{I}_0$.

The equilibria of (26) are given by:

$$F(X,\mathcal{I}) = 0 \tag{27}$$

By the implicit function theorem, we can solve (27) near $(X_0, \mathcal{I}_0)$ to obtain a map $X : \mathbf{R}^k \to \mathbf{R}^m$ such that

$$F(X(\mathcal{I}),\mathcal{I}) \equiv 0 \tag{28}$$

where $X(\mathcal{I}_0) = X_0$. Let

$$X(\mathcal{I}) = (Y(\mathcal{I}), x_o(\mathcal{I}))$$

Definition 11.1. *The* input-output map *of* (27) *near* $(X_0, \mathcal{I}_0)$ *is* $x_o : \mathbf{R}^k \to \mathbf{R}$.

Definition 11.2. *The point* $\mathcal{I}_0$ *is an* infinitesimal homeostasis point *of* x_o *if the derivative*

$$D_{\mathcal{I}} x_o(\mathcal{I}_0) = 0 \tag{29}$$

In particular, $\mathcal{I}_0$ *is a singularity—that is, the derivative of* x_o *is singular there—but the vanishing of* all *first derivatives selects a special subclass of singularities, said to have 'full corank'.*

The interpretation of an infinitesimal homeostasis point is that $x_o(\mathcal{I})$ differs from $x_o(\mathcal{I}_0)$ in a manner that depends quadratically (or to higher order) on $|\mathcal{I} - \mathcal{I}_0|$. This makes the graph of $x_o(\mathcal{I})$ flatter than any growth rate with a nonzero linear term. This condition motivates for the condition (29) rather than merely $D_{\mathcal{I}} x_o(\mathcal{I}_0)$ being singular.

Definition 11.2 places the study of homeostasis in the context of singularity theory, and we follow the standard line of development in that subject. A detailed discussion of singularity theory would be too extensive for this paper. A brief summary is given in [21] in the context of homeostasis, accessible descriptions can be found in [36,43], and full technical details are in [18,30] and many other sources.

Following Nijhout *et al.* [33] we define:

Definition 11.3. A *plateau* is a region of $\mathcal{I}$ over which $X(\mathcal{I})$ is approximately constant.

Remark 11.4. Universal unfolding theory implies that small perturbations of x_o (that is, variation of the suppressed parameters) change the plateau region only slightly. This point was explored for the chair singularity in [21]. It follows that for sufficiently small perturbations plateaus of singularities depend mainly on the singularity itself and not on its universal unfolding.

Remark 11.5. In this section we focus on how singularities in the input-output map shape plateaus, and we use the normal form and unfolding theorems of elementary catastrophe theory to do this. We remark that typically the variables other than x_o, the manipulated variables Y, can vary substantially while the output variable is held approximately constant. See, for example, Fig. 3 in [1].

11.1 Catastrophe Theory Classification

The results of [21] reduce the classification of homeostasis points for a single node to that of singularities of input-output maps $\mathbf{R}^k \to \mathbf{R}$. As mentioned in Sect. 9.2, this is precisely the abstract set-up for elementary catastrophe theory [4,18,36, 43]. The case $k = 1$ is discussed there.

We now consider the next case $k = 2$. Table 1 summarizes the classification when $k = 2$, so $\mathcal{I} = (\mathcal{I}_1, \mathcal{I}_2) \in \mathbf{R}^2$. Here the list is restricted to codimension ≤ 3. The associated geometry, especially for universal unfoldings, is described in [4,18,36] up to codimension 4. Singularities of much higher codimension have also been classified, but the complexities increase considerably. For example Arnold [2] provides an extensive classification up to codimension 10 (for the complex analog).

Table 1. Classification of singularities of input-output maps $\mathbf{R}^2 \to \mathbf{R}$ of codimension ≤ 3.

Name	Normal form	Codim	Universal unfolding
Morse (simple homeostasis)	$\pm I_1^2 \pm I_2^2$	0	$\pm I_1^2 \pm I_2^2$
Fold (chair)	$I_1^3 \pm I_2^2$	1	$I_1^3 + aI_1 \pm I_2^2$
Cusp	$\pm I_1^4 \pm I_2^2$	2	$\pm I_1^4 + aI_1^2 + bI_1 \pm I_2^2$
Swallowtail	$I_1^5 \pm I_2^2$	3	$I_1^5 + aI_1^3 + bI_1^2 + cI_1 \pm I_2^2$
Hyperbolic umbilic	$I_1^3 + I_2^3$	3	$I_1^3 + I_2^3 + aI_1I_2 + bI_1 + cI_2$
Elliptic umbilic	$I_1^3 - 3I_1I_2^2$	3	$I_1^3 - 3I_1I_2^2 + a(I_1^2 + I_2^2) + bI_1 + cI_2$

Remark 11.6. Because $k = 2$, the normal forms for $k = 1$ appear again, but now there is an extra quadratic term $\pm I_2^2$. This term is a consequence of the splitting lemma in singularity theory, arising here when the second derivative $D^2 x_o$ has rank 1 rather than rank 0 (corank 1 rather than corank 2). See [4,36,43]. The presence of the $\pm I_2^2$ term affects the range over which $x_o(\mathcal{I})$ changes when I_2 varies, but not when I_1 varies.

11.2 Normal Forms and Plateaus

The standard geometric features considered in catastrophe theory focus on the gradient of the function $x_o(\mathcal{I})$ in normal form. In contrast, what matters here is the function itself. Specifically, we are interested in the region in the $\mathcal{I}$-plane where the function x_o is approximately constant.

More specifically, for each normal form $x_o(\mathcal{I})$ we choose a small $\delta > 0$ and form the set

$$P_\delta = \{\mathcal{I} \in \mathbf{R}^2 : |x_o(\mathcal{I})| \leq \delta\}. \tag{30}$$

This is the *plateau* region on which $x_o(\mathcal{I})$ is approximately constant, where δ specifies how good the approximation is. If $x_o(\mathcal{I})$ is perturbed slightly, P_δ varies continuously. Therefore we can compute the approximate plateau by focusing on the singularity, rather than on its universal unfolding.

This observation is important because the universal unfolding has many zeros of the gradient of $x_o(\mathcal{I})$, hence 'homeostasis points' near which the value of $x_o(\mathcal{I})$ varies more slowly than linear. However, this structure seems less important when considering the relationship of infinitesimal homeostasis with homeostasis. See the discussion of the unfolding of the chair summarized in [21, Figure 3].

The 'qualitative' geometry of the plateau—that is, its differential topology and associated invariants—is characteristic of the singularity. This offers one way to infer the probable type of singularity from numerical data; it also provides information about the region in which the system concerned is behaving homeostatically. We do not develop a formal list of invariants here, but we indicate a few possibilities.

The main features of the plateaus associated with the six normal forms are illustrated in Table 1. Figure 10 plots, for each normal form, a sequence of contours from $-\delta$ to δ; the union is a picture of the plateaus. By unfolding theory, these features are preserved by small perturbations of the model, and by the choice of δ in (30) provided it is sufficiently small. Graphical plots of such perturbations (not shown) confirm this assertion. Again, we do not attempt to make these statements precise in this paper.

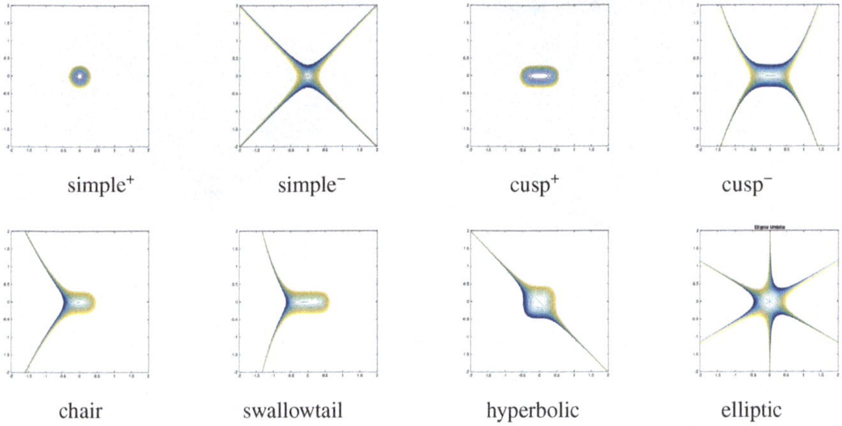

Fig. 10. Plateaus shown by contour plots for each singularity in Table 1. Reproduced from [23, Fig. 4]. 200 equally spaced contours for δ from −0.2 to 0.2.

11.3 The Hyperbolic Umbilic

As we have discussed, homeostasis can occur when one variable is held approximately constant on variation of two or more input parameters. For example, body temperature can be homeostatic with respect to both external temperature *and* amount of exercise. A biological network example is Fig. 3, where there is homeostasis of extracellular dopamine (eDA) in response to variation in the activities of the enzyme tyrosine hydroxylase (TH) and the dopamine transporters (DAT), Best *et al.* [3]. These authors derive a differential equation model for this biochemical network. They fix reasonable values for all parameters in the model with the exception of the concentrations of TH and DAT. Figure 11 (left) shows the equilibrium value of eDA as a function of TH and DAT in their model. The white dots indicate the predicted eDA values for the observationally determined values of TH and DAT in the wild type genotype (large white disk) and the polymorphisms observed in human populations (small white disks). Their result is scientifically important because almost all of the white disks lie on the plateau (the region where the surface is almost horizontal) that indicates homeostasis of eDA. Note that the flat region contains a line from left to right at about eDA = 0.9. In this respect the surface graph in Fig. 11 (left) appears to resemble that of a nonsingular perturbed hyperbolic umbilic (see Table 1) in Fig. 11 (right). See also the level contours of the hyperbolic umbilic in Fig. 10. This figure shows that the hyperbolic umbilic is the only low codimension singularity that contains a single line in its zero set.

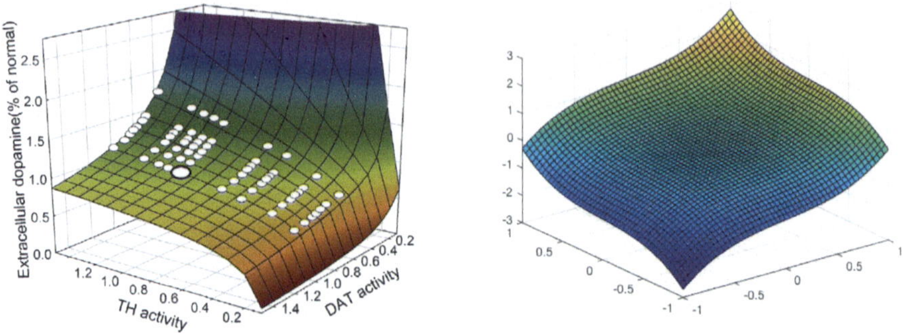

Fig. 11. (Left): Nijhout *et al.* [33, Fig. 8] and Reed *et al.* [37, Fig. 14]. At equilibrium there is homeostasis of eDA as a function of TH and DAT. There is a plateau around the wild-type genotype (large white disk). Smaller disks indicate positions of polymorphisms of TH and DAT found in human populations. (Right): Graph of surface of perturbed hyperbolic umbilic without singularities: $Z(I_1, I_2) = I_1^3 + I_2^3 + I_1 + I_2/2$.

The number of curves ('whiskers') forming the zero-level contour of the plateau is a characteristic of the plateau. For example, Fig. 11 appears to have one curve in the plateau. This leads us to conjecture that the hyperbolic umbilic is the singularity that organizes the homeostatic region of eDA in the example discussed in [3]. It may be the case however, that there is no infinitesimal homeostasis in this example, and the cause is more global. We have discussed in Sect. 10 why the chair and the hyperbolic umbilic are the singularities that might be expected to organize two output homeostasis.

Theorem 11.7. *Consider input-output functions with two inputs and one output. Then the only singularities of codimension ≤ 3 that have perturbations with no infinitesimal homeostasis are the fold (chair), swallowtail, and hyperbolic umbilic.*

The proof of this theorem is in [23].

Remark 11.8. In Sect. 10 we note that a system of equations that evolves toward infinitesimal homeostasis does so by transitioning through a singularity that has unfolding parameters with no infinitesimal homeostasis. It follows from Theorems 10.2 and 11.7 that the most likely ways to transition to homeostasis in systems with one input variable is through the chair and in systems with two input variables the hyperbolic umbilic and the two variable chair.

12 Gene Regulatory Networks and Housekeeping Genes

Antoneli et al. [1] use infinitesimal homeostasis to find regions of homeostasis in a differential equation model for the gene regulatory network (GRN) that is believed to regulate the production of the protein PGA2 in *Escherichia coli* and yeast. Specifically, in this model the input parameter is an *external* parameter $\mathcal{I}$ that represents the collective influence of other gene proteins on this specific GRN. We find regions of homeostasis that gives a plausible explanation of how the level of the PGA2 protein might be held approximately constant while other reactions are taking place.

Gene expression is a general name for a number of sequential processes, the most well known and best understood being *transcription* and *translation*. These processes control the level of gene expression and ultimately result in the production of a specific quantity of a target protein.

The genes, regulators, and the regulatory connections between them forms a *gene regulatory network (GRN)*. A gene regulatory network can be represented pictorially by a directed graph where the genes correspond to network nodes, incoming arrows to transcription factors, and outgoing arrows to levels of gene expression (protein concentration).

12.1 Gene Regulatory Networks and Homeostasis

Numerous terms are used to describe types of genes according to how they are regulated. A *constitutive gene* is a gene that is transcribed continually as opposed to a *facultative gene* that is transcribed only when needed. A *housekeeping gene* is a gene that is required to maintain basic cellular function and so is typically expressed in all cell types of an organism. Some housekeeping genes are transcribed at a 'relatively constant rate' in most non-pathological situations and are often used as reference points in experiments to measure the expression rates of other genes.

Even though this scheme is more or less universal among all life forms, from uni-cellular to multi-cellular organisms, there are some important differences according to whether the cell possesses a nucleus (eukaryote) or not (prokaryote). In single-cell organisms, gene regulatory networks respond to changes in the external environment adapting the cell at a given time for survival in this environment. For example, a yeast cell, finding itself in a sugar solution, will turn on genes to make enzymes that process the sugar to alcohol.

Recently, there has been an ongoing effort to map out the GRNs of some the most intensively studied single-cell *model organisms*: the prokaryote *E. coli* and the eukaryote *Saccharomyces cerevisiae*, a species of *yeast*. A hypothesis that has emerged from these efforts is that the GRN has evolved into a modular structure in terms of small sub-networks appearing as recurrent patterns in the GRN, called *network motifs*. Moreover, experiments on the dynamics generated by network motifs in living cells indicate that they have characteristic dynamical functions. This suggests that network motifs may serve as building blocks in modeling gene regulatory networks.

Much experimental work has been devoted to understanding network motifs in gene regulatory networks of single-cell model organisms. The GRNs of *E. coli* and yeast, for example, contain three main motif families that make up almost the entire network. Some well-established network motifs and their corresponding functions in the GRN of *E. coli* and yeast include negative (or inhibitory) self-regulation, positive (or excitatory) self-regulation and several types of feed-forward loops. Nevertheless, most analyses of motif function are carried out looking at the motif operating in isolation. There is, however, mounting evidence that network context, that is, the connections of the motif with the rest of the network, are important when drawing inferences on characteristic dynamical functions of the motif.

In this context, an interesting question is how the GRN of a single-cell organism is able to sustain the production rates of the housekeeping genes and at same time be able to quickly respond to environmental changes, by turning on and off the appropriate facultative genes. If we assume that the dynamics of gene expression is modeled by coupled systems of differential equations then this question can be formulated as the existence of a homeostatic mechanism associated to some types of network motifs imbedded in the GRN.

Latest estimates on the number of feedforward loops in the GRN of *S. cerevisiae* assert that there are least 50 feedforward loops (not all of the same type) potentially controlling 240 genes. One example of such a feedforward loop is shown in Fig. 12. The three genes in this network are considered constitutive.

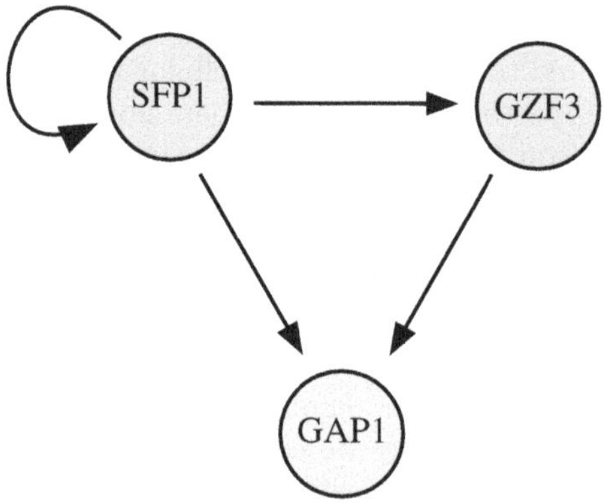

Fig. 12. An example of feedforward regulation network from the GRN of *S. cerevisiae*, involving the genes SFP1, CIN5 and PGA2. The PGA2 gene produces an essential protein involved in protein trafficking (null mutants have a cell separation defect). The CIN5 gene is a basic leucine zipper (bZIP) transcription factor. The SPF1 gene regulates transcription of ribosomal protein and biogenesis genes.

12.2 Basic Structural Elements of GRNs

The fundamental building block or *node* in a gene regulatory network is a *gene* that is composed of two parts: *transcription* and *translation*. The transcription part produces messenger RNA (mRNA) and the translation part produces the protein. The system of ODEs associated to one gene has the form

$$\dot{x} = f(x, t_J, \mathcal{I}) \qquad x = (x^R, x^P) \in \mathbf{R}^2$$

where x^R is the *mRNA concentration*, x^P is the *protein concentration* and the t_J are the coupling protein concentrations of *transcription factors* that regulate the gene and are produced by other genes in the network. The parameter $\mathcal{I}$ represents the effect of upstream transcription factors that regulate the gene but are not part of the network. The vector field f has the form

$$f = (f^R + \mathcal{I}, f^P) \in \mathbf{R}^2,$$

where f^R models the dynamics of mRNA concentration and f^P models the dynamics of the protein concentration.

When the gene is not self-regulated the system has the form

$$\begin{aligned} \dot{x}^R &= f^R(x^R, t_J) + \mathcal{I} \\ \dot{x}^P &= f^P(x^R, x^P) \end{aligned}$$

and when the gene is self-regulated the system of two scalar equations has the form

$$\begin{aligned} \dot{x}^R &= f^R(x^R, x^P, t_J) + \mathcal{I} \\ \dot{x}^P &= f^P(x^R, x^P) \end{aligned}$$

In both cases the gene output is the scalar variable x_P.

12.3 The Gene Regulatory Network for PGA2

Consider the network consisting of three genes (and six nodes) shown in Fig. 13, where the dashed lines represent inhibitory coupling (repression or negative control) and the solid lines represent excitatory coupling (activation or positive control).

Observe that the six-node network in Fig. 13 has two simple paths:

$$x^R \longrightarrow x^P \longrightarrow z^R \longrightarrow z^P \qquad \text{and} \qquad x^R \longrightarrow x^P \longrightarrow y^R \longrightarrow y^P \longrightarrow z^R \longrightarrow z^P$$

There are two possible Haldane homeostasis arrows $x^R \longrightarrow x^P$ and $z^R \longrightarrow z^P$, and one structural homeostasis of degree three consisting of two paths $x^P \longrightarrow y^R \longrightarrow y^P \longrightarrow z^R$ and $x^P \longrightarrow z^R$. To verify this we compute the homeostasis matrix.

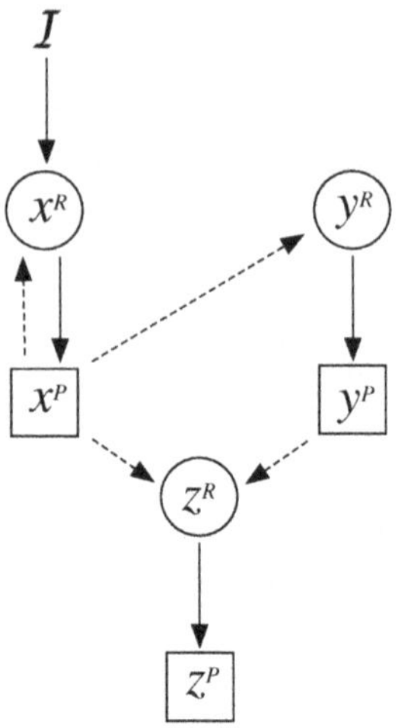

Fig. 13. Example of a 3-gene six-cell network. All arrows are different, but for simplicity this is not made explicit in the figure. Circles stand for mRNA concentration and squares for protein concentration. Solid lines indicate excitatory coupling and dashed lines indicate inhibitory coupling.

The steady-state equations associated with the network in Fig. 13 have the form:

$$\begin{aligned} f^R(x^R, x^P) + \mathcal{I} &= 0 \\ f^P(x^R, x^P) &= 0 \\ g^R(x^P, y^R) &= 0 \\ g^P(y^R, y^P) &= 0 \\ h^R(x^P, y^P, z^R) &= 0 \\ h^P(z^R, z^P) &= 0 \end{aligned} \tag{31}$$

where the input parameter $\mathcal{I}$ represents the action of all upstream transcription factors that affect the x-gene and do not come from the y- and z-genes. Our goal is to find regions of homeostasis in the steady-state protein concentration z^P as a function of the input parameter $\mathcal{I}$. To do this we compute det(H), where H is the 5×5 homeostasis matrix.

$$H = \begin{bmatrix} f^P_{x^R} & f^P_{x^P} & 0 & 0 & 0 \\ 0 & g^R_{x^P} & g^R_{y^R} & 0 & 0 \\ 0 & 0 & g^P_{y^R} & g^P_{y^P} & 0 \\ 0 & h^R_{x^P} & 0 & h^R_{y^P} & h^R_{z^R} \\ 0 & 0 & 0 & 0 & h^P_{z^R} \end{bmatrix} \tag{32}$$

A short calculation shows that

$$\det(H) = f^P_{x^R}\, h^P_{z^R} \left(g^R_{y^R} g^P_{y^P} h^R_{x^P} + h^R_{y^P} g^P_{y^R} g^R_{x^P} \right)$$

Therefore structural homeostasis is found by solving $h = h' = 0$, where

$$h(\mathcal{I}) \equiv g^R_{yR} g^P_{yP} h^R_{xP} + h^R_{yP} g^P_{yR} g^R_{xP} \tag{33}$$

This equation is analysed in Antoneli et al. [1], who show that standard ODE models for gene regulation, when inserted into a feedforward loop motif, do indeed lead to chair structural homeostasis in the output protein housekeeping genes. In [1] this cubic expression was obtained by direct calculation and its appearance was somewhat mysterious; here it emerges from the general theory of homeostasis matrices in Sect. 7.

Acknowledgments. We thank Janet Best, Tony Nance, and Mike Reed, for helpful conversations. This research was supported in part by the National Science Foundation Grant DMS-1440386 to the Mathematical Biosciences Institute. FA and MG were supported in part by the OSU-FAPESP Grant 2015/50315-3.

References

1. Antoneli, F., Golubitsky, M., Stewart, I.: Homeostasis in a feed forward loop gene regulatory network motif. J. Theor. Biol. **445**, 103–109 (2018). https://doi.org/10.1016/j.jtbi.2018.02.026
2. Arnold, V.I.: Local normal forms of functions. Invent. Math. **35**, 87–109 (1976)
3. Best, J., Nijhout, H.F., Reed, M.: Homeostatic mechanisms in dopamine synthesis and release: a mathematical model. Theor. Biol. Med. Model. **6** (2009). https://doi.org/10.1186/1742-4682-6-21
4. Bröcker, Th., Lander, L.: Differentiable Germs and Catastrophes. LMS Lect. Notes, vol. 17. Cambridge University Press, Cambridge (1975)
5. Brualdi, R.A., Ryser, H.J.: Combinatorial Matrix Theory. Cambridge University Press, Cambridge (1991)
6. Brualdi, R.A., Cvetkoić, D.M.: A Combinatorial Approach to Matrix Theory and its Applications. Chapman & Hall/CRC Press, Boca Raton (2009)
7. Coates, C.L.: Flow graph solutions of linear algebraic equations. IRE Trans. Circuit Theory **CT–6**, 170–187 (1959)
8. Cvetković, D.M.: The determinant concept defined by means of graph theory. Mat. Vesnik **12**(27), 333–336 (1975)
9. Dellnitz, M.: Hopf-Verzweigung in Systemen mit Symmetrie und deren Numerische Behandlung. Uni. Diss, Hamburg (1988)
10. Dellnitz, M.: A computational method and path following for periodic solutions with symmetry. In: Roose, D., De Dier, B., Spence, A. (eds.) Continuation and Bifurcations: Numerical Techniques and Applications, pp. 153–167. Kluwer, Dordrecht (1990)
11. Dellnitz, M.: Computational bifurcation of periodic solutions in systems with symmetry. IMA J. Numer. Anal. **12**, 429–455 (1992)
12. Dellnitz, M., Junge, O.: On the approximation of complicated dynamical behavior. SIAM J. Numer. Anal. **36**, 491–515 (1999)
13. Dellnitz, M., Junge, O., Thiere, B.: The numerical detection of connecting orbits. Discret. Continuous Dyn. Syst. - B **1**, 125–135 (2001)
14. Desoer, C.A.: The optimum formula for the gain of a flow graph or a simple derivation of Coates' formula. Proc. IRE **48**, 883–889 (1960)

15. Donovan, G.M.: Biological version of Braess' paradox arising from perturbed homeostasis. Phys. Rev. E **98**, 062406-1 (2018)
16. Donovan, G.M.: Numerical discovery and continuation of points of infinitesimal homeostasis. Math. Biosci. **311**, 62–67 (2019)
17. Ferrell, J.E.: Perfect and near perfect adaptation in cell signaling. Cell Syst. **2**, 62–67 (2016)
18. Gibson, C.: Singular Points of Smooth Mappings. Research Notes in Math, vol. 25. Pitman, London (1979)
19. Golubitsky, M.: An introduction to catastrophe theory and its applications. SIAM Rev. **20**(2), 352–387 (1978)
20. Golubitsky, M., Schaeffer, D.G.: Singularities and Groups in Bifurcation Theory I. Applied Mathematics Series, vol. 51. Springer, New York (1985)
21. Golubitsky, M., Stewart, I.: Symmetry methods in mathematical biology. São Paulo J. Math. Sci. **9**, 1–36 (2015)
22. Golubitsky, M., Stewart, I.: Homeostasis, singularities and networks. J. Math. Biol. **74**, 387–407 (2017). https://doi.org/10.1007/s00285-016-1024-2
23. Golubitsky, M., Stewart, I.: Homeostasis with multiple inputs. SIAM J. Appl. Dyn. Syst. **17**, 1816–1832 (2018)
24. Golubitsky, M., Stewart, I., Schaeffer, D.G.: Singularities and Groups in Bifurcation Theory II. Applied Mathematics Series, vol. 69. Springer, New York (1988)
25. Golubitsky, M., Wang, Y.: Infinitesimal homeostasis in three-node input-output networks. J. Math. Biol. **80**, 1163–1185 (2020). https://doi.org/10.1007/s00285-019-01457-x
26. Harary, F.: The determinant of the adjacency matrix of a graph. SIAM Rev. **4**(3), 202–210 (1962)
27. Jepson, A.D., Spence, A.: The numerical solution of nonlinear equations having several parameters, I: scalar equations. SIAM J. Numer. Anal. **22**, 736–759 (1985)
28. Jepson, A.D., Spence, A., Cliffe, K.A.: The numerical solution of nonlinear equations having several parameters, III: equations with $\mathbf{Z}_2$ symmetry. SIAM J. Numer. Anal. **28**, 809–832 (1991)
29. Ma, W., Trusina, A., El-Samad, H., Lim, W.A., Tang, C.: Defining network topologies that can achieve biochemical adaptation. Cell **138**, 760–773 (2009)
30. Martinet, J.: Singularities of Smooth Functions and Maps. London Mathematical Society Lecture Notes Series, vol. 58. Cambridge University Press, Cambridge (1982)
31. Moore, G., Garratt, T.J., Spence, A.: The numerical detection of Hopf bifurcation points. In: Roose, D., De Dier, B., Spence, A. (eds.) Continuation and Bifurcations: Numerical Techniques and Applications, pp. 227–246. Kluwer, Dordrecht (1990)
32. Morrison, P.R.: Temperature regulation in three Central American mammals. J. Cell Comp. Physiol. **27**, 125–137 (1946)
33. Nijhout, H.F., Best, J., Reed, M.C.: Escape from homeostasis. Math. Biosci. **257**, 104–110 (2014)
34. Nijhout, H.F., Reed, M.C.: Homeostasis and dynamic stability of the phenotype link robustness and plasticity. Integr. Comp. Biol. **54**(2), 264–275 (2014). https://doi.org/10.1093/icb/icu010
35. Nijhout, H.F., Reed, M., Budu, P., Ulrich, C.: A mathematical model of the folate cycle: new insights into folate homeostasis. J. Biol. Chem. **226**, 55008–55016 (2004)
36. Poston, T., Stewart, I.: Catastrophe Theory and Its Applications. Surveys and Reference Works in Math, vol. 2. Pitman, London (1978)

37. Reed, M., Best, J., Golubitsky, M., Stewart, I., Nijhout, F.: Analysis of homeostatic mechanisms in biochemical networks. Bull. Math. Biol. **79**, 2534–2557 (2017). https://doi.org/10.1007/s11538-017-0340-z
38. Reed, M.C., Lieb, A., Nijhout, H.F.: The biological significance of substrate inhibition: a mechanism with diverse functions. BioEssays **32**, 422–429 (2010)
39. Savageau, M.A., Jacknow, G.: Feedforward inhibition in biosynthetic pathways: inhibition of the aminoacyl-tRNA synthetase by intermediates of the pathway. J. Theor. Biol. **77**, 405–425 (1979)
40. Schechter, M.: Modern Methods in Partial Differential Equations. McGraw-Hill, New York (1977)
41. Tang, Z.F., McMillen, D.R.: Design principles for the analysis and construction of robustly homeostatic biological networks. J. Theor. Biol. **408**, 274–289 (2016)
42. Wang, Y., Huang, Z., Antoneli, F., Golubitsky, M.: The structure of infinitesimal homeostasis in input-output networks, preparation
43. Zeeman, E.C.: Catastrophe Theory: Selected Papers 1972–1977. Addison-Wesley, London (1977)
44. www.biology-online.org/4/1_physiological_homeostasis.htm (2000, updated)

The Approximation of Invariant Sets in Infinite Dimensional Dynamical Systems

Raphael Gerlach(✉) and Adrian Ziessler

Paderborn University, Warburger Straße 100, 33098 Paderborn, Germany
rgerlach@math.upb.de, adrianziessler@googlemail.com

Abstract. In this work we review the novel framework for the computation of finite dimensional invariant sets of infinite dimensional dynamical systems developed in [6] and [36]. By utilizing results on embedding techniques for infinite dimensional systems we extend a classical subdivision scheme [8] as well as a continuation algorithm [7] for the computation of attractors and invariant manifolds of finite dimensional systems to the infinite dimensional case. We show how to implement this approach for the analysis of delay differential equations and partial differential equations and illustrate the feasibility of our implementation by computing the attractor of the Mackey-Glass equation and the unstable manifold of the one-dimensional Kuramoto-Sivashinsky equation.

1 Introduction

In order to understand the long term behavior of complicated nonlinear dynamical systems, a promising approach is to study invariant sets such as the global attractor and invariant manifolds. To numerically approximate these sets set-oriented methods have been developed [7–9,17]. The underlying idea is to cover the set of interest by outer approximations that are generated by multilevel subdivision or continuation methods. They have been used successfully in various application areas such as molecular dynamics [30], astrodynamics [11] or ocean dynamics [18].

Until recently, the applicability of the subdivision scheme and the continuation method was restricted to finite dimensional dynamical systems, i.e., ordinary differential equations or finite dimensional discrete systems. In this work we show how to extend these algorithms for the computation of attractors as well as invariant manifolds to the infinite dimensional context, e.g. delay differential equations with small delays [3,12] and dissipative partial differential equations such as the Kuramoto-Sivashinsky equation [25,32], the Ginzburg-Landau equation and reaction-diffusion equations [22]. For all these systems a finite dimensional so-called *inertial manifold* exists to which trajectories are attracted exponentially fast, e.g., [4,15,34].

The novel approach utilizes infinite dimensional *embedding results* [21,28] that allow the reconstruction of finite dimensional invariant sets of infinite

O. Junge et al. (Eds.): SON 2020, SSDC 304, pp. 66–85, 2020.
https://doi.org/10.1007/978-3-030-51264-4_3

dimension dynamical systems. These results extend the results of Takens [33] and Sauer et al. [29] to the infinite dimensional context. The so–called *observation map* which consists of observations of the systems dynamics produces – in a generic sense – a one-to-one image of the underlying (infinite dimensional) dynamics provided the number of observations is large enough. This observation map and its inverse are then used for the construction of the *core dynamical system* (CDS), i.e., a continuous dynamical system on a state space of finite dimension. By construction the CDS possesses topologically the same dynamical behavior on the invariant set as the original infinite dimensional system. Then, application of the subdivision scheme and the continuation method allow us to compute the reconstructed invariant set of the CDS. The general numerical approach is in principle applicable to infinite dimensional dynamical systems described by a Lipschitz continuous operator on a Banach space. However, here we will restrict our attention to delay differential equations with constant delay and partial differential equations for the numerical realization. We note that in [35] the approach has been generalized to delay differential equations with state dependent time delay.

A detailed outline of the article is as follows. In Sect. 2 we briefly summarize the infinite dimensional embedding theory introduced in [21,28]. In Sect. 3 we employ this embedding technique for the construction of the CDS. Then in Sect. 4 we review the adapted subdivision scheme and continuation method for infinite dimensional systems developed in [6,36]. A numerical realization of the CDS for DDEs and PDEs is given in Sect. 5. Finally, in Sect. 6 we illustrate the efficiency of our methods for the Mackey-Glass delay differential equation and for the one-dimensional Kuramoto-Sivashinsky equation.

2 Infinite Dimensional Embedding Techniques

We consider dynamical systems of the form

$$u_{j+1} = \Phi(u_j), \quad j = 0, 1, \dots, \tag{1}$$

where $\Phi : Y \to Y$ is Lipschitz continuous on a Banach space Y. Moreover, we assume that Φ has an invariant compact set $\mathcal{A}$, that is

$$\Phi(\mathcal{A}) = \mathcal{A}.$$

In order to approximate the set $\mathcal{A}$ or invariant subsets of $\mathcal{A}$ we combine classical subdivision and continuation techniques for the computation of such objects in a finite dimensional space with infinite dimensional embedding results (cf. [21,28]). For the statement of the main result of [28] we require three particular notions: *prevalence* [29], *upper box counting dimension* and *thickness exponent* [21].

Definition 1

(a) A Borel subset S of a normed linear space V is *prevalent* if there is a finite dimensional subspace E of V (the 'probe space') such that for each $v \in V$, $v+e$ belongs to S for (Lebesgue) almost every $e \in E$.

Following a remark made in [29] we will say that "almost every" map in a function space V satisfies a certain property if the set of such maps is prevalent, even in the infinite dimensional case. Then this property will be called *generic* (in the sense of prevalence).

(b) Let Y be a Banach space, and let $\mathcal{A} \subset Y$ be compact. For $\varepsilon > 0$, denote by $N_Y(\mathcal{A}, \varepsilon)$ the minimal number of balls of radius ε (in the norm of Y) necessary to cover the set $\mathcal{A}$. Then

$$d(\mathcal{A}; Y) = \limsup_{\varepsilon \to 0} \frac{\log N_Y(\mathcal{A}, \varepsilon)}{-\log \varepsilon} = \limsup_{\varepsilon \to 0} \; -\log_\varepsilon N_Y(\mathcal{A}, \varepsilon)$$

denotes the upper box counting dimension of $\mathcal{A}$.

(c) Let Y be a Banach space, and let $\mathcal{A} \subset Y$ be compact. For $\varepsilon > 0$, denote by $d_Y(\mathcal{A}, \varepsilon)$ the minimal dimension of all finite dimensional subspaces $V \subset Y$ such that every point of $\mathcal{A}$ lies within distance ε of V; if no such V exists, $d_Y(\mathcal{A}, \varepsilon) = \infty$. Then

$$\sigma(\mathcal{A}, Y) = \limsup_{\varepsilon \to 0} \; -\log_\varepsilon d_Y(\mathcal{A}, \varepsilon)$$

is called the *thickness exponent* of $\mathcal{A}$ in Y.

These notions are essential in addressing the question when a delay embedding technique applied to an invariant subset $A \subset Y$ will generically work. More precisely, the results are as follows.

Theorem 1 ([21, Theorem 3.9]). *Let Y be a Banach space and $\mathcal{A} \subset Y$ compact, with upper box counting dimension $d(\mathcal{A}; Y) =: d$ and thickness exponent $\sigma(\mathcal{A}, Y) =: \sigma$. Let $N > 2d$ be an integer, and let $\alpha \in \mathbb{R}$ with*

$$0 < \alpha < \frac{N - 2d}{N \cdot (1 + \sigma)}.$$

Then for almost every (in the sense of prevalence) bounded linear map $\mathcal{L} : Y \to \mathbb{R}^N$ there is $C > 0$ such that

$$C \cdot \|\mathcal{L}(x - y)\|^\alpha \geq \|x - y\| \quad \textit{for all } x, y \in \mathcal{A}.$$

Note that this result implies that - if N is large enough - almost every (in the sense of prevalence) bounded linear map $\mathcal{L} : Y \to \mathbb{R}^N$ will be one-to-one on $\mathcal{A}$. Using this theorem, the following result concerning embedding techniques can be proven.

Theorem 2 ([28, Theorem 5.1]). *Let Y be a Banach space and $\mathcal{A} \subset Y$ a compact, invariant set, with upper box counting dimension d, and thickness exponent σ. Choose an integer $k > 2(1 + \sigma)d$ and suppose further that the set $\mathcal{A}_p$ of p-periodic points of Φ satisfies $d(\mathcal{A}_p; Y) < p/(2 + 2\sigma)$ for $p = 1, \ldots, k$. Then for* *almost every (in the sense of prevalence) Lipschitz map $f : Y \to \mathbb{R}$ the observation map $D_k[f, \Phi] : Y \to \mathbb{R}^k$ defined by*

$$D_k[f, \Phi](u) = (f(u), f(\Phi(u)), \ldots, f(\Phi^{k-1}(u)))^\top$$

is one-to-one on $\mathcal{A}$.

Remark 1

(a) Following an observation already made in [29, Remark 2.9], we note that this result may be generalized to the case where several distinct observables are evaluated. More precisely, for almost all (in the sense of prevalence) Lipschitz maps $f_i : Y \to \mathbb{R}$, $i = 1, \ldots, q$, the observation map $D_k[f_1, \ldots, f_q, \Phi] : Y \to \mathbb{R}^k$,

$$u \mapsto (f_1(u), \ldots, f_1(\Phi^{k_1-1}(u)), \ldots, f_q(u), \ldots, f_q(\Phi^{k_q-1}(u)))^\top$$

is also one-to-one on $\mathcal{A}$, provided that

$$k = \sum_{i=1}^{q} k_i > 2(1+\sigma) \cdot d \quad \text{and} \quad d(\mathcal{A}_p) < p/(2+2\sigma) \quad \forall p \le \max(k_1, \ldots, k_q).$$

(b) If the thickness exponent σ is unknown, a *worst-case* embedding dimension $k > 2(1+d)d$ can always be chosen since the thickness exponent is bounded by the (upper) box counting dimension (cf. [21]).
(c) In [27] it is suspected that many of the attractors arising in dynamical systems defined by the evolution equations of mathematical physics have thickness exponent zero. In addition, in [16] it is shown that the thickness exponent is essentially inversely proportional to smoothness. This result does not rely on the dynamics associated with the set $\mathcal{A}$ or the form of the underlying equations, but only on assumptions on the smoothness of functions in $\mathcal{A}$. Thus, it is reasonable to assume $\sigma = 0$, i.e., an embedding dimension $k > 2d$ is sufficient in most cases.

3 The Core Dynamical System

In this section we show how the results in Sect. 2 lay the theoretical foundation for the construction of the so–called *core dynamical system* (CDS). This finite dimensional dynamical system then allows us to approximate invariant sets of an infinite dimensional dynamical system. For details on the construction and corresponding proofs we refer to [6].

Let $\mathcal{A}$ be a compact invariant set of an infinite dynamical systems (1) on a Banach space Y and suppose $k \in \mathbb{N}$ is large enough such that the embedding result (Theorem 1 or 2) is valid. Suppose $R : Y \to \mathbb{R}^k$ is the corresponding *observation map*, i.e., $R = \mathcal{L}$ or $R = D_k[f, \Phi]$, respectively. We denote by A_k the image of $\mathcal{A} \subset Y$ under the observation map, that is,

$$A_k = R(\mathcal{A}).$$

The *core dynamical system* (CDS)

$$x_{j+1} = \varphi(x_j), \quad j = 0, 1, 2, \ldots,$$

with $\varphi : \mathbb{R}^k \to \mathbb{R}^k$ is then constructed as follows: Since R is invertible as a mapping from $\mathcal{A}$ to A_k there is a unique continuous map $\widetilde{E} : A_k \to Y$ satisfying

$$(\widetilde{E} \circ R)(u) = u \quad \forall u \in \mathcal{A} \quad \text{and} \quad (R \circ \widetilde{E})(x) = x \quad \forall x \in A_k. \tag{2}$$

Thus, as a first step this allows us to define φ solely on A_k via

$$\varphi = R \circ \Phi \circ \widetilde{E}.$$

For the extension of φ to $\mathbb{R}^k$ we need to extend the map $\tilde{E}$ to a continuous map $E : \mathbb{R}^k \to Y$. By employing a generalization of the well-known Tietze extension theorem [14, I.5.3] found by Dugundji [13, Theorem 4.1] we obtain a continuous map $E : \mathbb{R}^k \to Y$ with $E_{|A_k} = \widetilde{E}$ and we define the CDS by

$$\varphi = R \circ \Phi \circ E,$$

see Fig. 1 for an illustration. Observe that by this construction A_k is an invariant set for φ, and that the dynamics of φ on A_k is topologically conjugate to that of Φ on $\mathcal{A}$.

Proposition 1 ([6, Propostion 1])
There is a continuous map $\varphi : \mathbb{R}^k \to \mathbb{R}^k$ *satisfying*

$$\varphi(R(u)) = R(\Phi(u)) \text{ for all } u \in \mathcal{A}.$$

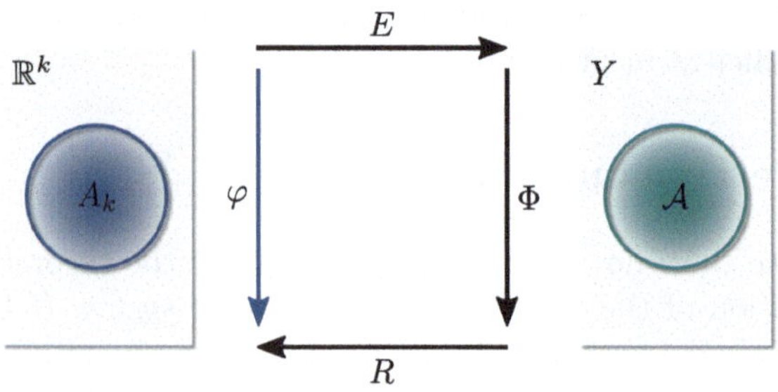

Fig. 1. Definition of the CDS φ (Figure adapted from [36]).

Note that the arguments stated above only guarantee the existence of the continuous map E and provide no guideline on how to design or approximate it. In fact, the particular realization of the map E will depend on the actual application (see Sect. 5).

4 Computation of Embedded Invariant Sets

We are now in the position to approximate the embedded invariant set A_k or invariant subsets such as the invariant manifold of a steady state via the core dynamical system

$$\varphi = R \circ \Phi \circ E.$$

To this end, we employ the subdivision and continuation schemes as defined in [8] and [7].

4.1 A Subdivision Scheme for the Approximation of Embedded Attractors

In this section we give a brief review of the adapted subdivision scheme developed in [6] that allows us to approximate the set A_k.

Let $Q \subset \mathbb{R}^k$ be a compact set and suppose $A_k \subset Q$ for simplicity. The *embedded global attractor relative to* Q is defined by

$$A_Q = \bigcap_{j \geq 0} \varphi^j(Q).$$

The aim is to approximate this set with a subdivision procedure. Given an initial finite collection $\mathcal{B}_0$ of compact subsets of $\mathbb{R}^k$ such that

$$Q = \bigcup_{B \in \mathcal{B}_0} B,$$

we recursively obtain $\mathcal{B}_\ell$ from $\mathcal{B}_{\ell-1}$ for $\ell = 1, 2, \ldots$ in two steps such that the diameter

$$\operatorname{diam}(\mathcal{B}_\ell) = \max_{B \in \mathcal{B}_\ell} \operatorname{diam}(B)$$

converges to zero for $\ell \to \infty$.

Algorithm 1. The subdivision method for embedded global attractors

---

Initialization: Given $k > 2(1 + \sigma)d$ choose a compact set $Q \subset \mathbb{R}^k$ and a finite collection $\mathcal{B}_0$, such that $A_k \subset Q$ and $Q = \bigcup_{B \in \mathcal{B}_0} B$. Fix $0 < \theta_{\min} \leq \theta_{\max} < 1$.

1) *Subdivision:* Construct a new collection $\hat{\mathcal{B}}_\ell$ such that

$$\bigcup_{B \in \hat{\mathcal{B}}_\ell} B = \bigcup_{B \in \mathcal{B}_{\ell-1}} B$$

and

$$\operatorname{diam}(\hat{\mathcal{B}}_\ell) = \theta_\ell \operatorname{diam}(\mathcal{B}_{\ell-1}),$$

where $0 < \theta_{\min} \leq \theta_\ell \leq \theta_{\max} < 1$.

2) *Selection:* Define the new collection $\mathcal{B}_\ell$ by

$$\mathcal{B}_\ell = \left\{ B \in \hat{\mathcal{B}}_\ell : \exists \hat{B} \in \hat{\mathcal{B}}_\ell \text{ such that } \varphi^{-1}(B) \cap \hat{B} \neq \emptyset \right\}. \quad (3)$$

Remark 2

a) A numerical implementation of Algorithm 1 is included in the software package `GAIO` *(Global Analysis of Invariant Objects)* [5, 10]. Here, the sets B constituting the collections $\mathcal{B}_\ell$ are realized by generalized k-dimensional rectangles ("boxes") of the form
$$B(c,r) = \left\{y \in \mathbb{R}^k : |y_i - c_i| \leq r_i \text{ for } i = 1,\ldots,k\right\},$$
where $c, r \in \mathbb{R}^k$, $r_i > 0$ for $i = 1, \ldots, k$, are the center and the radii, respectively. In each subdivision step each box of the current collection is subdivided by bisection with respect to the j-th coordinate, where j is varied cyclically. Therefore, these collections can easily be stored in a binary tree.
b) Given a collection $\hat{\mathcal{B}}_\ell$ the selection step is realized as follows: At first φ is evaluated for a large number of test points $x \in B'$ for each box $B' \in \hat{\mathcal{B}}_\ell$. Then a box B is kept in the collection $\mathcal{B}_\ell$ if there is a least one $x \in B'$ such that $\varphi(x) \in B$. We note that the binary tree structure implemented in `GAIO` allows a fast identification of the boxes that are not discarded.

The subdivision step results in decreasing box diameters with increasing ℓ. In fact, by construction
$$\operatorname{diam}(\mathcal{B}_\ell) \leq \theta_{\max}^\ell \operatorname{diam}(\mathcal{B}_0) \to 0 \quad \text{for } \ell \to \infty.$$

In the selection step each subset whose preimage does neither intersect itself nor any other subset in $\hat{\mathcal{B}}_\ell$ is removed. Denote by Q_ℓ the collection of compact subsets obtained after ℓ subdivision steps, that is
$$Q_\ell = \bigcup_{B \in \mathcal{B}_\ell} B.$$

Since the Q_ℓ's define a nested sequence of compact sets, that is, $Q_{\ell+1} \subset Q_\ell$ we conclude for each m
$$Q_m = \bigcap_{\ell=1}^{m} Q_\ell.$$

Then by considering
$$Q_\infty = \bigcap_{\ell=1}^{\infty} Q_\ell$$
as the limit of the Q_ℓ's the selection step accounts for the fact that Q_m approaches the relative global attractor.

Proposition 2 ([6, Proposition 2])
Suppose A_Q satisfies $\varphi^{-1}(A_Q) \subset A_Q$. Then
$$A_Q = Q_\infty.$$

We note that we can, in general, not expect that $A_k = A_Q$. In fact, by construction A_Q may contain several invariant sets and related heteroclinic connections. However, if $\mathcal{A}$ is an attracting set equality can be proven (see [6]).

4.2 A Continuation Technique for the Approximation of Embedded Unstable Manifolds

In [36] the classical continuation method of [7] was extended to the approximation of embedded unstable manifolds. In the following we state the main result of this scheme. Let us denote by

$$\mathcal{W}^u_\Phi(u^*) \subset \mathcal{A}$$

the unstable manifold of $u^* \in \mathcal{A}$, where u^* is a steady state solution of the infinite dimensional dynamical system Φ (cf. (1)). Furthermore, let us define the *embedded unstable manifold* $W^u(p)$ by

$$W^u(p) = R(\mathcal{W}^u_\Phi(u^*)) \subset A_k,$$

where $p = R(u^*) \in \mathbb{R}^k$ and R is the observation map introduced in Sect. 3. We now choose a compact set $Q \subset \mathbb{R}^k$ containing p and we assume for simplicity that Q is large enough so that it contains the entire closure of the embedded unstable manifold, i.e.,

$$\overline{W^u(p)} \subset Q.$$

For the purpose of initializing the developed algorithm we define a partition $\mathcal{P}$ of Q to be a finite family of compact subsets of Q such that

$$\bigcup_{B \in \mathcal{P}} B = Q \quad \text{and} \quad \text{int} B \cap \text{int} B' = \emptyset, \text{ for all } B, B' \in \mathcal{P},\ B \neq B'.$$

We consider a nested sequence $\mathcal{P}_s$, $s \in \mathbb{N}$, of successively finer partitions of Q, requiring that for all $B \in \mathcal{P}_s$ there exist $B_1, \ldots, B_m \in \mathcal{P}_{s+1}$ such that $B = \cup_i B_i$ and $\text{diam}(B_i) \leq \theta \,\text{diam}(B)$ for some $0 < \theta < 1$. A set $B \in \mathcal{P}_s$ is said to be of *level* s.

The aim of the continuation method is to approximate subsets $W_j \subset W^u(p)$ where $W_0 = W^u_{loc}(p) = R(\overline{\mathcal{W}^u_{\Phi,loc}(u^*)})$ is the *local embedded unstable manifold* and

$$W_{j+1} = \varphi(W_j) \quad \text{for } j = 0, 1, 2, \ldots$$

in two steps:

At first we approximate $W^u_{loc}(p)$ by applying Algorithm 1 on a compact neighborhood $C \subset A_k$ of p such that $p \in$ int C and $W^u_{loc}(p) \subset C$ in order to compute the relative global attractor A_C. In fact, if the steady state $u^* \in \mathcal{A}$ is hyperbolic $W^u_{loc}(p)$ is identical to A_C [36, Proposition 3.1 (b)]. In the second step this covering of $W^u_{loc}(p)$ is then to globalized to obtain an approximation of the compact subsets $W_j \subset W^u(p)$ or even the entire closure $\overline{W^u(p)}$.

Algorithm 2. The continuation method for embedded unstable manifolds

Initialization: Given $k > 2(1+\sigma)d$ choose an initial box $Q \subset \mathbb{R}^k$ such that $A_k \subset Q$. Choose a partition $\mathcal{P}_s$ of Q and a set $C \in \mathcal{P}_s$ such that $p = R(u^*) \in C$.

1) Perform ℓ steps of Algorithm 1 on $\mathcal{B}_0 = \{C\}$ to obtain a covering $\mathcal{B}_\ell \subset \mathcal{P}_{s+\ell}$ of $W^u_{loc}(p)$. Set $C_0^{(\ell)} = \mathcal{B}_\ell$.
2) *Continuation:* For $j = 0, 1, 2, \dots$ define

$$C_{j+1}^{(\ell)} = \left\{ B \in \mathcal{P}_{s+\ell} : \exists B' \in C_j^{(\ell)} \text{ such that } B \cap \varphi(B') \neq \emptyset \right\}. \tag{4}$$

Remark 3

a) Algorithm 2 is also implemented within the software package GAIO. In fact, the binary tree structure encodes a nested sequence of finer partitions of Q.
b) Numerically the continuation step is realized as follows: At first φ is evaluated for a large number of test points $x \in B'$ for each box $B' \in C_j^{(\ell)}$. Then a box $B \in \mathcal{P}_{s+\ell}$ is added to the collection $C_{j+1}^{(\ell)}$ if there is a least one $x \in B'$ such that $\varphi(x) \in B$.

Observe that the unions

$$C_j^{(\ell)} = \bigcup_{B \in C_j^{(\ell)}} B$$

form a nested sequence in ℓ, i.e.,

$$C_j^{(0)} \supset C_j^{(1)} \supset \dots \supset C_j^{(\ell)} \dots .$$

In fact, it is also a nested sequence in j, i.e.,

$$C_0^{(\ell)} \subset C_1^{(\ell)} \dots \subset C_j^{(\ell)} \dots .$$

Due to the compactness of Q the continuation step of Algorthm 2 will terminate after finitely many, say J_ℓ, steps. We denote the corresponding box covering obtained by the continuation method by

$$G_\ell = \bigcup_{j=0}^{J_\ell} C_j^{(\ell)} = C_{J_\ell}^{(\ell)}.$$

In [36] it was proven that increasing ℓ eventually leads to convergence of $C_j^{(\ell)}$ to the subsets W_j and assuming that the closure of the embedded unstable manifold $\overline{W^u(p)}$ is attractive G_ℓ converges to $\overline{W^u(p)}$.

Proposition 3 ([36, Proposition 5])

(a) The sets $C_j^{(\ell)}$ *are coverings of* W_j *for all* $j, \ell = 0, 1, \ldots$. *Moreover, for fixed* j, *we have*

$$\bigcap_{\ell=0}^{\infty} C_j^{(\ell)} = W_j.$$

(b) Suppose that $\overline{W^u(p)}$ *is linearly attractive, i.e., there is a* $\lambda \in (0, 1)$ *and a neighborhood* $U \supset Q \supset \overline{W^u(p)}$ *such that*

$$dist\left(\varphi(y), \overline{W^u(p)}\right) \leq \lambda \; dist\left(y, \overline{W^u(p)}\right) \quad \forall y \in U.$$

Then the box coverings obtained by Algorithm 2 converge to the closure of the embedded unstable manifold $\overline{W^u(p)}$. *That is,*

$$\bigcap_{\ell=0}^{\infty} G_\ell = \overline{W^u(p)}.$$

5 Numerical Realization of the CDS

As discussed in the introduction, dynamical systems with infinite dimensional state space, but finite dimensional attractors arise in particular in two areas of applied mathematics, namely dissipative partial differential equations and delay differential equations with small constant delay. In this section we show how to numerically realize the CDS for both cases. From now on we assume that upper bounds for both the box counting dimension d and the thickness exponent σ are available. This allows us to fix $k > 2(1 + \sigma)d$ according to Theorem 2.

In order to numerically realize the construction of the map $\varphi = R \circ \Phi \circ E$ described in Sect. 3, we have to address three tasks: the implementation of E, of R, and of the *time-T-map*, denoted by Φ, respectively. For the latter we will rely on standard methods for forward time integration of DDEs [1] and PDEs, e.g., a fourth-order time stepping method for the one-dimensional Kuramoto-Sivashinsky equation [24]. The map R will be realized on the basis of Theorem 2 and Remark 1 by an appropriately chosen observables. For the numerical construction of the continuous map E we will employ a bootstrapping method that re-uses results of previous computations. This way we will in particular guarantee that the identities in (2) are at least approximately satisfied.

5.1 Delay Differential Equations

We consider equations of the form

$$\dot{y}(t) = g(y(t), y(t - \tau)), \tag{5}$$

where $y(t) \in \mathbb{R}^n$, $\tau > 0$ is a constant time delay and $g : \mathbb{R}^n \times \mathbb{R}^n \to \mathbb{R}^n$ is a smooth map. Here, we will only consider the one-dimensional case, that is $n = 1$, and refer to [6] for $n > 1$. Following [19], we denote by $Y = C([-\tau, 0], \mathbb{R}^n)$ the (infinite dimensional) state space of the dynamical system (5). Observe that equipped with the maximum norm Y is indeed a Banach space. We set $T > 0$ to be a natural fraction of τ, that is

$$T = \frac{\tau}{K} \text{ for } K \in \mathbb{N}. \tag{6}$$

5.1.1 Numerical Realization of *R*

For the definition of R we have to specify the time span T and appropriate corresponding observables. In the case of a scalar equation we choose the observable f to be

$$f(u) = u(-\tau).$$

Thus, in this case the restriction R is simply given by

$$R = D_k[f, \Phi](u) = (u(-\tau), \Phi(u)(-\tau), \ldots, \Phi^{k-1}(u)(-\tau))^\top.$$

Observe that a natural choice for K in (6) would be $K = k - 1$ for $k > 1$. That is, for each evaluation of R the observable would be applied to a function $u : [-\tau, 0] \to \mathbb{R}$ at k equally distributed time steps within the interval $[-\tau, 0]$.

5.1.2 Numerical Realization of *E*

In the application of the subdivision scheme Algorithm 1 as well as the continuation method Algorithm 2 the CDS has to be evaluated for a set of test points (see Remark 2 and Remark 3). Thus, for the evaluation of $\varphi = R \circ \Phi \circ E$ at a test point x we need to define the image $E(x)$, i.e., we need to generate adequate initial conditions for the forward integration of the DDE (5).

In the first selection or continuation step, when no information on $\mathcal{A}$ or $\mathcal{W}^u_\Phi(u^*)$, respectively, is available, we construct a piecewise linear function $u = E(z)$, where

$$u(t_i) = z_i,$$

for $t_i = -\tau + i \cdot T$, $i = 0, \ldots, k - 1$. Observe that by this choice of E and R the condition $(R \circ E)(x) = x$ is satisfied for each test point x (see (2)). In the following steps we make use of the following observations for both schemes:

Remark 4. If a box $B \in \mathcal{B}_\ell$ (resp. $B \in C^{(\ell)}_{j+1}$), then, by the subdivision (resp. selection) step, there must have been a $\hat{B} \in \mathcal{B}_{\ell-1}$ (resp. $\hat{B} \in C^{(\ell)}_j$) such that $\bar{x} = R(\Phi(E(\hat{x}))) \in B$ for at least one test point $\hat{x} \in \hat{B}$. Therefore, we can use the information from the computation of $\Phi(E(\hat{x}))$ to construct an appropriate $E(x)$ for each test point $x \in B$ in both cases.

More concretely, in every step of the procedures, for every set $B \in \mathcal{B}_\ell$ (resp. $B \in C^{(\ell)}_{j+1}$) we keep additional information about the trajectories $\Phi(E(\hat{z}))$ that were mapped into B by R in the previous step. For simplicity, we store $k_i \geq 1$ additional equally distributed function values for each interval $(-\tau + (i-1)T, -\tau + iT)$ for $i = 1, \ldots, k-1$.

When $\varphi(B)$ is to be computed using test points from B, we first use the points in B for which additional information is available and generate the corresponding initial value functions via spline interpolation. Note that the more information we store, the smaller the error $\|\Phi(E(\hat{x})) - E(x)\|$ becomes for $x = R(\Phi(E(\hat{x})))$. That is, we enforce an approximation of the identity $(E \circ R)(u) = u$ for all $u \in \mathcal{A}$ (see (2)). If the additional information is available only for a few points in B, we generate new test points in B at random and construct corresponding trajectories by piecewise linear interpolation.

5.2 Partial Differential Equations

We will consider explicit differential equations of the form

$$\frac{\partial}{\partial t} u(y,t) = F(y,u), \; u(y,0) = u_0(y), \tag{7}$$

where $u : \mathbb{R}^n \times \mathbb{R} \to \mathbb{R}^n$ is in some Banach space Y and F is a (nonlinear) differential operator. We assume that the dynamical system (7) has a well-defined semiflow on Y.

5.2.1 Numerical Realization of *R*

In the previous section for delay differential equations R is defined by the delay coordinate map. In principle it would also be possible to observe the evolution of a partial differential equation by a delay coordinate map. However, from a computational point of view this would be impractical. The reason is that for the realization of the map $E : \mathbb{R}^k \to Y$ one would have to reconstruct functions (i.e., the space-dependent state) from time delay coordinates (of scalar, e.g., point-wise observations). Thus, for each point in observation space one would essentially have to store the entire corresponding function.

To overcome this problem, we will present a different approach. In what follows, we will assume that the function $u \in Y$ can be represented in terms of an orthonormal basis $\{\Psi_i\}_{i=1}^{\infty}$, i.e.,

$$u(y,t) = \sum_{i=1}^{\infty} x_i(t)\Psi_i(y),$$

where the Ψ_i are elements from a Hilbert space. Then our observation map R will be defined by projecting a function onto k coefficients x_i of its Galerkin expansion. The function u can then be approximated within the (truncated) linear subspace spanned by the basis $\{\Psi_i\}_{i=1}^{k}$:

$$u(y,t) \approx \sum_{i=1}^{k} x_i(t)\Psi_i(y).$$

For the computation of the basis, we use the well-known Proper Orthogonal Decomposition (POD), cf. [2,20,31]. The reason is that for each basis size k, POD yields the *optimal* basis, i.e., the basis with the minimal L^2 projection error. In order to compute such a basis $\{\Psi_i\}_i$ we use the so-called *method of snapshots* (cf. [20] for details). To this end, we construct the so–called snapshot matrix $S_M \in \mathbb{R}^{n_x \times r}$, where each column consists of the (discretized) state at equidistant time instances obtained from a single long-time integration of the underlying PDE (7). Then we perform a singular value decomposition (SVD) of the matrix S_M and obtain $S_M = U\Sigma V^\top$, where $U \in \mathbb{R}^{n_x \times n_x}, \Sigma \in \mathbb{R}^{n_x \times r}$ and $V \in \mathbb{R}^{r \times r}$. The columns of U give us a discrete representation of the POD modes Ψ_i. Using the fact that this basis is orthogonal, we then define the observation map by choosing k different observables

$$f_i(u) = \langle u, \Psi_i \rangle = x_i \quad \text{for } i = 1, \ldots, k.$$

This yields

$$R(u) = (f_1(u), \ldots, f_k(u))^\top = (x_1, \ldots, x_k)^\top.$$

Observe that R is linear and bounded and hence, for k sufficiently large, Theorem 1 and Remark 1, respectively, guarantee that generically (in the sense of prevalence) R will be a one-to-one map on $\mathcal{A}$.

5.2.2 Numerical Realization of *E*

Since the state space for the CDS φ is given by points $x \in \mathbb{R}^k$ where $x_1, \ldots, x_k$ are the POD coefficients we simply construct initial conditions $u = E(x)$ by defining the map E as

$$E(x) = \sum_{i=1}^{k} x_i \Psi_i,$$

if no additional information is available. Observe again that by this choice the condition $(R \circ E)(x) = x$ is satisfied for each test point x.

However, the linear space spanned by the first k POD modes is not invariant under the dynamics of Φ if k is not sufficiently large. Thus, $(E \circ R)(\bar{u}) = \bar{u}$ with $\bar{u} = \Phi(E(x))$ will in general not be satisfied anymore. This is not acceptable since the requirement $(E \circ R)(u) = u$ for all $u \in \mathcal{A}$ (see (2)) is crucial in order to compute reliable coverings.

To enforce this equality at least approximately we extend the expansion and construct initial functions by

$$E(x) = \sum_{i=1}^{k} x_i \Psi_i + \sum_{l=k+1}^{S} x_l \Psi_l, \tag{8}$$

where $S \gg k$. To address the choice of S we note that the singular values σ_i of the snapshot matrix S_M, i.e. the diagonal elements of Σ, determine the amount of information that is neglected by truncating the basis $\{\Psi_i\}_i$ to size $S < r$ [31].

Thus, we choose S such that

$$\varepsilon(S) := \frac{\sum_{i=1}^{S} \sigma_i}{\sum_{j=1}^{r} \sigma_j} \approx 1.$$

In (8) only the first k POD coefficients are given by the coordinates of points inside $B \subset \mathbb{R}^k$. Thus, it remains to discuss how to choose the POD coefficients $x_{k+1}, \ldots, x_S$. The idea is to use a heuristic strategy that utilizes statistical information. By Remark 4 we can compute the POD coefficients $\bar{x}_{k+1}, \ldots, \bar{x}_S$ for all these points $\bar{x}$ by

$$\bar{x}_i = \langle \Phi(E(\hat{x})), \Psi_i \rangle, \quad i = k+1, \ldots, S.$$

Then we sample the box B with all points $\bar{x}$ for which additional information is available. However, the number of these points $\bar{x}$ might be too small, such that B is not discretized sufficiently well and we have to generate additional test points. For this, we first choose a certain number of points $\widetilde{x} \in B$ at random. Then we extend these points to elements in $\mathbb{R}^S$ as follows: We first compute componentwise the mean value μ_i and the variance σ_i^2 of all POD coefficients $\bar{x}_i$, for $i = k+1, \ldots, S$. This allows us to make a Monte Carlo sampling for the additional coefficients of $\widetilde{x}_i$ for $i = k+1, \ldots, S$, i.e., $\widetilde{x}_i \sim \mathcal{N}(\mu_i, \sigma_i^2)$ for $i = k+1, \ldots, S$. Finally, we compute initial functions of the form

$$E(\widetilde{x}) = \sum_{i=1}^{S} \widetilde{x}_i \Psi_i.$$

By this construction we expect to generate initial functions that at least approximately satisfy the identity $(E \circ R)(u) = u$ for all $u \in \mathcal{A}$.

6 Numerical Results

In this section we present results of computations carried out for the Mackey-Glass delay differential equation and the Kuramoto-Sivashinsky equation, respectively.

6.1 The Mackey-Glass Equation

As in [6], we consider the well-known delay differential equation introduced by Mackey and Glass in 1977 [26], namely

$$\dot{u}(t) = \beta \frac{u(t-\tau)}{1 + u(t-\tau)^\eta} - \gamma u(t), \tag{9}$$

where we choose $\beta = 2, \gamma = 1$, $\eta = 9.65$, and $\tau = 2$. This equation is a model for blood production, where $u(t)$ represents the concentration of blood at time t, $\dot{u}(t)$ represents production at time t and $u(t-\tau)$ is the concentration at an earlier time.

Direct numerical simulations indicate that the dimension of the corresponding attracting set is approximately $d = 2$. Thus, we choose the embedding dimension $k = 7$, and approximate the relative global attractor A_Q for $Q = [0, 1.5]^7 \subset \mathbb{R}^7$.

In Fig. 2 (a) to (c), we show projections of the coverings obtained after $\ell = 28, 42$ and 63 subdivision steps. In order to investigate the effect of using information retained from prior integration runs in the implementation of the map E (see Sect. 5.1.2), we show in Fig. 2 (d) a projection of the coverings obtained after 63 subdivision steps with the map E using only piecewise linear functions – that is, no additional information from previous time integration is used. The results indicate that in this case the influence on the quality of the approximation of A_k is only marginal.

6.2 The Kuramoto-Sivashinsky Equation

The well-known Kuramoto-Sivashinsky equation in one spatial dimension is given by

$$\begin{aligned} &u_t + 4u_{yyyy} + \mu \left[u_{yy} + \frac{1}{2}(u_y)^2 \right] = 0, \quad 0 \le y \le 2\pi, \\ &u(y, 0) = u_0(y), \quad u(y + 2\pi, t) = u(y, t). \end{aligned} \tag{10}$$

Here, the parameter is $\mu = L^2/4\pi^2$, where L denotes the size of a typical pattern scale. As in [36] we are interested in computing the unstable manifold of the trivial unstable steady state $u^* = 0$ for $\mu = 15$.

In what follows, the observation space is defined through projections onto the first k POD coefficients, and thus, $p = R(u^*) = 0 \in \mathbb{R}^k$. We compute the POD basis (cf. Sect. 5.2.1) by using the snapshot matrix obtained through a long-time integration with the initial condition

$$u_0(y) = 10^{-4} \cdot \cos(y) \cdot (1 + \sin(y)).$$

For $\mu = 15$ the Kuramoto-Sivashinsky equation has two stable traveling waves (see. Fig. 3 (a)) traveling in opposite directions due to the symmetry imposed by the periodic boundary conditions. In the observation space this corresponds to two stable limit cycles that are symmetric in the first POD coefficient a_1 (see. Fig. 3 (b)). We assume that the dimension of the embedded unstable manifold is approximately two since different initial conditions result in trajectories in observation space that are rotations of each other about the origin. Therefore, assuming that the thickness exponent is zero, we have to choose $k \ge 5$ in order to obtain a one-to-one image of $\mathcal{W}^u_\Phi(u^*)$. To allow for a larger dimension or thickness exponent we choose the embedding dimension $k = 7$ in the following. We choose $Q = [-8, 8]^7$ and initialize a fine partition $\mathcal{P}_s$ of Q for $s = 21, 35, 49, 63$. Next we set $T = 200$. In addition, we define a finite time grid $\{t_0, \ldots, t_N\}$, where

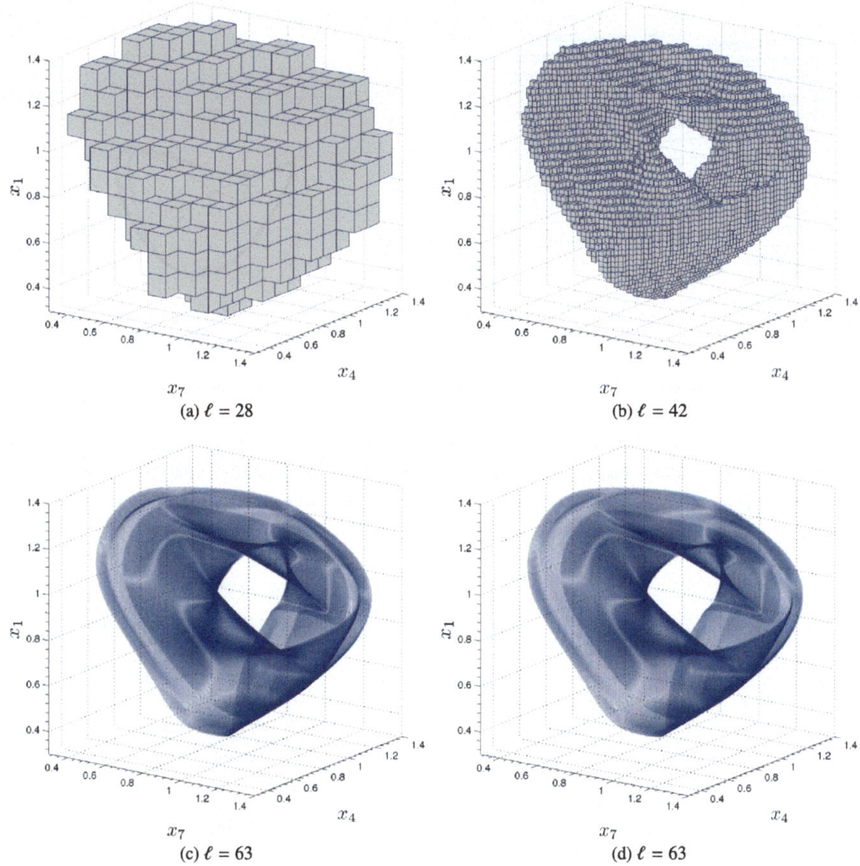

Fig. 2. (a)–(c) Successively finer coverings of a relative global attractor after ℓ subdivision steps for the Mackey-Glass equation (9). (d) Embedding E using only piecewise linear interpolation.

$t_N = T$, and add all boxes that are hit in any of these time steps t_i (a similar approach has been used in [23]). In Fig. 4 (a) to (d) we illustrate successively finer box coverings of the unstable manifold as well as a transparent box covering depicting the complex internal structure of the unstable manifold. Observe that – as mentioned above – the boundary of the unstable manifold consists of two limit cycles which are symmetric in the first POD coefficient x_1. This is due to the fact that the Kuramoto-Sivashinsky equation with periodic boundary conditions (10) possesses $O(2)$-symmetry.

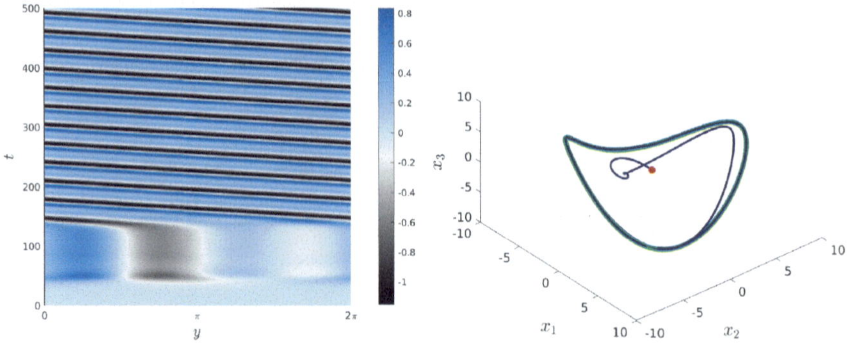

(a) Direct simulation
(b) Observation space

Fig. 3. (a) Direct simulation of the Kuramoto-Sivashinsky equation for $\mu = 15$. The initial value is attracted to a traveling wave solution; (b) Corresponding embedding in observation space. Here, the red dot depicts the unstable steady state. As expected the CDS possesses a limit cycle (green).

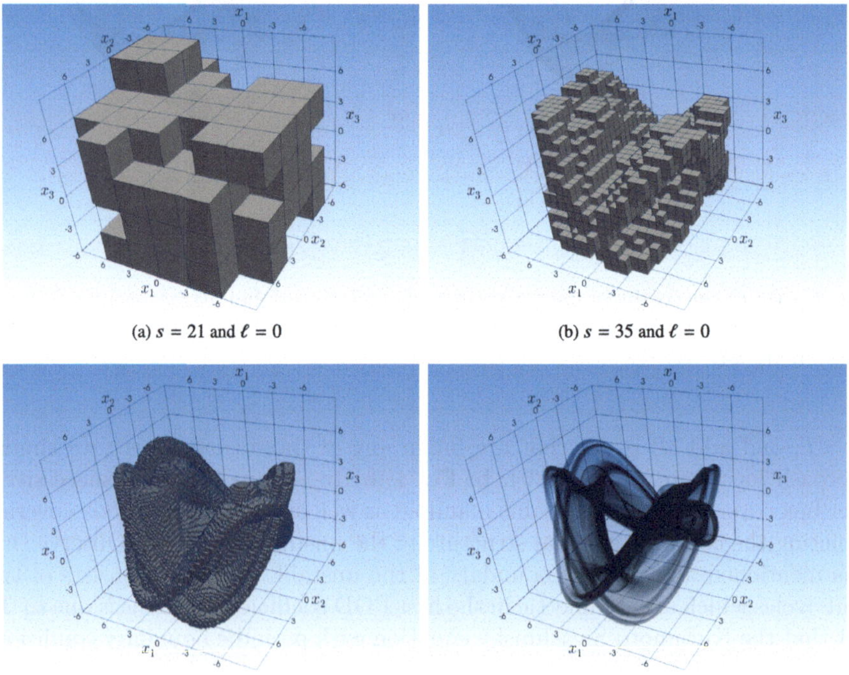

(a) $s = 21$ and $\ell = 0$
(b) $s = 35$ and $\ell = 0$
(c) $s = 49$ and $\ell = 0$
(d) $s = 63$ and $\ell = 0$

Fig. 4. (a)–(d) Successively finer box-coverings of the unstable manifold for $\mu = 15$. (d) Transparent box covering for $s = 63$ and $\ell = 0$ depicting the internal structure of the unstable manifold.

7 Conclusion

In this work we review the contents of [6] and [36], where infinite dimensional embedding results have extended to the numerical analysis of infinite dimensional dynamical systems. To this end, a continuous dynamical system, a finite dimensional *core dynamical system* (CDS) is constructed to obtain a one-to-one representation of the underlying dynamics. For the numerical realization of this system we also identify suitable observables for delay differential and partial differential equations. This finite dimensional system then employed in the subdivision scheme for the computation of relative global attractors and the continuation method for the approximation of invariant manifolds feasible for infinite dimensional systems. The applicability of this novel framework is illustrated by the computation of the attractor of the Mackey-Glass delay differential equation and the unstable manifold of the one-dimensional Kuramoto-Sivashinsky equation.

The numerical effort of the methods proposed in this work essentially depends on the dimension of the object to be computed, and not on the dimension of the observation space of the CDS. However, note that for the numerical realization of the selection step 3 and the continuation step 4 we have to evaluate the CDS for each box and each test point $x \in B'$. Therefore, for each test point we also have to evaluate the underlying infinite dimensional dynamical system which may result in a prohibitively large computational effort. For this reason data-based local *reduced order models* can be used in order to significantly reduce the number of CDS evaluations [35].

Acknowledgments. We would like to acknowledge Michael Dellnitz for developing the underlying ideas as well as the theoretical foundations of this work.

References

1. Bellen, A., Zennaro, M.: Numerical Methods for Delay Differential Equations. Oxford University Press, Oxford (2013)
2. Berkooz, G., Holmes, P., Lumley, J.L.: The proper orthogonal decomposition in the analysis of turbulent flows. Annu. Rev. Fluid Mech. **25**(1), 539–575 (1993)
3. Chicone, C.: Inertial and slow manifolds for delay equations with small delays. J. Differ. Equ. **190**(2), 364–406 (2003)
4. Constantin, P., Foias, C., Nicolaenko, B., Temam, R.: Integral Manifolds and Inertial Manifolds for Dissipative Partial Differential Equations. Springer, Heidelberg (1988)
5. Dellnitz, M., Froyland, G., Junge, O.: The algorithms behind GAIO—set oriented numerical methods for dynamical systems. In: Ergodic Theory, Analysis, and Efficient Simulation of Dynamical Systems, pp. 145–174. Springer, Heidelberg (2001)
6. Dellnitz, M., Hessel-von Molo, M., Ziessler, A.: On the computation of attractors for delay differential equations. J. Comput. Dyn. **3**(1), 93–112 (2016)
7. Dellnitz, M., Hohmann, A.: The computation of unstable manifolds using subdivision and continuation. In: Nonlinear Dynamical Systems and Chaos, pp. 449–459. Birkhäuser Basel (1996)

8. Dellnitz, M., Hohmann, A.: A subdivision algorithm for the computation of unstable manifolds and global attractors. Numer. Math. **75**, 293–317 (1997)
9. Dellnitz, M., Junge, O.: On the approximation of complicated dynamical behavior. SIAM J. Numer. Anal. **36**(2), 491–515 (1999)
10. Dellnitz, M., Junge, O.: Set oriented numerical methods for dynamical systems. Handb. Dyn. Syst. **2**, 221–264 (2002)
11. Dellnitz, M., Junge, O., Lo, M., Marsden, J.E., Padberg, K., Preis, R., Ross, S., Thiere, B.: Transport of Mars-crossing asteroids from the quasi-Hilda region. Phys. Rev. Lett. **94**(23), 231102 (2005)
12. Driver, R.D.: On Ryabov's asymptotic characterization of the solutions of quasilinear dfferential equations with small delays. SIAM Rev. **10**(3), 329–341 (1968)
13. Dugundji, J.: An extension of Tietze's theorem. Pacific J. Math. **1**(3), 353–367 (1951)
14. Dunford, N., Schwartz, J.T.: Linear Operators. Part I: General theory. Wiley-Interscience (1988)
15. Foias, C., Jolly, M.S., Kevrekidis, I.G., Sell, G.R., Titi, E.S.: On the computation of inertial manifolds. Phys. Lett. A **131**(7), 433–436 (1988)
16. Friz, P.K., Robinson, J.C.: Smooth attractors have zero 'thickness'. J. Math. Anal. Appl. **240**(1), 37–46 (1999)
17. Froyland, G., Dellnitz, M.: Detecting and locating near-optimal almost invariant sets and cycles. SIAM J. Sci. Comput. **24**(6), 1839–1863 (2003)
18. Froyland, G., Horenkamp, C., Rossi, V., Santitissadeekorn, N., Sen Gupta, A.: Three-dimensional characterization and tracking of an Agulhas ring. Ocean Model. **52–53**, 69–75 (2012)
19. Hale, J.K., Lunel, S.M.V.: Introduction to Functional Differential Equations, vol. 99. Springer, Heidelberg (2013)
20. Holmes, P., Lumley, J.L., Berkooz, G., Rowley, C.W.: Turbulence, Coherent Structures, Dynamical Systems and Symmetry. Cambridge University Press, Cambridge (2012)
21. Hunt, B.R., Kaloshin, V.Y.: Regularity of embeddings of infinite-dimensional fractal sets into finite-dimensional spaces. Nonlinearity **12**(5), 1263–1275 (1999)
22. Jolly, M.S.: Explicit construction of an inertial manifold for a reaction diffusion equation. J. Diff. Equ. **78**(2), 220–261 (1989)
23. Junge, O.: Mengenorientierte Methoden zur numerischen Analyse dynamischer Systeme. Shaker Verlag (1999)
24. Kassam, A.K., Trefethen, L.N.: Fourth-order time-stepping for stiff pdes. SIAM J. Sci. Comput. **26**(4), 1214–1233 (2005)
25. Kuramoto, Y., Tsuzuki, T.: Persistent propagation of concentration waves in dissipative media far from thermal equilibrium. Progress Theor. Phys. **55**(2), 356–369 (1976)
26. Mackey, M.C., Glass, L.: Oscillation and chaos in physiological control systems. Science **197**(4300), 287–289 (1977)
27. Ott, W., Hunt, B., Kaloshin, V.: The effect of projections on fractal sets and measures in Banach spaces. Ergodic Theory Dyn. Syst. **26**(3), 869–891 (2006)
28. Robinson, J.C.: A topological delay embedding theorem for infinite-dimensional dynamical systems. Nonlinearity **18**, 2135–2143 (2005)
29. Sauer, T., Yorke, J.A., Casdagli, M.: Embedology. J. Stat. Phys. **65**(3–4), 579–616 (1991)
30. Schütte, C., Huisinga, W., Deuflhard, P.: Transfer operator approach to conformational dynamics in biomolecular systems. In: Ergodic Theory, Analysis, and Efficient Simulation of Dynamical Systems, pp. 191–223. Springer, Heidelberg (2001)

31. Sirovich, L.: Turbulence and the dynamics of coherent structures part I: coherent structures. Q. Appl. Math. **45**(3), 561–571 (1987)
32. Sivashinsky, G.: Nonlinear analysis of hydrodynamic instability in laminar flames - I. Derivation of basic equations. Acta Astronautica **4**(11–12), 1177–1206 (1977)
33. Takens, F.: Detecting strange attractors in turbulence. In: Dynamical Systems and Turbulence, Warwick 1980, pp. 366–381. Springer, Heidelberg (1981)
34. Temam, R.: Infinite-Dimensional Dynamical Systems in Mechanics and Physics, Applied Mathematical Sciences, vol. 68. Springer, Heidelberg (1997)
35. Ziessler, A.: Analysis of infinite dimensional dynamical systems by set-oriented numerics. Ph.D. thesis, Paderborn University (2018)
36. Ziessler, A., Dellnitz, M., Gerlach, R.: The numerical computation of unstable manifolds for infinite dimensional dynamical systems by embedding techniques. SIAM J. Appl. Dyn. Syst. **18**(3), 1265–1292 (2019)

Set-Oriented and Finite-Element Study of Coherent Behavior in Rayleigh-Bénard Convection

Anna Klünker[1], Christiane Schneide[1], Gary Froyland[2], Jörg Schumacher[3], and Kathrin Padberg-Gehle[1(✉)]

[1] Institute of Mathematics and its Didactics, Leuphana Universität Lüneburg, Universitätsallee 1, 21335 Lüneburg, Germany
{anna.kluenker,christiane.schneide,padberg}@leuphana.de
[2] School of Mathematics and Statistics, University of New South Wales, Sydney, NSW 2052, Australia
g.froyland@unsw.edu.au
[3] Department of Mechanical Engineering, Technische Universität Ilmenau, P.O. Box 100 565, 98684 Ilmenau, Germany
joerg.schumacher@tu-ilmenau.de

Abstract. Transfer operator methods have been recognized as powerful tools for the computational study of transport and mixing processes in nonautonomous dynamical systems. The main applications in this context have been geophysical flows with large-scale and long-lived isolated vortical coherent flow structures such as eddies or gyres. The present paper aims to demonstrate the applicability of set-oriented and finite-element frameworks to more complex systems. To this end, we study coherent behavior in turbulent Rayleigh-Bénard convection in two- and three-dimensional settings.

1 Introduction

Transport and mixing processes have been widely studied in dynamical systems. Of key interest are regions in the phase space of an autonomous or a nonautonomous dynamical system that remain coherent, or minimally dispersive, under the action of the flow. Over the last two decades, a number of different concepts have been proposed that describe the notion of Lagrangian coherent behavior. For discussions and comparisons of the major current approaches we refer to [1,2].

Among these established concepts are transfer operator-based methods within a set-oriented numerical framework. Similar in spirit to cell-mapping techniques [3,4], Dellnitz and Hohmann [5] developed a set-oriented approach for the outer approximation of attracting sets in dissipative dynamical systems. Dellnitz and Junge [6,7] extended these ideas to approximate natural invariant measures as well as almost-invariant and almost-cyclic sets of the underlying dynamical

O. Junge et al. (Eds.): SON 2020, SSDC 304, pp. 86–108, 2020.
https://doi.org/10.1007/978-3-030-51264-4_4

system. Almost-invariant sets [6,7] and their finite-time counterparts [8,9] are spatially fixed regions in phase space with the property that trajectories leave such a set only with a relatively small probability over a given time span. Hence, almost-invariant sets mitigate transport between their interior and the rest of phase space.

The key mathematical tool for this probabilistic approach is the Perron-Frobenius operator (transfer operator) or, for continuous-time dynamics, its generator [10]. In a set-oriented framework, the Ulam approximation [11] of the transfer operator produces a stochastic matrix and there are many results in the literature, dating back to [12], concerning the convergence of the leading eigenvector of the resulting stochastic matrix to a physical invariant measure (when one exists). Eigenvectors corresponding to real eigenvalues close to one contain information about almost-invariant sets [7]. This latter concept has been applied to many different dynamical systems, including molecular dynamics [13,14], astrodynamics [15,16], and ocean dynamics [17,18]. A special type of (almost-) invariant sets are attracting sets in dissipative systems and their basins, which can also be identified based on leading eigenvectors of the numerical transfer operator [4,19,20] or the generator [21].

The almost-invariant set framework was extended by Froyland and co-workers to the identification of mobile regions that move about with minimal dispersion under the time-asymptotic [22] and finite-time [23–25] action of a nonautonomous dynamical system. In the finite-time setting, subdominant singular vectors of numerically approximated transfer operators are used to determine the phase space structures of interest. The relation between almost-invariant sets and finite-time coherent sets was discussed in [25]. In [26] the existence of coherent sets over long time spans was linked to the existence of almost-invariant sets for small time spans, given that the coherent sets move sufficiently slowly. A study of coherent sets for the Fokker-Planck equation is in [27] and recent generator-based approaches remove the need for trajectory integration in periodically driven [28] and aperiodically driven [29] flows.

The set-oriented approach to identifying finite-time coherent sets relies on the addition of small amount of diffusion to create the necessary spectral gap [24]; in fact this reliance is also present for almost-invariant sets. In practice, this diffusion is usually provided by the numerical diffusion inherent in the set-oriented numerics. By formally sending this added diffusion to zero, one arrives at a second-order differential operator called the dynamic Laplace operator [30]. The dominant eigenfunctions of the dynamic Laplacian correspond to the dominant singular vectors of the transfer operator. A finite-element approach [31] to approximating the dynamic Laplacian provides a robust, numerical scheme for sparse trajectory data. The cluster-based approach [32] for the estimation of finite-time coherent sets from sparse trajectory data with possibly missing trajectory elements has been followed by several other data-based methods [33–36].

Transfer operators can also be employed to estimate finite-time expansive behavior along trajectories in autonomous and nonautonomous dynamical systems. Finite-time entropy (FTE) captures nonlinear stretching directly from the entropy growth experienced by a small localized density evolved by the transfer operator. An approximation of the FTE field can be obtained very efficiently within the set-oriented framework. It gives very similar results to finite-time Lyapunov exponent [37] calculations, which many of the geometric approaches for the identification of Lagrangian coherent structures are based on [38]. The FTE-concept has been introduced in [39], see also [40] for related previous work.

In this chapter we consider Rayleigh–Bénard convection (RBC), which is an idealized model of thermal convection in natural systems. In RBC a fluid layer placed between two solid horizontal plates is uniformly heated from below and cooled from above [41]. This model setting contains already many of the properties which can be observed in natural convection flows. One is the formation of large-scale coherent patterns when RBC is investigated in horizontally extended domains [42–47]. These coherent sets, which have been detected in the Eulerian frame of reference, are termed turbulent superstructures as the characteristic horizontal scale extends the height of the convection layer. In thermal convection flows, they consist of convection rolls and cells that are concealed in instantaneous velocity fields by turbulent fluctuations. However, they show up prominently after time averaging of the velocity or temperature fields.

In this paper, we will extend our previous Lagrangian investigations of coherent behavior in turbulent Rayleigh-Bénard convection flows [48,49]. We begin by discussing transport phenomena in nonautonomous systems and the transfer operator framework for the identification of coherent flow behavior in Sect. 2. In Sect. 3 the numerical approximation of such operators within a set-oriented approach is described, and in Sect. 4 the finite-element approach is outlined. The discretized transfer operator and dynamic Laplace operator are the fundamental tools for the extraction of coherent sets and transport barriers and we will introduce the respective approaches. In Sect. 5 we will apply these methods to turbulent Rayleigh-Bénard convection flows in two and three dimensions. In particular, we will extract turbulent superstructures of convection in terms of dominant convection roles. We conclude with a short discussion and outlook in Sect. 6.

2 Nonautonomous Dynamics, Transfer Operators and Transport

We consider a nonautonomous differential equation

$$\dot{\mathbf{x}} = \mathbf{u}(\mathbf{x}, t) \tag{1}$$

with state $\mathbf{x} \in M \subset \mathbb{R}^d$, time $t \in \mathbb{R}$ and sufficiently smooth right-hand side $\mathbf{u}$ such that the flow map $\Phi(\mathbf{x}, t; \tau) : M \times \mathbb{R} \times \mathbb{R} \to M$, $M \subset \mathbb{R}^d$ exists. Here τ denotes the flow time and t the initial time. We aim at identifying almost-invariant and coherent subsets of M, i.e. mobile regions in M that minimally mix with the surrounding phase space. Frequently used indicators for barriers of transport and hence boundaries of coherent regions are ridges in the **finite-time Lyapunov exponent** (FTLE) field [37]

$$FTLE(\mathbf{x}, t; \tau) = \frac{1}{2|\tau|} \log \left(\lambda_{max}[D_{\mathbf{x}}\Phi(\mathbf{x}, t; \tau)^{\top} D_{\mathbf{x}}\Phi(\mathbf{x}, t; \tau)] \right). \tag{2}$$

They are the basis of some geometric approaches for the identification of Lagrangian coherent structures [38]. In this work, we follow a probabilistic approach, which considers the evolution of sets, or, more abstractly, probability measures. Later, when discussing the dynamic Laplacian, we follow a geometric approach related to finding persistently small set boundaries.

A set $A \subset M$ is called **Φ-invariant** over $[t, t+\tau]$ if $\Phi(A, t+s; -s) = A$ for all $0 \le s \le \tau$. That is, the set A remains unchanged under the evolution of Φ. Almost-invariant sets obey an approximate invariance principle $\Phi(A, t+s; -s) \approx A$ for all $0 \le s \le \tau$. To be more precise, given a probability measure μ on M, we call a set $A \subset M$ with $\mu(A) \ne 0$ **almost-invariant** [6] over $[t, t+\tau]$ if

$$\rho(A) := \frac{\mu(A \cap \Phi(A, t+\tau; -\tau))}{\mu(A)} \approx 1. \tag{3}$$

If $A \subset M$ is almost-invariant over the interval $[t, t+\tau]$, then the probability (according to μ) of a trajectory leaving A at some time in $[t, t+\tau]$ and not returning to A by time $t+\tau$ is relatively small.

Unlike almost-invariant sets, **coherent sets** are allowed to move in phase space under the evolution of the time-dependent system. Given a reference probability measure μ on M at time t, one seeks to find pairs of sets $(A_t, A_{t+\tau})$ [23] such that

$$\rho(A_t, A_{t+\tau}) = \frac{\mu(A_t \cap \Phi(A_{t+\tau}, t+\tau; -\tau))}{\mu(A_t)} \approx 1. \tag{4}$$

Equation (4) measures the proportion of the set A_t at time t that is mapped to the set $A_{t+\tau}$ at time $t+\tau$ and one seeks to find sets such that $A_{t+\tau} \approx \Phi(A_t, t; \tau)$. Under set-oriented discretisation, optimal almost-invariant [8] and coherent [23] sets maximize (3) and (4).

The NP-hard discrete optimization problems can then be approximately solved by considering the Perron-Frobenius operator $\mathbf{P}_{t,\tau} : L^1(M, m) \to L^1(M, m)$ associated with the flow map Φ, where m denotes Lebesgue measure. The transfer operator is defined by

$$\mathbf{P}_{t,\tau} f(\mathbf{x}) = \frac{f(\Phi(\mathbf{x}, t+\tau; -\tau))}{|\det D\Phi(\Phi(\mathbf{x}, t+\tau; -\tau), t; \tau)|} \tag{5}$$

The interpretation is that if f is a density and $f(\mathbf{x})$ the density value in $\mathbf{x}$ at time t, then $\mathbf{P}_{t,\tau} f(\mathbf{x})$ describes the density value in $\Phi(\mathbf{x}, t; \tau)$ at time $t+\tau$ induced by the flow map. In [24,25] it was shown that maximizing ρ in (3) and (4) can be described in the framework of optimizing an inner product involving a compact self-adjoint operator obtained from $\mathbf{P}_{t,\tau}$. In order to avoid the technical functional analytic description underlying [24,25], we will briefly recall the concept of finite-time coherent sets in the finitary setting [23] in Sect. 3.2 based on a finite-rank approximation of $\mathbf{P}_{t,\tau}$ introduced in Sect. 3.1.

A stretching measure, similar to FTLE in (2), has been derived using the evolution of $\mathbf{P}_{t,\tau}$ [39]. It is based on the concept of differential entropy $h(f) = -\int_\Omega f \log f \, dm$, where Ω is the support of the density f. For a given initial condition $\mathbf{x}_0$, let$f_{\epsilon,\mathbf{x}_0} := \frac{1}{m(B_\epsilon(\mathbf{x}_0))}\mathbf{1}_{B_\epsilon(\mathbf{x}_0)}$ denote a uniform density supported on $B_\epsilon(\mathbf{x}_0)$, a ball of radius ϵ about $\mathbf{x}_0$. An ϵ-smoothing operator is then defined by

$$\mathbf{A}_\epsilon f(\mathbf{x}) := \frac{1}{m(B_\epsilon(\mathbf{x}))}\int_{B_\epsilon(\mathbf{x})} f\, dm.$$

The rate of increase in entropy experienced in the ϵ-neighborhood of $\mathbf{x}_0$ over the time span $[t, t+\tau]$ of the ϵ-perturbed dynamics can now be described by

$$FTE_\epsilon(\mathbf{x}_0, t; \tau) := \frac{1}{|\tau|}[h(\mathbf{A}_\epsilon \mathbf{P}_{t,\tau} f_{\epsilon,\mathbf{x}_0}) - h(f_{\epsilon,\mathbf{x}_0})]. \tag{6}$$

In [39] several properties of FTE_ϵ and its deterministic limit $\lim_{\epsilon\to 0} FTE_\epsilon$ have been derived. In particular, FTE_ϵ measures nonlinear stretching and can be compared with finite-time Lyapunov exponents (2) in the deterministic limit. In Sect. 3.3 we will outline a very efficient set-oriented approximation of the FTE-field.

We denote by $\mathbf{P}_{t,\tau,\epsilon} := \mathbf{A}_\epsilon \mathbf{P}_{t,\tau} \mathbf{A}_\epsilon$ the slightly mollified transfer operator. As mentioned above, finite-time coherent sets are extracted from the dominant singular vectors of the normalised L^2-compact operator $\mathbf{L}_{t,\tau,\epsilon} := \mathbf{P}_{t,\tau,\epsilon}/(\mathbf{P}_{t,\tau,\epsilon}\mathbf{1})$; see [24], also [25].

One could equivalently consider the dominant eigenvectors[1] of $\mathbf{L}^*_{t,\tau,\epsilon}\mathbf{L}_{t,\tau,\epsilon}$, and in the pure advection limit of $\epsilon \to 0$, one obtains

$$\lim_{\epsilon\to 0} \frac{\mathbf{L}^*_{t,\tau,\epsilon}\mathbf{L}_{t,\tau,\epsilon} - I}{\epsilon^2} = \Delta^D_{[t,t+\tau]},$$

where $\Delta^D_{[t,t+\tau]}$ is the dynamic Laplace operator [30], a self-adjoint, elliptic, second-order differential operator.

Extending ideas from isoperimetric theory, which concern sets of minimal boundary size relative to volume (the Cheeger ratio), one can create a *dynamic* isoperimetric theory [30] and prove connections between the spectrum of $\Delta^D_{[t,t+\tau]}$ and sets with persistently small boundary size relative to evolved volume (the dynamic Cheeger ratio). These sets with persistently small boundary size relative to evolved volume are excellent candidates for finite-time coherent sets because their boundaries resist filamentation and in the presence of small diffusion, diffusive flux across the boundary is minimised. In analogy to the second singular value of $\mathbf{L}_{t,\tau,\epsilon}$ bounding the mixing factor over $[t, t+\tau]$ of all nonequilibrium distributions (Theorem 2 [24], Theorems 3 and 4 [25]), the second singular value of $\Delta^D_{[t,t+\tau]}$ bounds the smallest Cheeger ratio taken over all smooth decompositions of the domain (Corollary 3.6 [30]).

3 Set-Oriented Numerical Framework

We now describe a set-oriented numerical framework for the approximation of the nonautonomous Perron-Frobenius operator in terms of a transition matrix

[1] In the following expression $\mathbf{L}^*_{t,\tau,\epsilon}$ is the adjoint of $\mathbf{L}_{t,\tau,\epsilon}$ between its domain and codomain; see [24] for details.

of a finite-state Markov chain. The discretized transfer operator is the basis for extracting coherent sets (Sect. 3.2) as well as for the computation of FTE-fields (Sect. 3.3).

3.1 Approximation of Transfer Operator

Following [23] we consider some compact subset $X \subset M$ and a small neighborhood Y of $\Phi(X, t; \tau)$. Let $\{B_1, \ldots, B_k\}$ be a partition of X, $\{C_1, \ldots, C_n\}$ a partition of Y. The partition elements are typically generalized rectangles, but other settings are possible. Applying Ulam's method [11] a finite-rank approximation of $\mathbf{P}_{t,\tau} : L^1(X, m) \to L^1(Y, m)$ is given via the transition matrix

$$P_{ij} = \frac{m(B_i \cap \Phi(C_j, t + \tau; -\tau))}{m(B_i)}, \; i = 1, \ldots, k, \; j = 1, \ldots, n \tag{7}$$

where we drop the t and τ-dependence of P for brevity. In practice the entries P_{ij} of the transition matrix P are estimated via

$$P_{ij} \approx \frac{\#\{r : \Phi(\mathbf{z}_{i,r}, t; \tau) \in C_j\}}{R}. \tag{8}$$

with uniformly distributed sample points $\mathbf{z}_{i,r}, r = 1, \ldots, R$ chosen in each partition element B_i, $i = 1, \ldots, k$. P is a sparse, row-stochastic matrix and thus all its eigenvalues are contained in the unit circle. For the efficient computation of the transition matrix P we use the software package GAIO [50] (available at http://github.com/gaioguy/GAIO).

The interpretation of the P-induced dynamics is that if $\mathbf{p} \geq 0$ (componentwise) is a probability vector ($\sum_i p_i = 1$), then $\mathbf{p}' = \mathbf{p}P$ is the push-forward of $\mathbf{p}$ under the discretized action of $\Phi(\cdot, t; \tau)$. Note that the numerical scheme introduces diffusion – which is also theoretically needed for robust results [24,25].

3.2 Extracting Finite-Time Coherent Sets

Consider a reference probability measure μ on X at time t, which is discretely represented as a probability vector $\mathbf{p}$ with $p_i = \mu(B_i)$, $i = 1, \ldots, k$. The image probability vector on Y at time $t + \tau$ is then simply computed as $\mathbf{q} = \mathbf{p}P$. We assume both $\mathbf{p} > 0$ and $\mathbf{q} > 0$ (component-wise) and form a normalized matrix L via

$$L_{ij} = \frac{p_i P_{ij}}{q_j}. \tag{9}$$

This matrix has the property that $\mathbf{1}_{\mathbb{R}^k} L = \mathbf{1}_{\mathbb{R}^n}$. In [23,24] it was shown that (under some technical assumptions) the problem of finding optimal coherent sets can be approximated by considering the left eigenvectors $\mathbf{w}_2 \in \mathbb{R}^k$ of LL^* and $\hat{\mathbf{w}}_2 \in \mathbb{R}^n$ of L^*L to the second largest eigenvalue $\lambda_2 < 1$. Here $L^* = P^\top$. Note that these two eigenvalue problems can be turned into the task of finding leading singular values and corresponding left and right singular vectors of a sparse matrix (see [23] for the exact construction), which can be very efficiently computed by iterative schemes (e.g. `svds` in MATLAB). The signed vector entries of $\mathbf{w}_2$ and $\hat{\mathbf{w}}_2$ can be interpreted as relaxations of indicator functions of the sets A_t and $A_{t+\tau}$ and their complements. Thus the vector $\mathbf{w}_2$ defines fuzzy coherent sets on X, whereas $\hat{\mathbf{w}}_2$ represents their image on Y. Optimal partitions of X and Y into finite-time coherent pairs can be approximated via a line search in $\mathbf{w}_2$ and $\hat{\mathbf{w}}_2$ [23,25]. However, this approach is restricted to finding two-partitions in terms of a coherent set and its complement. In practice, there are often $k > 2$ singular values close to one (followed by a spectral gap) whose corresponding singular vectors highlight the location of coherent sets. In this case, one can postprocess the singular vectors by a k-means clustering to obtain a hard partition into k coherent sets. Alternatively, to preserve the eigenspace structure, one can project the singular vectors to a sparse basis (SEBA) [51], where the entries of each vector denote probabilities that the underlying box B_i belongs to a specific coherent set. Hard assignment of boxes to sets may then be performed by thresholding (see [51]) to form (i) a subpartition of unity, (ii) the largest collection of disjoint sets, or (iii) by maximum likelihood.

3.3 Set-Oriented Computation of FTE

In the discrete context, densities (which are central to the FTE-construction in Eq. (6)) are now represented by discrete probability measures μ and the entropy of a probability vector $\mathbf{p}$ with $p_i = \mu(B_i)$, $i = 1, \ldots, k$, is simply $H(\mathbf{p}) = -\sum_{i=1}^{n} p_i \log p_i$. Under the assumption that all partition elements $\{B_1, \ldots, B_k\}$ are of equal volume let δ_i be a k-vector with a 1 in the ith position and 0 elsewhere. Then the discrete FTE of a partition set B_i is given by

$$FTE(B_i, t; \tau) = \frac{1}{|\tau|} H(\delta_i P) = -\frac{1}{|\tau|} \sum_{j=1}^{n} P_{ij} \log P_{ij}. \tag{10}$$

Note that once the transition matrix P has been computed, the FTE field (6) can be very quickly approximated by application of Eq. (10). In particular, we do not require to explicitly push forward probability densities with P. In addition, stretching rates for differing box volumes as well as for the backward-time dynamics can be conveniently computed, see [39] for more details.

4 Finite-Element Framework

A set-oriented approach for approximating $\Delta^D_{[t,t+\tau]}$ is possible and effective [30]. The ingredients are Ulam approximation(s) of the dynamics and a finite-difference approximation of the standard Laplace operator. As the dynamic Laplace operator is an elliptic second-order differential operator, it turns out to be more efficient to adapt the well-worn finite-element approximation approach to our dynamic setting [31]. The main advantages are the ability to produce good results with dramatically decreased trajectory numbers, and the increased smoothness of the estimates (continuous, piecewise-affine approximations rather than the discontinuous, piecewise-constant approximations from Ulam). One of the main advantages of the set-oriented framework, namely sparsity, is retained. The finite-element approach detailed in [31] has other structural benefits, such as preservation of the symmetry of the true operator, and the ability to have incomplete trajectories.

Briefly, in the time interval $[t, t+\tau]$, one creates a sequence of n time instances $t = t_1 < t_2 < \cdots < t_n = t + \tau$ at which one has N trajectory data points $\mathbf{x}_{i,k} := \Phi(\mathbf{x}_i, t; t_k - t)$, $i = 1, \ldots, N$, $k = 1, \ldots, n$. At the time instance t_k, the trajectory points are meshed and a basis of N piecewise-affine nodal hat functions $\phi_{i,k} : M \to \mathbb{R}$ are defined with $x_{i,k}$ the node of the ith hat function. The function $\phi_{i,k}$ is locally supported on mesh elements with $x_{i,k}$ as a vertex and $\phi_{i,k}(x_{i,k}) = 1$. The usual stiffness and mass matrices are computed on each mesh and averaged across time.

$$D = \frac{1}{n}\sum_{k=1}^{n} D_k, \qquad D_{k,lm} = \int_{\Phi(M,t;t_k-t)} \nabla\phi_{l,k} \cdot \nabla\phi_{m,k} \tag{11}$$

$$M = \frac{1}{n}\sum_{k=1}^{n} M_k, \qquad M_{k,lm} = \int_{\Phi(M,t;t_k-t)} \phi_{l,k} \cdot \phi_{m,k} \tag{12}$$

The discrete representation of the eigenproblem $\Delta^D_{[t,t+\tau]} f = \lambda f$ is $Dv = \lambda M v$, which is immediately solved in e.g. MATLAB. The approximate solution f is then recovered as $f = \sum_{i=1}^{N} v_i \phi_{i,1}$. An example computation is shown in Fig. 15 for the dominant 17 eigenfunctions.

4.1 Disentangling Multiple Features with SEBA

In both the transfer operator approach of Sect. 3.2 and the dynamic Laplace approach considered here, it is frequently the case that multiple finite-time coherent sets are encoded in several dominant approximate singular vectors of $\mathbf{L}_{t,\tau,\epsilon}$ and eigenfunctions of $\Delta^D_{[t,t+\tau]}$, respectively. This is illustrated in Fig. 15 for the 17 dominant eigenfunctions of $\Delta^D_{[t,t+\tau]}$. In order to disentangle individual finite-time coherent sets, we seek a rotation of the eigendata so that each rotated vector contains a single set. We use sparsity as the means to drive the rotation towards this individual feature separation, because sparse[2] vectors imply a small total feature support in each vector. In more detail, if each singular vector or eigenvector $v^b, b = 1, \ldots, B$ is a column vector in $\mathbb{R}^N$ and $V := [v^1|v^2|\cdots|v^B]$ is an $N \times B$ array, we wish to find a sparse array $S = [s^1|s^2|\cdots|s^B]$ for which $\mathrm{span}\{s^1, s^2, \ldots, s^B\} \approx \mathrm{span}\{v^1, v^2, \ldots, v^B\}$. This is carried out using the SEBA (Sparse EigenBasis Approximation) algorithm [51], which finds a locally optimal $B \times B$ rotation matrix R with $V \approx SR$ small and S sparse; see Sect. 3 [51] for further details. Figure 16 shows the conversion of the eigenbasis displayed

[2] A sparse vector (resp. array) is a vector (resp. array) with a high proportion of zero elements.

in Fig. 15 to a new approximate sparse eigenbasis, with each vector isolating a single feature. The same approach is employed in the three-dimensional results in Fig. 18.

5 Application to Rayleigh-Bénard Convection

Turbulent convection flows in nature are often organized in regular large-scale patterns which evolve gradually compared to the typical convective time unit and arranged on spatial scales which are much larger than the layer height H. Prominent examples are cloud streets in atmospheric or granulation and super-granulation patterns in solar convection. This order in a fully developed turbulent flow is termed *turbulent superstructure of convection* in the following. Pandey et al. [47] reported their appearance in turbulent RBC flows with very different molecular dissipation properties which are characterized by the dimensionless Prandtl number Pr. A second dimensionless parameter of RBC which measures the vigor of convective turbulence is the Rayleigh number Ra. They are defined as

$$\mathrm{Ra} = \frac{\alpha g \Delta T H^3}{\nu\kappa}, \tag{13}$$

$$\mathrm{Pr} = \frac{\nu}{\kappa}, \tag{14}$$

where α, ν, and κ are the isobaric expansion coefficient, the kinematic viscosity, and the thermal diffusivity of the fluid, respectively. The wall-to-wall temperature difference is given by $\Delta T = T_{\mathrm{bottom}} - T_{\mathrm{top}}$. The acceleration vector due to gravity is given by $\boldsymbol{g} = (0, 0, -g)$. The Prandtl number is extremely small in stellar or solar convection with $\mathrm{Pr} \lesssim 10^{-6}$; it is $\mathrm{Pr} \approx 0.7$ for atmospheric turbulence, and $\mathrm{Pr} \approx 7$ for convective motion in the oceans. The large-scale structure formation in turbulent RBC became recently accessible in direct numerical simulations (DNS), which can now resolve all involved scales of turbulence in simulations in horizontally extended domains with a large aspect ratio [42–45].

Here, we study RBC in two different settings. Our first setting is a two-dimensional DNS of a RBC system with a larger Prandtl number $\mathrm{Pr} = 10$ close to convection in water at a small aspect ratio of $\Gamma = 4$ as in [49]. We restrict here to a two-dimensional model as it has been previously shown that for large Prandtl numbers the large- and small-scale quantities show similar scalings in two- and three-dimensional systems. The second setting is a DNS of three-dimensional RBC with a smaller Prandtl number $\mathrm{Pr} = 0.7$ corresponding to convection in air at a larger aspect ratio of $\Gamma = 16$ as in [48].

5.1 2D System

We consider the same two-dimensional RBC system as in [49], given in the Boussinesq approximation [41], in non-dimensional form by

$$\frac{\partial \boldsymbol{u}}{\partial t} + \boldsymbol{u} \cdot \boldsymbol{\nabla} \boldsymbol{u} = -\nabla p + T\boldsymbol{e}_z + \sqrt{\frac{\mathrm{Pr}}{\mathrm{Ra}}} \boldsymbol{\nabla}^2 \boldsymbol{u}, \tag{15}$$

$$\frac{\partial \theta}{\partial t} + \boldsymbol{u} \cdot \boldsymbol{\nabla} \theta = u_z + \frac{1}{\sqrt{\mathrm{PrRa}}} \boldsymbol{\nabla}^2 \theta, \tag{16}$$

$$\boldsymbol{\nabla} \cdot \boldsymbol{u} = 0, \tag{17}$$

where $\boldsymbol{u} = (u_x, u_z)$, θ, and p are the velocity, temperature fluctuation, and pressure fluctuation fields, respectively. The temperature fluctuations θ are deviations from the linear conductive (equilibrium) profile and related to the total temperature field T via

$$T(x, z, t) = T_{\mathrm{bottom}} - \frac{\Delta T}{H} z + \theta(x, z, t), \tag{18}$$

where T_{bottom} is the temperature at the bottom plate. Equations (15–17) are nondimensionalized using the height H of the convective layer as the length scale, the free-fall velocity $u_f = \sqrt{\alpha g \Delta T H}$ as the velocity scale, and the temperature difference ΔT as the temperature scale. Stress-free boundary conditions for the velocity field are applied at all walls. The side walls have Neumann boundary conditions, $\partial T / \partial n = 0$. Top and bottom walls are held at constant temperatures (as already mentioned before). Consequently, $\theta = 0$ at the top and bottom. Equations (15–17) are solved for $\mathrm{Pr} = 10$ and $\mathrm{Ra} = 10^6$ in a two-dimensional box of aspect ratio $\Gamma = L_x/H = 4$ subject to appropriate boundary conditions. The computational details are described in [49].

We start our simulation with random velocity and temperature fields as the initial condition and continued until a statistically steady state is reached. The steady state flow structure exhibits a pair of counter-rotating circulation rolls. Hot fluid rises in the central region whereas cold fluid falls near the sidewalls. The velocity and temperature fields at all the grid points were written to output files at every $0.1\, t_f$, with $t_f = H/u_f$ being the free-fall time (which is taken as the convective time unit).

The flow map required for setting up the transition matrix P is obtained from numerical advection of massless particles with coordinates $\boldsymbol{x}$ in the computed velocity field corresponding to

$$\frac{d\boldsymbol{x}}{dt} = \boldsymbol{u}(\boldsymbol{x}, t). \tag{19}$$

Time integration is done by the RK4 method and spatial interpolation of the velocities by cubic splines.

We consider a box covering $\{B_1, \ldots, B_n\}$ of the simulation domain $X = [0, 4] \times [0, 1]$ by 2^{12} or 2^{16} square boxes. As the system is closed we can choose $Y = X$ and thus use the same box covering for the initial and the final time.

For the computation of the transition matrix a 4×4 uniform grid of test points is initialized in each box B_i and advected by the flow map $\Phi(\cdot, 2000; \tau)$. As in [49] we consider the two different cases $\tau = 20t_f$ and $\tau = 200t_f$. Note that the average turnover time for a tracer is $20t_f$ for this setting [49].

In the following, we will compare different Lagrangian methods for coherent sets in the two-dimensional flow at hand. These are finite-time entropy (FTE), finite time Lyapunov exponents (FTLE), transfer operator method, and the sparse eigenbasis approximation (SEBA).

5.1.1 Short Flow Time $\tau = 20$

In order to visualize the major transport barrier separating the two convection roles, we compute the forward time FTE field from P as described in Sect. 3.3. As shown in Fig. 1, the FTE field has large values in the box center where hot fluid rises and also at the boundaries, where cold fluid falls, and thus where the main heat transport takes place. This result is in agreement with the computed FTLE field shown in Fig. 2.

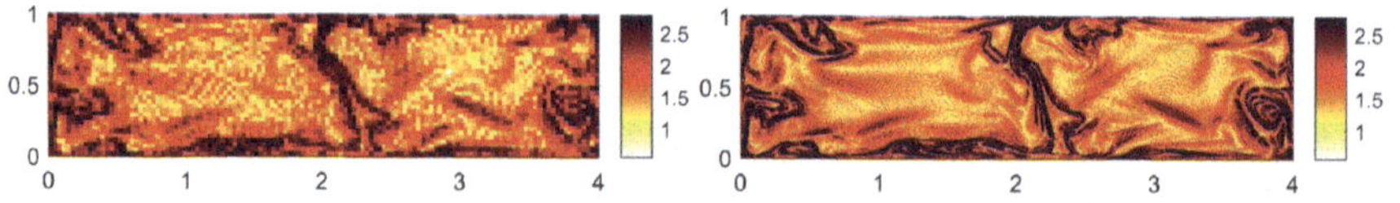

Fig. 1. FTE fields $FTE(\cdot, 2000; 20)$ computed over the time interval [2000, 2020] and plotted with respect to initial positions for two different box coverings (left: 2^{12} boxes; right: 2^{16} boxes). Dark regions are characterized by large stretching and correspond to dominant transport barriers.

For the extraction of coherent sets, we compute the leading left and right singular vectors as described in Sect. 3.2 for the coarse (Fig. 3) and the finer box discretization (Fig. 4). The second singular vectors identify the left-right division induced by the major transport barrier and highlight the two different gyres (Figs. 3 and 4, left columns). The third singular vectors (right columns) distinguish the two gyre cores from the background flow. Further singular vectors (not shown) subdivide the gyre cores into smaller structures. This has also been

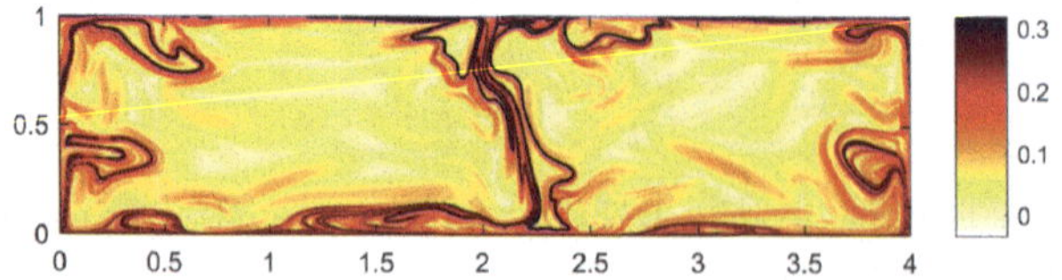

Fig. 2. FTLE field $FTLE(\cdot, 2000; 20)$ computed over the time interval [2000, 2020]. As in Fig. 1 dark regions correspond to dominant transport barriers.

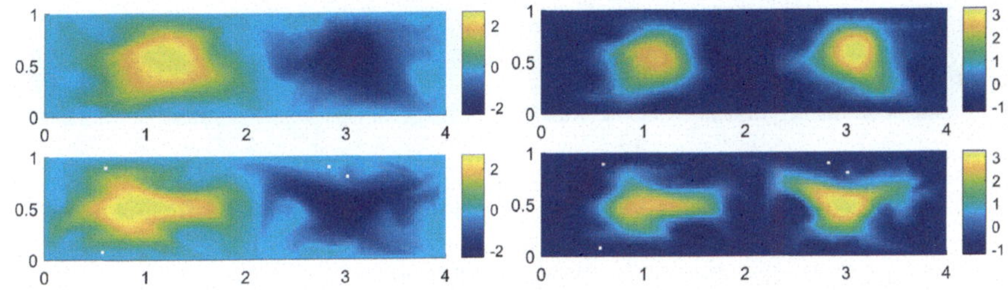

Fig. 3. Left singular vectors $\mathbf{w}_2$ and $\mathbf{w}_3$ (top row) and corresponding right singular vectors $\hat{\mathbf{w}}_2$ and $\hat{\mathbf{w}}_3$ (bottom row) obtained from the modified transition matrix (9) highlight coherent sets at initial and final time of the computation over the short time interval [2000, 2020]. Here 2^{12} boxes are used for setting up the transition matrix.

observed in [49]. The results for the two different box coverings are very similar, indicating that the computational results are very robust.

There are spectral gaps after the second singular values in both settings. However, in order to extract the apparently three dominant coherent sets (two gyres and background) from the leading singular vectors, we apply a standard k-means algorithm to the three leading left singular vectors. The results for both box coverings, which are again very similar, are shown in Fig. 5. This approach separates the two gyre cores from the background flow, where most of the heat transport takes place.

As an alternative to the hard-clustering resulting from k-means, we aim to find a sparse basis representation of the space spanned by the leading three singular vectors. Using SEBA [51] as briefly explained at the end of Sect. 4, two of the resulting sparse vectors are supported on each the gyre cores (Fig. 6, top), and the third sparse vector is supported on the background flow region (Fig. 6, bottom, left). A superposition of the three vectors (Fig. 6, bottom, right) reveals in dark blue a particularly incoherent (well mixing) region separating the two

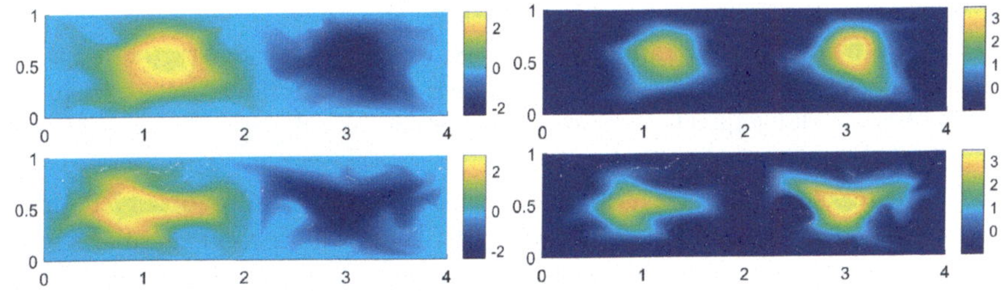

Fig. 4. Left singular vectors $\mathbf{w}_2$ and $\mathbf{w}_3$ (top row) and corresponding right singular vectors $\hat{\mathbf{w}}_2$ and $\hat{\mathbf{w}}_3$ (bottom row) obtained from the modified transition matrix (9) highlight coherent sets at initial and final time of the computation over the short time interval [2000, 2020]. Here 2^{16} boxes are used for setting up the transition matrix.

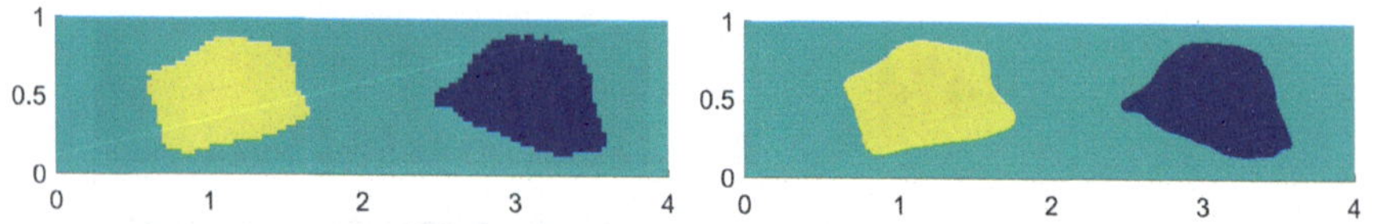

Fig. 5. Extracted coherent sets via an application of the standard k-means algorithm on the first three left singular vectors based discretizations with 2^{12} boxes (left) and 2^{16} boxes (right) for the short time interval $[2000, 2020]$.

gyres from the background flow. We note that the results are comparable to those in [49].

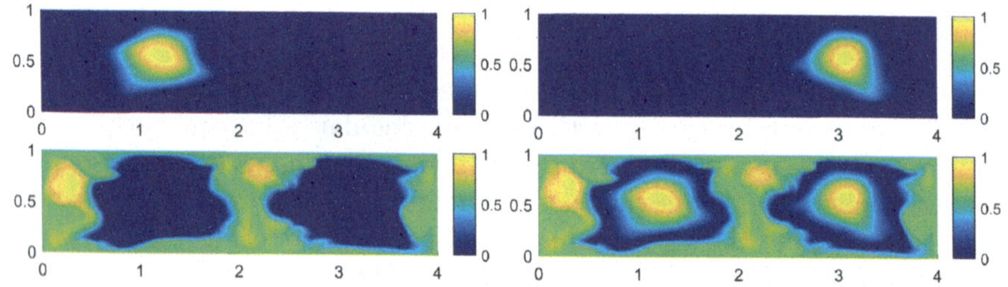

Fig. 6. SEBA applied to the first three left singular vectors, 2^{16} boxes, time interval $[2000, 2020]$. The upper row shows two of the output sparse vectors. Lower left shows the third output sparse vector and lower right shows the superposition of the three sparse vectors, revealing an incoherent region in dark blue.

5.1.2 Long Flow Time $\tau = 200$

For the long flow time $\tau = 200$ we have studied again the system using the coarse and the fine box covering. As these results are again very similar, we will show only the results for the finer box covering of 2^{16} boxes.

The FTE field (Fig. 7) highlights extended regions of strong stretching, which fill the space apart from the gyre cores, which appear to have decreased in size considerably and have developed into more filamentary shapes. This is confirmed by the FTLE field shown in Fig. 8.

The shrinking and filamentation of the gyre cores is also observed in the leading left and right singular vectors (Fig. 9). While the second singular vectors (left column) are analogous to those of the short time study, the third singular vector (right column) appears to further subdivide the right gyre. This has also been observed in our previous studies [49].

As there is a spectral gap after the fourth singular value, we use the corresponding four leading left singular vectors for postprocessing. Applying k-means (Fig. 10, left) and SEBA (Fig. 10, right) results in the identification of three very small gyre cores and the background flow.

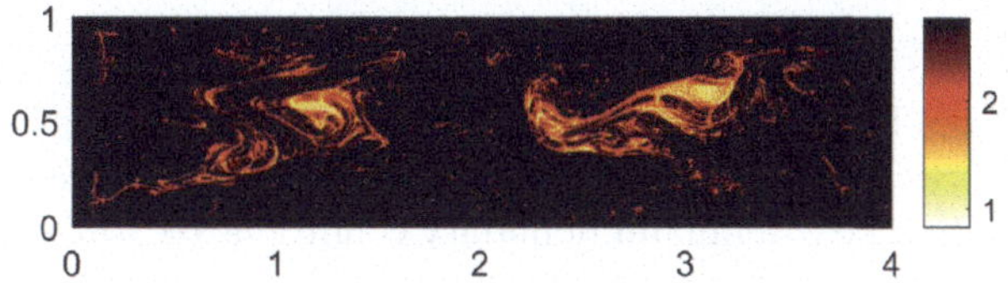

Fig. 7. Forward time FTE field computed over the long time interval [2000, 2200] and plotted with respect to initial positions for a box covering consisting of 2^{16} boxes. Dark regions are characterized by large stretching and correspond to dominant transport barriers.

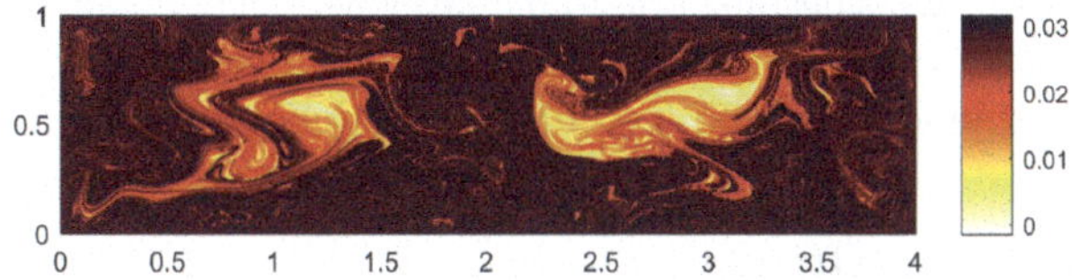

Fig. 8. FTLE field computed over the long time interval [2000, 2200]. Dark regions are characterized by large stretching and correspond to dominant transport barriers.

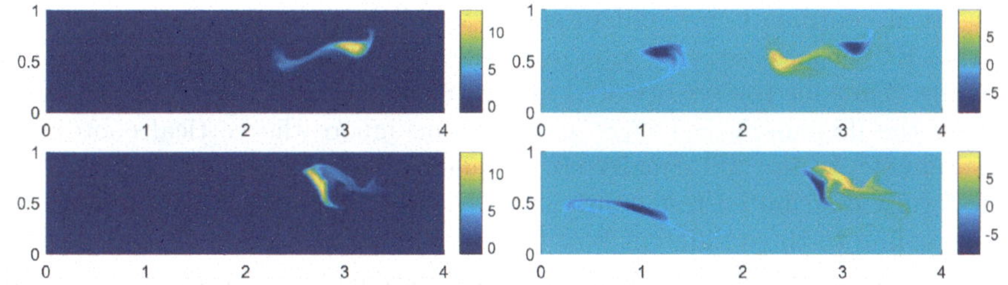

Fig. 9. Left singular vectors $\mathbf{w}_2$ and $\mathbf{w}_3$ (top row) and corresponding right singular vectors $\hat{\mathbf{w}}_2$ and $\hat{\mathbf{w}}_3$ (bottom row) obtained from the modified transition matrix (9) highlight coherent sets at initial and final time of the computation over the long time interval [2000, 2200]. Here 2^{16} boxes are used for setting up the transition matrix.

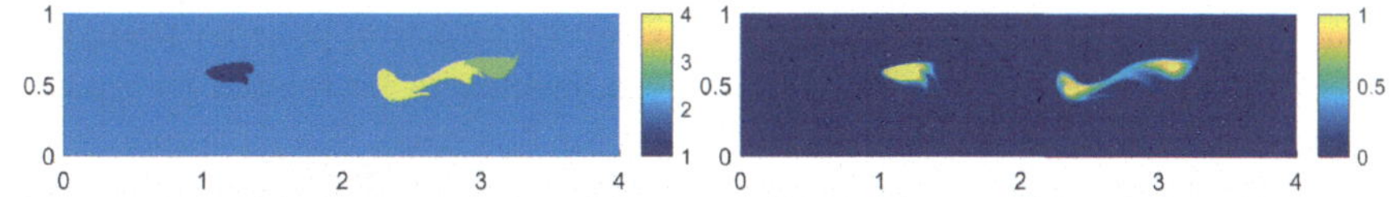

Fig. 10. Extracted coherent sets from the leading four left singular vectors for the long time interval [2000, 2200] via k-means (left) and SEBA (right).

5.2 3D System

Here, we solve the three-dimensional version of RBC which is again given in dimensionless form by Eqs. (15–17). No-slip boundary conditions for the velocity field $\boldsymbol{u} = (u_x, u_y, u_z)$ are applied at all walls, i.e., $\boldsymbol{u} = 0$. The side walls are again thermally insulated, i.e., Neumann boundary conditions $\partial T/\partial n = 0$ are applied. At the top and bottom walls, a constant dimensionless temperature of $T = 0$ and 1 is maintained again. Following ref. [48], we solve these equations numerically for $Ra = 10^5$ and $Pr = 0.7$ in a closed three dimensional box of aspect ratio $\Gamma = L_x/H = L_y/H = 16$, i.e. $M = [-8, 8] \times [-8, 8] \times [0, 1]$. For more details on the DNS we refer to [48]. During the simulation, the trajectories required for setting up the transition matrix are approximated. For this 512^2 points (tracers are initialized on a regular grid at a height of $z = 0.03$ above the bottom plate which is well inside the thermal boundary layer δ_T (that has a mean thickness of about 0.12). The tracers are advected by a 3-step explicit Adams-Bashforth scheme. The interpolation of the velocity field is done spectrally.

5.2.1 Quasi-2D Set-Oriented Study

As the convection cell is very flat ($\Gamma \gg 1$) the large-scale structures are expected to be arranged in horizontal patterns. This is clearly visible in the time-averaged temperature fields, see Fig. 11. We therefore restrict to a quasi-two-dimensional set-oriented study and take the temperature field in the midplane as the reference. As discussed in [47], an average of T at a given time t with respect to the vertical coordinate z would provide basically the same information. Note also that RBC has a statistical up-down reflection symmetry with respect to the midplane. Thus for the set-oriented approximation of the transfer operator we consider the domain $X = [-8, 8] \times [-8, 8]$ and ignore the vertical coordinate. We subdivide X into 2^{14} equally sized square boxes, hence each box contains 16 uniformly distributed test points initially. We set up the transition matrices corresponding to three different flow times $t = 2.6, 5.2, 10.5t_f$. We also incorporate a small amount of explicit diffusion as the dynamics is very dissipative at the beginning. Note that the flow times are short in comparison to the average turnover time of a tracer in the layer. This time is on average $t \approx 19t_f$ for this parameter setting [48].

At the beginning of the simulation, the tracers are attracted to the regions where the hot fluid rises from the bottom to the top of the convection cell, which correspond to attracting sets, at least on a finite-time span. We first study this particular behavior by means of the FTE field, see Fig. 12 and compare with Fig. 11. Regions of strong stretching are observed which appear to bound the different basins of attraction. Note that these basins can be related to a pair of convection rolls. In particular, the FTE field for flow time $5.2t_f$ compares very nicely to the time-averaged temperature fields in Fig. 11. For longer flow times the picture becomes increasingly fuzzy due to turbulent dispersion. The same behavior was also observed for the FTLE field in previous work [48], see also Fig. 13.

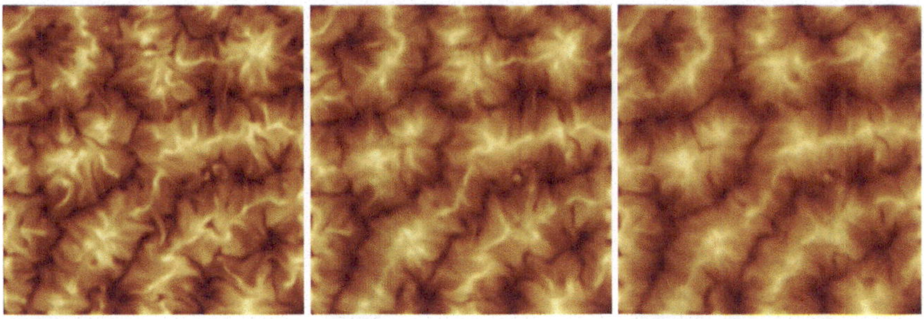

Fig. 11. Time-averaged temperature fields at mid-plane for different time spans $2.6, 5.2, 10.5t_f$. Light regions correspond to hot rising fluid, dark areas to cold descending fluid.

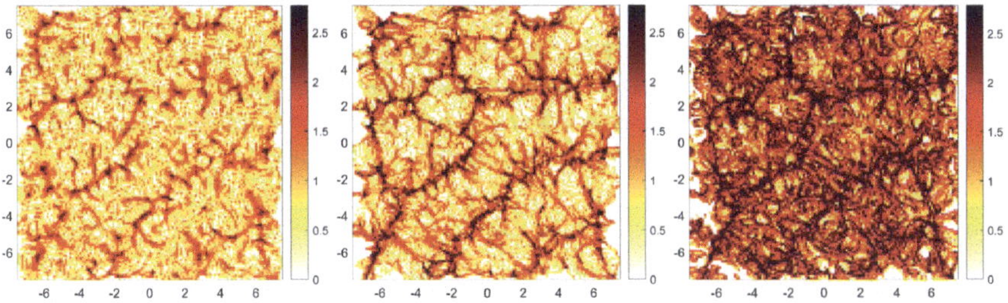

Fig. 12. FTE fields obtained from the transition matrices for flow times $2.6, 5.2, 10.5t_f$. Dark colors indicate to regions of strong stretching, which compare well with the structures formed by descending cold fluid in Fig. 11.

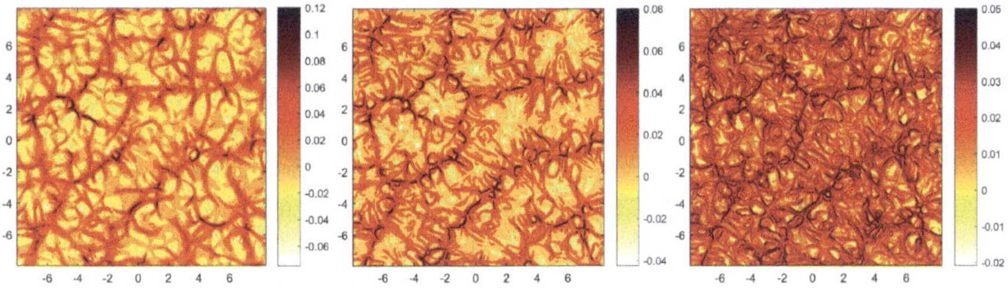

Fig. 13. FTLE fields for flow times $2.6, 5.2, 10.5t_f$. Dark colors indicate to regions of strong stretching, which compare well with the structures formed by descending cold fluid in Fig. 11 and also with the corresponding FTE fields in Fig. 12.

We also extract coherent sets based from the numerical transfer operator for the three different time spans. After inspecting the spectra we use the 13 leading left singular vectors for the settings with flow times $2.6t_f$ and $5.2t_f$ and 19 for the longer flow time. The results after a k-means postprocessing are shown in Fig. 14. The results for the two smaller flow times are very similar and compare again well to the temperature contours in Fig. 11. In particular, coherent sets appear to be made up of pairs of convection roles. The picture becomes more fuzzy for flow time $10.5t_f$ due to turbulent dispersion. The results compare well to our previous data-based studies in [48].

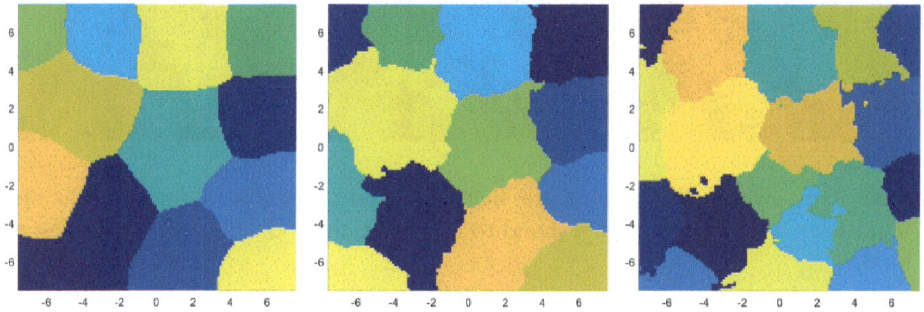

Fig. 14. Coherent sets extracted from the leading left singular vectors of the corresponding transition matrices for the three different flow times using k-means clustering. 13 clusters are obtained for flow times $2.6t_f$ (left) and $5.2t_f$ (middle), and 19 clusters for the long flow time $10.5t_f$ (right).

5.2.2 3D Finite-Element Study

For the remaining experiments we populate the entire three-dimensional domain with test points distributed throughout the domain. Further, we reduce the number of test points by more than six-fold to 40,000 and employ the dynamic Laplace approach of Sect. 4.

Before presenting the fully three-dimensional results, we remove the z-coordinate and investigate the flow for the longest time duration of $10.5t_f$. Triangulating the 40,000 points creates meshes of around 80,000 triangles; we compute the matrices D and M from (11) and (12) and solve the eigenproblem $Dv = \lambda Mv$. Using SEBA and the sparse vector heuristic in Sect. 4.2.2 [51], we choose 17 eigenvectors as a strong local minimum of the envelope produced by the MATLAB function `MinValStackedPlot.m` in Appendix A.6 [51], and illustrated in Fig. 11 [51]. Figure 15 shows the first 17 eigenfunctions of the dynamic Laplacian. These dominant 17 eigenvectors are input to SEBA in order to extract 17 individual coherent features. These 17 sparse vectors, representing the likelihoods of points belonging to individual coherent sets, are shown in Fig. 16. Maxima of the likelihoods can be plotted to create a single "hot-spot" image, and hard thresholded; see Fig. 17.

We now include the z-coordinate and begin our fully three-dimensional experiments. The 40,000 points are meshed into around 265,000 tetrahedra, and the matrices D and M are computed using the three-dimensional version of FEMDL (see [31] and http://github.com/gaioguy/FEMDL for examples and code). We again look for local minima in the `MinValStackedPlot.m` output and for visualisation purposes, choose a slightly smaller number of vectors, namely we use the most robust 10 SEBA vectors from the most dominant 12 eigenvectors. As in Fig. 16, each of the 10 vectors provides a pointwise likelihood that a point belongs

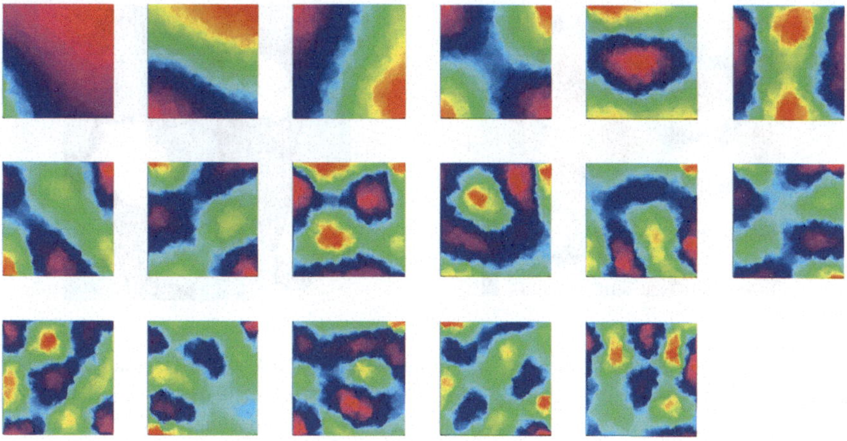

Fig. 15. Dominant 17 eigenvectors of the dynamic Laplacian for flow time $10.5t_f$. The colourmap is chosen so that bright pink and red values are extreme values that correspond to coherent features.

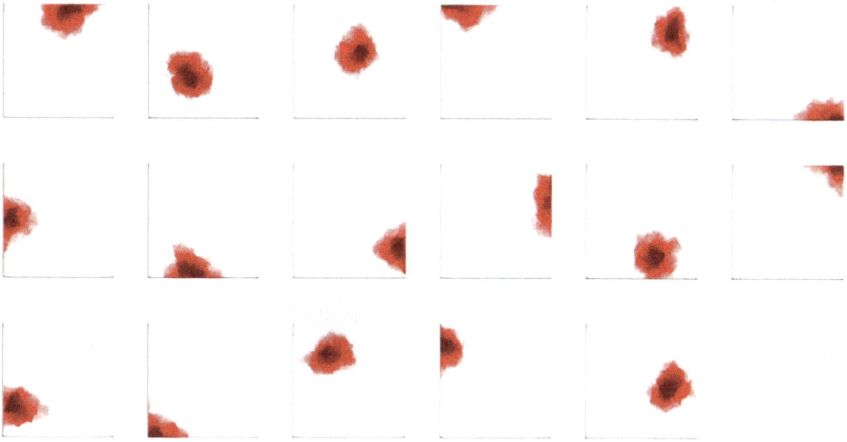

Fig. 16. Output of SEBA applied to the 17 dominant eigenvectors shown in Fig. 15. Negative parts have been removed. The colour scale ranges from 0 (white) to 1 (red); the intensity of red indicates the likelihood that a point belongs to each individual coherent feature.

to a particular coherent set. To clearly visualise these 10 three-dimensional coherent sets, we plot the isosurface[3] at value 1/2. As the likelihood increases toward the centre of the sets (we verified this visually, not shown), one can interpret

---

[3] Recall a SEBA vector defines a continuous, piecewise-linear function, affine on each tetrahedron. MATLAB's isosurface function requires a regular grid, so we reinterpolate to approximately the same number of points and the approximately the same density on a regular $80 \times 80 \times 10$ grid.

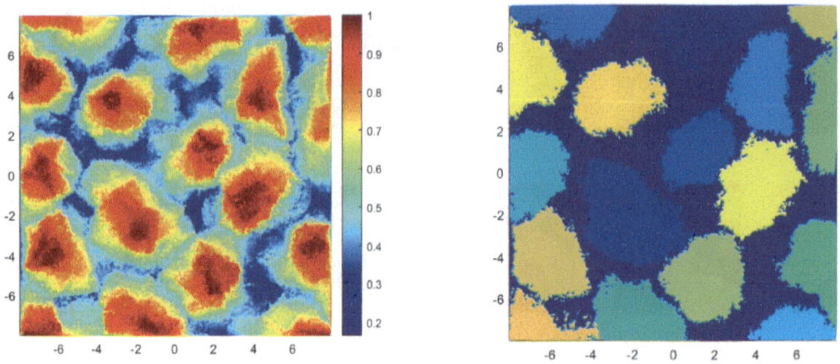

Fig. 17. Left: Maximum of intensities of individual SEBA vectors shown in Fig. 16, representing total probability to belong to one of the 17 identified coherent sets. Right: Maximum likelihood coherent sets created from the left image.

these surfaces as containing all points that are more than 50% likely to belong to each coherent set; one could call them the "cores" of the coherent sets. Figure 18 upper left shows the 10 three-dimensional cores. Note that each of these cores extends almost all of the way from the bottom to the top of the domain, consistent with the overall nature of the flow, where fluid mostly rises from the bottom of the domain to the top, before overturning and heading back toward the bottom of the domain. This full vertical extent feature is extremely robust to the number of eigenvectors and SEBA vectors used. Figure 18 lower left displays the same image as the upper left, but with commensurate lengths, emphasising that the domain is much shorter in the vertical direction. Figure 18 right is again the

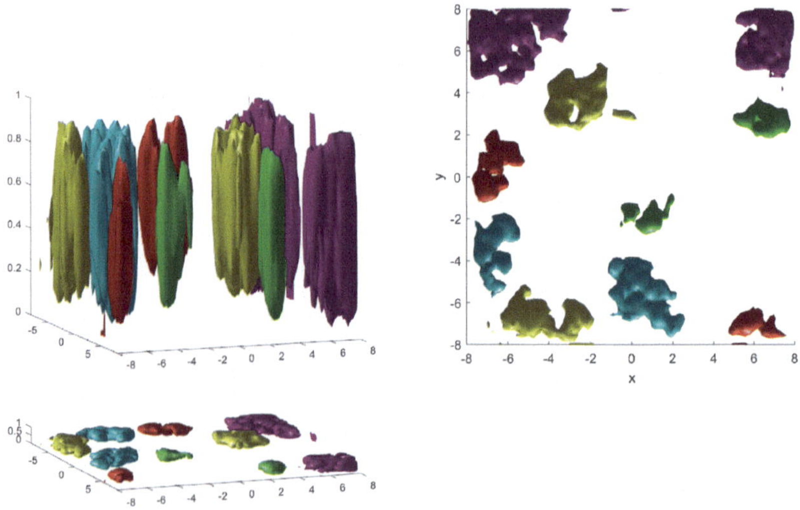

Fig. 18. Cores of 10 three-dimensional coherent sets for flow time $10.5t_f$.

same image, but viewing directly from above. Figure 18 displays cores roughly of the same size, and in some cases in similar locations, as the coherent sets in Fig. 17, although with fewer plotted for clearer three-dimensional visualisation.

6 Conclusion and Outlook

In this work, we have applied transfer operator based numerical frameworks for analyzing coherent behavior in nonautonomous systems to Rayleigh-Bénard convection in two- and three-dimensional settings. To this end, we used set-oriented approximations of the transfer operator and finite-element approximations of the dynamic Laplacian. It turns out that these general frameworks reliably identify the core regions of the various convection rolls as the regions that contribute least to the turbulent heat transfer from the bottom to the top. The two-dimensional results compare well with those of previous studies [48,49]. Future work will address the long-term evolution of the turbulent superstructures of convection and as well as their impact on transport properties.

Acknowledgements. KPG is thankful for the many fruitful discussions with Michael Dellnitz over the past more than twenty years. His fundamental contributions to the transfer operator based analysis of transport processes in dynamical systems have had a huge impact on the development of the overall field and on her own research. GF thanks Michael Dellnitz for countless stimulating mathematical discussions, for introducing him to the idea of almost-invariant sets, and for introducing him to his German colleagues in the late 90s, many of whom remain collaborators (and friends). Michael Dellnitz has also been one of the drivers of the Priority Programme SPP 1881 Turbulent Superstructures of the Deutsche Forschungsgemeinschaft by which the present work is supported and the authors are grateful for this. GF additionally thanks Universities Australia and the DAAD for a joint research travel grant, which partially supported his visit to Germany.

References

1. Allshouse, M.R., Peacock, T.: Lagrangian based methods for coherent structure detection. Chaos **25**, 097617 (2015)
2. Hadjighasem, A., Farazmand, M., Blazevski, D., Froyland, G., Haller, G.: A critical comparison of Lagrangian methods for coherent structure detection. Chaos **27**, 053104 (2017)
3. Kreuzer, E.: Numerische Untersuchung nichtlinearer dynamischer Systeme. Springer (1987)
4. Hsu, C.S.: Cell-to-Cell Mapping: A Method of Global Analysis for Nonlinear Systems. Springer, New York (1987)
5. Dellnitz, M., Hohmann, A.: A subdivision algorithm for the computation of unstable manifolds and global attractors. Numer. Math. **75**, 293–317 (1997)
6. Dellnitz, M., Junge, O.: Almost invariant sets in Chua's circuit. Int. J. Bifurc. Chaos **7**, 2475–2485 (1997)
7. Dellnitz, M., Junge, O.: On the approximation of complicated dynamical behaviour. SIAM J. Numer. Anal. **36**(2), 491–515 (1999)

8. Froyland, G.: Statistically optimal almost-invariant sets. Physica D **200**, 205–219 (2005)
9. Froyland, G., Padberg, K.: Almost-invariant sets and invariant manifolds - connecting probabilistic and geometric descriptions of coherent structures in flows. Physica D **238**, 1507–1523 (2009)
10. Froyland, G., Junge, O., Koltai, P.: Estimating long-term behavior of flows without trajectory integration: the infinitesimal generator approach. SIAM J. Numer. Anal. **51**, 223–247 (2013)
11. Ulam, S.: Problems in Modern Mathematics. Wiley, New York (1964)
12. Li, T.Y.: Finite approximation for the Frobenius-Perron operator. A solution to Ulam's conjecture. J. Approx. Theory **17**, 177–186 (1976)
13. Deuflhard, P., Dellnitz, M., Junge, O., Schütte, C.: Computation of essential molecular dynamics by subdivision techniques. In: Deuflhard, P., Hermans, J., Leimkuhler, B., Mark, A.E., Reich, S., Skeel, R.D. (eds.) Computational Molecular Dynamics: Challenges, Methods, Ideas, pp. 98–115. Springer, Heidelberg (1999)
14. Schütte, C., Huisinga, W., Deuflhard, P.: Transfer operator approach to conformational dynamics in biomolecular systems. In: Fiedler, B. (ed.) Ergodic Theory, Analysis, and Efficient Simulation of Dynamical Systems, pp. 191–223. Springer, Heidelberg (2001)
15. Dellnitz, M., Junge, O., Koon, W., Lekien, F., Lo, M., Marsden, J., Padberg, K., Preis, R., Ross, S., Thiere, B.: Transport in dynamical astronomy and multibody problems. Int. J. Bifurc. Chaos **15**, 699–727 (2005)
16. Dellnitz, M., Junge, O., Lo, M.W., Marsden, J.E., Padberg, K., Preis, R., Ross, S.D., Thiere, B.: Transport of Mars-crossing asteroids from the quasi-Hilda region. Phys. Rev. Lett. **94**, 231102 (2005)
17. Froyland, G., Padberg, K., England, M.H., Treguier, A.M.: Detection of coherent oceanic structures via transfer operators. Phys. Rev. Lett. **98**, 224503 (2007)
18. Dellnitz, M., Froyland, G., Horenkamp, C., Padberg-Gehle, K., Sen Gupta, A.: Seasonal variability of the subpolar gyres in the Southern Ocean: a numerical investigation based on transfer operators. Nonlinear Process. Geophys. **16**, 655–663 (2009)
19. Neumann, N., Goldschmidt, S., Wallaschek, J.: On the application of set-oriented numerical methods in the analysis of railway vehicle dynamics. Proc. Appl. Math. Mech. **4**, 578–579 (2004)
20. Froyland, G., Stuart, R.M., van Sebille, E.: How well connected is the surface of the global ocean? Chaos **24**, 033126 (2014)
21. Koltai, P.: A stochastic approach for computing the domain of attraction without trajectory simulation. Discrete Contin. Dyn. Syst. Suppl. **2011**, 854–863 (2011)
22. Froyland, G., Lloyd, S., Santitissadeekorn, N.: Coherent sets for nonautonomous dynamical systems. Physica D **239**, 1527–1541 (2010)
23. Froyland, G., Santitissadeekorn, N., Monahan, A.: Transport in time-dependent dynamical systems: finite-time coherent sets. Chaos **20**, 043116 (2010)
24. Froyland, G.: An analytic framework for identifying finite-time coherent sets in time-dependent dynamical systems. Physica D **250**, 1–19 (2013)
25. Froyland, G., Padberg-Gehle, K.: Almost-invariant and finite-time coherent sets: directionality, duration, and diffusion. In: Bahsoun, W., Bose, C., Froyland, G. (eds.) Ergodic Theory, Open Dynamics, and Coherent Structures, pp. 171–216. Springer, Heidelberg (2014)
26. Dellnitz, M., Horenkamp, C.: The efficient approximation of coherent pairs in non-autonomous dynamical system. Discrete Contin. Dyn. Syst. A **32**, 3029–3042 (2012)

27. Denner, A., Junge, O., Matthes, D.: Computing coherent sets using the Fokker-Planck equation. J. Comput. Dyn. **3**, 3–18 (2016)
28. Froyland, G., Koltai, P.: Estimating long-term behavior of periodically driven flows without trajectory integration. Nonlinearity **30**, 1948 (2017)
29. Froyland, G., Koltai, P., Plonka, M.: Computation and optimal perturbation of finite-time coherent sets for aperiodic flows without trajectory integration. *arXiv preprint* arXiv:1902.09263 (2019)
30. Froyland, G.: Dynamic isoperimetry and the geometry of Lagrangian coherent structures. Nonlinearity **28**, 3587 (2015)
31. Froyland, G., Junge, O.: Robust FEM-based extraction of finite-time coherent sets using scattered, sparse, and incomplete trajectories. SIAM J. Appl. Dyn. Syst. **17**, 1891–1924 (2018)
32. Froyland, G., Padberg-Gehle, K.: A rough-and-ready cluster-based approach for extracting finite-time coherent sets from sparse and incomplete trajectory data. Chaos **25**, 087406 (2015)
33. Hadjighasem, A., Karrasch, D., Teramoto, H., Haller, G.: Spectral-clustering approach to Lagrangian vortex detection. Phys. Rev. E **93**, 063107 (2016)
34. Schlueter-Kuck, K.L., Dabiri, J.O.: Coherent structure colouring: identification of coherent structures from sparse data using graph theory. J. Fluid Mech. **811**, 468 (2017)
35. Banisch, R., Koltai, P.: Understanding the geometry of transport: diffusion maps for Lagrangian trajectory data unravel coherent sets. Chaos **27**, 035804 (2017)
36. Padberg-Gehle, K., Schneide, C.: Network-based study of Lagrangian transport and mixing. Nonlinear Process. Geophys. **24**, 661 (2017)
37. Pierrehumbert, R.T., Yang, H.: Global chaotic mixing on isentropic surfaces. J. Atmos. Sci. **50**, 2462–2480 (1993)
38. Haller, G.: Lagrangian coherent structures. Ann. Rev. Fluid Mech. **47**, 137 (2015)
39. Froyland, G., Padberg-Gehle, K.: Finite-time entropy: a probabilistic approach for measuring nonlinear stretching. Physica D **241**, 1612–1628 (2012)
40. Padberg, K., Thiere, B., Preis, R., Dellnitz, M.: Local expansion concepts for detecting transport barriers in dynamical systems. Commun. Nonlinear Sci. Numer. Simul. **14**, 4176–4190 (2009)
41. Chillà, F., Schumacher, J.: New perspectives in turbulent Rayleigh-Bénard convection. Eur. Phys. J. E **35**, 58 (2012)
42. Hartlep, T., Tilgner, A., Busse, F.H.: Large scale structures in Rayleigh-Bénard convection at high Rayleigh numbers. Phys. Rev. Lett. **91**, 064501 (2003)
43. von Hardenberg, J., Parodi, A., Passoni, G., Provenzale, A., Spiegel, E.A.: Large-scale patterns in Rayleigh-Bénard convection. Phys. Lett. A **372**, 2223 (2008)
44. Bailon-Cuba, J., Emran, M.S., Schumacher, J.: Aspect ratio dependence of heat transfer and large-scale flow in turbulent convection. J. Fluid Mech. **655**, 152 (2010)
45. Emran, M.S., Schumacher, J.: Large-scale mean patterns in turbulent convection. J. Fluid Mech. **776**, 96 (2015)
46. Stevens, R.A.J.M., Blass, A., Zhu, X., Verzicco, R., Lohse, D.: Turbulent thermal superstructures in Rayleigh-Bénard convection. Phys. Rev. Fluids **3**, 041501(R) (2018)
47. Pandey, A., Scheel, J.D., Schumacher, J.: Turbulent superstructures in Rayleigh-Bénard convection. Nat. Commun. **9**, 2118 (2018)
48. Schneide, C., Pandey, A., Padberg-Gehle, K., Schumacher, J.: Probing turbulent superstructures in Rayleigh-Bénard convection by Lagrangian trajectory clusters. Phys. Rev. Fluids **3**, 113501 (2018)

49. Schneide, C., Stahn, M., Pandey, A., Junge, O., Koltai, P., Padberg-Gehle, K., Schumacher, J.: Lagrangian coherent sets in turbulent Rayleigh-Bénard convection. Phys. Rev. E **100**, 053103 (2019)
50. Dellnitz, M., Froyland, G., Junge, O.: The algorithms behind GAIO - set oriented numerical methods for dynamical systems. In: Fiedler, B. (ed.) Ergodic Theory, Analysis, and Efficient Simulation of Dynamical Systems, pp. 145–174. Springer, Heidelberg (2001)
51. Froyland, G., Rock, C.P., Sakellariou, K.: Sparse eigenbasis approximation: multiple feature extraction across spatiotemporal scales with application to coherent set identification. Commun. Nonlinear Sci. Numer. Simul. **77**, 81–107 (2019)

Singular Value Decomposition of Operators on Reproducing Kernel Hilbert Spaces

Mattes Mollenhauer[1], Ingmar Schuster[2], Stefan Klus[1(✉)], and Christof Schütte[1,3]

[1] Department of Mathematics and Computer Science, Freie Universität Berlin, Berlin, Germany
{mattes.mollenhauer,stefan.klus,christof.schuette}@fu-berlin.de
[2] Zalando Research, Zalando SE, Berlin, Germany
ingmar.schuster@zalando.de
[3] Zuse Institute Berlin, Berlin, Germany

Abstract. Reproducing kernel Hilbert spaces (RKHSs) play an important role in many statistics and machine learning applications ranging from support vector machines to Gaussian processes and kernel embeddings of distributions. Operators acting on such spaces are, for instance, required to embed conditional probability distributions in order to implement the kernel Bayes rule and build sequential data models. It was recently shown that transfer operators such as the Perron–Frobenius or Koopman operator can also be approximated in a similar fashion using covariance and cross-covariance operators and that eigenfunctions of these operators can be obtained by solving associated matrix eigenvalue problems. The goal of this paper is to provide a solid functional analytic foundation for the eigenvalue decomposition of RKHS operators and to extend the approach to the singular value decomposition. The results are illustrated with simple guiding examples.

1 Introduction

A majority of the characterizing properties of a linear map such as range, null space, numerical condition, and different operator norms can be obtained by computing the *singular value decomposition* (SVD) of the associated matrix representation. Furthermore, the SVD is used to optimally approximate matrices under rank constraints, solve least squares problems, or to directly compute the *Moore–Penrose pseudoinverse*. Applications range from solving systems of linear equations and optimization problems and to a wide variety of methods in statistics, machine learning, signal processing, image processing, and other computational disciplines.

Although the matrix SVD can be extended in a natural way to compact operators on Hilbert spaces [1], this infinite-dimensional generalization is not

O. Junge et al. (Eds.): SON 2020, SSDC 304, pp. 109–131, 2020.
https://doi.org/10.1007/978-3-030-51264-4_5

as multifaceted as the finite-dimensional case in terms of numerical applications. This is mainly due to the complicated numerical representation of infinite-dimensional operators and the resulting problems concerning the computation of their SVD. As a remedy, one usually considers finite-rank operators based on finite-dimensional subspaces given by a set of fixed basis elements. The SVD of such finite-rank operators will be the main focus of this paper. We will combine the theory of the SVD of finite-rank operators with the concept of the reproducing kernel Hilbert space (RKHS), a specific type of function space. A significant part of the theory of RKHSs was originally developed in a functional analytic setting [2] and made its way into pattern recognition and statistics [3–5]. RKHSs are often used to derive nonlinear extensions of linear computational methods. This is typically achieved by representing observational data in terms of RKHS elements and rewriting the methods based on the inner product of the RKHS. This strategy is known as the *kernel trick* [3]. The approach of embedding a finite number of observations into the RKHS can be generalized to the embedding of probability distributions associated with random variables into the RKHS [6]. The theory of the resulting *kernel mean embedding* (see [7] for a comprehensive review), *conditional mean embedding* [8–11] and *Kernel Bayes rule* [12,13] spawned a wide range of nonparametric approaches to problems in statistics and machine learning. Recent advances based on the conditional mean embedding show that data-driven methods in various fields such as transfer operator theory, time series analysis, and image and text processing naturally give rise to a spectral analysis of finite-rank RKHS operators [14,15].

Practical applications of these spectral analysis techniques include the identification of the slowest relaxation processes of dynamical systems, e.g., conformational changes of complex molecules or slowly evolving coherent patterns in fluid flows, but also dimensionality reduction and blind source separation. The eigendecomposition, however, is beneficial only in the case where the underlying system is ergodic with respect to some density. If this is not the case, however, i.e., the stochastic process is time-inhomogeneous, eigendecompositions can be replaced by singular value decompositions in order to obtain similar information about the global dynamics [16]. Moreover, outside of the context of stochastic processes, the conditional mean embedding operator has been shown to be the solution of certain vector-valued regression problems [9,11]. Contrary to the transfer operator setting, input and output space can differ fundamentally (e.g., the input space could be text) and the constraint that the RKHS for input and output space must be identical is too restrictive. The SVD of RKHS operators does not require this assumption and is hence a more general analysis tool applicable to operators that solve regression problems and to transfer operators associated with more general stochastic processes.

In this paper, we will combine the functional analytic background of the Hilbert space operator SVD and the theory of RKHSs to develop a self-contained and rigorous mathematical framework for the SVD of finite-rank operators acting on RKHSs and show that the SVD of such operators can be computed numerically by solving an auxiliary matrix eigenvalue problem. The remainder of the paper is structured as follows: Sect. 2 briefly recapitulates the theory of compact operators. In Sect. 3, RKHS operators and their eigendecompositions and singular value decompositions will be described. Potential applications are discussed in Sect. 4, followed by a brief conclusion and a delineation of open problems in Sect. 5.

2 Preliminaries

We recall the most important properties of compact operators on Hilbert spaces. For details, we refer the reader to [1,17]. In what follows, let H be a real Hilbert space, $\langle \cdot, \cdot \rangle_H$ its inner product, and $\|\cdot\|_H$ the induced norm. For a Hilbert space H, we call a set $\{h_i\}_{i \in I} \subseteq H$ with an index set I an *orthonormal system* if $\langle h_i, h_j \rangle_H = \delta_{ij}$ for all $i, j \in I$. If additionally $\operatorname{span}\{h_i\}_{i \in I}$ is dense in H, then we call $\{h_i\}_{i \in I}$ a *complete orthonormal system.* If H is separable, then the index set I of every complete orthonormal system of H is countable. Given a complete orthonormal system, every $x \in H$ can be expressed by the series expansion $x = \sum_{i \in I} \langle h_i, x \rangle_H h_i$.

Definition 1. Given two Hilbert spaces H and F and nonzero elements $x \in H$ and $y \in F$, we define the *tensor product operator* $y \otimes x \colon H \to F$ by $(y \otimes x)h = \langle x, h \rangle_H y$.

Remark 1. When $H = \mathbb{R}^m$ and $F = \mathbb{R}^n$ and both spaces are equipped with the Euclidean inner product, the tensor product operator $y \otimes x = y^\top x \in \mathbb{R}^{n \times m}$ reduces to the standard outer product of vectors in $x \in H$ and $y \in F$.

Note that tensor product operators are bounded linear operators. Boundedness follows from the Cauchy–Schwarz inequality on H. We define $\mathcal{E} := \operatorname{span}\{y \otimes x \mid x \in H, y \in F\}$ and call the completion of $\mathcal{E}$ with respect to the inner product

$$\langle y_1 \otimes x_1, y_2 \otimes x_2 \rangle := \langle y_1, y_2 \rangle_F \langle x_1, x_2 \rangle_H$$

the *tensor product* of the spaces F and H, denoted by $F \otimes H$. It follows that $F \otimes H$ is again a Hilbert space. It is well known that, given a self-adjoint compact operator $A \colon H \to H$, there exists an *eigendecomposition* of the form

$$A = \sum_{i \in I} \lambda_i (e_i \otimes e_i),$$

where I is either a finite or countably infinite ordered index set, $\{e_i\}_{i \in I} \subseteq H$ an orthonormal system, and $\{\lambda_i\}_{i \in I} \subseteq \mathbb{R} \setminus \{0\}$ the set of nonzero eigenvalues. If the index set I is not finite, then the resulting sequence $(\lambda_i)_{i \in I}$ is a null sequence. Similarly, given a compact operator $A \colon H \to F$, there exists a *singular value decomposition* given by

$$A = \sum_{i \in I} \sigma_i (u_i \otimes v_i),$$

where I is again an either finite or countably infinite ordered index set, $\{v_i\}_{i\in I} \subseteq H$ and $\{u_i\}_{i\in I} \subseteq F$ two orthonormal systems, and $\{\sigma_i\}_{i\in I} \subseteq \mathbb{R}_{>0}$ the set of singular values. As for the eigendecomposition, the sequence $(\sigma_i)_{i\in I}$ is a null sequence if I is not finite. Without loss of generality, we assume the singular values of compact operators to be ordered in non-increasing order, i.e., $\sigma_i \geq \sigma_{i+1}$. We additionally write $\sigma_i(A)$ for the ith singular value of a compact operator A if we want to emphasize to which operator we refer. The following result shows the connection of the eigendecomposition and the SVD of compact operators.

Lemma 1 (cf. [1]). *Let $A\colon H \to F$ be compact and let $\{\lambda_i\}_{i\in I}$ denote the set of nonzero eigenvalues of A^*A counted with their multiplicities and $\{v_i\}_{i\in I}$ the corresponding normalized eigenfunctions of A^*A, then, for $u_i := \lambda_i^{-1/2} A v_i$, the singular value decomposition of A is given by*

$$A = \sum_{i\in I} \lambda_i^{1/2} (u_i \otimes v_i).$$

A bounded operator $A\colon H \to F$ is said to be r-dimensional if $\operatorname{rank}(A) = r$. If $r < \infty$, we say that A is *finite-rank.*

Theorem 1 (cf. [17]). *Let H and F be two Hilbert spaces and $A\colon H \to F$ a linear operator. The operator A is finite-rank with* $\operatorname{rank}(A) = r$ *if and only if there exist linearly independent sets $\{h_i\}_{1\leq i\leq r} \subseteq H$ and $\{f_i\}_{1\leq i\leq r} \subseteq F$ such that $A = \sum_{i=1}^r f_i \otimes h_i$. Furthermore, then $A^* = \sum_{i=1}^r h_i \otimes f_i$.*

The class of finite-rank operators is a dense subset of the class of compact operators with respect to the operator norm.

Definition 2. Let H and F be Hilbert spaces and $\{h_i\}_{i\in I} \subseteq H$ be a complete orthonormal system. An operator $A\colon H \to F$ is called a *Hilbert–Schmidt operator* if $\sum_{i\in I} \|A h_i\|_F^2 < \infty$.

The space of Hilbert–Schmidt operators from H to F is itself a Hilbert space with the inner product $\langle A, B\rangle_{\mathrm{HS}} := \sum_{i\in I} \langle A h_i, B h_i\rangle_F$. Furthermore, it is isomorphic to the tensor product space $F \otimes H$. The space of finite-rank operators is a dense subset of the Hilbert–Schmidt operators with respect to the Hilbert–Schmidt norm. Furthermore, every Hilbert–Schmidt operator is compact and therefore admits an SVD.

Remark 2. Based on the definitions of the operator norm and the Hilbert–Schmidt norm, we have $\|A\| = \sigma_1(A)$ for any compact operator and $\|A\|_{\mathrm{HS}} = \left(\sum_{i\in I} \sigma_i(A)^2\right)^{1/2}$ for any Hilbert–Schmidt operator.

We will now derive an alternative characterization of the SVD of compact operators by generalizing a classical block-matrix decomposition approach to compact operators. For the matrix version of this result, we refer the reader to [18]. For two Hilbert spaces H and F, we define the *external direct sum*

$F \oplus H$ as the Hilbert space of tuples of the form (f, h), where $h \in H$ and $f \in F$, with the inner product

$$\langle (f, h), (f', h') \rangle_\oplus := \langle h, h' \rangle_H + \langle f, f' \rangle_F \,.$$

If $A\colon H \to F$ is a compact operator, then the operator $T\colon F \oplus H \to F \oplus H$, with

$$(f, h) \mapsto (Ah, A^* f) \tag{1}$$

is compact and self-adjoint with respect to $\langle \cdot, \cdot \rangle_\oplus$. By interpreting the elements of $F \oplus H$ as column vectors and generalizing algebraic matrix operations, we may rewrite the action of the operator T on (f, h) in a block operator notation as

$$\begin{bmatrix} & A \\ A^* & \end{bmatrix} \begin{bmatrix} f \\ h \end{bmatrix} = \begin{bmatrix} Ah \\ A^* f \end{bmatrix}.$$

We remark that the block operator notation should be applied with caution since vector space operations amongst $h \in H$ and $f \in F$ in terms of the matrix multiplication are only defined if $F \oplus H$ is an internal direct sum.

Lemma 2. *Let $A\colon H \to F$ be a compact operator and $T\colon F \oplus H \to F \oplus H$ be the block-operator given by* (1)*. If A admits the SVD*

$$A = \sum_{i \in I} \sigma_i (u_i \otimes v_i) \tag{2}$$

then T admits the eigendecomposition

$$T = \sum_{i \in I} \sigma_i \left[\tfrac{1}{\sqrt{2}} (u_i, v_i) \otimes \tfrac{1}{\sqrt{2}} (u_i, v_i) \right] - \sigma_i \left[\tfrac{1}{\sqrt{2}} (-u_i, v_i) \otimes \tfrac{1}{\sqrt{2}} (-u_i, v_i) \right]. \tag{3}$$

A proof of this lemma can be found in Appendix A.1.

Corollary 1. *Let $A\colon H \to F$ be a compact operator. If $\sigma > 0$ is an eigenvalue of the block-operator $T\colon F \oplus H \to F \oplus H$ given by* (1) *with the corresponding eigenvector $(u, v) \in F \otimes H$, then σ is a singular value of A with the corresponding left and right singular vectors $\|u\|_F^{-1} u \in F$ and $\|v\|_H^{-1} v \in H$.*

3 Decompositions of RKHS Operators

We will first introduce reproducing kernel Hilbert spaces, and then consider empirical operators defined on such spaces. The main results of this section are a basis orthonormalization technique via a kernelized QR decomposition in Sect. 3.3 and the eigendecomposition and singular value decomposition of empirical RKHS operators in Sect. 3.4 and Sect. 3.5 via auxiliary problems, respectively. The notation is adopted from [7,14] and summarized in Table 1.

3.1 RKHS

The following definitions are based on [3,5]. In order to distinguish reproducing kernel Hilbert spaces from standard Hilbert spaces, we will use script style letters for the latter, i.e., $\mathscr{H}$ and $\mathscr{F}$.

Table 1. Overview of notation.

Random variable	X	Y
Domain	$\mathbb{X}$	$\mathbb{Y}$
Observation	x	y
Kernel function	$k(x, x')$	$l(y, y')$
Feature map	$\phi(x)$	$\psi(y)$
Feature matrix	$\Phi = [\phi(x_1), \ldots, \phi(x_m)]$	$\Psi = [\psi(y_1), \ldots, \psi(y_n)]$
Gram matrix	$G_\phi = \Phi^\top \Phi$	$G_\psi = \Psi^\top \Psi$
RKHS	$\mathscr{H}$	$\mathscr{F}$

Definition 3 (Reproducing kernel Hilbert space, [3]). Let $\mathbb{X}$ be a set and $\mathscr{H}$ a space of functions $f : \mathbb{X} \to \mathbb{R}$. Then $\mathscr{H}$ is called a *reproducing kernel Hilbert space (RKHS)* with corresponding inner product $\langle \cdot, \cdot \rangle_{\mathscr{H}}$ if a function $k : \mathbb{X} \times \mathbb{X} \to \mathbb{R}$ exists such that

(i) $\langle f, k(x, \cdot) \rangle_{\mathscr{H}} = f(x)$ for all $f \in \mathscr{H}$ and
(ii) $\mathscr{H} = \overline{\operatorname{span}\{k(x, \cdot) \mid x \in \mathbb{X}\}}$.

The function k is called *reproducing kernel* and the first property the *reproducing property*. It follows in particular that $k(x, x') = \langle k(x, \cdot), k(x', \cdot) \rangle_{\mathscr{H}}$. The canonical feature map $\phi : \mathbb{X} \to \mathscr{H}$ is given by $\phi(x) := k(x, \cdot)$. Thus, we obtain $k(x, x') = \langle \phi(x), \phi(x') \rangle_{\mathscr{H}}$. It was shown that an RKHS has a unique symmetric and positive definite kernel with the reproducing property and, conversely, that a symmetric positive definite kernel k induces a unique RKHS with k as its reproducing kernel [2]. We will refer to the set $\mathbb{X}$ as the corresponding *observation space*.

3.2 RKHS Operators

Finite-rank operators can be defined by a finite number of fixed basis elements in the corresponding RKHSs. In practice, finite-rank RKHS operators are usually estimates of infinite-dimensional operators based on a set of empirical observations. We later refer to this special type of finite-rank operator as *empirical RKHS operator* although the concepts in this section are more general and do not need the assumption of the data in the observation space being given by random events.

Let $\mathscr{H}$ and $\mathscr{F}$ denote RKHSs based on the observation spaces $\mathbb{X}$ and $\mathbb{Y}$, respectively, with kernels k and l and feature maps ϕ and ψ. Given $x_1, \dots, x_m \in \mathbb{X}$ and $y_1, \dots, y_n \in \mathbb{Y}$, we call

$$\Phi := [\phi(x_1), \dots, \phi(x_m)] \quad \text{and} \quad \Psi := [\psi(y_1), \dots, \psi(y_n)]$$

their associated *feature matrices.* Note that feature matrices are technically not matrices but row vectors in $\mathscr{H}^m$ and $\mathscr{F}^n$, respectively. Since the embedded observations in the form of $\phi(x_i) \in \mathscr{H}$ and $\psi(y_j) \in \mathscr{F}$ can themselves be interpreted as (possibly infinite-dimensional) vectors, the term *feature matrix* is used. In what follows, we assume that feature matrices contain linearly independent elements. This is, for example, the case if $k(\cdot, \cdot)$ is a radial basis kernel and the observations $x_1, \dots, x_m \in \mathbb{X}$ consist of pairwise distinct elements. We adopt the commonly used notation $\Phi^\top v := [\langle \phi(x_1), v \rangle_{\mathscr{H}}, \dots, \langle \phi(x_m), v \rangle_{\mathscr{H}}]^\top$ for all $v \in \mathscr{H}$, which we also use to express pairwise kernel evaluations between objects in two feature matrices. Given the feature matrices Φ and Ψ, we can define the corresponding Gram matrices by $G_\Phi = \Phi^\top \Phi \in \mathbb{R}^{m \times m}$ and $G_\Psi = \Psi^\top \Psi \in \mathbb{R}^{n \times n}$. That is, $[G_\Phi]_{ij} = k(x_i, x_j)$ and $[G_\Psi]_{ij} = l(y_i, y_j)$. We will now analyze operators $S\colon \mathscr{H} \to \mathscr{F}$ of the form $S = \Psi B \Phi^\top$, where $B \in \mathbb{R}^{n \times m}$. Given $v \in \mathscr{H}$, we obtain

$$Sv = \Psi B \Phi^\top v = \sum_{i=1}^{n} \psi(y_i) \sum_{j=1}^{m} b_{ij} \langle \phi(x_j), v \rangle_{\mathscr{H}}.$$

We will refer to operators S of this form as *empirical RKHS operators.* Examples of such operators are described in Sect. 4.

Remark 3. If the rows of B are linearly independent in $\mathbb{R}^m$, then the elements of $B\Phi^\top$ are linearly independent in $\mathscr{H}$. The analogue statement holds for linearly independent columns of B and elements of ΨB in $\mathscr{F}$.

Proposition 1. *The operator S defined above has the following properties:*

(i) S *is a finite-rank operator. In particular,* $\operatorname{rank}(S) = \operatorname{rank}(B)$.
(ii) $S^* = \Phi B^\top \Psi^\top$.
(iii) *Let* $B = W \Sigma Z^\top$ *be the singular value decomposition of* B, *where* $W = [\mathbf{w}_1, \dots, \mathbf{w}_n]$, $\Sigma = \operatorname{diag}(\sigma_1, \dots, \sigma_r, 0, \dots, 0)$, *and* $Z = [\mathbf{z}_1, \dots, \mathbf{z}_m]$, *then*

$$\|S\| \leq \sum_{i=1}^{r} \sigma_i \|\Psi \mathbf{w}_i\|_{\mathscr{F}} \|\Phi \mathbf{z}_i\|_{\mathscr{H}}.$$

Proof. The linearity of S follows directly from the linearity of the inner product in $\mathscr{H}$. We now show that properties (i)–(iii) can directly be obtained from Theorem 1. Using $B = W \Sigma Z^\top$, we can write $S = (\Psi W)\Sigma(Z^\top \Phi^\top)$ and obtain

$$Sv = \sum_{i=1}^{r} \sigma_i \Psi \mathbf{w}_i \langle \Phi \mathbf{z}_i, v \rangle_{\mathscr{H}} \quad \text{for all } v \in \mathscr{H}. \tag{4}$$

Since the elements in Φ and Ψ are linearly independent, we see that ΦZ and ΨW are also feature matrices containing the linearly independent elements $\Phi \mathbf{z}_i \in \mathscr{H}$ and $\Psi \mathbf{w}_i \in \mathscr{F}$ as stated in Remark 3. Therefore, (4) satisfies the assumptions in Theorem 1 if we choose $\{\Phi \mathbf{z}_i\}_{1 \leq i \leq r} \subseteq \mathscr{H}$ and $\{\sigma_i \Psi \mathbf{w}_i\}_{1 \leq i \leq r} \subseteq \mathscr{F}$ to be the required linearly independent sets. Theorem 1 directly yields all the desired statements. □

Note that the characterization (4) is in general not a singular value decomposition of S since the given basis elements in ΦZ and ΨW are not necessarily orthonormal systems in $\mathscr{H}$ and $\mathscr{F}$, respectively.

3.3 Basis Orthonormalization and Kernel QR Decomposition

When we try to perform any type of decomposition of the operator $S = \Psi B \Phi^\top$, we face the problem that the representation matrix B is defined to work on feature matrix entries of Ψ and Φ, which are not necessarily orthonormal systems in the corresponding RKHSs. This leads to the fact that we can not simply decompose B with standard numerical routines based on the Euclidean inner products and expect a meaningful equivalent decomposition of S in terms of RKHS inner products. We therefore orthonormalize the feature matrices with respect to the RKHS inner products and capture these transformations in a new representation matrix $\tilde{B}$ which allows using matrix decompositions to obtain operator decompositions of S. We now generalize the matrix QR decomposition to feature matrices, which is essentially equivalent to a kernelized Gram–Schmidt procedure [19]. By expressing empirical RKHS operators with respect to orthonormal feature matrices, we can perform operator decompositions in terms of a simple matrix decomposition.

Proposition 2 (Kernel QR decomposition). *Let $\Phi \in \mathscr{H}^m$ be a feature matrix. Then there exists a unique upper triangular matrix $R \in \mathbb{R}^{m \times m}$ with strictly positive diagonal elements and a feature matrix $\tilde{\Phi} \in \mathscr{H}^m$, such that*

$$\Phi = \tilde{\Phi} R$$

and $\tilde{\Phi}^\top \tilde{\Phi} = I_m$.

Proof. We have assumed that elements in feature matrices are linearly independent. Therefore $\Phi^\top \Phi$ is strictly positive definite. We have a Cholesky decomposition $\Phi^\top \Phi = R^\top R$ for a unique upper triangular matrix $\mathbb{R}^{m \times m}$ with positive diagonal entries. By setting $\tilde{\Phi} := \Phi R^{-1}$ and observing that $\tilde{\Phi}^\top \tilde{\Phi} = (\Phi R^{-1})^\top \Phi R^{-1} = I_m$, the claim follows. □

By using Proposition 2, we can express empirical operators in orthonormalized basis elements. Given an empirical RKHS operator $S = \Psi B \Phi^\top$ and the two corresponding kernel QR decompositions $\Phi = \tilde{\Phi} R_\Phi$ and $\Psi = \tilde{\Psi} R_\Psi$, we can rewrite

$$S = (\tilde{\Psi} R_\Psi^{-1}) B (\tilde{\Phi} R_\Phi^{-1})^\top = \tilde{\Psi} (R_\Psi^{-1} B (R_\Phi^{-1})^\top) \tilde{\Phi}^\top = \tilde{\Psi} \tilde{B} \tilde{\Phi}^\top. \tag{5}$$

We can now simply perform any type of matrix decomposition on the new representation matrix $\tilde{B} := R_\Psi^{-1} B (R_\Phi^{-1})^\top$ to obtain an equivalent decomposition of the operator S. As examples, we give the SVD and the eigendecomposition of S.

Corollary 2 (Singular value decomposition). *Let $S = \tilde{\Psi}\tilde{B}\tilde{\Phi}^\top : \mathscr{H} \to \mathscr{F}$ be given by orthonormalized basis elements as above. If $\tilde{B} = \sum_{i=1}^r \sigma_i u_i v_i^\top$ is the singular value decomposition of $\tilde{B}$, then*

$$S = \sum_{i=1}^{r} \sigma_i (\tilde{\Psi} u_i \otimes \tilde{\Phi} v_i)$$

is the singular value decomposition of S.

For the eigendecomposition, we require the operator to be a mapping from $\mathscr{H}$ to itself. We will assume that both the domain and the range of S are defined via the same feature matrix Φ. We consider the self-adjoint case, that is B (or equivalently $\tilde{B}$) is symmetric.

Corollary 3 (Eigendecomposition). *Let $S = \tilde{\Phi}\tilde{B}\tilde{\Phi}^\top : \mathscr{H} \to \mathscr{H}$ be given by orthonormalized basis elements as above. Let $\tilde{B}$ be symmetric. If $\tilde{B} = \sum_{i=1}^r \lambda_i v_i v_i^\top$ is the eigendecomposition of $\tilde{B}$, then*

$$S = \sum_{i=1}^{r} \lambda_i (\tilde{\Phi} v_i \otimes \tilde{\Phi} v_i) \tag{6}$$

is the eigendecomposition of S.

In particular, the matrix $\tilde{B}$ and the operator S share the same singular values (or eigenvalues, respectively) potentially up to zero. In practice, computing the singular value decomposition of S by this approach needs two kernel QR decompositions (which numerically results in Cholesky decompositions of the Gram matrices), inversions of the triangular matrices R_Ψ and R_Φ and the final decomposition of $\tilde{B}$. For the eigendecomposition we need a single kernel QR decomposition and inversion before performing the eigendecomposition of $\tilde{B}$. Since this may numerically be costly, we give an overview of how eigendecompositions and singular value decompositions of empirical RKHS operators can be performed by solving a single related auxiliary problem.

Remark 4. Representation (5) makes it possible to compute classical matrix decompositions such as Schur decompositions, LU-type decompositions, or polar decompositions on $\tilde{B}$ and obtain a corresponding decomposition of the operator S. Note however that when S approximates an operator for $n, m \to \infty$, it is not necessarily given that these empirical decompositions of S converge to a meaningful infinite-rank concept that is equivalent. For the eigendecomposition and the singular value decomposition, this reduces to classical operator perturbation theory [20].

3.4 Eigendecomposition via Auxiliary Problem

The eigendecomposition of RKHS operators via an auxiliary problem was first considered in [14]. For the sake of completeness, we will briefly recapitulate the main result and derive additional properties. For the eigendecomposition, we again require the operator to be a mapping from $\mathscr{H}$ to itself. For this section, we define a new feature matrix by $\Upsilon = [\phi(x_1'), \dots, \phi(x_m')]$. Note that the sizes of Φ and Υ have to be identical.

Proposition 3 (cf. [14]). *Let $S\colon \mathscr{H} \to \mathscr{H}$ with $S = \Upsilon B \Phi^\top$ and $B \in \mathbb{R}^{m \times m}$ be an empirical RKHS operator. Then the following statements hold:*

(i) If λ is an eigenvalue of $B\Phi^\top\Upsilon \in \mathbb{R}^{m\times m}$ with corresponding eigenvector $\mathbf{w} \in \mathbb{R}^m$, then $\Upsilon\mathbf{w} \in \mathscr{H}$ is an eigenfunction of S corresponding to λ.
(ii) Conversely, if $\lambda \neq 0$ is an eigenvalue of S corresponding to the eigenfunction $v \in \mathscr{H}$, then $B\Phi^\top v \in \mathbb{R}^m$ is an eigenvector of $B\Phi^\top\Upsilon \in \mathbb{R}^{m\times m}$ corresponding to the eigenvalue λ.

In particular, the operator S and the matrix $B\Phi^\top\Upsilon$ share the same nonzero eigenvalues.

Proof. For the sake of completeness, we briefly reproduce the gist of the proof.

(i) Let $\mathbf{w} \in \mathbb{R}^m$ be an eigenvector of the matrix $B\Phi^\top\Upsilon$ corresponding to the eigenvalue λ. Using the associativity of feature matrix multiplication and kernel evaluation, we have

$$S(\Upsilon\mathbf{w}) = \Upsilon(B\Phi^\top\Upsilon\mathbf{w}) = \lambda\Upsilon\mathbf{w}.$$

Furthermore, since $\mathbf{w} \neq 0 \in \mathbb{R}^m$ and the elements in Υ are linearly independent, we have $\Upsilon\mathbf{w} \neq 0 \in \mathscr{H}$. Therefore, $\Upsilon\mathbf{w}$ is an eigenfunction of S corresponding to λ.

(ii) Let v be an eigenfunction of S associated with the eigenvalue $\lambda \neq 0$. By assumption, we then have

$$\Upsilon B\Phi^\top v = \lambda v.$$

By "multiplying" both sides from the left with $B\Phi^\top$ and using the associativity of the feature matrix notation, we obtain

$$(B\Phi^\top\Upsilon)B\Phi^\top v = \lambda B\Phi^\top v.$$

Furthermore, $B\Phi^\top v$ cannot be the zero vector in $\mathbb{R}^m$ as we would have $\Upsilon(B\Phi^\top v) = Sv = 0 \neq \lambda v$ otherwise since λ was assumed to be a nonzero eigenvalue. Therefore, $B\Phi^\top v$ is an eigenvector of the matrix $B\Phi^\top\Upsilon$. □

Remark 5. Eigenfunctions of empirical RKHS operators may be expressed as a linear combination of elements contained in the feature matrices. However, there exist other formulations of this result [14]. We can, for instance, define the alternative auxiliary problem

$$\Phi^\top \Upsilon B \mathbf{w} = \lambda \mathbf{w}.$$

For eigenvalues λ and eigenvectors $\mathbf{w} \in \mathbb{R}^m$ satisfying this equation, we see that $\Upsilon B v \in \mathscr{H}$ is an eigenfunction of S. Conversely, for eigenvalues $\lambda \neq 0$ and eigenfunctions $v \in \mathscr{H}$ of S, the auxiliary matrix has the eigenvector $\Phi^\top v \in \mathbb{R}^m$.

Example 1. The eigendecomposition of RKHS operators can be used to obtain an approximation of the Mercer feature space representation[1] of a kernel. Let us consider the domain $\mathbb{X} = [-2, 2] \times [-2, 2]$ equipped with the Lebesgue measure and the kernel $k(x, x') = \left(1 + x^\top x'\right)^2$. The associated feature space is in this case six-dimensional.[2] The nonzero eigenvalues and eigenfunctions of the integral operator $\mathcal{E}_k$ defined by

$$\mathcal{E}_k f(x) = \int k(x, x') f(x') \mathrm{d}\mu(x')$$

are given by

$$\begin{aligned}
\lambda_1 &= \tfrac{269+\sqrt{60841}}{90} \approx 5.72, & e_1(x) &= c_1 \left(\tfrac{-179+\sqrt{60841}}{120} + x_1^2 + x_2^2\right),\\
\lambda_2 &= \tfrac{32}{9} \approx 3.56, & e_2(x) &= c_2 x_1 x_2,\\
\lambda_3 &= \tfrac{8}{3} \approx 2.67, & e_3(x) &= c_3 x_1,\\
\lambda_4 &= \tfrac{8}{3} \approx 2.67, & e_4(x) &= c_4 x_2,\\
\lambda_5 &= \tfrac{64}{45} \approx 1.42, & e_5(x) &= c_5 \left(x_1^2 - x_2^2\right),\\
\lambda_6 &= \tfrac{269-\sqrt{60841}}{90} \approx 0.24, & e_6(x) &= c_6 \left(\tfrac{-179-\sqrt{60841}}{120} + x_1^2 + x_2^2\right),
\end{aligned}$$

where $c_1, \ldots, c_6$ are normalization constants so that $\|e_i\|_\mu = 1$. Defining $\phi = [\phi_1, \ldots, \phi_6]^\top$, with $\phi_i = \sqrt{\lambda_i} e_i$, we thus obtain the Mercer feature space representation of the kernel, i.e., $k(x, x') = \langle \phi(x), \phi(x') \rangle$. Here, $\langle \cdot, \cdot \rangle$ denotes the standard inner product in $\mathbb{R}^6$. For $f \in \mathscr{H}$, it holds that $\mathcal{E}_k f = \mathcal{C}_{XX} f$, where $\mathcal{C}_{XX}$ is the covariance operator.[3] We now compute eigenfunctions of its empirical

[1] Given a continuous kernel k on a compact domain, Mercer's theorem allows for a series representation of the form $k(x, x') = \sum_{i \in I} \lambda_i e_i(x) \, e_i'(x)$, see, e.g., [5]. In particular, $\{\sqrt{\lambda_i}\, e_i\}_{i \in I}$ forms an (at most countable) orthonormal system in $\mathscr{H}$. The Mercer feature space can be constructed by computing eigenfunctions of the operator $\mathcal{E}_k$ introduced below.

[2] For a d-dimensional state space, the polynomial kernel with degree p spans a $\binom{p+d}{p}$-dimensional feature space [19].

[3] For a detailed introduction of covariance and cross-covariance operators, see Sect. 4.

estimate $\widehat{\mathcal{C}}_{XX}$ with the aid of the methods described above. That is, $B = \frac{1}{m} I_m$. Drawing $m = 5000$ test points from the uniform distribution on $\mathbb{X}$, we obtain the eigenvalues and (properly normalized) eigenfunctions shown in Fig. 1. The eigenfunctions are in good agreement with the analytically computed ones. We evaluate the analytically and numerically computed eigenfunctions in the midpoints of a regular 50×50 box discretization and the average relative error is approximately 2.9% for the first eigenfunction. Note that the eigenspace corresponding to the eigenvalues λ_3 and λ_4 is only determined up to basis rotations. The computed eigenvalues λ_i for $i > 6$ are numerically zero. This indicates that the feature space is, as expected, only six-dimensional. ▲

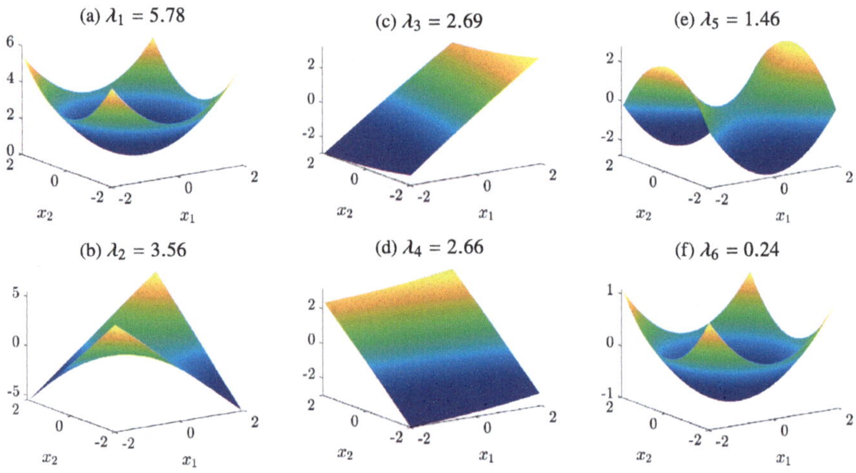

Fig. 1. Numerically computed eigenvalues and eigenfunctions of $\widehat{\mathcal{C}}_{XX}$ associated with the second-order polynomial kernel on $\mathbb{X} = [-2, 2] \times [-2, 2]$.

While we need the assumption that the eigenvalue λ of S is nonzero to infer the eigenvector of the auxiliary matrix from the eigenfunction from S, this assumption is not needed the other way around. This has the simple explanation that a rank deficiency of B always introduces a rank deficiency to $S = \Upsilon B \Phi^\top$. On the other hand, if $\mathscr{H}$ is infinite-dimensional, S as a finite-rank operator *always* has a natural rank deficiency, even when B has full rank. In this case, S has the eigenvalue 0 while B does not.

In order to use Proposition 3 as a consistent tool to compute eigenfunctions of RKHS operators, we must ensure that all eigenfunctions corresponding to nonzero eigenvalues of empirical RKHS operators can be computed. In particular, we have to be certain that eigenvalues with a higher geometric multiplicity allow to capture a full set of linearly independent basis eigenfunctions in the associated eigenspace.

Lemma 3. *Let $S\colon \mathscr{H} \to \mathscr{H}$ with $S = \Upsilon B \Phi^\top$ be an empirical RKHS operator. Then it holds:*

(i) If $\mathbf{w}_1 \in \mathbb{R}^m$ and $\mathbf{w}_2 \in \mathbb{R}^m$ are linearly independent eigenvectors of $B\Phi^\top\Upsilon$, then $\Upsilon\mathbf{w}_1 \in \mathscr{H}$ and $\Upsilon\mathbf{w}_2 \in \mathscr{H}$ are linearly independent eigenfunctions of S.
(ii) If v_1 and v_2 are linearly independent eigenfunctions belonging to the eigenvalue $\lambda \neq 0$ of S, then $B\Phi^\top v_1 \in \mathbb{R}^m$ and $B\Phi^\top v_2 \in \mathbb{R}^m$ are linearly independent eigenvectors of $B\Phi^\top\Upsilon$.

In particular, if $\lambda \neq 0$, then we have $\dim\ker(B\Phi^\top\Upsilon - \lambda\mathcal{I}_m) = \dim\ker(S - \lambda\mathcal{I}_{\mathscr{H}})$.

Proof. The eigenvalue-eigenfunction correspondence is covered in Proposition 3, it therefore remains to check the linear independence in statements (i) and (ii). Part (i) follows from Remark 3. We show part (ii) by contradiction: Let v_1 and v_2 be linearly independent eigenfunctions associated with the eigenvalue $\lambda \neq 0$ of S. Then assume for some $\alpha \neq 0 \in \mathbb{R}$, we have $B\Phi^\top v_1 = \alpha B\Phi^\top v_2$. Applying Υ from the left to both sides, we obtain

$$\Upsilon B\Phi^\top v_1 = Sv_1 = \lambda v_1 = \alpha\lambda v_2 = \alpha S v_2 = \Upsilon\alpha B\Phi^\top v_2,$$

which contradicts the linear independence of v_1 and v_2. Therefore, $B\Phi^\top v_1$ and $B\Phi^\top v_2$ have to be linearly independent in $\mathbb{R}^m$.

From (i) and (ii), we can directly infer $\dim\ker(B\Phi^\top\Upsilon - \lambda\mathcal{I}_m) = \dim\ker(S - \lambda\mathcal{I}_{\mathscr{H}})$ by contradiction: Let $\lambda \neq 0$ be an eigenvalue of S and $B\Phi^\top\Upsilon$. We assume that $\dim\ker(B\Phi^\top\Upsilon - \lambda I_m) > \dim\ker(S - \lambda\mathcal{I}_{\mathscr{H}})$. This implies that there exist two eigenvectors $\mathbf{w}_1, \mathbf{w}_2 \in \mathbb{R}^m$ of $B\Phi^\top\Upsilon$ that generate two linearly dependent eigenfunctions $\Upsilon\mathbf{w}_1, \Upsilon\mathbf{w}_2 \in \mathscr{H}$, contradicting statement (i). Hence, we must have $\dim\ker(B\Phi^\top\Upsilon - \lambda I_m) \leq \dim\ker(S - \lambda\mathcal{I}_{\mathscr{H}})$. Analogously, applying the same logic to statement (ii), we obtain $\dim\ker(B\Phi^\top\Upsilon - \lambda I_m) \geq \dim\ker(S - \lambda\mathcal{I}_{\mathscr{H}})$, which concludes the proof. □

Corollary 4. *If $S = \Upsilon B\Phi^\top$ is an empirical RKHS operator and $\lambda \in \mathbb{R}$ is nonzero, it holds that* $\{\Upsilon\mathbf{w} \mid B\Phi^\top\Upsilon\mathbf{w} = \lambda\mathbf{w}\} = \ker(S - \lambda\mathcal{I}_{\mathscr{H}})$.

The corollary justifies to refer to the eigenvalue problems $Sv = \lambda v$ as *primal problem* and $B\Phi^\top\Upsilon\mathbf{w} = \lambda\mathbf{w}$ as *auxiliary problem*, respectively.

3.5 Singular Value Decomposition via Auxiliary Problem

We have seen that we can compute eigenfunctions corresponding to nonzero eigenvalues of empirical RKHS operators. This can be extended in a straightforward fashion to the singular value decomposition of such operators.

3.5.1 Standard Derivation

We apply the eigendecomposition to the self-adjoint operator S^*S to obtain the singular value decomposition of S.

Proposition 4. *Let $S\colon \mathscr{H} \to \mathscr{F}$ with $S = \Psi B \Phi^\top$ be an empirical RKHS operator, where $\Phi = [\phi(x_1), \ldots, \phi(x_m)]$, $\Psi = [\psi(y_1), \ldots, \psi(y_n)]$, and $B \in \mathbb{R}^{n \times m}$. Assume that the multiplicity of each singular value of S is 1. Then the SVD of S is given by*

$$S = \sum_{i=1}^{r} \lambda_i^{1/2} (u_i \otimes v_i),$$

where

$$v_i := (\mathbf{w}_i^\top G_\Phi \mathbf{w}_i)^{-1/2}\, \Phi \mathbf{w}_i,$$
$$u_i := \lambda_i^{-1/2} S v_i,$$

with the nonzero eigenvalues $\lambda_1, \ldots, \lambda_r \in \mathbb{R}$ of the matrix

$$M G_\Phi \in \mathbb{R}^{m \times m} \quad \text{with} \quad M := B^\top G_\Psi B \in \mathbb{R}^{m \times m}$$

counted with their multiplicities and corresponding eigenvectors $\mathbf{w}_1, \ldots, \mathbf{w}_r \in \mathbb{R}^m$.

Proof. Using Proposition 1, the operator

$$S^* S = \Phi (B^\top G_\Psi B) \Phi^\top = \Phi M \Phi^\top$$

is an empirical RKHS operator on $\mathscr{H}$. Naturally, S^*S is also positive and self-adjoint. We apply Corollary 4 to calculate the normalized eigenfunctions

$$v_i := \|\Phi \mathbf{w}_i\|_{\mathscr{H}}^{-1} \Phi \mathbf{w}_i = (\mathbf{w}_i^\top G_\Phi \mathbf{w}_i)^{-1/2}\, \Phi \mathbf{w}_i$$

of S^*S by means of the auxiliary problem

$$M G_\Phi \mathbf{w}_i = \lambda_i \mathbf{w}_i, \quad \mathbf{w}_i \in \mathbb{R}^m,$$

for nonzero eigenvalues λ_i. We use Lemma 1 to establish the connection between the eigenfunctions of S^*S and singular functions of S and obtain the desired form for the SVD of S. □

Remark 6. Whenever the operator S possesses singular values with multiplicities larger than 1, a Gram-Schmidt procedure may need to be applied to the resulting singular functions in order to ensure that they form an orthonormal system in the corresponding eigenspaces of S^*S and SS^*.

Remark 7. As described in Remark 5, several different auxiliary problems to compute the eigendecomposition of S^*S can be derived. As a result, we can reformulate the calculation of the SVD of S for every possible auxiliary problem.

Example 2. We define a probability density on $\mathbb{R}^2$ by

$$p(x, y) = \frac{1}{2}\big(p_1(x) p_2(y) + p_2(x) p_1(y)\big),$$

with

$$p_1(x) = \frac{1}{\sqrt{2\pi\rho^2}} e^{-\frac{(x-1)^2}{2\rho^2}} \quad \text{and} \quad p_2(x) = \frac{1}{\sqrt{2\pi\rho^2}} e^{-\frac{(x+1)^2}{2\rho^2}},$$

see Fig. 2(a), and draw $m = n = 10000$ test points (x_i, y_i) from this density as shown in Fig. 2(b). Let us now compute the singular value decomposition of $\widehat{\mathcal{C}}_{YX} = \frac{1}{m}\Psi\Phi^\top$, i.e., $B = \frac{1}{m}I_m$. That is, we have to compute the eigenvalues and eigenvectors of the auxiliary matrix $\frac{1}{m^2}G_\psi G_\phi$. Using the normalized Gaussian kernel with bandwidth 0.1 results in singular values $\sigma_1 \approx 0.47$ and $\sigma_2 \approx 0.43$ and the corresponding right and left singular functions displayed in Fig. 2(c) and Fig. 2(d). The subsequent singular values are close to zero. Thus, we can approximate $\widehat{\mathcal{C}}_{YX}$ by a rank-two operator of the form $\widehat{\mathcal{C}}_{YX} \approx \sigma_1(u_1 \otimes v_1) + \sigma_2(u_2 \otimes v_2)$, see also Fig. 2(e) and Fig. 2(f). This is due to the decomposability of the probability density $p(x, y)$. ▲

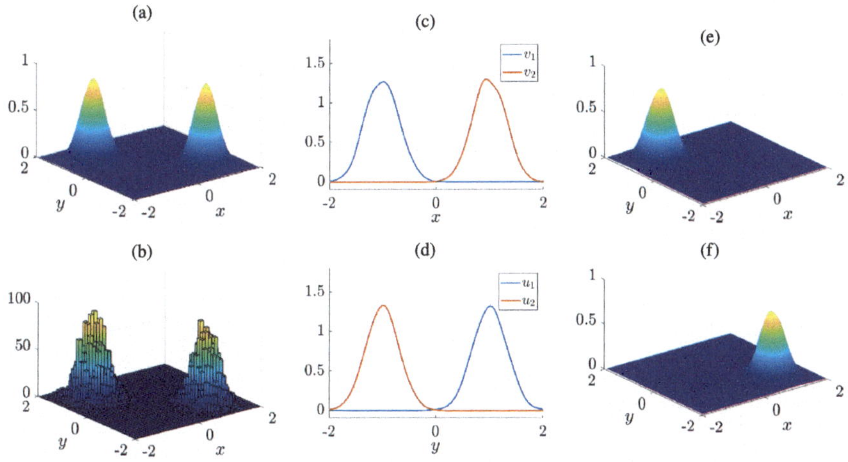

Fig. 2. Numerically computed singular value decomposition of $\widehat{\mathcal{C}}_{YX}$. (a) Joint probability density $p(x, y)$. (b) Histogram of the 10000 sampled data points. (c) First two right singular functions. (d) First two left singular functions. (e) $\sigma_1(u_1 \otimes v_1)$. (f) $\sigma_2(u_2 \otimes v_2)$.

With the aid of the singular value decomposition, we are now, for instance, able to compute low-rank approximations of RKHS operators—e.g., to obtain more compact and smoother representations—or their pseudoinverses. This will be described below. First, however, we show an alternative derivation of the decomposition. Proposition 4 gives a numerically computable form of the SVD of the empirical RKHS operator S. Since the auxiliary problem of the eigendecomposition of S^*S involves several matrix multiplications, the problem might become ill-conditioned.

3.5.2 Block-Operator Formulation

We now employ the relationship described in Corollary 1 between the SVD of the empirical RKHS operator $S\colon \mathscr{H} \to \mathscr{F}$ and the eigendecomposition of the block-operator $T\colon \mathscr{F} \oplus \mathscr{H} \to \mathscr{F} \oplus \mathscr{H}$, with $(f, h) \mapsto (Sh, S^* f)$.

Theorem 2. *The SVD of the empirical RKHS operator $S = \Psi B \Phi^\top$ is given by*

$$S = \sum_{i \in I}^{r} \sigma_i \left[\left(\|\Psi \mathbf{w}_i\|_{\mathscr{F}}^{-1} \Psi \mathbf{w}_i \right) \otimes \left(\|\Phi \mathbf{z}_i\|_{\mathscr{H}}^{-1} \Phi \mathbf{z}_i \right) \right],$$

where σ_i are the strictly positive eigenvalues and $\left[\begin{smallmatrix} \mathbf{w}_i \\ \mathbf{z}_i \end{smallmatrix}\right] \in \mathbb{R}^{n+m}$ the corresponding eigenvectors of the auxiliary matrix

$$\begin{bmatrix} 0 & BG_\Phi \\ B^\top G_\Psi & 0 \end{bmatrix} \in \mathbb{R}^{(n+m)\times(n+m)}. \tag{7}$$

Proof. The operator T defined above can be written in block form as

$$T \begin{bmatrix} f \\ h \end{bmatrix} = \begin{bmatrix} & S \\ S^* & \end{bmatrix} \begin{bmatrix} f \\ h \end{bmatrix} = \begin{bmatrix} Sh \\ S^* f \end{bmatrix}. \tag{8}$$

By introducing the block feature matrix $\Lambda := [\,\Psi\ \Phi\,]$, we may rewrite (8) as the empirical RKHS operator

$$\Lambda \begin{bmatrix} 0 & B \\ B^\top & 0 \end{bmatrix} \Lambda^\top.$$

Invoking Corollary 4 yields the auxiliary problem

$$\begin{bmatrix} 0 & B \\ B^\top & 0 \end{bmatrix} \Lambda^\top \Lambda = \begin{bmatrix} 0 & B \\ B^\top & 0 \end{bmatrix} \begin{bmatrix} G_\Psi & 0 \\ 0 & G_\Phi \end{bmatrix} = \begin{bmatrix} 0 & BG_\Phi \\ B^\top G_\Psi & 0 \end{bmatrix} \in \mathbb{R}^{(n+m)\times(n+m)}$$

for the eigendecomposition of T. We again emphasize that the block-operator notation has to be used with caution since $\mathscr{F} \oplus \mathscr{H}$ is an external direct sum. We use Corollary 1 to obtain the SVD of S from the eigendecomposition of T. □

Remark 8. In matrix analysis and numerical linear algebra, one often computes the SVD of a matrix $A \in \mathbb{R}^{n \times m}$ through an eigendecomposition of the matrix $\left[\begin{smallmatrix} 0 & A \\ A^\top & 0 \end{smallmatrix}\right]$. This leads to a symmetric problem, usually simplifying iterative SVD schemes [18]. The auxiliary problem (7), however, is in general not symmetric.

4 Applications

In this section, we describe different operators of the form $S = \Psi B \Phi^\top$ or $S = \Phi B \Psi^\top$, respectively, and potential applications. All of the presented examples are empirical estimates of Hilbert–Schmidt RKHS operators. Therefore, the

SVD of the given empirical RKHS operators converges to the SVD of their analytical counterparts. For results concerning the convergence and consistency of the estimators, we refer to [9,11–13]. Note that in practice the examples below may bear additional challenges such as ill-posed inverse problems and regularization of compact operators, which we will not examine in detail. We will also not cover details such as measurability of feature maps and properties of related integral operators in what follows. For these details, the reader may consult, for example [5].

4.1 Low-Rank Approximation, Pseudoinverse and Optimization

With the aid of the SVD it is now also possible to compute low-rank approximations of RKHS operators. This well-known result is called *Eckart–Young theorem* or *Eckart–Young–Mirsky theorem*, stating that for every compact operator A with SVD $A = \sum_{i \in I} \sigma_i (u_i \otimes v_i)$ and $k \leq \operatorname{rank}(A)$, the operator given by the truncated SVD

$$A_k := \sum_{i=1}^{k} \sigma_i (u_i \otimes v_i)$$

satisfies the optimality property

$$A_k = \arg\min_{\operatorname{rank}(B)=k} \|A - B\|_{\mathrm{HS}},$$

see [21] for details. Another application is the computation of the (not necessarily globally defined) *pseudoinverse* or *Moore–Penrose inverse* [22] of operators, defined as $A^+ \colon \mathscr{F} \supseteq \operatorname{dom}(A^+) \to \mathscr{H}$, with

$$A^+ := \sum_{i \in I} \sigma_i^{-1} (v_i \otimes u_i).$$

We can thus obtain the solution $x \in H$ of the—not necessarily well-posed—inverse problem $Ax = y$ for $y \in \operatorname{dom}(A^+)$ through the Moore–Penrose pseudoinverse, i.e.,

$$A^+ y = \arg\min_{x \in \mathscr{H}} \|Ax - y\|_{\mathscr{F}},$$

where $A^+ y$ in $\mathscr{H}$ is the unique minimizer with minimal norm. For the connection to regularized least-squares problems and the theory of inverse problems, see [22].

4.2 Kernel Covariance and Cross-Covariance Operator

Let X and Y be random variables with values in $\mathbb{X}$ and $\mathbb{Y}$ defined on a probability space $(\Omega, \mathcal{F}, \mathbb{P})$. The *kernel covariance operator* $\mathcal{C}_{XX} \colon \mathscr{H} \to \mathscr{H}$ and the *kernel cross-covariance operator* [23] $\mathcal{C}_{YX} \colon \mathscr{H} \to \mathscr{F}$ are defined by

$$\begin{aligned} \mathcal{C}_{XX} &= \int \phi(X) \otimes \phi(X) \, \mathrm{d}\mathbb{P}(X) &&= \mathbb{E}_X[\phi(X) \otimes \phi(X)], \\ \mathcal{C}_{YX} &= \int \psi(Y) \otimes \phi(X) \, \mathrm{d}\mathbb{P}(Y, X) &&= \mathbb{E}_{YX}[\psi(Y) \otimes \phi(X)], \end{aligned}$$

assuming that the second moments (in the Bochner integral sense) of the embedded random variables $\phi(X), \psi(Y)$ exist. Kernel (cross-)covariance operators can be regarded as generalizations of (cross-)covariance matrices and are frequently used in nonparametric statistical methods, see [7] for an overview. Given training data $\mathbb{D}_{XY} = \{(x_1, y_1), \ldots, (x_n, y_n)\}$ drawn i.i.d. from the joint probability distribution $\mathbb{P}(X, Y)$, we can estimate these operators by

$$\widehat{\mathcal{C}}_{XX} = \frac{1}{n}\sum_{i=1}^{n} \phi(x_i) \otimes \phi(x_i) = \frac{1}{n}\Phi\Phi^\top \quad \text{and} \quad \widehat{\mathcal{C}}_{YX} = \frac{1}{n}\sum_{i=1}^{n} \psi(y_i) \otimes \phi(x_i) = \frac{1}{n}\Psi\Phi^\top.$$

Thus, $\widehat{\mathcal{C}}_{XX}$ and $\widehat{\mathcal{C}}_{YX}$ are empirical RKHS operators with $B = \frac{1}{n}I_n$, where $\Psi = \Phi$ for $\widehat{\mathcal{C}}_{XX}$. Decompositions of these operators are demonstrated in Example 1 and Example 2, respectively, where we show that we can compute approximations of the Mercer feature space and obtain low-rank approximations of operators.

4.3 Conditional Mean Embedding

The conditional mean embedding is an extension of the mean embedding framework to conditional probability distributions. Under some technical assumptions, the RKHS embedding of a conditional distribution can be represented as a linear operator [8]. We will not cover the technical details here and refer the reader to [10] for the mathematical background. We note that alternative interpretations of the conditional mean embedding exist in a least-squares context which needs less assumptions than the operator-theoretic formulation [9,11].

Remark 9. For simplicity, we write $\mathcal{C}_{XX}^{-1}$ for the inverse covariance operator in what follows. However, note that $\mathcal{C}_{XX}^{-1}$ does in general not exist as a globally defined bounded operator – in practice, a Tikhonov-regularized inverse (i.e., $(\mathcal{C}_{XX} + \epsilon \mathrm{Id})^{-1}$ for some $\epsilon > 0$) is usually considered instead (see [22] for details), leading to regularized matrices in the empirical versions.

The conditional mean embedding operator of $\mathbb{P}(Y \mid X)$ is given by

$$\mathcal{U}_{Y|X} = \mathcal{C}_{YX}\mathcal{C}_{XX}^{-1}.$$

Note that when the joint distribution $\mathbb{P}(X, Y)$ and hence $\mathcal{C}_{XX}$ and $\mathcal{C}_{YX}$ are unknown, we can not compute $\mathcal{U}_{Y|X}$ directly. However, if the training data $\mathbb{D}_{XY} = \{(x_1, y_1), \ldots, (x_n, y_n)\}$ is drawn i.i.d. from the probability distribution $\mathbb{P}(X, Y)$, it can be estimated as

$$\widehat{\mathcal{U}}_{Y|X} = \Psi G_\phi^{-1} \Phi^\top.$$

This is an empirical RKHS operator, where $B = G_\phi^{-1}$. The conditional mean operator is often used for nonparametric models, for example in state-space models [8], filtering and Bayesian inference [12,13], reinforcement learning [24–26], and density estimation [27].

4.4 Kernel Transfer Operators

For this example, we consider a (stochastic) dynamical system $X = (X_t)_{t \in T}$. Transfer operators associated with X such as the Perron–Frobenius operator $\mathcal{P}$ and Koopman operator $\mathcal{K}$ are frequently used for the analysis of the global behaviour of molecular dynamics and fluid dynamics problems but also for model reduction and control [28–30]. Approximations of these operators in RHKSs are strongly related to the conditional mean embedding framework [14]. The kernel-based variants $\mathcal{P}_k$ and $\mathcal{K}_k$ are defined by

$$\mathcal{P}_k = \mathcal{C}_{XX}^{-1} \mathcal{C}_{YX} \quad \text{and} \quad \mathcal{K}_k = \mathcal{C}_{XX}^{-1} \mathcal{C}_{XY},$$

where $Y = (X_{t+\tau})_{t \in T}$ is a time-lagged version of X for a fixed time lag τ. The empirical estimates of $\mathcal{P}_k$ and $\mathcal{K}_k$ are given by

$$\widehat{\mathcal{P}}_k = \Psi G_{\Phi\Psi}^{-1} G_{\Phi}^{-1} G_{\Phi\Psi} \Phi^\top \quad \text{and} \quad \widehat{\mathcal{K}}_k = \Phi G_{\Phi}^{-1} \Psi^\top.$$

Here, we use the feature matrices

$$\Phi := [\phi(x_1), \ldots, \phi(x_m)] \quad \text{and} \quad \Psi := [\phi(y_1), \ldots, \phi(y_n)]$$

with data x_i and $y_i = \Xi^\tau(x_i)$, where Ξ^τ denotes the flow map associated with the dynamical system X with time step τ. Note that in particular $\mathscr{H} = \mathscr{F}$. Both operators $\mathcal{P}_k$ and $\mathcal{K}_k$ can be written as empirical RKHS operators, with $B = G_{\Phi\Psi}^{-1} G_{\Phi}^{-1} G_{\Phi\Psi}$ and $B = G_{\Phi}^{-1}$, respectively, where $G_{\Phi\Psi} = \Phi^\top \Psi$ is a time-lagged Gram matrix. Examples pertaining to the eigendecomposition of kernel transfer operators associated with molecular dynamics and fluid dynamics problems as well as text and video data can be found in [14]. The eigenfunctions and corresponding eigenvalues of kernel transfer operators contain information about the dominant slow dynamics and their implied time-scales. Moreover, the singular value decomposition of kernel transfer operators is known to be connected to *kernel canonical correlation analysis* [31] and the detection of coherent sets in dynamical systems [15]. In particular, the singular value decomposition of the operator

$$S := \widehat{\mathcal{C}}_{YY}^{-1/2} \widehat{\mathcal{C}}_{YX} \widehat{\mathcal{C}}_{XX}^{-1/2}$$

solves the kernel CCA problem. This operator can be written as

$$S = \Psi B \Phi^\top,$$

where $B = G_{\Psi}^{-1/2} G_{\Phi}^{-1/2}$. For the derivation, see Appendix A.2. We will give an example in the context of coherent sets to illustrate potential applications.

Example 3. Let us consider the well-known periodically driven double gyre flow

$$\begin{aligned}
\dot{x}_1 &= -\pi A \sin(\pi f(x_1, t)) \cos(\pi x_2), \\
\dot{x}_2 &= \pi A \cos(\pi f(x_1, t)) \sin(\pi x_2) \frac{\partial f}{\partial x}(x_1, t),
\end{aligned}$$

with $f(y,t) = \delta \sin(\omega t)\, y^2 + (1 - 2\delta \sin(\omega t))\, y$ and parameters $A = 0.25$, $\delta = 0.25$, and $\omega = 2\pi$, see [32] for more details. We choose the lag time $\tau = 10$ and define the test points x_i to be the midpoints of a regular 120×60 box discretization of the domain $[0,2] \times [0,1]$. To obtain the corresponding data points $y_i = \Xi^\tau(x_i)$, where Ξ^τ denotes the flow map, we use a Runge–Kutta integrator with variable step size. We then apply the singular value decomposition to the operator described above using a Gaussian kernel with bandwidth $\sigma = 0.25$. The resulting right singular functions are shown in Fig. 3. ▲

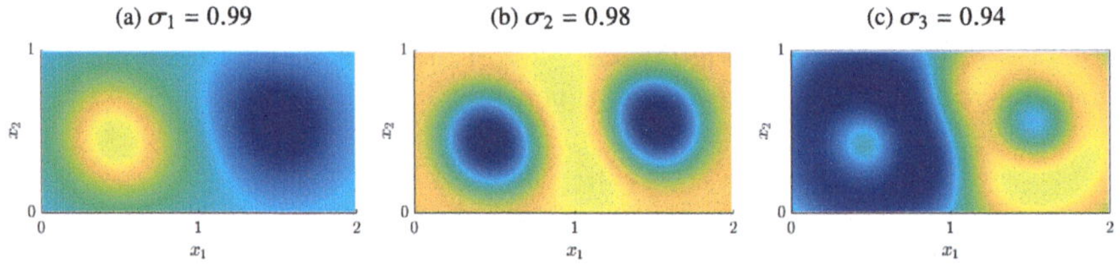

Fig. 3. Numerically computed singular values and right singular functions of $\widehat{\mathcal{C}}_{YY}^{-1/2}\widehat{\mathcal{C}}_{YX}\widehat{\mathcal{C}}_{XX}^{-1/2}$ associated with the double gyre flow.

5 Conclusion

We showed that the eigendecomposition and singular value decomposition of empirical RKHS operators can be obtained by solving associated matrix eigenvalue problems. To underline the practical importance and versatility of RKHS operators, we listed potential applications concerning kernel covariance operators, conditional mean embedding operators, and kernel transfer operators. While we provide the general mathematical theory for the spectral decomposition of RKHS operators, the interpretation of the resulting eigenfunctions or singular functions depends strongly on the problem setting. The eigenfunctions of kernel transfer operators, for instance, can be used to compute conformations of molecules, coherent patterns in fluid flows, slowly evolving structures in video data, or topic clusters in text data [14]. Singular value decompositions of transfer operators might be advantageous for non-equilibrium dynamical systems. Furthermore, the decomposition of the aforementioned operators can be employed to compute low-rank approximations or their pseudoinverses, which might open up novel opportunities in statistics and machine learning. Future work includes analyzing connections to classical methods such as kernel PCA, regularizing finite-rank RKHS operators by truncating small singular values, solving RKHS operator regression problems with the aid of the pseudoinverse, and optimizing numerical schemes to compute the operator SVD by applying iterative schemes and symmetrization approaches.

Acknowledgements. M. M., S. K., and C. S were funded by Deutsche Forschungsgemeinschaft (DFG) through grant CRC 1114 (Scaling Cascades in Complex Systems, project ID: 235221301) and through Germany's Excellence Strategy (MATH+: The Berlin Mathematics Research Center, EXC-2046/1, project ID: 390685689). We would like to thank Ilja Klebanov for proofreading the manuscript and valuable suggestions for improvements.

A Appendix

A.1 Proof of Block SVD

Proof ***(Lemma*** 2***).*** Let A admit the SVD given in (2). Then by the definition of T, we have

$$T(\pm u_i, v_i) = (Av_i, A^* u_i) = \pm\sigma_i(\pm u_i, v_i)$$

for all $i \in I$. For any element $(f, h) \in \operatorname{span}\{(\pm u_i, v_i)\}_{i\in I}^{\perp}$, we can immediately deduce

$$0 = \langle (f, h),\, (\pm u_i, v_i)\rangle_{\oplus} = \pm\langle f,\, u_i\rangle_F + \langle h,\, v_i\rangle_H$$

for all $i \in I$ and hence $f \in \operatorname{span}\{u_i\}_{i\in I}^{\perp}$ and $h \in \operatorname{span}\{v_i\}_{i\in I}^{\perp}$. Using the SVD of A in (2), we therefore have

$$T\Big|_{\operatorname{span}\{(\pm u_i, v_i)\}_{i\in I}^{\perp}} = 0.$$

It now remains to show that $\left\{\frac{1}{\sqrt{2}}(\pm u_i, v_i)\right\}_{i\in I}$ is an orthonormal system in $F \oplus H$, which is clear since $\langle (\pm u_i, v_i),\, (\pm u_j, v_j)\rangle_{\oplus} = 2\,\delta_{ij}$ and $\langle (-u_i, v_i),\, (u_j, v_j)\rangle_{\oplus} = 0$ for all $i, j \in I$. Concluding, T has the form (3) as claimed. □

A.2 Derivation of the Empirical CCA Operator

The claim follows directly when we can show the identity

$$\Phi^\top(\Phi\Phi^\top)^{-1/2} = G_{\Phi}^{-1/2}\Phi^\top$$

and its analogue for the feature map Ψ. Let $G_{\Phi} = U\Lambda U^\top$ be the eigendecomposition of the Gramian. We know that in this case we have the SVD of the operator $\Phi\Phi^\top = \sum_{i\in I}\lambda_i(\lambda_i^{-1/2}\Phi u_i)\otimes(\lambda_i^{-1/2}\Phi u_i)$, since

$$\left\langle \lambda_i^{-1/2}\Phi u_i,\, \lambda_j^{-1/2}\Phi u_j\right\rangle_{\mathscr{H}} = \lambda_i^{-1/2} u_i G_{\Phi} u_j \lambda_j^{-1/2} = \delta_{ij}.$$

We will write this operator SVD for simplicity as $\Phi\Phi^\top = (\Phi U\Lambda^{-1/2})\Lambda(\Lambda^{-1/2} U\Phi^\top)$ with an abuse of notation. Note that we can express the inverted operator square root elegantly in this form as $(\Phi\Phi^\top)^{-1/2} = (\Phi U\Lambda^{-1/2})\Lambda^{-1/2} (\Lambda^{-1/2}U\Phi^\top) = (\Phi U)\Lambda^{-3/2}(U\Phi^\top)$. Therefore, we immediately get

$$\begin{aligned}
\Phi^\top(\Phi\Phi^\top)^{-1/2} &= \Phi^\top(\Phi U\Lambda^{-3/2}U^\top\Phi^\top)\\
&= G_{\Phi}U\Lambda^{-3/2}U^\top\Phi^\top\\
&= U\Lambda U^\top U\Lambda^{-3/2}U^\top\Phi^\top\\
&= U\Lambda^{-1/2}U^\top\Phi^\top = G_{\Phi}^{-1/2}\Phi^\top,
\end{aligned}$$

which proves the claim. In the regularized case, all operations work the same with an additional ϵ-shift of the eigenvalues, i.e., the matrix Λ is replaced with the regularized version $\Lambda + \epsilon \mathrm{I}$.

References

1. Reed, M., Simon, B.: Methods of Mathematical Physics I: Functional Analysis, 2nd edn. Academic Press Inc., Cambridge (1980)
2. Aronszajn, N.: Theory of reproducing kernels. Trans. Am. Math. Soc. **68**(3), 337–404 (1950)
3. Schölkopf, B., Smola, A.J.: Learning with Kernels: Support Vector Machines, Regularization, Optimization and Beyond. MIT Press, Cambridge (2001)
4. Berlinet, A., Thomas-Agnan, C.: Reproducing Kernel Hilbert Spaces in Probability and Statistics. Kluwer Academic Publishers, Berlin (2004)
5. Steinwart, I., Christmann, A.: Support Vector Machines. Springer, Heidelberg (2008)
6. Smola, A., Gretton, A., Song, L., Schölkopf, B.: A Hilbert space embedding for distributions. In: Proceedings of the 18th International Conference on Algorithmic Learning Theory, pp. 13–31. Springer (2007)
7. Muandet, K., Fukumizu, K., Sriperumbudur, B., Schölkopf, B.: Kernel mean embedding of distributions: a review and beyond. Found. Trends Mach. Learn. **10**(1–2), 1–141 (2017)
8. Song, L., Huang, J., Smola, A., Fukumizu, K.: Hilbert space embeddings of conditional distributions with applications to dynamical systems. In: Proceedings of the 26th Annual International Conference on Machine Learning, pp. 961–968 (2009)
9. Grünewälder, S., Lever, G., Baldassarre, L., Patterson, S., Gretton, A., Pontil, M.: Conditional mean embeddings as regressors. In: International Conference on Machine Learning, vol. 5 (2012)
10. Klebanov, I., Schuster, I., Sullivan, T.J.: A rigorous theory of conditional mean embeddings (2019)
11. Park, J., Muandet, K.: A measure-theoretic approach to kernel conditional mean embeddings (2020)
12. Fukumizu, K., Song, L., Gretton, A.: Kernel Bayes' rule: Bayesian inference with positive definite kernels. J. Mach. Learn. Res. **14**, 3753–3783 (2013)
13. Fukumizu, K.: Nonparametric Bayesian inference with kernel mean embedding. In: Peters, G., Matsui, T. (eds.) Modern Methodology and Applications in Spatial-Temporal Modeling (2017)
14. Klus, S., Schuster, I., Muandet, K.: Eigendecompositions of transfer operators in reproducing kernel Hilbert spaces. J. Nonlinear Sci. **30**, 283–315 (2019)
15. Klus, S., Husic, B.E., Mollenhauer, M., Noé, F.: Kernel methods for detecting coherent structures in dynamical data. Chaos Interdiscip. J. Nonlinear Sci. **29**(12), 123112 (2019)
16. Koltai, P., Wu, H., Noé, F., Schütte, C.: Optimal data-driven estimation of generalized Markov state models for non-equilibrium dynamics. Computation **6**(1), 22 (2018)
17. Weidmann, J.: Lineare Operatoren in Hilberträumen, 3rd edn. Teubner, Stuttgart (1976)
18. Golub, G.H., Van Loan, C.F.: Matrix Computations, 4th edn. John Hopkins University Press, Baltimore (2013)

19. Shawe-Taylor, J., Christianini, N.: Kernel Methods for Pattern Analysis. Cambridge University Press, Cambridge (2004)
20. Kato, T.: Perturbation Theory for Linear Operators. Springer, Berlin (1980)
21. Eubank, R., Hsing, T.: Theoretical Foundations of Functional Data Analysis with an Introduction to Linear Operators, 1st edn. Wiley, New York (2015)
22. Engl, H., Hanke, M., Neubauer, A.: Regularization of Inverse Problems. Kluwer, Berlin (1996)
23. Baker, C.: Joint measures and cross-covariance operators. Trans. Am. Math. Soc. **186**, 273–289 (1973)
24. Lever, G., Shawe-Taylor, J., Stafford, R., Szepesvári, C.: Compressed conditional mean embeddings for model-based reinforcement learning. In: Association for the Advancement of Artificial Intelligence (AAAI), pp. 1779–1787 (2016)
25. Stafford, R., Shawe-Taylor, J.: ACCME: actively compressed conditional mean embeddings for model-based reinforcement learning. In: European Workshop on Reinforcement Learning 14 (2018)
26. Gebhardt, G.H.W., Daun, K., Schnaubelt, M., Neumann, G.: Learning robust policies for object manipulation with robot swarms. In: IEEE International Conference on Robotics and Automation (2018)
27. Schuster, I., Mollenhauer, M., Klus, S., Muandet, K.: Kernel conditional density operators. In: The 23rd International Conference on Artificial Intelligence and Statistics (2020, accepted for publication)
28. Lasota, A., Mackey, M.C.: Chaos, Fractals, and Noise: Stochastic Aspects of Dynamics. Applied Mathematical Sciences, vol. 97, 2nd edn. Springer, Heidelberg (1994)
29. Mezić, I.: Spectral properties of dynamical systems, model reduction and decompositions. Nonlinear Dyn. **41**(1), 309–325 (2005)
30. Klus, S., Nüske, F., Koltai, P., Wu, H., Kevrekidis, I., Schütte, C., Noé, F.: Data-driven model reduction and transfer operator approximation. J. Nonlinear Sci. **28**, 985–1010 (2018)
31. Melzer, T., Reiter, M., Bischof, H.: Nonlinear feature extraction using generalized canonical correlation analysis. In: Dorffner, G., Bischof, H., Hornik, K. (eds.) Artificial Neural Networks – ICANN 2001, pp. 353–360. Springer, Heidelberg (2001)
32. Froyland, G., Padberg-Gehle, K.: Almost-invariant and finite-time coherent sets: directionality, duration, and diffusion. In: Bahsoun, W., Bose, C., Froyland, G. (eds.) Ergodic Theory, Open Dynamics, and Coherent Structures, pp. 171–216. Springer, New York (2014)

A Weak Characterization of Slow Variables in Stochastic Dynamical Systems

Andreas Bittracher[1(✉)] and Christof Schütte[1,2]

[1] Department of Mathematics and Computer Science, Freie Universität Berlin, Berlin, Germany
bittracher@mi.fu-berlin.de, Christof.Schuette@fu-berlin.de
[2] Zuse Institute Berlin, Berlin, Germany

Abstract. We present a novel characterization of slow variables for continuous Markov processes that provably preserve the slow timescales. These slow variables are known as reaction coordinates in molecular dynamical applications, where they play a key role in system analysis and coarse graining. The defining characteristics of these slow variables is that they parametrize a so-called transition manifold, a low-dimensional manifold in a certain density function space that emerges with progressive equilibration of the system's fast variables. The existence of said manifold was previously predicted for certain classes of metastable and slow-fast systems. However, in the original work, the existence of the manifold hinges on the pointwise convergence of the system's transition density functions towards it. We show in this work that a convergence in average with respect to the system's stationary measure is sufficient to yield reaction coordinates with the same key qualities. This allows one to accurately predict the timescale preservation in systems where the old theory is not applicable or would give overly pessimistic results. Moreover, the new characterization is still constructive, in that it allows for the algorithmic identification of a good slow variable. The improved characterization, the error prediction and the variable construction are demonstrated by a small metastable system.

1 Introduction

The ability and practice to perform all-atom molecular simulations of more and more complex biochemical systems has led to an unprecedented increase in the available amount of dynamical data about those systems. This has exponentiated the importance to identify good chemical reaction coordinates (RCs), low-dimensional observables of the full system that are associated with the relevant, often slowly-progressing sub-processes. For one, a meaningful RC permits insight into the essential mechanisms and parameters of a reaction, by acting as a filter

O. Junge et al. (Eds.): SON 2020, SSDC 304, pp. 132–150, 2020.
https://doi.org/10.1007/978-3-030-51264-4_6

for the overwhelming complexity of the data. As an example, computing the free energy (also known as the potential of mean force) along such a coordinate is typically used for identifying energy barriers and associated transition states [10,37]. RCs are also essential for the development of accurate reduced dynamical models. The Mori-Zwanzig formalism and related schemes [18,27,42,44] can be used to derive approximate closed equations of motion of the dynamics projected onto the image space of the RC. Depending on the chosen RC, the essential dynamical properties of the reduced model — such as transition rates between reactant and product — may or may not resemble those of the original system [43]. Finally, accelerated sampling schemes such as metadynamics [20], Blue Moon sampling [8] and umbrella sampling [38] also rely heavily on an accurate RC to guide them efficiently into unexplored territory.

In each of those applications, the result depends crucially on the "quality" of the RC, an elusive measure for how well the RC suits the specified task. In most cases, this quality can be brought down to how well the RC "captures the essential dynamics", in particular the rates of transitions between reactant and product state (see also [28] for an in-depth review on the effect of poorly chosen RCs on different classic rate theories). Due to this ambiguity, the search for universal and mathematically rigorous optimality criteria for RCs remains an active field of research, and numerous new approaches have been suggested during the last decade. For reactions involving one clearly defined reactant and product state, a in multiple ways ideal RC is the committor function [3,23], a one-dimensional observable that in each point describes the probability to hit the product state before returning to the reactant state. As the committor function is notoriously hard to compute, advanced numerical schemes have been developed to either approximate it efficiently [12], or find RCs that are equivalent by certain metrics [29]. Still, the computation of committor-like RCs often remains out of reach for high-dimensional systems.

For systems where the relevant behavior involves transitions between more than tw states [36], where the reaction is not adequately described by a transition between isolated states [35], or where the states are not known or cannot be computed, other optimality criteria must be employed. Here one common approach is to demand the preservation of the system's longest (equilibration) time scales under projection of the dynamics onto the RC. This leads naturally to a characterization of RCs in terms of the eigenvalues of the system's transfer operator, a widely used mathematical tool for time scale analysis in molecular dynamics and beyond [7,11,19,34,41]. It is in this setting where the authors and coworkers have previously proposed a novel mathematical framework for the characterization and numerical computation of ideal RCs [6]. The proposed theory builds on the insight that in many systems, the equilibration of the fast sub-processes over time manifests as the convergence of the system's transition density functions towards a certain low-dimensional manifold in density space, the so-called transition manifold (TM). This convergence is observed even if there is no equivalent low-dimensional structure in state space, such as a transition pathway between isolated states. Any parametrization of the TM then can in theory be used to construct an ideal RC.

The framework demands that the convergence towards the TM must occur for *all* transition density functions, i.e., for every conceivable starting state. In practice however, this rather strong condition is often violated for starting states with high potential energy, as the associated transition density functions may stay far away from any sensible candidate TM for all times. The probability to encounter these states in the canonical ensemble is however exponentially low, and thus should not contribute significantly to the shape of the RC. Indeed, the numerical methods built around parametrizing the TM are able to successfully deal with this problem by heuristically ignoring sparse outliers by tuning the manifold learning algorithm [4,5].

Still, a rigorous argument for why those outliers can be safely ignored was lacking so far, a gap that the present article aims to fill. In short, we show that the distance to the TM does not need to be uniformly low for all transition density functions, but that the distance is permitted to scale with the potential energy of the starting state. The RC received by parametrizing the TM is then of the same quality as in the uniform distance case. This extension to the TM theory will therefore allow to measure the quality of given RCs, and the numerical computation of ideal RCs in systems that been previously deemed unsuitable for the theory.

This paper is structured as follows: Section 2 reviews the time scale-based definition of good RCs. Section 3 presents the main contribution of this article, weakened but sufficient conditions for the existence of good RCs. In Sect. 4 we give an example of a metastable toy system that fulfills the relaxed but not the original reducibility condition, and demonstrate how the new characterization can improve the quality of error bounds for the dominant timescales. In Sect. 5, concluding remarks and an outlook on future work are given.

2 Good Reaction Coordinates

Before introducing the (generalized) transition manifold framework, we first revisit the fundamental time scale-based definition of good reaction coordinates.

2.1 Timescales of Molecular Dynamics

We consider a time- and space-continuous, reversible and ergodic Markov process $\mathbf{X}_t$ on a state space $\mathbb{X} \subset \mathbb{R}^n$. In a molecular dynamical system consisting of N atoms, $\mathbb{X}$ often is the Euclidean space describing the three-dimensional positions of all atoms, i.e., $\mathbb{X} = \mathbb{R}^{3N}$ (or $\mathbb{X} = \mathbb{R}^{6N}$ if the atom's momenta are also included). In this case, $\mathbf{X}_t$ is typically described by a thermostated Hamiltonian dynamics or Langevin dynamics.

$\mathbf{X}_t$ is fully characterized by its stochastic transition functions $p^t(x,\cdot) : \mathbb{X} \to \mathbb{R}^+$, or, equivalently, by its family of *transfer operators* $\mathcal{T}^t : L^1_\mu \to L^1_\mu$, $t \geq 0$,

$$\mathcal{T}^t u(x) = \int_{\mathbb{X}} \frac{\rho(x')}{\rho(x)} p^t(x', x) u(x')\, dx'.$$

Here, ρ is the system's (positive) stationary density, which is unique due to the ergodicity of $\mathbf{X}_t$, and μ is the associated invariant measure. Operating on L^1_μ, $\mathcal{T}^t$ can be understood as the evolution operator of densities with respect to μ under the dynamics.

On L^1_μ, $\mathcal{T}^t$ is a linear Markov operator, [21, Chap. 3], and in particular non-expansive. Hence, no eigenvalue of $\mathcal{T}^t$ has absolute value greater than 1. Due to the uniqueness of the stationary density, the eigenvalue $\lambda_0^t := 1$ is single; the associated unique eigenfunction is $\varphi_0 \equiv 1$.

Furthermore, $\mathcal{T}^t$ is well-defined as an operator $\mathcal{T}^t : L^p_\mu \to L^p_\mu$ for any $1 \leq p \leq \infty$ [2]. We understand $\mathcal{T}^t$ as an operator on L^2_μ from now on, where we will be able to exploit the additional Hilbert space structure. In particular, $\mathcal{T}^t$ is self-adjoint with respect to the inner product on L^2_μ [33], hence its point spectrum is real and therefore confined to the interval $(-1, 1]$. Note that $\mathcal{T}^t$ cannot possess the eigenvalue -1, as this would imply the existence of an eigenfunction $\widetilde{\varphi}_0 \neq \varphi_0$ of $\mathcal{T}^{2t}$ to eigenvalue 1. This however contradicts the uniqueness of φ_0 as the only eigenfunction to eigenvalue 1 of $\mathcal{T}^t$ for all t.

In the following we will always order the eigenvalues so that

$$1 = \lambda_0^t > \lambda_1^t \geq \lambda_2^t \geq \cdots .$$

The associated eigenfunctions φ_i of $\mathcal{T}^t$ form an orthonormal basis of L^2_μ. Hence, on L^2_μ, $\mathcal{T}^t$ admits the decomposition

$$\mathcal{T}^t = \sum_{i=0}^{\infty} \lambda_i^t \langle \varphi_i, \cdot \rangle_\mu \varphi_i,$$

which lets us examine the behavior of $\mathbf{X}_t$ on different time scales. The i-th *relaxation rate*, i.e., the exponential rate with which the i-th eigenfunction φ_i of $\mathcal{T}^t$ decays, is given by

$$\sigma_i = -\log(\lambda_i^t)/t, \quad i = 0, 1, 2, \ldots, \tag{1}$$

independent of t. These rates, as well as their inverse, the *relaxation time scales* $t_i = 1/\sigma_i$, $i = 0, 1, 2, \ldots$, measure the influence of the different φ_i on the long time density transport under $\mathcal{T}^t$, and hence are central quantities of the system.

2.2 Reaction Coordinates

A reaction coordinate (RC) now is a continuous map $\xi : \mathbb{X} \to \mathbb{Y} \subset \mathbb{R}^r$, where typically $r \ll n$. Note that the term "reaction coordinate" does not imply that ξ describes a reaction of some sort, it simply is a continuous map. For $y \in \mathbb{Y}$, let $\Sigma_\xi(y)$ be the y-level set of ξ, i.e.,

$$\Sigma_\xi(y) = \{x \in \mathbb{X} \mid \xi(x) = y\}.$$

Following [22], we now define the *coordinate projection operator* $\Pi_\xi : L^1_\mu \to L^1_\mu$ for a RC ξ by

$$\begin{aligned}(\Pi_\xi u)(x) &= \int_{\Sigma_\xi(\xi(x))} u(x')d\mu_{\xi(x)}(x') \\ &= \frac{1}{\Gamma(\xi(x))}\int_{\Sigma_\xi(\xi(x))} u(x')\rho(x')\det\left(\nabla\xi(x')^\intercal\nabla\xi(x')\right)^{-1/2} d\sigma_{\xi(x)}(x'),\end{aligned}$$

where $\Gamma(y)$ is a normalization constant given by

$$\Gamma(y) = \int_{\Sigma_\xi(y)} \rho(x')\det\left(\nabla\xi(x')^\intercal\nabla\xi(x')\right)^{-1/2} d\sigma_y(x'),$$

and $d\sigma_y$ denotes the surface measure on $\Sigma_\xi(y)$. μ_y can be understood as the invariant measure μ conditioned on $\Sigma_\xi(y)$, and formally is induced by the density

$$\rho_y = \frac{\rho}{\Gamma(y)}\left[\det\left(\nabla\xi^\intercal\nabla\xi\right)\right]^{-1/2}.$$

As $L^2_\mu \subset L^1_\mu$ due to Hölder's inequality, Π_ξ is defined on L^2_μ as well. Informally, Π_ξ has the effect of averaging an input function u over each level set $\Sigma_\xi(y)$ with respect to μ_y.

It has been shown in [6] that Π_ξ is indeed a projection operator. Moreover, Π_ξ is equivalent to the Zwanzig projection operator, described in detail in [17], although the latter is typically constructed so that its image are functions over the reduced space $\mathbb{Y}$. For our presentation, however, it is advantageous to define Π_ξ to project onto a true subspace of L^2_μ (namely the subspace of functions that are constant on each $\Sigma_\xi(y)$, $y \in \mathbb{Y}$).

The *effective transfer operator* $\mathcal{T}^t_\xi : L^2_\mu \to L^2_\mu$ associated with the RC ξ is now defined by

$$\mathcal{T}^t_\xi = \Pi_\xi \circ \mathcal{T}^t \circ \Pi_\xi.$$

Originally considered in [42], $\mathcal{T}^t_\xi$ has been shown to again be self-adjoint and bounded in L^2_μ-norm by 1 [6]. Hence, the eigenvalues $\lambda^t_{\xi,i}$, $i = 0, 1, 2, \ldots$ of $\mathcal{T}^t_\xi$ are also confined to the interval $[-1, 1]$.

2.3 Preservation of Time Scales

Our characterization of *good* RCs — originally proposed in [6] — now revolves around the central assumption that the relevant part of the dynamics (the "reaction") occurs on the slowest time scales of $\mathbf{X}_t$. Moreover, we assume that the time scales of the reaction are well-separated from non-reactive time scales, i.e., $t_0 > t_1 \geq \cdots \geq t_d \gg t_{d+1}$ for some $d \in \mathbb{N}$. This is a sensible and commonly made assumption [26,31,32,34], as it holds true for many difference classes of chemical and molecular reactions. However, there are relevant molecular systems whose effective behavior cannot be explained by its slowest timescales alone [25,40], and hence valid criticism of the general equivalence of the slow with the relevant time scales exist. Nevertheless, we assume that the reaction in question is associated with the d dominant time scales.

With the goal of preserving the dominant time scales under projection onto the RC, and the close connection between those time scales and the dominant transfer operator eigenvalues (1), we use the following definition of good RCs:

Definition 1 (Good reaction coordinates [6]**).** Let λ_i^t, $i = 0, 1, 2, \ldots$ and $\lambda_{\xi,i}^t$, $i = 0, 1, 2, \ldots$ denote the eigenvalues of $\mathcal{T}^t$ and $\mathcal{T}_\xi^t$, respectively. Let t_d be the last time scale of the system that is relevant to the reaction. Let $\varepsilon > 0$.

An RC $\xi : \mathbb{X} \to \mathbb{Y}$ is called a *ε-good RC*, if for all $t > 0$ holds

$$|\lambda_i^t - \lambda_{\xi,i}^t| \leq \varepsilon, \quad i = 0, 1, \ldots, d. \tag{2}$$

Informally, we will call ξ a *good RC* if it is ε-good for small ε.

Alternatively, the following sufficient condition characterizes good RC by the projection error of the dominant eigenfunctions under Π_ξ:

Theorem 1 ([6], Corollary 3.6). *Let (λ_i^t, φ_i), $i = 1, 2, \ldots$ denote the eigenpairs of $\mathcal{T}^t$. For any given i, if*

$$\|\Pi_\xi \varphi_i - \varphi_i\|_{L^2_\mu} \leq \varepsilon,$$

then there is an eigenvalue $\lambda_{\xi,i}^t$ of $\mathcal{T}_\xi^t$ such that

$$|\lambda_i^t - \lambda_{\xi,i}^t| \leq \frac{\varepsilon}{\sqrt{1-\varepsilon^2}}.$$

Remark 1. By the above theorem, choosing the d dominant eigenfunctions as the d components of ξ results in a "perfect" RC. However, this approach may lead to redundancy if the φ_i, $i = 1, \ldots, d$ are strongly correlated and can be parametrized by a common, lower-dimensional ξ. For example, a system with d metastable sets along a common, one-dimensional transition pathway would possess d dominant eigenfunctions, but a one-dimensional good RC that parametrizes the transition pathway (see [6, Sect. 5.2] for a detailed example).

Using eigenfunctions as RCs was also promoted by Froyland et al. [14,15], for the special case where the timescale separation stems from a pointwise local separation of the dynamics into a slow and a fast part. Just like for the transition manifold approach presented in Sect. 3, the short-time equilibration of the dynamics again plays an important part, but unlike in our approach it is assumed to take place on certain "fast fibers" of state space. The transition manifold framework can therefore be considered a generalization of the approach of Froyland et al.

3 Weak Reducibility of Stochastic Systems

Definition (2) is not constructive, in that it allow one to check the quality of a given RC, but does not indicate how to find a good RC algorithmically. To this end, we will now derive a reducibility condition that binds the existence of good RCs to the existence of a certain low-dimensional structure in the space of transition density functions. This structure, called the *transition manifold*, can be interpreted as the backbone of the essential dynamics, can be visualized, and ultimately can be used to numerically compute good RCs.

3.1 Condition for Good Reaction Coordinates Based on Transfer Operator Eigenfunctions

It was shown in [6] that if for some functions $\hat{\varphi}_i : \mathbb{Y} \to \mathbb{R}$ the condition

$$\|\varphi_i - \hat{\varphi}_i \circ \xi\|_\infty \leq \varepsilon, \quad i = 0, 1, \dots, d \tag{3}$$

holds, then ξ is a $\frac{\varepsilon}{\sqrt{1-\varepsilon^2}}$-good RC by Theorem 1. In other words, if the dominant eigenfunctions are pointwise almost constant along the level sets of ξ, then ξ is a good RC.

It turns out, however, that condition (3) is unnecessarily strong. To be precise, the pointwise approximation implied by the $\|\cdot\|_\infty$-norm can be replaced by the following weaker condition. This was already observed previously [6, Remark 4.3], but has not been proven formally.

Theorem 2. *Assume that for an RC $\xi : \mathbb{X} \to \mathbb{Y}$ and some functions $\hat{\varphi}_i : \mathbb{Y} \to \mathbb{R}$, $i = 0, 1, \dots, d$ holds*

$$\int_{\Sigma_\xi(y)} \big|\varphi_i(x') - \hat{\varphi}_i(y)\big| \, d\mu_y(x') \leq \varepsilon \tag{4}$$

for all level sets $\Sigma_\xi(y)$ of ξ. Then

$$\|\Pi_\xi \varphi_i - \varphi_i\|_{L^2_\mu} \leq 2\varepsilon.$$

Remark 2. In words, for a specific value $y \in \mathbb{Y}$, the dominant eigenfunctions φ_i do not need to be almost constant everywhere on $\Sigma_\xi(y)$, but only the average deviation of φ_i from some value $\hat{\varphi}(y)$ along $\Sigma_\xi(y)$, weighted by μ_y, must be small. Hence, ξ may be a good RC even if $\varphi_i(x')$ substantially deviates from the value $\hat{\varphi}(y)$, as long as it is in regions where the measure μ_y is small. These are precisely the regions of state space that are lowly-populated in the canonical ensemble, and thus are statistically irrelevant.

Proof ***(Proof of Theorem** 2).* The projection error is

$$\|\Pi_\xi \varphi_i - \varphi_i\|_{L^2_\mu} \leq \|\Pi_\xi \varphi_i - (\hat{\varphi}_i \circ \xi)\|_{L^2_\mu} + \|(\hat{\varphi}_i \circ \xi) - \varphi_i\|_{L^2_\mu}.$$

For the first summand, consider

$$\begin{aligned}(\Pi_\xi \varphi_i)(x) &= \int_{\Sigma_\xi(\xi(x))} \varphi_i(x') d\mu_{\xi(x)}(x') \\ &= \int_{\Sigma_\xi(\xi(x))} \Big(\hat{\varphi}_i\big(\underbrace{\xi(x')}_{=\xi(x)}\big) + \varphi_i(x') - \hat{\varphi}_i\big(\xi(x')\big)\Big) d\mu_{\xi(x)}(x') \\ &= \hat{\varphi}_i\big(\xi(x)\big) + \int_{\Sigma_\xi(\xi(x))} \Big(\varphi_i(x') - \hat{\varphi}_i\big(\xi(x')\big)\Big) d\mu_{\xi(x)}(x'),\end{aligned}$$

and hence

$$\begin{aligned}\|\Pi_\xi \varphi_i - (\hat{\varphi}_i \circ \xi)\|_{L^2_\mu}^2 &\leq \int_{\mathbb{X}} \underbrace{\Big(\int_{\Sigma_\xi(\xi(x))} \big|\varphi_i(x') - \hat{\varphi}_i\big(\xi(x')\big)\big| d\mu_{\xi(x)}(x')\Big)^2}_{\leq \varepsilon} d\mu(x) \\ &\leq \varepsilon^2 \int_{\mathbb{X}} d\mu(x) = \varepsilon^2.\end{aligned}$$

For the second summand, we get with the co-area formula [13]

$$\begin{aligned}\|(\hat{\varphi}_i \circ \xi) - \varphi_i\|_{L^2_\mu}^2 &= \int_{\mathbb{Y}} \int_{\Sigma_\xi(y)} \big|\hat{\varphi}_i(\xi(x')) - \varphi_i(x')\big|^2 d\mu_y(x') \Gamma(y)\, dy \\ &\leq \int_{\mathbb{Y}} \underbrace{\Big(\int_{\Sigma_\xi(y)} \big|\hat{\varphi}_i(\xi(x')) - \varphi_i(x')\big| d\mu_y(x')\Big)^2}_{\leq \varepsilon} \Gamma(y)\, dy \\ &\leq \varepsilon^2 \int_{\mathbb{Y}} \Gamma(y)\, dy = \varepsilon^2.\end{aligned}$$

3.2 Weak Reducibility and Weak Transition Manifolds

From the abstract condition (4) of good RCs, one can now derive a constructive condition for the existence of a good RC. We will also repeat the strong version of this condition, based on (3), which was originally derived in [6].

The parametrizations of certain manifolds will play a central role in our constructions. Specifically, we consider the special class of manifolds $\mathbb{M} \subset L^1$ for which a compact and connected set $\mathbb{Y} \subset \mathbb{R}^r$, as well as a homeomorphism $\mathcal{E} : \mathbb{M} \to \mathbb{Y}$ exists, such that

$$\mathbb{M} = \mathcal{E}^{-1}(\mathbb{Y}). \tag{5}$$

$\mathbb{Y}$ will later become the image space of our constructed RC.

For a fixed lag time $\tau > 0$, we now call the set of functions

$$\widetilde{\mathbb{M}} = \big\{p^\tau(x, \cdot) \mid x \in \mathbb{X}\big\} \subset L^1$$

the *fuzzy transition manifold*. Note that $\widetilde{\mathbb{M}}$ is not a manifold; the reason behind the choice of name will however soon become clear. Now, for any manifold $\mathbb{M} \subset \widetilde{\mathbb{M}}$ of form (5), define the projection onto $\mathbb{M}$ by

$$\mathcal{Q} : \mathbb{X} \to \mathbb{M}, \quad x \mapsto \arg\min_{f \in \mathbb{M}} \|f - p^\tau(x, \cdot)\|_{L^2_{1/\mu}}. \tag{6}$$

Definition 2. We call the system *strongly* (ε, r, τ)*-reducible*, if there exists a manifold $\mathbb{M} \subset \widetilde{\mathbb{M}}$ of form (5) so that for all $x \in \mathbb{X}$

$$\left\| \mathcal{Q}(x) - p^\tau(x, \cdot) \right\|_{L^2_{1/\mu}} \leq \varepsilon. \tag{7}$$

We call any such $\mathbb{M}$ a *strong transition manifold.*

We call the system *weakly* (ε, r, τ)*-reducible*, if there exists a manifold $\mathbb{M} \subset \widetilde{\mathbb{M}}$ of form (5) so that for all $x \in \mathbb{X}$

$$\int_{\Sigma_{\mathcal{Q}}(\mathcal{Q}(x))} \left\| \mathcal{Q}(x') - p^\tau(x', \cdot) \right\|_{L^2_{1/\mu}} d\mu_{\mathcal{Q}(x)}(x') \leq \varepsilon, \tag{8}$$

where $\Sigma_{\mathcal{Q}}(f)$ is the f-level set of $\mathcal{Q}$. We call any such $\mathbb{M}$ a *weak transition manifold.*

Example 1. As an illustration of the core idea behind the TM construction, we give a simple example of a metastable system with a strong TM, originally published in [5].

Consider a two-dimensional system described by the overdamped Langevin equation

$$d\mathbf{X}_t = -\nabla V(\mathbf{X}_t)\, dt + \sqrt{2\beta^{-1}} d\mathbf{W}_t, \tag{9}$$

where V is the potential energy function and $\mathbf{W}_t$ is a Wiener diffusion process scaled by the inverse temperature $\beta \in \mathbb{R}^+$. Now suppose that V possesses two local energy wells, connected by a linear, one-dimensional transition path, such as in Fig. 1(left). The "reaction" in this system is the rare transition from one well to the other. Hence, an intuitively good RC is the horizontal coordinate of a point, $\xi(x) = x_1$, as it describes the progress of x along the transition pathway.

The key insight now is that, if the lag time τ was chosen long enough for a typical trajectory to move to one of the metastable sets, then the transition densities $p^\tau(x, \cdot) \in L^1$ also essentially depend only on the progress of x along the transition path. The reason is that the $p^\tau(x, \cdot)$ are essentially convex combinations of two Gaussians[1] centered in the energy minima A and B,

$$p^\tau(x, \cdot) \approx c(x)\rho_A(\cdot) + (1 - c(x))\rho_B(\cdot)$$

with the convex factor $c(x)$ determined by the progress of the starting point x along the transition path. This is represented in Fig. 1(right) by the fact that the transition densities for each gray and white starting point, respectively, concentrate around one point each in L^1. Hence, overall, the fuzzy TM $\widetilde{\mathbb{M}}$ concentrates around a one-dimensional manifold in L^1. This manifold is therefore a strong TM.

An example of a system with only a weak TM will be discussed in detail in Sect. 4.

[1] To be precise, the $p^\tau(x, \cdot)$ are approximately convex combinations of the quasi-stationary densities [16] of the metastable sets, that here however resemble Gaussians.

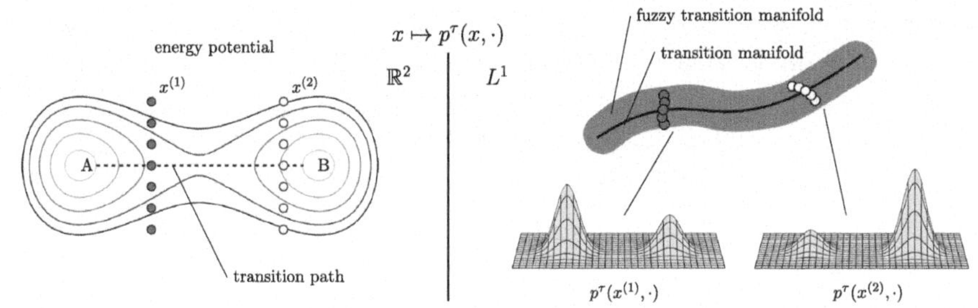

Fig. 1. Illustration of the transition manifold concept for metastable systems. Left: energy potential of a two-dimensional metastable system. Right: Sketch of the (fuzzy) TM for this system. Starting points x with the same progress along the transition path get mapped to approximately the same density under the map $x \mapsto p^\tau(x,\cdot)$. Geometrically, this means that the fuzzy TM concentrates around a one-dimensional manifold in L^1.

Remark 3. Note that we slightly deviate from the original definition of the transition manifold in [6] by requiring that $\mathbb{M} \subset \widetilde{\mathbb{M}}$ instead of only $\mathbb{M} \subset L^1$. Also note that $\mathcal{Q}$ is now defined on $\mathbb{X}$ and not on $\widetilde{\mathbb{M}}$ as originally in [6]. The interpretation of $\mathcal{Q}$ as "closest point projection onto $\mathbb{M}$" is still valid, however.

Condition (7) indicates whether the fuzzy TM $\widetilde{\mathbb{M}}$ clusters ε-closely around an actual manifold $\mathbb{M}$ with respect to the $L^2_{1/\rho}$-norm. Again, condition (8) represents a relaxation of this condition, as the integral introduces a weighting with respect to $d\mu_{\mathcal{Q}(x)}$. Informally speaking, for points x' with $\rho(x') = \mathcal{O}(\varepsilon)$, a distance $\big\|\mathcal{Q}(x') - p^\tau(x',\cdot)\big\|_{L^2_{1/\mu}} = \mathcal{O}(1)$ is now permitted without violating the reducibility condition.

It was shown in [6] that strongly reducible systems possess good RCs. The following theorem now shows that weakly reducible systems still possess good RCs. It characterizes $\mathcal{Q}$ as a good "$\mathbb{M}$-valued RC" (cf. (4)):

Theorem 3. *Let the system be weakly (ε, r, τ)-reducible. Then for each eigenpair $(\lambda_i^\tau, \varphi_i)$ of the transfer operator $\mathcal{T}^\tau$ there exists a map $\tilde{\varphi}_i : \mathbb{M} \to \mathbb{R}$ so that for all $x \in \mathbb{X}$*

$$\int_{\Sigma_{\mathcal{Q}}(\mathcal{Q}(x))} \big|\varphi_i(x') - \tilde{\varphi}_i\big(\mathcal{Q}(x')\big)\big|\, d\mu_{\mathcal{Q}(x)}(x') \leq \frac{\varepsilon}{|\lambda_i^\tau|}.$$

Proof. As $\mathbb{M} \subset \widetilde{\mathbb{M}}$, for $x \in \mathbb{X}$ we can choose $q(x) \in \mathbb{X}$ so that $\mathcal{Q}(x) = p^t\big(q(x),\cdot\big)$. Let $\tilde{\varphi}_i : \mathbb{M} \to \mathbb{R}$ be defined by

$$\tilde{\varphi}_i\big(\mathcal{Q}(x)\big) = \varphi_i\big(q(x)\big).$$

Then

$$\begin{aligned}\int_{\Sigma_{\mathcal{Q}}(\mathcal{Q}(x))} \big|\varphi_i(x') - \tilde{\varphi}_i\big(\mathcal{Q}(x')\big)\big|\, d\mu_{\mathcal{Q}(x)}(x') &= \int_{\Sigma_{\mathcal{Q}}(\mathcal{Q}(x))} \big|\varphi_i(x') - \tilde{\varphi}_i\big(\mathcal{Q}(x)\big)\big|\, d\mu_{\mathcal{Q}(x)}(x') \\ &= \int_{\Sigma_{\mathcal{Q}}(\mathcal{Q}(x))} \big|\varphi_i(x') - \varphi_i\big(q(x)\big)\big|\, d\mu_{\mathcal{Q}(x)}(x') =: (\star)\end{aligned}$$

As the system is reversible, the detailed balance condition $\rho(x)p^\tau(x,x'') = \rho(x'')p^\tau(x'',x)$ holds. Hence, the eigenfunctions φ_i of $\mathcal{T}^\tau$ have the property

$$\lambda_i^\tau \varphi_i = \mathcal{T}^\tau \varphi_i = \int_{\mathbb{X}} \frac{\rho(x'')}{\rho(x)} p^\tau(x'',\cdot)\varphi_i(x'')\, dx'' = \int_{\mathbb{X}} \varphi_i(x'')p^\tau(\cdot,x'')\, dx'',$$

and thus

$$(\star) = \int_{\Sigma_{\mathcal{Q}}(\mathcal{Q}(x))} \frac{1}{|\lambda_i^\tau|} \left| \int_{\mathbb{X}} \varphi_i(x'')\Big(p^\tau(x',x'') - p^\tau\big(q(x),x''\big)\Big)\, dx'' \right| d\mu_{\mathcal{Q}(x)}(x').$$

Swapping integrals gives

$$(\star) \leq \frac{1}{|\lambda_i^\tau|} \int_{\mathbb{X}} |\varphi_i(x'')| \int_{\Sigma_{\mathcal{Q}}(\mathcal{Q}(x))} \left| p^\tau(x',x'') - p^\tau\big(q(x),x''\big)\right| d\mu_{\mathcal{Q}(x)}(x')\, dx'',$$

and with Hölder's inequality, $\|fg\|_{L^1} \leq \|f\|_{L^2_\mu} \|g\|_{L^2_{1/\mu}}$, we get

$$\leq \frac{1}{|\lambda_i^\tau|} \underbrace{\|\varphi_i\|_{L^2_\mu}}_{=1} \left\| \int_{\Sigma_{\mathcal{Q}}(\mathcal{Q}(x))} \left| p^\tau(x',\cdot) - p^\tau\big(q(x),\cdot\big)\right| d\mu_{\mathcal{Q}(x)}(x') \right\|_{L^2_{1/\mu}}.$$

Applying triangle inequality and using $p^\tau\big(q(x),\cdot\big) = \mathcal{Q}(x)$ gives

$$\begin{aligned}(\star) &\leq \frac{1}{|\lambda_i^\tau|} \int_{\Sigma_{\mathcal{Q}}(\mathcal{Q}(x))} \left\| p^t(x',\cdot) - p^t\big(q(x),\cdot\big)\right\|_{L^2_{1/\mu}} d\mu_{\mathcal{Q}(x)}(x') \\ &= \frac{1}{|\lambda_i^\tau|} \int_{\Sigma_{\mathcal{Q}}(\mathcal{Q}(x))} \| p^t(x',\cdot) - \underbrace{\mathcal{Q}(x)}_{=\mathcal{Q}(x')} \|_{L^2_{1/\mu}} d\mu_{\mathcal{Q}(x)}(x').\end{aligned}$$

By our assumption, this integral is at most ε. Hence,

$$(\star) \leq \frac{\varepsilon}{|\lambda_i^\tau|}.$$

As the last step, we can now construct from $\mathcal{Q}$ an r-dimensional RC that meets the condition (2):

Corollary 1. *Let the system be weakly (ε, r, τ)-reducible. Let $\mathcal{E} : \mathbb{M} \to \mathbb{R}^r$ be any parametrization of the transition manifold $\mathbb{M}$. Then for the RC*

$$\xi : \mathbb{X} \to \mathbb{R}^r, \quad x \mapsto \mathcal{E}\big(\mathcal{Q}(x)\big) \tag{10}$$

and the eigenpairs $(\lambda_i^\tau, \varphi_i)$ of $\mathcal{T}^\tau$ holds

$$\|\Pi_\xi \varphi_i - \varphi_i\|_{L^2_\mu} \leq \frac{2\varepsilon}{|\lambda_i^\tau|}. \tag{11}$$

Proof. Let $\tilde{\varphi}_i : \mathbb{M} \to \mathbb{R}$ as in the proof of Theorem 3, and define $\hat{\varphi}_i : \mathbb{Y} \to \mathbb{R}$ via

$$\hat{\varphi}_i(y) := \tilde{\varphi}_i\big(\mathcal{E}^{-1}(y)\big).$$

Note that for any $x \in \mathbb{X}$ holds $\Sigma_\xi\big(\xi(x)\big) = \Sigma_{\mathcal{Q}}(\mathcal{Q}(x))$. Thus,

$$\begin{aligned}\int_{\Sigma_\xi(\xi(x))} \big|\varphi_i(x') - (\hat{\varphi}_i \circ \xi)(x')\big|\, d\mu_y(x') &= \int_{\Sigma_{\mathcal{Q}}(\mathcal{Q}(x))} \big|\varphi_i(x') - (\tilde{\varphi}_i \circ \mathcal{Q})(x')\big|\, d\mu_{\mathcal{Q}(x)}(x') \\ &\leq \frac{\varepsilon}{|\lambda_i^\tau|},\end{aligned}$$

where the last inequality is Theorem 3. The assertion now follows from Theorem 2. □

If $(\lambda_i^\tau, \varphi_i)$ is dominant, i.e., $\lambda_i^\tau \approx 1$, then the projection error (11) is small. In that case, $\xi : x \mapsto \mathcal{E}\big(\mathcal{Q}(x)\big)$ is indeed a good RC, by Theorem 1.

Remark 4. Any RC of form (10) is called an *ideal RC* [6]. As in practice, however, neither the projection $\mathcal{Q}$ nor the parametrization $\mathcal{E}$ of $\mathbb{M}$ are known, this RC cannot be computed analytically. Instead, for strongly reducible systems, an approximate parametrization of $\mathbb{M}$ is computed by applying manifold learning methods to a finite sample of the fuzzy TM $\widetilde{\mathbb{M}}$ [4–6]. Our ongoing efforts to extend these techniques to the newly-identified weak reducibility condition will be discussed in the outlook in Sect. 5.

4 Numerical Example: A Weakly Reducible System

In order to compare the strong and weak reducibility condition, we consider a simple two-dimensional metastable system that possesses a one-dimensional RC. This system, originally considered in [22], is governed by an overdamped Langevin equation of form (9), where the potential energy function V is given by

$$V(x) = (x_1^2 - 1)^2 + 10(x_1^2 + x_2 - 1)^2.$$

We choose the inverse temperature $\beta = 1$, and consider the system on the domain $\mathbb{X} = [-2, 2] \times [-2, 2]$ (though no boundary conditions have been enforced in the following computations). The potential V, depicted in Fig. 2(a), possesses two local minima in the states $A = (-1, 0)$ and $B = (1, 0)$. The reaction in question hence is the transition from the area around one minimum (without loss of generality state A) to the other (state B). The minimum energy pathway (MEP) [24], which in the zero temperature limit supports almost all reactive trajectories [30], is indicated by the white dashed line.

The spectrum of $\mathcal{T}^\tau$ for $\tau = 0.5$, computed by a Ulam method [39] from a long, equilibrated trajectory of the system, exhibits a spectral gap after λ_1 (Fig. 2(b)). The relevant reaction, i.e., the transition between the two metastable sets, is associated primarily with the process on the dominant timescale t_1.

The (MEP) of the potential is given by the set

$$A_{\text{MEP}} = \{(x_1, x_2) \in \mathbb{X} \mid x_2 = 1 - x_1^2\}.$$

Intuitively, the manifold

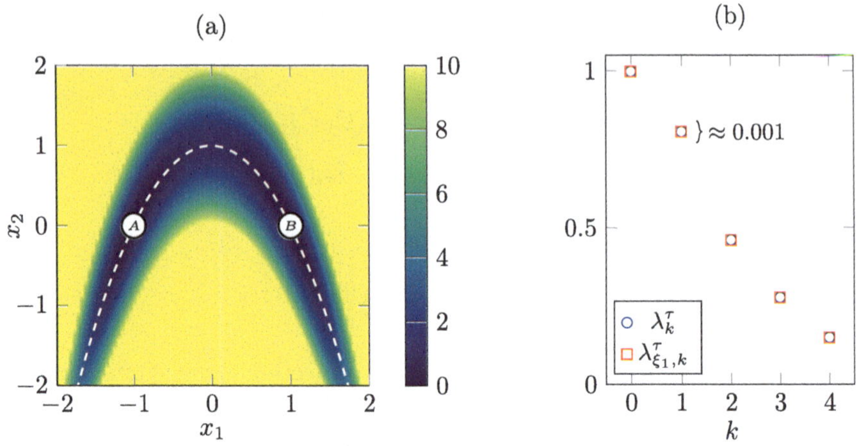

Fig. 2. (a) Energy potential of a two-dimensional drift-diffusion system. The reaction of interest here is the transition between the two local minima. (b) Eigenvalues of the full transfer operator $\mathcal{T}^\tau$ and of the effective transfer operator $\mathcal{T}_{\xi_1}^\tau$ projected onto the computed RC ξ_1.

$$\mathbb{M}_{\text{MEP}} = \{p^\tau(x,\cdot) \mid x \in A_{\text{MEP}}\}$$

should constitute a good TM. This statement should come with a warning: The intuition that the MEP allows to construct a good TM is wrong in general. There are many cases where the relevant transition pathways are completely different from the MEPs of the underlying system, mainly because for finite temperatures all statistically relevant transition paths concentrate in regions not close to the MEP and only converge to the MEP in the limit of zero temperature. In the case considered herein, however, relevant transition paths concentrate around the MEP even for finite temperatures.

Before quantitatively assessing whether or not $\mathbb{M}_{\text{MEP}}$ is indeed is a good TM, we visualize the fuzzy TM of the system, i.e., the set $\widetilde{\mathbb{M}} = \{p^\tau(x,\cdot) \mid x \in \mathbb{X}\}$. As $\widetilde{\mathbb{M}}$ lies in the function space L^1, it first needs to be embedded into a (finite-dimensional) Euclidean space. This is done by computing the mean of every $p^\tau(x,\cdot) \in \widetilde{\mathbb{M}}$ via the function $\mathbf{m} : L^1 \to \mathbb{R}^2$,

$$\mathbf{m}\big(p^\tau(x,\cdot)\big) := \int_{\mathbb{X}} x' p^\tau(x,x')\,dx'. \tag{12}$$

The set $\mathbf{m}(\widetilde{\mathbb{M}})$ then serves as the Euclidean embedding[2] of $\widetilde{\mathbb{M}}$.

[2] While for general dynamics $\mathbf{m}$ is not an embedding of the fuzzy TM in the strict topological sense, we conjecture that in this system, no two transition densities $p^\tau(x_1,\cdot), p^\tau(x_2,\cdot)$ possess the same mean, and hence that $\mathbf{m}$ is homeomorphic on $\widetilde{\mathbb{M}}$ and its image. Still, we neither formally confirm this, nor assess the distortion of $\widetilde{M}$ under $\mathbf{m}$, and hence $\mathbf{m}(\widetilde{\mathbb{M}})$ as a replacement for $\widetilde{M}$ should be handled with care.

Furthermore, as $\mathbf{m}(\widetilde{\mathbb{M}})$ is an infinite set, only a finite subsample can be visualized. For this we draw a large number, specifically $N = 8000$, of starting points $\{x_1, \ldots, x_N\}$ uniformly from $\mathbb{X}$ and for each x_k compute $\mathbf{m}_k := \mathbf{m}(p^\tau(x_k, \cdot))$. Here the integral in (12) is approximated via Monte Carlo quadratur, i.e., for $M \gg 1$,

$$\mathbf{m}(p^\tau(x_k, \cdot)) \approx \frac{1}{M} \sum_{l=1}^{M} z_k^{(l)}, \tag{13}$$

where the $z_k^{(l)}$ are samples of the density $p^\tau(x_k, \cdot)$. These were computed numerically by an Euler-Maruyama integrator of (9), starting in x_k, with a different random seed for each $l = 1, \ldots, M$.

The points $\mathbf{m}_k$ are shown in Fig. 3. We observe that most of the $\mathbf{m}_k$ lie close to a parabola-like structure, though there appear to exist systematic outliers, associated with starting points from the high energy regions in the lower part of $\mathbb{X}$. The maximum distance is assumed by the starting point $x^* = (0, -2)$. The parabola is exactly the Euclidean embedding of $\mathbb{M}_{\text{MEP}}$, which is also shown in Fig. 3.

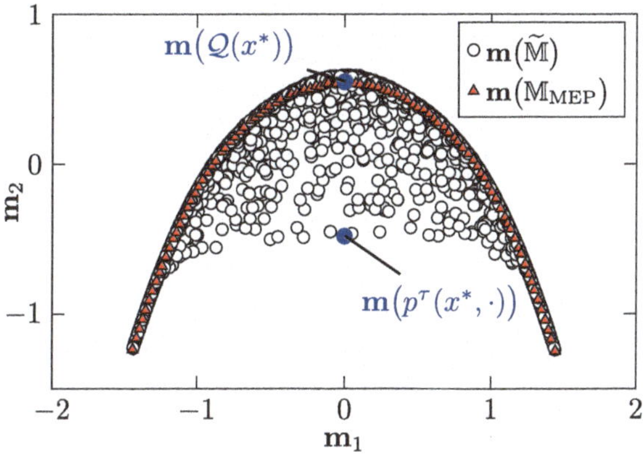

Fig. 3. Euclidean embeddings via the mean embedding function $\mathbf{m}$ of the fuzzy TM $\widetilde{\mathbb{M}}$, and the TM based on the minimum energy pathway, $\mathbb{M}_{\text{MEP}}$. Shown are $N = 8000$ sample points of $\mathbf{m}(\widetilde{\mathbb{M}})$, and $N = 100$ sample points of $\mathbf{m}(\mathbb{M}_{\text{MEP}})$. $\mathbf{m}(\widetilde{\mathbb{M}})$ appears to cluster around $\mathbf{m}(\mathbb{M}_{\text{MEP}})$, except for outliers from the high energy regions below the MEP.

However, the outliers prevent $\mathbb{M}_{\text{MEP}}$ from being a good strong TM by Definition 2. To be precise, for the point $x^* = (0, -3)$, we get for the distance in (7)

$$\left\| \mathcal{Q}(x^*) - p^t(x^*, \cdot) \right\|_{L^2_{1/\mu}} \approx 2.5, \tag{14}$$

where again finite samples of $\widetilde{\mathbb{M}}$ and $\mathbb{M}_{\text{MEP}}$, and kernel density estimations of the $p^t(x, \cdot)$ were used in the computation. Using (14) as a lower bound for the eigenvalue approximation (2) via Theorems 2 and 1 is of course worthless, hence $\mathbb{M}_{\text{MEP}}$ is not a strong TM.

On the other hand, for the defining condition (8) of weak reducibility holds

$$\int_{\Sigma_{\mathcal{Q}}(\mathcal{Q}(x^*))} \left\| \mathcal{Q}(x') - p^\tau(x', \cdot) \right\|_{L^2_{1/\mu}} d\mu_{\mathcal{Q}(x^*)}(x') \approx 0.02 \tag{15}$$

for the problematic point x^*. Assuming this value is indeed an upper bound for (8), the system is weakly reducible with parameter $\varepsilon = 0.06$, and $\mathbb{M}_{\text{MEP}}$ is the corresponding weak TM. The eigenvalue error for λ_1^τ predicted by Theorems 2 and 1 then is

$$|\lambda_1^\tau - \lambda_{\xi,1}^\tau| \leq 0.06, \tag{16}$$

for any RC ξ of the form (10).

To confirm this error bound, we now construct such an RC. For this, a parametrization $\mathcal{E}$ of $\mathbb{M}_{\text{MEP}}$ must be chosen. Any such parametrization is sufficient, for simplicity we choose

$$\mathcal{E}\big(p^\tau(x, \cdot)\big) := x_1,$$

i.e., the map of $p^\tau(x, \cdot)$ onto the first component x_1 of its starting point x. Next, the projection $\mathcal{Q}$ of $\widetilde{\mathbb{M}}$ onto the TM $\mathbb{M}_{\text{MEP}}$ is required. In order to avoid the costly calculation of kernel density estimates for the large number of starting points, and to avoid the badly-conditioned scaling by the factor $1/\rho$, we replace the $L^2_{1/\rho}$ distance in (6) by the Euclidean distance between the mean-embedded densities, i.e., utilize

$$\widetilde{\mathcal{Q}}(x) = \operatorname*{arg\,min}_{f \in \mathbb{M}_{\text{MEP}}} \left\| \mathbf{m}(f) - \mathbf{m}\big(p^\tau(x, \cdot)\big) \right\|_2.$$

Numerically, this projection is approximated by choosing from the 100 sample points of $\mathbf{m}(\mathbb{M}_{\text{MEP}})$ that are shown in Fig. 3 the point of minimum distance from $\mathbf{m}(p^\tau(x, \cdot))$. The point $\mathbf{m}(p^\tau(x, \cdot))$ is here again computed via (13). While using the projection $\widetilde{\mathcal{Q}}$ instead of $\mathcal{Q}$ might slightly distort the computed RC, it will have a negative impact on the quality of the RC, so if the bound (16) holds for $\widetilde{\mathcal{Q}}$, it will hold for $\mathcal{Q}$ as well. Moreover, it has been shown in [5] that the $L^2_{1/\rho}$ distance is equivalent to the distance in certain embedding spaces.

The final RC is then given by $\xi_1 : x \mapsto \mathcal{E}\big(\widetilde{\mathcal{Q}}(p^\tau(x, \cdot))\big)$. By numerically evaluating ξ_1 at the 8000 sample points (where the $p^\tau(x, \cdot)$ are again approximated by finite samples) and interpolating the resulting values bilinearly, we receive a continuous RC on $\mathbb{X}$. Figure 4 shows the level plot of ξ_1. We see that the level sets of ξ_1 are essentially identical to those of the dominant eigenfunction φ_1, also shown in Fig. 4. This is not surprising, as ξ_1 is constructed to fulfill the requirements of Theorem 1, i.e., the dominant eigenfunctions are required to be almost invariant under averaging over the level sets of ξ_1. As there is only one dominant eigenfunction φ_1, and ξ_1 is also one-dimensional, this implies that the level sets of ξ_1 and φ_1 are almost identical. Note however that the precise ranges

(a)

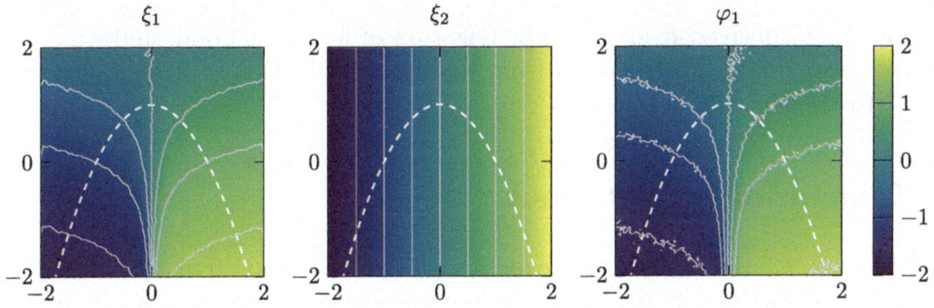

Fig. 4. Level plots of the RCs ξ_1 computed by the TM method, a naively-constructed RC ξ_2, as well as the dominant eigenfunction φ_1 of $\mathcal{T}^\tau$. We see that the level sets of ξ_1 and φ_1 are essentially identical.

of ξ and φ_1 are not necessarily identical, but strongly depend on the chosen parametrization $\mathcal{E}$.

The effective transfer operator $\mathcal{T}^\tau_{\xi_1}$ associated with ξ_1 can again be approximated by an Ulam method. Its leading eigenvalues, shown in Fig. 3(b), approximate the eigenvalues of the full transfer operator $\mathcal{T}^\tau$ very well. In particular, for the second dominant eigenvalue holds

$$|\lambda_1^\tau - \lambda_{\xi,1}^\tau| \approx 0.001.$$

As a consequence, the relaxation rate of the projected system $\xi_1(X_t)$, denoted σ_{ξ_1} and computed from $\lambda_{\xi,1}$ via (1), also approximate the rate of the full system σ_{full} very well; we have $\sigma_{\xi_1} \approx 0.43$, $\sigma_{\text{full}} \approx 0.43$. In contrast, projections onto other, naively chosen RCs, such as

$$\xi_2(x) := x_1,$$

seem to systematically over-estimate the equilibration rate, hence under-estimates the metastability of the system. Specifically, we have $\sigma_{\xi_2} \approx 0.46$. Reduced models built based on ξ_2 would therefore run the risk of equilibrating quicker than the full model by artificially increasing the number of transitions.

That said, the difference between $|\sigma_{\xi_1} - \sigma_{\xi_2}| \approx 0.03$ is rather small, so the naive RC ξ_2 can already be considered quite good. The reason is that at low temperatures the dynamics concentrates near the MEP, and here for each level set of ξ_2 there exists a level set of ξ_1 that is close (in the sense that the minimum pairwise point distance is small), and the RCs are both smooth. Still, the difference is measurable, and this causes the discrepancy.

Overall, this example confirms that

(1) the RC ξ_1 derived from a parametrization of $\mathbb{M}_{\text{MEP}}$ is good, and
(2) the error bound (16) derived from the characterization of $\mathbb{M}_{\text{MEP}}$ as a weak TM is reasonably accurate.

5 Conclusion and Outlook

In this work, we derived an improved and generalized characterization of good reaction coordinates for timescale-separated stochastic processes. We built upon a recently developed framework that constructs good RCs from parametrizations of the so-called transition manifold, a potentially low-dimensional manifold in the space of probability densities. We have shown that the criteria on the underlying system to possess such a manifold were overly strict, in the sense that certain systems with demonstrated good reaction coordinates do not possess a transition manifold by the old definition. We thus provided an alternative, relaxed definition of the transition manifold that is applicable to a larger class of systems, while still allowing the construction of good reaction coordinates.

One natural next step would be to implement the novel definition of weak TMs into a data-driven algorithm for the identification of good RCs. Unlike in the toy example from Sect. 4, the parametrization of the transition manifold (or of a suitable candidate) is not known analytically in practice. Instead, an approximate parametrization is identified by applying a nonlinear manifold learning algorithm to a large sample of $\widetilde{\mathbb{M}}$ (or a suitable embedding thereof) [4]. Many manifold learning algorithms, such as the diffusion maps algorithm [9] can be tuned to ignore outliers, which can be seen as a heuristic way weighing with respect to the invariant measure μ. A more rigorous approach however would be to directly implement the weighted distance (8) into the diffusion maps algorithm. This could be achieved by using the target measure-extension of diffusion maps [1], which at the same time allows one to estimate the in general unknown measure μ from data.

Acknowledgements. This research has been funded by Deutsche Forschungsgemeinschaft (DFG) through grant CRC 1114 "Scaling Cascades in Complex Systems", Project B03 "Multilevel coarse graining of multi-scale problems".

References

1. Banisch, R., Trstanova, Z., Bittracher, A., Klus, S., Koltai, P.: Diffusion maps tailored to arbitrary non-degenerate itô processes. Appl. Comput. Harmonic Anal. **48**(1), 242–265 (2020)
2. Baxter, J.R., Rosenthal, J.S.: Rates of convergence for everywhere-positive Markov chains. Stat. Probab. Lett. **22**(4), 333–338 (1995)
3. Best, R.B., Hummer, G.: Reaction coordinates and rates from transition paths. Proc. Natl. Acad. Sci. **102**(19), 6732–6737 (2005)

4. Bittracher, A., Banisch, R., Schütte, C.: Data-driven computation of molecular reaction coordinates. J. Chem. Phys. **149**(15), 154103 (2018)
5. Bittracher, A., Klus, S., Hamzi, B., Koltai, P., Schütte, C.: Dimensionality reduction of complex metastable systems via kernel embeddings of transition manifolds. arXiv eprint 1904.08622 (2019)
6. Bittracher, A., Koltai, P., Klus, S., Banisch, R., Dellnitz, M., Schütte, C.: Transition manifolds of complex metastable systems: theory and data-driven computation of effective dynamics. J. Nonlinear Sci. **28**(2), 471–512 (2017)
7. Chodera, J.D., Noé, F.: Markov state models of biomolecular conformational dynamics. Curr. Opin. Struct. Biol. **25**, 135–144 (2014)
8. Ciccotti, G., Kapral, R., Vanden-Eijnden, E.: Blue moon sampling, vectorial reaction coordinates, and unbiased constrained dynamics. Chem. Phys. Chem. **6**(9), 1809–1814 (2005)
9. Coifman, R.R., Lafon, S.: Diffusion maps. Appl. Comput. Harmonic Anal. **21**(1), 5–30 (2006)
10. Daldrop, J.O., Kappler, J., Brünig, F.N., Netz, R.R.: Butane dihedral angle dynamics in water is dominated by internal friction. Proc. Natl. Acad. Sci. **115**(20), 5169–5174 (2018)
11. Dellnitz, M., Junge, O.: On the approximation of complicated dynamical behavior. SIAM J. Numer. Anal. **36**(2), 491–515 (1999)
12. Elber, R., Bello-Rivas, J.M., Ma, P., Cardenas, A.E., Fathizadeh, A.: Calculating Iso-committor surfaces as optimal reaction coordinates with milestoning. Entropy **19**(5) (2017)
13. Evans, L.C., Gariepy, R.F.: Measure Theory and Fine Properties of Functions. Chapman and Hall/CRC, New York (2015)
14. Froyland, G., Gottwald, G., Hammerlindl, A.: A computational method to extract macroscopic variables and their dynamics in multiscale systems. SIAM J. Appl. Dyn. Syst. **13**(4), 1816–1846 (2014)
15. Froyland, G., Gottwald, G.A., Hammerlindl, A.: A trajectory-free framework for analysing multiscale systems. Physica D **328**, 34–43 (2016)
16. Gesùa, G.D., Lelièvre, T., Peutreca, D.L., Nectouxa, B.: Jump markov models and transition state theory: the quasi-stationary distribution approach. Faraday Discuss. **195**, 469–495 (2016)
17. Givon, D., Kupferman, R., Stuart, A.: Extracting macroscopic dynamics: model problems and algorithms. Nonlinearity **17**(6), R55–R127 (2004)
18. Kappler, J., Daldrop, J.O., Bruenig, F.N., Boehle, M.D., Netz, R.R.: Memory-induced acceleration and slowdown of barrier crossing. J. Chem. Phys. **148**, 014903 (2018)
19. Klus, S., Nüske, F., Koltai, P., Wu, H., Kevrekidis, I., Schütte, C., Noé, F.: Data-driven model reduction and transfer operator approximation. J. Nonlinear Sci. **28**, 985–1010 (2018)
20. Laio, A., Gervasio, F.L.: Metadynamics: a method to simulate rare events and reconstruct the free energy in biophysics, chemistry and material science. Rep. Prog. Phys. **71**(12), 126601 (2008)
21. Lasota, A., Mackey, M.C.: Chaos, Fractals, and Noise: Stochastic Aspects of Dynamics, vol. 97. Springer, Berlin (2013)
22. Legoll, F., Lelièvre, T.: Effective dynamics using conditional expectations. Nonlinearity **23**(9), 2131 (2010)
23. Ma, A., Dinner, A.R.: Automatic method for identifying reaction coordinates in complex systems. J. Phys. Chem. B **109**(14), 6769–6779 (2005)

24. Maragliano, L., Fischer, A., Vanden-Eijnden, E., Ciccotti, G.: String method in collective variables: minimum free energy paths and isocommittor surfaces. J. Phys. Chem. **125**(2), 024106 (2006)
25. McGibbon, R.T., Husic, B.E., Pande, V.S.: Identification of simple reaction coordinates from complex dynamics. J. Phys. Chem. **146**(4), 44109 (2017)
26. Noé, F., Nüske, F.: A variational approach to modeling slow processes in stochastic dynamical systems. Multiscale Model. Simul. **11**(2), 635–655 (2013)
27. Pavliotis, G.A., Stuart, A.M.: Multiscale Methods: Averaging and Homogenization. Springer, New York (2008)
28. Peters, B.: Reaction coordinates and mechanistic hypothesis tests. Annu. Rev. Phys. Chem. **67**(1), 669–690 (2016). PMID: 27090846
29. Peters, B., Trout, B.L.: Obtaining reaction coordinates by likelihood maximization. J. Chem. Phys. **125**(5), 054108 (2006)
30. Ren, W.: Higher order string method for finding minimum energy paths. Commun. Math. Sci. **1**(2), 377–384 (2003)
31. Sarich, M., Noé, F., Schütte, C.: On the approximation quality of Markov state models. Multiscale Model. Simul. **8**(4), 1154–1177 (2010)
32. Sarich, M., Schütte, C.: Approximating selected non-dominant timescales by Markov state models. Commun. Math. Sci. **10**(3), 1001–1013 (2012)
33. Schervish, M.J., Carlin, B.P.: On the convergence of successive substitution sampling. J. Comput. Graph. Stat. **1**(2), 111–127 (1992)
34. Schütte, C., Fischer, A., Huisinga, W., Deuflhard, P.: A direct approach to conformational dynamics based on hybrid Monte Carlo. J. Comput. Phys. **151**(1), 146–168 (1999)
35. Sengupta, U., Carballo-Pacheco, M., Strodel, B.: Automated markov state models for molecular dynamics simulations of aggregation and self-assembly. J. Chem. Phys. **150**(11), 115101 (2019)
36. Sirur, A., De Sancho, D., Best, R.B.: Markov state models of protein misfolding. J. Chem. Phys. **144**(7), 075101 (2016)
37. Smith, P.E.: The alanine dipeptide free energy surface in solution. J. Chem. Phys. **111**(12), 5568–5579 (1999)
38. Torrie, G.M., Valleau, J.P.: Nonphysical sampling distributions in Monte Carlo free-energy estimation: umbrella sampling. J. Comput. Phys. **23**(2), 187–199 (1977)
39. Ulam, S.: A Collection of Mathematical Problems. Interscience Tracts in Pure and Applied Mathemtics, vol. 8. Interscience Publishers, New York (1960)
40. Wedemeyer, W.J., Welker, E., Scheraga, H.A.: Proline cis-trans isomerization and protein folding. Biochemistry **41**(50), 14637–14644 (2002)
41. Williams, M.O., Kevrekidis, I.G., Rowley, C.W.: A data-driven approximation of the koopman operator: extending dynamic mode decomposition. J. Nonlinear Sci. **25**(6), 1307–1346 (2015)
42. Zhang, W., Hartmann, C., Schütte, C.: Effective dynamics along given reaction coordinates, and reaction rate theory. Faraday Discuss. **195**, 365–394 (2016)
43. Zhang, W., Schuette, C.: Reliable approximation of long relaxation timescales in molecular dynamics. Entropy **19**(7), 367 (2017)
44. Zwanzig, R.: Memory effects in irreversible thermodynamics. Phys. Rev. **124**, 983–992 (1961)

Analysis and Simulation of Extremes and Rare Events in Complex Systems

Meagan Carney[1], Holger Kantz[1], and Matthew Nicol[2](✉)

[1] Max Planck Institute for the Physics of Complex Systems, Nöthnitzer Str. 38, 01187 Dresden, Germany
{meagan,kantz}@pks.mpg.de

[2] Department of Mathematics, University of Houston, Houston, TX 77204-3008, USA
nicol@math.uh.edu

Abstract. Rare weather and climate events, such as heat waves and floods, can bring tremendous social costs. Climate data is often limited in duration and spatial coverage, and so climate forecasting has often turned to simulations of climate models to make better predictions of rare weather events. However very long simulations of complex models, in order to obtain accurate probability estimates, may be prohibitively slow. It is an important scientific problem to develop probabilistic and dynamical techniques to estimate the probabilities of rare events accurately from limited data. In this paper we compare four modern methods of estimating the probability of rare events: the generalized extreme value (GEV) method from classical extreme value theory; two importance sampling techniques, geneaological particle analysis (GPA) and the Giardina-Kurchan-Lecomte-Tailleur (GKLT) algorithm; as well as brute force Monte Carlo (MC). With these techniques we estimate the probabilities of rare events in three dynamical models: the Ornstein-Uhlenbeck process, the Lorenz '96 system and PlaSim (a climate model). We keep the computational effort constant and see how well the rare event probability estimation of each technique compares to a gold standard afforded by a very long run control. Somewhat surprisingly we find that classical extreme value theory methods outperform GPA, GKLT and MC at estimating rare events.

1 Extremes and Rare Event Computation.

Rare weather and climate events such as heat waves, floods, hurricanes, and the like, have enormous social and financial consequences. It is important to be able to estimate as accurately as possible the probability of the occurrence and duration of such extreme events. However the time series data available to predict rare events is usually too short to assess with reasonable confidence the probability of events with very long recurrence times, for example on the

O. Junge et al. (Eds.): SON 2020, SSDC 304, pp. 151–182, 2020.
https://doi.org/10.1007/978-3-030-51264-4_7

order of decades or centuries. In this regard, one may consider return levels of exceedances which represent the level that is expected to be exceeded on average, say, once every 100 years by a process. For example, a 100-year return level estimate of a time series of temperature or precipitation data would tell us the temperature or amount of precipitation that is expected to be observed only once in 100 years. It is common, however, that the amount of weather data available is limited in spatial density and time range. As a result, climate forecasting has often turned to simulations of climate models to make better predictions of rare weather events. These simulations are not without limitations; a more accurate model requires a large amount of inputs to take into account most of the environmental factors which impact weather. With these more complex models, very long simulations may be required to obtain probability estimates of rare events with long return times. These simulations may be very slow and have motivated the study of statistical techniques which allow for more accurate rare event probability estimates with lower computational cost.

One approach to estimate the probability of rare events or extremes is to use classical extreme value theory, perhaps aided by clustering techniques or other statistical approaches suitable for the application at hand. Other techniques to accurately estimate the probabilities of rare events include importance sampling (IS) methods. In general, importance sampling is a probabilistic technique which allows us to choose those trajectories or paths in a random or deterministic model which will most likely end in an extreme event. This reduces the number of long trajectories that are required to obtain an estimate on the tail probabilities of extremes and essentially changes the sampling distribution to make rare events less rare. The goal of importance sampling is not only to estimate probabilities of rare events with less computational cost, but also more accurately in that the ratio of the likely error in estimation to the probability of the event is lessened.

Importance sampling algorithms have been successfully applied in many fields, especially in chemical and statistical physics [3,26,28]. Recently these techniques have been applied to dynamical systems and dynamical climate models [27,29]. In this paper we will consider two similar types of IS techniques, geneaological particle analysis (GPA) and the Giardina-Kurchan-Lecomte-Tailleur (GKLT) algorithm. The GKLT algorithm is designed to estimate probabilities of events such as heatwaves as it considers time-averaged quantities. GKLT is motivated by ideas from large deviations theory, though in its implementation it does not explicitly require calculation of large deviation quantities such as rate functions.

The main goal of this paper is to compare the performance of the generalized extreme value (GEV) method with GPA, GKLT and brute force Monte Carlo (MC) at estimating rare events of our test models: the Ornstein-Uhlenbeck process, the Lorenz '96 system and PlaSim (a climate model). We keep the computational effort constant and see how well the rare event probability estimation of each technique compares to a gold standard afforded by a very long run control. Somewhat surprisingly we find that GEV outperforms GPA, GKLT and MC at estimating rare events. Perhaps this advantage comes from the fact that

GEV methods are parametric and maximum likelihood estimation, in practice, results in close to optimal parameters and confidence intervals.

2 The Four Methods

Extreme value theory is a well-established branch of statistics [5,8,23]. Over the last ten years or so the theory has been investigated in the setting of chaotic dynamics, for a state of the art review see [2, Chapters 4 and 6]. The goal of extreme value theory is to estimate probabilities associated to rare events. Another way to approach this problem is via importance sampling. Recently ideas from importance sampling have been successfully applied to several dynamical models (a non-exhaustive list includes [16,17,19,20]). How do the methods compare, for a given computational cost, at accurately determining the probabilities of rare events? We now describe the four methods we investigate in this paper.

2.1 Generalized Extreme Value Distribution (GEV)

There are two main approaches for classical extreme value theory: peaks over threshold; and the block maxima method. They are equivalent mathematically [5], but more research has been done on the block maxima method in the setting of deterministic models (for a treatment of this topic and further references see [2, Chapters 4 and 6]). We will use the block maxima method in this paper. In the context of modeling extremes in dynamical models, Galfi et al. [14] have used the peaks over threshold method to benchmark their large deviations based analysis of heat-waves and cold spells in the PUMA model of atmospheric circulation. Given a sequence of iid random variables $\{X_1, X_2, \ldots, X_n, \ldots\}$ it is known that the maxima process $M_n = \max\{X_1, X_2, \ldots, X_n\}$ has only three possible non-degenerate limit distributions under linear scaling: Types I (Gumbel), II (Fréchet) and III (Weibull) [13], no matter the distribution of X_1. By linear scaling we mean the choice of a sequence of constants A_n, B_n such that $P(A_n(M_n - B_n) \leq y) \to H(y)$ for a nondegenerate distribution H. The extreme value distributions are universal and play a similar role to that of the Gaussian distribution in explaining a wide variety of phenomena. These three distributions can be subsumed into a Generalized Extreme Value (GEV) distribution

$$G(x) = \exp\Big(- [1 + \zeta(\frac{x - \mu}{\sigma})]^{\frac{-1}{\zeta}} \Big) \ (*)$$

defined for $\{x : 1 + \zeta(\frac{x-\mu}{\sigma}) > 0\}$ with three parameters $-\infty < \mu < \infty$, $\sigma > 0$, $-\infty < \zeta < \infty$. The parameter μ is the location parameter, σ the scale and ζ the shape parameter (the most important parameter as ζ determines the tail behavior). A type I distribution corresponds to the limit as $\zeta \to 0$, while Type II corresponds to $\zeta > 0$ and Type III to $\zeta < 0$. The three types differ in the behavior of the tail of the distribution function F for the underlying process

(X_i). For type III the X_i are essentially bounded, while the tail of F decays exponentially for Type I and polynomially (fat tails) for Type II.

The advantage of using GEV over brute force fitting a tail distribution by simulation or data collection is that a statistical distribution is assumed, and only three parameters need to be determined (like fitting a normal distribution, where only 2 parameters need to be estimated). This has enormous advantages over methods which try to determine an a priori unknown form of distribution. The GEV parameters may be estimated, for example, by the method of maximum likelihood. Once the parameters are known $G(x)$ can be used to make predictions about extremes. This is done for a time series of observations in the following way. A sequence of observations are taken X_1, X_2, ... and grouped into blocks of length m (for example it could be daily rainfall amounts clumped into blocks of one year length). This gives a series of block maxima $M_{m,1}$, $M_{m,2}$, ... where $M_{m,\ell}$ is the maximum of the observations in block ℓ (which consists of m observations). Using parameter estimation like maximum likelihood, the GEV model is fitted to the sequence of $M_{m,\ell}$ to yield μ, σ and ζ. The probability of certain return levels of exceedance for the maximum of time-series of length m are obtained by inverting (*) and subtracting from 1. For example, m could correspond to a length of one year made of $m = 365$ daily rainfall data points, then the result is the level of rainfall a that the yearly maximum is expected to exceed once every $1/(1 - G(a))$ years.

One issue in the implementation of GEV is the possibly slow rate of convergence to the limiting distribution. There are some results [12,22] on rates of convergence to an extreme distribution for chaotic systems, but even in the iid case rates of convergence may be slow [21]. Another is the assumption of independence. Time-series from weather readings, climate models or deterministic dynamical systems are usually highly correlated. There are conditions in the statistical literature [6,11,15,23] under which the GEV distributional limit holds for maxima M_n of observables $\phi(X_j)$ which are "weakly dependent" i.e. the underlying X_j are correlated, and which ensure that M_n has the same extreme value limit law as an iid process with the same distribution function. Usually two conditions are given, called Condition D_2 (a decay of correlations requirement), and Condition D' (which quantifies short returns) which need to be checked. Collet [6] first used variants of Condition D_2 and Condition D' to establish return time statistics and extremes for certain dynamical systems. Recent results [2] have shown that maxima of time-series of Hölder observables on a wide variety of chaotic dynamical systems (Lorenz models, chaotic billiard systems, logistic-type maps and other classical systems) satisfy classical extreme value laws. The development of extreme value theory for deterministic dynamical systems has been an intensive area of research. For the current state of knowledge we refer to "Extremes and Recurrence in Dynamical Systems" [2, Chapters 4 and 6].

Even using a parametric model like GEV there is still an issue of having enough data. There are several approaches to extract the most information possible from given measurements. For example, in [1,4] sophisticated clustering techniques based on information theory ideas were used to group measurements from different spatial locations and amplify time-series of temperature recordings to improve the validity of GEV estimates for annual summer temperature extremes in Texas and Germany.

Despite these caveats this paper shows that GEV works very well in estimating probabilities of rare events in realistic models such as PlaSim, performing better at the same computational cost than MC and the two IS techniques we investigate.

2.2 Brute Force Monte Carlo

Given a random variable X distributed according to a distribution $\rho(x)$, we want to estimate the probability of a rare event,

$$\gamma_A = P(X \in A) \ll 1$$

As a naive approach, one could do this by a brute force Monte Carlo (MC) estimate,

$$\hat{\gamma_A}(N) = \frac{1}{N} \sum_{i=1}^{N} 1_A(X_i)$$

for some sequence of random variables X_i sampled from $\rho(x)$. Here, $E(\hat{\gamma_A}) = \gamma_A$ (as $\hat{\gamma_A}$ is an unbiased estimator) and for large enough N,

$$\sqrt{N}\hat{\gamma_A}(N) \sim \mathcal{N}(\gamma_A, \sigma^2(\hat{\gamma_A}))$$

by the central limit theorem (where valid). The relative error of an estimator is defined to be the standard deviation of the estimator divided by the estimated quantity. As

$$\sigma^2(1_A) = E((1_A(X) - \gamma_A)^2) = E(1_A(X)) - \gamma_A^2 = \gamma_A - \gamma_A^2 \approx \gamma_A$$

for small γ_A, and $\text{Var}\hat{\gamma_A}(N) = \frac{\text{Var}\gamma_A}{N}$, the relative error is estimated as

$$\sigma(\hat{\gamma_A})(N)/\gamma_A \approx \frac{1}{\sqrt{N\gamma_A}},$$

which is large for small γ_A. This analysis can be found in [16].

2.3 Importance Sampling Techniques

Importance sampling methods work to lower the relative error by a change of measure from ρ to another measure $\tilde{\rho}$. The idea is to change the distribution of X in a way that rare events are sampled more often under $\tilde{\rho}$ and if the steps in the algorithm by which we do this are accounted for, we obtain an accurate estimate of the probability of the rare event under ρ with a significantly decreased relative error in our estimator. In our applications X is a real-valued random variable (distance from the origin in Ornstein-Uhlenbeck process, energy level in the Lorenz'96 model and temperature or averaged temperature in the PlaSim model) and rare events will correspond to high values of X.

We alter the probability of rare events by using a weight function whose goal is to perform a change of measure. Provided X has tails which decay exponentially, the weight function can be chosen as an "exponential tilt". We now provide an illustration of the exponential tilt in the context of a normally distributed random variable. Details for the following estimates are provided in [16].

Suppose we want to estimate the probability γ_A of a rare event $A = \{X > a\}$ for $X \sim \mathcal{N}(0, 1)$ so that $\rho(X) = e^{-x^2/2}$. If we choose,

$$\tilde{\rho}(X) = \frac{\rho(X)e^{CX}}{E(e^{CX})} = \frac{1}{\sqrt{2\pi}}\exp[\frac{-(X-C)^2}{2}] \tag{1}$$

we obtain a shift of the average by C. The error of our estimate in the shifted distribution is given by its variance,

$$\sigma^2(\tilde{\gamma}_A) = P_{C,1}(X > a)e^{C^2} - \gamma_A^2$$

where $P_{C,1}$ denotes the probability under a normal distribution with mean C and variance 1. If we take $a = 2$ there is a unique minimum of the variance for a value of C close to 2. In this example a decrease of relative error by a factor of roughly 4 is produced. Because of the scaling $\frac{1}{\sqrt{N\gamma_A}}$ it would take a 16 times longer brute force run to achieve this result. We remark that this exponential tilt of the original distribution results in an optimal value of $C(a)$ for each threshold a for which $\gamma_A = P(X > a)$. Part of the finesse in using IS techniques is to tune the parameter C.

We now describe the two importance sampling techniques we investigate.

2.3.1 Genealogical Particle Analysis

Genealogical particle analysis (GPA) [16,17] is an importance sampling algorithm that uses weights to perform a change of measure, by a weight function $V(x)$ (in the previous example $V(x)$ was taken to be Cx but V(x) is application specific) applied to the original distribution of particles x_t under the dynamics. When we talk of particles we may mean paths in a Markov chain model or trajectories in a dynamical model such as the Ornstein-Uhlenbeck process or Lorenz'96. These weights can be thought of as measuring the performance of a particle's trajectory. If the particle is behaving as though it comes from the distribution tilted by the weight function $V(x)$ then it is cloned, otherwise it is killed and no longer used in the simulation. The act of killing or cloning based on weights is performed at specified time steps separated by length τ. We will refer to τ as the *resampling time*. In theory, the resampling time can chosen between the limits of the Lyapunov time, so as to not be too large that samples relax back to their original distribution and the decorrelation time, so as to not be too small that all clones remain close to each other. In practice, the decorrelation rate of a trajectory x_t under the dynamics is calculated as the autocorrelation taken over a time lag and the sampling time is then chosen as the smallest time lag that results in the autocorrelation of x_t being close to zero at a specified tolerance. A description of the algorithm is given below.

1. Initiate $n = 1, \ldots, N$ particles with different initial conditions.
2. For $i = 1, \ldots, T_f/\tau$ where T_f is the final integration time.
   2a. Iterate each trajectory from time $t_{i-1} = (i-1)\tau$ to time $t_i = i\tau$.
   2b. At time t_i, stop the simulation and assign a weight to each trajectory n given by,

$$W_{n,i} = \frac{\exp(V(x_{n,t_i}) - V(x_{n,t_{i-1}}))}{Z_i} \tag{2}$$

   where

$$Z_i = \frac{1}{N} \sum_{n=1}^{N} W_{n,i} \tag{3}$$

   is the normalizing factor that ensures the number of particles in each iteration remains constant.
   2c. Determine the number of clones produced by each trajectory,

$$c_{n,i} = \lfloor W_{n,i} + u_n \rfloor \tag{4}$$

   where $\lfloor \cdot \rfloor$ is the integer portion and u_n are random variables generated from a uniform distribution on $[0, 1]$.
   2d. The number of trajectories present after each iteration is given by,

$$N_i = \sum_{n=1}^{N} c_{n,i} \tag{5}$$

   Clones are used as inputs into the next iteration of the algorithm. For large N, the normalizing factor ensures the number of particles N_i remains constant; however, in practice the number of particles fluctuates slightly on each iteration i. To ensure N_i remains constant it is common to compute the difference $\Delta N_i = N_i - N$. If $\Delta N_i > 0$, then ΔN_i trajectories are randomly selected (without replacement) and killed. If $\Delta N_i < 0$, then ΔN_i trajectories are randomly selected (with replacement) and cloned.
3. Provided τ is chosen between the two bounds specified above, the final set of particles tends to the new distribution affected by $V(x)$ as $N \to \infty$,

$$\tilde{p}(x) = \frac{p(x)e^{V(x)}}{E(e^{V(x)})}. \tag{6}$$

   where $p(x)$ is the original distribution of the sequence of realizations $x_{n,0}$ and $\tilde{p}(x)$ is the distribution tilted by the weight function $V(x)$.

Probability estimates for rare events $\gamma_A = P(X > a)$ under $p(x)$ are obtained by the reversibility of the algorithm and dividing out the product of weight factors applied to the particles. Suppose A is the event $(X > a)$ for $X \sim p(x)$, then the expected value in the original distribution denoted by E_0 is given by [16],

$$\gamma_A = E_0(1_A) = \frac{1}{N}\sum_{n=1}^{N} 1_A(x_{n,T_f/\tau})e^{V(x_{n,0})}e^{-V(x_{n,T_f/\tau})}\prod_{i=1}^{T_f/\tau} Z_i \tag{7}$$

Since GPA weights consider the end distribution of particles, they result in a telescoping sum in the exponential where the final rare event estimate is a function of the first and last weight terms only. For a detailed proof of this equivalence, we refer the reader to [16]. For an illustration of this algorithm, see Fig. 1.

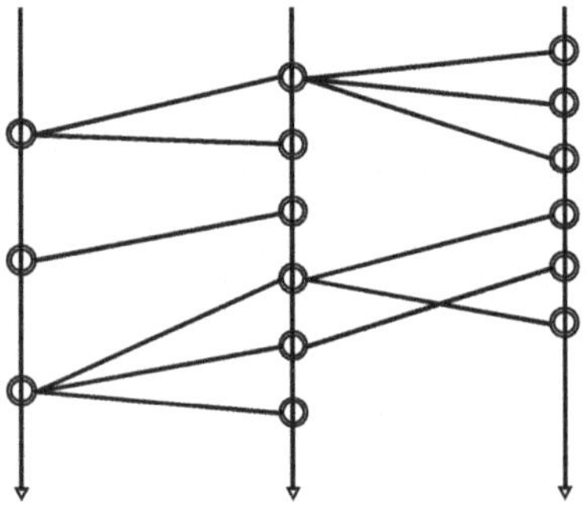

Fig. 1. Illustration of the GPA algorithm.

As seen above, the change of measure is completely determined by the choice of weight function $V(x)$ in the algorithm.

Furthermore, the algorithm can be applied to any observable ϕ by considering the continuous random variable $X_t = \phi(x_t)$ and defining

$$W_{n,i} = \frac{\exp(V(\phi(x_{n,t_i})) - V(\phi(x_{n,t_{i-1}}))}{Z_i}.$$

where $x_n(t)$ is one of our $n = 1, \ldots, N$ realizations and $x_{n,t_i} = x_n(t_i)$.

If we are interested in estimating rare event probabilities of a time-averaged quantity the weight $W_{n,i} = \frac{V(\int_{t_{i-1}}^{t_i} x_n(t)dt)}{Z_i}$ is given by an integral rather than the difference $W_{n,i} = \frac{\exp(V(x_{n,t_i})-V(x_{n,t_{i-1}}))}{Z_i}$ and the increments do not telescope. We next discuss a method, the GKLT algorithm, based on large deviations theory to estimate probabilities of rare events for time-averaged quantities in the next section. We note here that the GKLT algorithm in its implementation does not require explicit computation of large deviation quantities such as rate functions.

2.3.2 Giardina-Kurchan-Lecomte-Tailleur (GKLT) Algorithm

This technique was developed in a series of papers [7,19,20] and uses ideas from large deviations theory to make estimates of extremes for time-averaged quantities, for example heatwaves lasting a couple of months or more where the averaged maximal daily temperature over the two month period would be high. The advantage is that over long periods of averaging large deviation theory gives a method which works well, but a disadvantage is that the period of averaging needs to be long enough for the heuristic arguments involving the rate function and other quantities from large deviations theory to be valid. In practice, to calculate the probability of summer heatwave extremes in Europe, the duration of heatwaves has been set at the order of 90 to 120 days in the literature [14,18].

Suppose ϕ is an observable. We will consider time-averaged quantities $\frac{1}{T}\int_{t=jT}^{(j+1)T} \phi(x(t))\, dt$ over a fixed time-window of length T, $j = 1, \ldots, \lfloor T_f/T \rfloor$. We may choose to apply the weight function V to the integral of $n = 1, \ldots, N$ realizations $\phi(x_n(t))$ by defining the set of weights as,

$$W_{n,i} = \frac{V(\int_{t_{i-1}}^{t_i} \phi(x_n(t))\, dt)}{Z_i} \quad (8)$$

with normalizing factor,

$$Z_i = \frac{1}{N}\sum_{n=1}^{N} W_{n,i}$$

where the resampling time $\tau = t_{i-1} - t_i$ is chosen between the limits described in Sect. 2.3.1 and may differ from the choice of the time-average window length T.

Applying the method described in algorithm Sect. 2.3.1 equipped with Eq. 8 tilts the distribution of the integral $\int_{t_{i-1}}^{t_i} \phi(x(t))\, dt$ by $V(\cdot)$. As a result, the distribution of the T-time average trajectory $\frac{1}{T}\int_{t=jT}^{(j+1)T} \phi(x(t))\, dt$ is tilted in a similar way. For an illustration of this algorithm, see Fig. (2).

Since the weight is a function of segments of the trajectory (rather than the distribution of end particles), the telescoping property no longer holds and estimates in the original distribution require the reconstruction of N-backward trajectories $\hat{\phi}(x_n(t))$, $n = 1, \ldots, N$.

Let E_0 denote the expected value in the original distribution and suppose O is some functional of $\phi(x_n(t))$. Then it can be shown [18],

$$E_0(O(\{\phi(x_n(t))\}_{0\le t\le T_f})) \sim \frac{1}{N}\sum_{n=1}^{N} O(\{\hat{\phi}(x_n(t))\}_{0\le t\le T_f})e^{-V(\int_0^{T_f}\hat{\phi}(x_n(t))dt)}\prod_{i=1}^{T_f/\tau} Z_i. \quad (9)$$

Often, O in Eq. 9 is taken as some indicator function of a rare event so that, $E_0(O(\{\phi(x(t))\}_{0\le t\le T_f}))$ provides some rare event probability estimate. For example, to obtain the rare event probability estimate that the T-time averaged

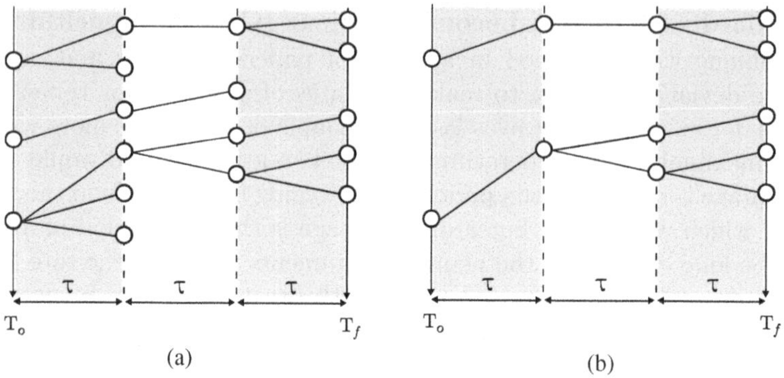

Fig. 2. (a) Illustration of the GKLT algorithm and (b) assembly of N backward trajectories. Although shifts in the distribution of the integral are defined by the resampling time τ, reconstruction of backward trajectories allows for estimates on T-time averaged trajectories after implementation of GKLT.

observable exceeds some threshold a, we may rewrite Eq. 9 as,

$$E_0\left(1_{\{\frac{1}{T}\int_{jT}^{(j+1)T}\phi(x(t))dt>a|0\leq j\leq\lfloor T_f/T\rfloor\}}(\phi(x(t)))\right)$$
$$\sim\frac{1}{N}\sum_{n=1}^{N}E\left(1_{\{\frac{1}{T}\int_{jT}^{(j+1)T}\hat{\phi}(x_n(t))dt>a|0\leq j\leq\lfloor T_f/T\rfloor\}}(\hat{\phi}(x_n(t)))\right)\cdot e^{-V(\int_0^{T_f}\hat{\phi}(x_n)(t)dt)}\prod_{i=1}^{T_f/\tau}Z_i \tag{10}$$

A consequence of Eq. 10 is that rare event probabilities $P(\Psi\circ\phi(x(t))>a)$ for any functional Ψ of the observed trajectory $\phi(x(t))$ can be calculated in a similar way.

Hence, rare event probabilities for longer time-averages can be estimated at no further computational expense. Different observables are considered in the next section. We end by remarking that a natural choice is to take $V(x)=Cx$, if the rare event consists of exceedance of a certain level.

3 Numerical Results

IS algorithms hinge on their ability to shift the sampling distribution of a system to increase the probability of the rare event. They open the possibility of reducing numerical cost while providing a more (or similarly) accurate estimate over a brute force method. Shifting of the sampling distribution relies on a convergence assumption to hold for a sufficiently large number N of initial particles. In [16] it is shown in certain models that the relative error (also a quantity relying on the number of initial particles N) is smaller for tail probability estimates obtained from IS methods if the shift is chosen optimally for a specific threshold. For a set of thresholds a_k, statistics on tail probabilities and return time estimates may be obtained by averaging over a set of trials, as in [18]. However, this requirement adds to the true numerical cost of the IS methods. Optimal values of a shift for any given threshold usually cannot be determined a priori. Moreover, the magnitude of a shift in the sampling distribution cannot be chosen arbitrarily because of its heavy dependence on the choice of observable, system and initial conditions. This dependence limits the algorithm in practice to smaller shift choices, larger errors and hence, less reliable return time estimates.

We compare numerical results from two well-known IS methods (GPA and GKLT) with GEV and MC under true numerical costs of obtaining statistical estimates for sequences of thresholds. In implementation of IS methods, we choose shifting values as large as possible to obtain accurate return-time estimates and illustrate the problems that occur with dependence on initial conditions. Following recent literature, we use the Ornstein-Uhlenbeck process as a benchmark for our work and expand to the more complex Lorenz'96 and PlaSim model. In all systems, we find that the GEV outperforms GPA, GKLT and MC under the same numerical cost.

3.1 The Generalized Extreme Value (GEV) Model for Numerical Comparison

3.1.1 GEV Model for Comparison to GPA Tail Estimates

Since the GPA algorithm considers only the distribution of end particles, tail probability estimates of a trajectory X_t are provided at a sampling rate of T_f intervals denoted $P(X_{T_f} > a_k)$ for a sequence of thresholds a_k. Recall that in the case of considering an observable under the dynamics, X_t can be seen as the random variable $X_t = \phi(x_t)$ where x_t is the trajectory under the dynamics at time t. To compare across methods, we use the same sampling rate for MC brute force and GEV modeling. Following standard literature, we may choose to consider one long trajectory X_t of length $\hat{N} \cdot T_f$, so that we obtain $\hat{N}$ samples taken at T_f intervals of X_t. From here, we define the subsequence of X_t taken at the sampling rate T_f to be $X_{\hat{j},T_f}$ for $\hat{j} = 1, \cdots, \hat{N}$. We may then define the block maxima over blocks of length m taken over our subsequence X_{i,T_f} by,

$$M_{\ell,m} = max_{\ell m \leq i \leq (\ell+1)m} \, X_{i,T_f}$$

such that the number of total block maxima is $\lfloor \hat{N}/m \rfloor$ and $\ell = 1, \cdots, \lfloor \hat{N}/m \rfloor$ and m is chosen at a length that ensures convergence of the block maxima. For the purposes of this paper, $m = 10$ and 100 were checked with m chosen as the value providing the best fit to the control.

Another option is to run many, say $\hat{N}$ again, trajectories $X_{\hat{i},t}$ for $\hat{i} = 1, \cdots, \hat{N}$ up to time T_f. We denote the sequence of end particles $X_{\hat{i},T_f}$ so that $X_{\hat{i},T_f}$ coincides with the appropriate fixed sampling rate T_f for each $\hat{i}$. Then, we may define the block maxima over blocks of length m by,

$$M_{\ell,m} = max_{\ell m \leq \hat{i} \leq (\ell+1)m} \, X_{\hat{i},T_f}$$

so that once again, $\ell = 1, \cdots, \lfloor \hat{N}/m \rfloor$ and the total number of block maxima is $\lfloor \hat{N}/m \rfloor$. In both cases, the distribution of $M_{\ell,m}$ is theoretically the same, however we choose the latter to lower numerical error which builds over long trajectories. An illustration of how the maxima are defined and their relationship to the GPA algorithm outcome can be seen in Fig. 3.

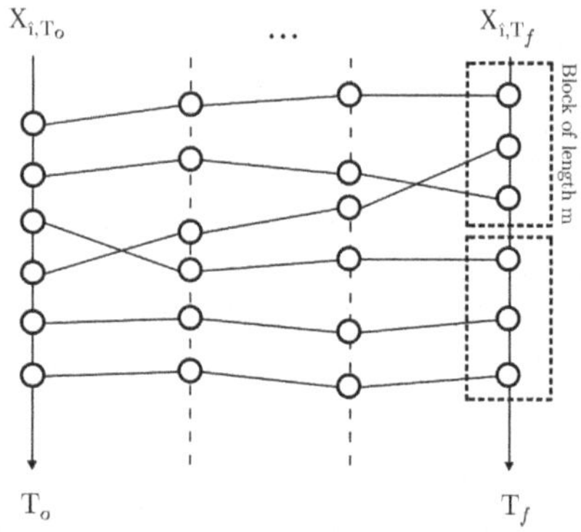

Fig. 3. Illustration of the block maxima for GEV to GPA comparison. Many trajectories are run under the dynamics up to the sampling time T_f and the final values are used to form the block maxima (indicated by dashed boxes).

Classical results for fitting a GEV to the sequence of block maxima $M_{\ell,m}$ require the sequence $X_{\hat{i},T_f}$ to be independent and stationary. The choice of $T_f \gg \tau$ ensures that samples taken at T_f intervals are nearly independent. We may fit the generalized extreme value (GEV) distribution $G(x)$ to the sequence $M_{\ell,m}$ by maximum likelihood estimation of the shape, ζ, scale σ, and location μ parameters [5, Section 3.3.2]. Independence assumptions on the sequence $X_{\hat{i},T_f}$ allows for reversibility of the probability estimates of the m-block maxima by the following relationship [5, Section 3.1.1],

$$G(x) = P(M_{\ell,m} \leq x) \approx (F(x))^m$$

where $G(x)$ is the GEV of the m-block maxima estimated by maximum likelihood estimation and $F(x)$ is the c.d.f. of the trajectory X_t sampled at a rate of T_f intervals. Hence,

$$P(X_{T_f} > x) \approx 1 - \sqrt[m]{G(x)} \tag{11}$$

In the event that independence of $X_{\hat{i},T_f}$ cannot be established, the dependence weaker conditions such as conditions D_2 and D', if valid, entail convergence of the sequence of m-block maxima to a GEV distribution.

3.1.2 GEV Model for Comparison to GKLT Tail Estimates

In the GKLT algorithm, we consider the distribution of the T-time averages created from the N-backward reconstructed trajectories $X_{n,t}$. That is, we consider the probability $P(A_T > a_k)$ that the T-time average, $A_T = \frac{1}{T}\int_0^T X(t)\,dt$ is greater than some threshold (or sequence of thresholds) a_k. Recall that $X_{n,t} = \phi(x_n(t))$ is some realization of a trajectory under the dynamics equipped with an observable ϕ. We run $\hat{N}$ trajectories under the dynamics up to time T_f and denote this sequence as $X_{\hat{i},t}$ for $0 \le t \le T_f$ and $\hat{i} = 1, \cdots, \hat{N}$. Then the sequence of (non-overlapping) T-time averages created from the set of trajectories $X_{\hat{i},t}$ is defined as,

$$A_{T,\hat{i},j} = \frac{1}{T}\int_{jT}^{(j+1)T} X_{\hat{i},t}\,dt$$

for $j = 1, \cdots, \lfloor T_f - T \rfloor$. For each fixed j, we define the sequence of maxima taken over blocks of length m

$$M_{h,j,m} = \max_{hm \le \hat{i} \le (h+1)m} A_{T,\hat{i},j}$$

for $h = 1, \cdots, \lfloor \lfloor T_f - T \rfloor / m \rfloor$ so that we have $\lfloor \lfloor T_f - T \rfloor / m \rfloor \cdot \hat{N}$ number of maxima in total. Defining the maxima over trajectories for every fixed time step j, rather than over time steps of a single (long) realization, allows us to keep the integration time small and minimize numerical error. Following our previous discussion, we may also choose to consider one long trajectory X_t, break it up into a sequence of non-overlapping T-time averages, and consider the sequence of maxima taken over blocks of length m taken from this long sequence of averages. Once again, we note that $T \ge \tau$ is chosen so that the sequence of averages is roughly independent. Hence, the GEV $G(x)$ can be fitted by maximum likelihood estimation to the sequence $M_{h,j,m}$. The independence of the sequence of T-time averages allows for reversibility of the probability estimates of the m-block maxima by,

$$G(x) = P(M_{h,j,m} \le x) \approx (F(x))^m$$

where $G(x)$ is the maximum likelihood estimate for the GEV model of the sequence of m-block maxima $M_{h,j,m}$ and $F(x)$ is the c.d.f. of the sequence of T-time averages taken from the trajectory X_t. Hence,

$$P(A_T > x) \approx 1 - \sqrt[m]{G(x)} \tag{12}$$

An illustration of how the block maxima in estimating the GEV are defined in terms of the sequence of T-time average trajectories for comparison to the GKLT algorithm can be found in Fig. 4.

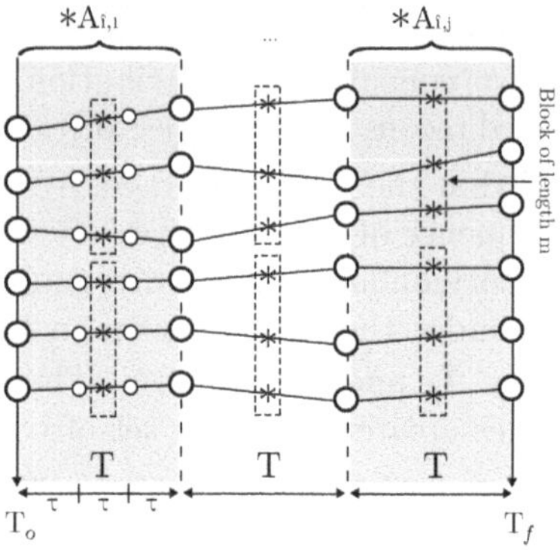

Fig. 4. Illustration of the block maxima for GEV to GKLT comparison. Many trajectories are run under the dynamics up to time T_f. T-time average sequences are calculated from the trajectories. For each fixed time step j, the block maxima (indicated by dashed boxes) are calculated. The τ interval is shown here to emphasize its difference to T and does **not** represent any weighting done to trajectories used in the GEV model.

3.1.3 Return Time Curves

We consider a long trajectory X_t such that X_t is sampled for over threshold probability estimates at time $T_f \geq \tau$ and a rare event threshold a such that $X_t < a$ for most times t. We define the return time $r(a)$ as the average waiting time between two statistically independent events exceeding the value a.

Following the scheme of [18] we divide the sequence X_t into pieces of duration ΔT and define $a_k = \max\{X_t | (k-1)\Delta T \leq t \leq k\Delta T\}$ and $s_k(a) = 1$ if $a_k > a$ and 0 otherwise. Then the number of exceedances of the maxima a_k over threshold a can be approximated by a Poisson process with rate $\lambda(a) = 1/r(a)$. Using the return time c.d.f. F_T^{-1} for the Poisson process, we have

$$F_T^{-1}(\frac{1}{K}\sum_{k=1}^{K} s_k(a)) = \frac{-\log(1 - \frac{1}{K}\sum_{k=1}^{K} s_k(a))}{\lambda(a)}$$

where $\frac{1}{K}\sum_{k=1}^{K} s_k(a) = F_T(\Delta T)$ is the probability of observing a return of the maxima a_k above threshold a in ΔT time steps. For any a_k we have an associated probability p_k. We denote the reordering of this sequence $(\hat{a}_k, \hat{p}_k)$ such that $\hat{a}_1 \geq \hat{a}_2 \geq \cdots \geq \hat{a}_K$. Then the return time is given by,

$$r(\hat{a}_k) = -\frac{1}{\log(1 - \sum_{k=m}^{K} \hat{p}_m)} \qquad (13)$$

where $\sum_{m=k}^{K} \hat{p}_m$ gives an approximation of the probability of exceeding threshold $\hat{a}_k$.

Return time plots estimated from outcomes of importance sampling methods are the result of first averaging return time estimates over a number of experiments for each C, then averaging over all C-return time plots. See Fig. (7) for an illustration. Only those return times corresponding to threshold values that fall within 1/2 standard deviation of the tilted distribution are used in this averaging. For the remainder of this paper, the term *experiment* will be used to describe a single run of an importance sampling algorithm.

3.2 Ornstein-Uhlenbeck Process

The Ornstein-Uhlenbeck process given by,

$$dx = -\lambda x dt + \sigma d\mathcal{W}$$

is a nice toy-example for importance sampling application because it is simple to implement, has low numerical cost, the distribution of position x is approximately Gaussian, and it's correlations decay exponentially. We use this process with $\lambda = 1$ and $\sigma = 1$ as a benchmark for the following numerical investigation.

3.2.1 GKLT

The GKLT importance sampling algorithm is performed on the Ornstein-Uhlenbeck process with $N = 100$ initial trajectories, resampling time $\tau = 0.1$, and a total integration time of $T_f = 2.0$. Here, the observable of interest is the position. At each time step of the algorithm, a new value of noise $\mathcal{W}$ is sampled from the standard normal distribution for each cloned trajectory to ensure divergence of the clones. Time average trajectories are calculated by averaging the $N = 100$ backward-reconstructed trajectories over time-windows of length $T = 0.25$ with step size equal to T so that no window has overlapping values.

Above threshold probabilities of the T-time average position $P(A_T > a_k)$ where $A_T = \frac{1}{T}\int_0^T x(t)\, dt$ are estimated for $C = [0.01, 0.03, 0.05, 0.07]$. We define the sequence of T-time averages obtained from realizations $\hat{\phi}(x_n(t))$ of the N-backward reconstructed trajectories as,

$$A_{n,j} = \frac{1}{T}\int_{jT}^{(j+1)T} \hat{\phi}(x_n(t))dt, \tag{14}$$

for $j = 1, \cdots, \lfloor T_f/T \rfloor$. Then the probability estimate for $P(A_T(t) > a)$ above a threshold a from Eq. 10 is given as,

$$E_0(1(x(t))_{\{A_T > a | 0 \le t \le T_f - T\}}) \sim \frac{1}{N}\sum_{n=1}^{N} E(1(\hat{\phi}(x_n(t))_{\{A_{n,j} > a | j=1 \cdots \lfloor T_f/T \rfloor\}})e^{-C(\int_0^{T_f} \hat{\phi}(x_n(t))dt)} \prod_{i=1}^{T_f/\tau} Z_i.$$

This approach results in a unique probability estimate for each predefined threshold a.

Return times are estimated for each value C and sequence of thresholds a_k by Eq. 13 resulting in four return time curves. We perform 100 experiments under these conditions for a total of 400 return time curves and average to obtain the result shown in Fig. (5). This process is illustrated in Fig. (7). The total numerical cost for this estimate is $4 \cdot 10^4$. Monte Carlo (MC) brute force and generalized extreme value (GEV) (Eq. 11) probability estimates are obtained through numerical costs of the same order. We find that GEV and MC brute force methods outperform GKLT by providing estimates of return times longer than $1 \cdot 10^6$.

Another option is to define the sequence corresponding to the maximum T-time average quantity of a single realization $\hat{\phi}(x_n)$ given by,

$$a_n(T) = \max_{1 \leq j \leq \lfloor T_f/T \rfloor} \frac{1}{T} \int_{jT}^{(j+1)T} \hat{\phi}(x_n(t))dt. \tag{15}$$

This results in a sequence of maximum thresholds $a_n(T)$, one per each realization of $\hat{\phi}(x_n(t))$. For each threshold $a_n(T)$, there exists an associated probability estimate,

$$p_n = \frac{1}{N} e^{-C \int_0^{T_f} \hat{\phi}(x_n(t))dt} \prod_{i=1}^{T_f/\tau} Z_i,$$

which is the result of plugging the threshold values of Eq. 15 into Eq. 10 and noting that,

$$E(1(x_n(t))_{\{\frac{1}{T} \int_{jT}^{(j+1)T} \hat{\phi}(x_n(t))dt > a_n(t,T) | 0 \leq t \leq T_f - T\}}) = 1.$$

The sequence $(a_n(T), p_n)$ for $1 \leq n \leq N$ is then reordered for decreasing values of a_n. We denote the ranked sequence $(\hat{a}_n(t), \hat{p}_n)$ where $\hat{a}_1 \geq \hat{a}_2 \geq \cdots \geq \hat{a}_N$ and associate a return time $r(\hat{a}_n)$ defined by Eq. 13 using the reordered sequence $\hat{p}_n$. We refer to [18] for more details on this approach. Return time curves are then obtained by linearly interpolating the pair $(\hat{a}_n(T), r_n(\hat{a}_n))$ over an equal spaced vector of return times. GKLT is run with the same initial conditions as stated above. We refer to Fig. (6) for this discussion. Choosing to calculate return time curves in this way allows for estimates of longer times; however, this tends to be at the expense of accuracy. Equation 14 allows for more control over the choice of range of thresholds included from the shifted distribution.

GEV and MC estimates are obtained through numerical costs of the same order. Deviation statistics for GKLT, GEV, and MC methods, represented by dashed lines in Fig. (6), are calculated by finding the minimum and maximum deviation in 100 experiments. Solid lines about the GEV represent the 95% confidence intervals coming from the likelihood function for the GEV estimated from the corresponding MC simulation. We compare all results against a long control run of order $1 \cdot 10^6$. We find that GEV and GKLT methods provide more accurate estimates of return times longer than $1 \cdot 10^5$ compared to the MC method. Moreover, the GEV outperforms the GKLT algorithm by providing surprisingly accurate return time estimates with smaller deviation for all thresholds except in a small fraction of cases.

A possible explanation for the poor performance of the GKLT algorithm comes from the fact that the tilting coefficient C cannot be chosen arbitrarily large to obtain longer return time estimates without some change in the initial conditions (e.g. integration time, number of starting trajectories). Large choices of C result in a lower number of parent trajectories (as many copies are killed) which causes the tilted distribution to breakdown Fig. (8). This breakdown results in increasingly inaccurate return time estimates, even for thresholds sitting close to the center of the tilted distribution.

3.2.2 GPA

The GPA importance sampling algorithm is performed with $N = 100$ starting trajectories, resampling time $\tau = 0.1$, and a total integration time of $T_f = 2.0$. The final trajectories X_{n,T_f} from GPA with tilting constants $C = [2, 3, 4]$ are used to estimate the above threshold probabilities $P(X_{T_f} > a_k)$ and return time curves. To begin, we perform 10 experiments, with the initial conditions described above, resulting in a total of 30 return time curves (10 experiments for each value of C) and average to obtain the result shown in Fig. (9). The total numerical cost for this estimate is $3 \cdot 10^3$ compared to the long control run of $1 \cdot 10^6$. We find that GPA and GEV methods provide nearly equivalent results for return times up to $1 \cdot 10^4$ with GPA and GEV methods outperform Monte Carlo brute force estimates for return times longer than $1 \cdot 10^4$. On average, GPA provides a slightly closer approximation to the control curve than that of the GEV method for longer return times; however, the deviation of this estimate is much larger than that of GEV.

Next, we consider larger values of C to test whether reliable estimates can be obtained for thresholds exceeding the control run. We run 30 experiments for 10 different values of $C = [1, 2, \ldots, 10]$ under the same initial conditions as stated above for a total numerical cost of $3 \cdot 10^4$. We average the resulting return time curves shown in Fig. (7) to obtain the final return time plot Fig. (10). As seen in the estimates for GKLT, higher values of C with unchanged initial conditions provide less accurate return-time results even for those thresholds which sit at the center (e.g. have the highest probability of occurrence) of the tilted distribution. On the other hand, GEV methods with the same numerical cost of $3 \cdot 10^4$ show surprisingly reasonable estimates for return times longer than the control method can provide at numerical costs of $1 \cdot 10^6$.

3.2.3 Relative Error Estimates

We now discuss relative error estimates on return probabilities across GPA, GEV, and MC methods. The relative error is estimated as $\sqrt{\sum_{j=1}^{K} \frac{1}{K}(\hat{\gamma} - \gamma)^2}/\gamma$ where $\hat{\gamma}$ is the estimate for each of $K = 100$ experiments and γ is the long control-run estimate. The relative error is essentially the average deviation of the tail probability estimate $\hat{\gamma}$ from the true value γ where it is assumed that $\hat{\gamma}$ follows a Gaussian distribution with mean γ [16,17] *for a sufficiently large number N of starting particles.* For lower values of N, the relative error calculated in this way has an underlying measurement error in the bias that is observed for $\hat{\gamma}$ in lower N values. Although this bias is often considered negligible, the sensitivity of long return times to small deviations in the tail probability estimate suggest otherwise. We first illustrate that the relative error cannot be used reliably for thresholds whose optimal tilting value is not approximately C. We calculate an estimate of the mean $\mu(\hat{\gamma}) = \frac{1}{K}\sum_{k=1}^{K}\hat{\gamma}_k$ for $K = 100$ experiments with $N = 1000$ and three different values of C. Then, we calculate the relative deviation of $\mu(\hat{\gamma})$ from the "true" mean γ by $\sqrt{(\mu(\hat{\gamma}) - \gamma)^2}/\gamma$ for each value of the threshold. Results in Fig. (11) show that this deviation is small only for thresholds whose tilting value C lies near the optimal value.

The effects of this deviation can be seen in return time estimates. We calculate the return time curves from 100 experiments of GPA and GEV methods with $N =$ 1000 (Fig. (13)) Clearly, GEV methods produce a larger standard deviation for return times. Under the assumptions above, the relative error for GEV methods would be larger than that of GPA; however, the mean of the tail probabilities obtained from GEV are nearly exactly those of the long control run. On the other hand, GPA produces a much smaller standard deviation (relative error) while the mean of the tail probabilities have accurate estimates only near thresholds for which the C value is chosen optimally.

We remark that for a single threshold and a close to optimal value of C, relative error estimates are reliable and GPA outperforms GEV and MC methods under relative error (Fig. (12)) while providing accurate return time estimates (Fig. (13)). These results are consistent with those of [16]. Interestingly, though not surprisingly, are the results on equivalent relative error for the GEV and MC methods for shorter return times. This equivalence suggests that the advantage of GEV over MC methods comes from its ability to estimate longer return times where MC methods fail to provide results.

3.3 Lorenz Model

The Lorenz 1996 model consists of J coupled sites x_l on a ring,

$$\dot{x}_l = x_{l-1}(x_{l+1} - x_{l-2}) + R - x_l$$

$l = 0, \ldots, J-1$ where the indices are in Z mod J. The parameter R is a forcing term and the dynamics is chaotic for $R \geq 8$ [24,25]. The energy $E(x) = \frac{1}{2J}\sum_{l=1}^{J} x_l^2$ is conserved and there is a repelling fixed hyperplane $x_l = R$, $l = 0, \ldots, J-1$. The extremes of interest investigated numerically in [16] and in our preliminary work were tail probabilities of the form $P(E(x(t)) > E_t)$. The energy observable on this system has an approximately Gaussian distribution.

3.3.1 GPA, GEV and MC

The weight function is taken to be the change ΔE of energy i.e. $E(x(t+1)) - E(t)$ for a single time step and from this an exponential weight function $W = \exp(C\Delta E)$ is constructed, depending on a single parameter C (large C makes tail probabilities greater). For this analysis, we choose $J = 32$ sites and a forcing coefficient $R = 64$.

The GPA importance sampling algorithm is performed with $N = 2000$ and 5000 starting trajectories, a resampling time $\tau = 0.08$, and a total integration time of $T_f = 1.28$. At each time-step of the algorithm, a random perturbation sampled from $[-\varepsilon, \varepsilon]$ where $\varepsilon = O(10^{-3})$ is added to the clones of the previous iteration to ensure divergence. The final trajectories from GPA with tilting constants $C = [3.2{\cdot}10^{-3}, 6.4{\cdot}10^{-3}]$ are used to calculate the above threshold probabilities and return time curves. The return time curve is calculated by averaging over 10 experiments. Return time curves from the GEV and MC methods are created from runs of equal numerical cost $4 \cdot 10^4$, and $1 \cdot 10^5$, respectively. All estimates are compared to a long control run of $1 \cdot 10^6$. For $N = 2000$ initial starting particles both GEV and MC methods outperform GPA by providing more accurate return time estimates for times longer than $1 \cdot 10^3$ (Fig. 14). GPA seems to provide more accurate estimates for returns longer than $1 \cdot 10^5$ for $N = 5000$; however, the deviation of the averaged return time curve is much larger than that of GEV or MC methods for all thresholds (Fig. 15).

The complexity of the Lorenz'96 highlights some of the major pitfalls in GPA. Intuitively, the choice of tilting value C is (roughly) the shift required for center of the distribution of the observable to lie directly over the threshold of interest. The Lorenz system provides an example of the difficulties involved in choosing this tilting value in practice. Similar to the OU system, the underlying dynamics of the Lorenz system equipped with the energy observable distorts the shifted distribution. Unlike the OU system, this occurs for very low values of C even though the observable range is much larger. As a result, the intuitive choice of C for thresholds in the tail of the distribution cannot be used. The values of C chosen here are taken from preliminary work related to [16].

A related issue is the number of initial particles required to give an accurate return time curve. Relative error arguments for GPA do not hold here both because the optimal tilting value C to threshold pair is nontrivial for complex systems and because the value C cannot be chosen arbitrarily large. An alternative to this issue is to choose a large enough number of initial particles N so that relative error is only affected by the standard deviation of the tail probability estimates $\hat{\gamma}$ (see. Sect. 3.2); however, this number is nontrivial as convergence depends on how far the optimal value is from the chosen tilting value.

GEV and GPA methods are able to estimate longer return times compared to MC brute force methods for the Lorenz 96 system. GEV has the advantage of maintaining the same relative error growth while difficulties in the optimal choice of C and initial values cause probability tail estimates from GPA to have much larger relative error. Furthermore, GEV likelihood estimation requires a single run to estimate the optimal return level plot with confidence intervals where relative error can be approximated by the standard brute force growth rate ($\approx 1/\sqrt{N}\gamma_A$). On the other hand, GPA requires many runs to estimate the relative error and return level plot for threshold values that do not correspond to the center (or near center) of the C-shifted distribution.

3.4 Planet Simulator (PlaSim)

We now describe a climate model on which our analysis will focus—*Planet Simulator* (**PlaSim**): a planet simulation model of intermediate complexity developed by the Universität Hamburg Meteorological Institute [10]. Like most atmospheric models, PlaSim is a simplified model derived from the Navier Stokes equation in a rotating frame of reference. The model structure is given by five main equations which allow for the conservation of mass, momentum, and energy. For a full list of the variables used in the following equations please see Table 1. The key equations are as follows:

- Vorticity Equation

$$\frac{\partial \zeta}{\partial t} = \frac{1}{1-\mu^2}\frac{\partial}{\partial \lambda}\mathcal{F}_v - \frac{\partial}{\partial \mu}\mathcal{F}_u - \frac{\xi}{\tau_F} - K(-1)^h \nabla^{2h} \xi \tag{16}$$

- Divergence Equation

$$\frac{\partial D}{\partial t} = \frac{1}{1-\mu^2}\frac{\partial}{\partial \lambda}\mathcal{F}_u + \frac{\partial}{\partial \mu}\mathcal{F}_v - \nabla^2\left(\frac{U^2+V^2}{2(1-\mu^2)} + \Phi + T_R \ln p_s\right) - \frac{D}{\tau_F} - K(-1)^h \nabla^{2h} D \tag{17}$$

- Thermodynamic Equation

$$\frac{\partial T'}{\partial t} = -\frac{1}{(1-\mu^2)}\frac{\partial}{\partial \lambda}(UT') - \frac{\partial}{\partial \mu}(VT') + DT' - \dot{\sigma}\frac{\partial T}{\partial \sigma} + \kappa\frac{T\omega}{p} + \frac{T_R - T}{\tau_R} - K(-1)^h \nabla^{2h} T' \tag{18}$$

- Continuity Equation

$$\frac{\partial(\ln p_s)}{\partial t} = -\frac{U}{1-\mu^2}\frac{\partial(\ln p_s)}{\partial \lambda} - V\frac{\partial(\ln p_s)}{\partial \mu} - D - \frac{\partial \dot{\sigma}}{\partial \sigma} \tag{19}$$

- Hydrostatic Equation

$$\frac{\partial \Phi}{\partial(\ln \sigma)} = -T \tag{20}$$

Here,

$$U = u\cos\phi - u\sqrt{1-\mu^2}, \quad V = v\cos\phi - v\sqrt{1-\mu^2},$$
$$\mathcal{F}_u = V\zeta - \dot{\sigma}\frac{\partial U}{\partial \sigma} - T'\frac{\partial(lnp_s)}{\partial \lambda}, \quad \mathcal{F}_v = -U\zeta - \dot{\sigma}\frac{\partial V}{\partial \sigma} - T'(1-\mu^2)\frac{\partial(\ln p_s)}{\partial \mu}.$$

The combination of vorticity (16) and divergence (17) equations ensure the conservation of momentum in the system while the continuity equation (19) ensures conservation of mass. The hydrostatic equation (18) describes air pressure at any height in the atmosphere while the thermodynamic equation (18) is essentially derived from the ideal gas law .

The equations above are solved numerically with discretization given by a (variable) horizontal Gaussian grid [9] and a vertical grid of equally spaced levels so that each grid-point has a corresponding latitude, longitude and depth triplet. (The default resolution is 32 latitude grid points, 64 longitude grid points and 5 levels.) At every fixed time step t and each grid point, the atmospheric flow is determined by solving the set of model equations through the spectral transform method which results in a set of time series describing the system; including temperature, pressure, zonal, meridional and horizontal wind velocity, among others. The resulting time series can be converted through the PlaSim interface into a readily accessible data file (such as netcdf) where further analysis can be performed using a variety of platforms. We refer to [10] for more information.

Table 1. List of variables used in PUMA.

ζ	Absolute vorticity	λ	Longitude
ξ	Relative vorticity	ϕ	Latitude
D	Divergence	μ	$\sin(\phi)$
Φ	Geopotential	κ	Adiabatic coefficient
ω	Vertical velocity	τ_R	Timescale of Newtonian cooling
p	Pressure	τ_F	Timescale of Rayleigh friction
p_s	Surface pressure	σ	Vertical coordinate p/p_s
K	Hyperdiffusion	$\dot{\sigma}$	Vertical velocity $d\sigma/dt$
u	Zonal wind	v	Meridional wind
h	Hyperdiffusion order	T_R	Restoration temperature
T	Temperature	T'	$T - T_R$

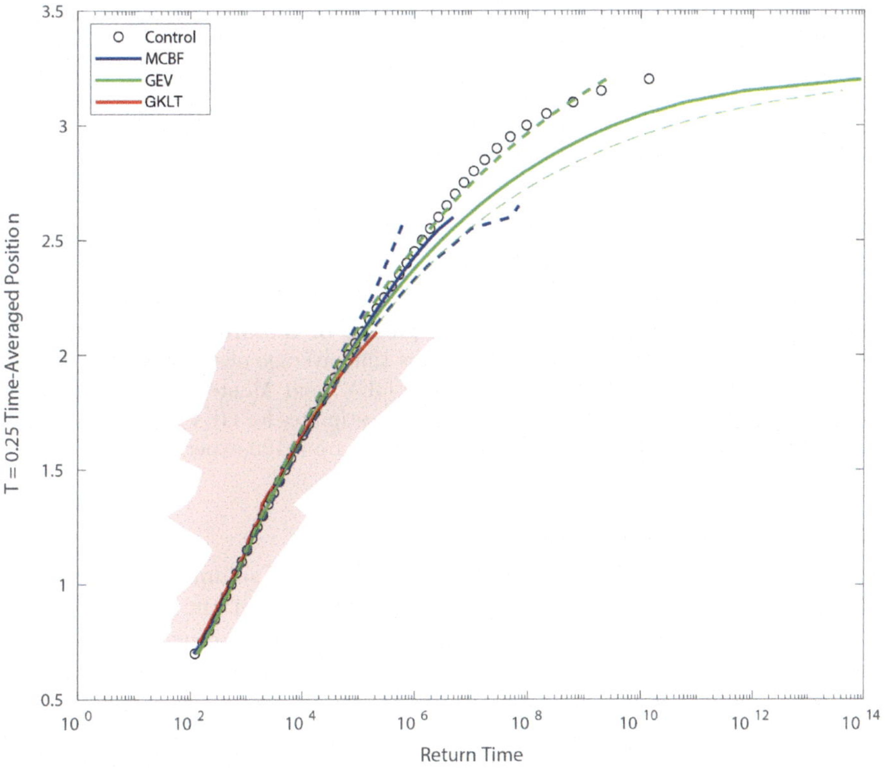

Fig. 5. Return time estimates for the Ornstein Uhlenbeck process time average observable using GKLT for 4 different C values and 100 experiments, GEV, and Monte Carlo brute forces methods with numerical cost $4 \cdot 10^4$. Relative error curves for MC brute force and GEV estimates are represented by dashed lines. Relative error estimated by 100 experiments of the GKLT process is represented by the shaded red region.

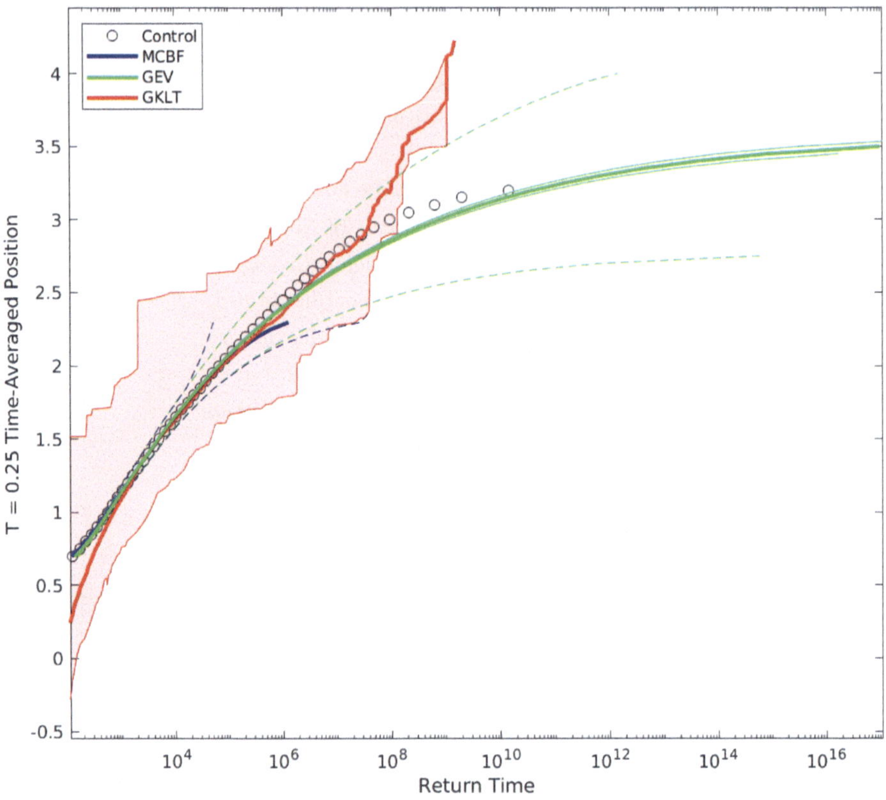

Fig. 6. Return time estimates **from the sequence of maxima taken over each trajectory** for the Ornstein Uhlenbeck process time average observable using GKLT for 4 different C values and 100 experiments, GEV, and Monte Carlo brute forces methods with numerical cost $4 \cdot 10^4$. Relative error estimates for GEV and MC methods (dashed lines) and GKLT (red region) are estimated from 100 experiments.

3.4.1 GKLT, GEV and MC

Our observable of interest in PlaSim is the time series of summer European spatial average temperature anomalies. For simplicity, we set the climate boundary data to consistent July 1st conditions and remove the diurnal and annual cycles. This allows for perpetual summer conditions and saves on computational time. We define the European spatial average as the average over the set of 2-dimensional latitude and longitude pairs on the grid located between $36°N$–$70°N$ and $11°W$–$25°E$. Spatial average values are taken at 6 h intervals. We subtract the long-run mean to obtain the sequence of summer European spatial average temperature anomalies used in this analysis.

We perform the GKLT algorithm on the European spatial averaged temperature time-series by considering initial values as the beginning of a year (360 days) to ensure each initial value is independent. It is important to note that

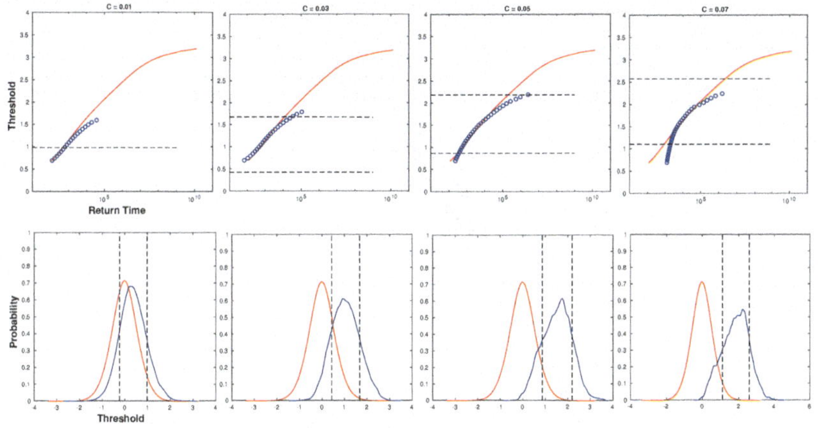

Fig. 7. Return time estimates for the Ornstein Uhlenbeck process time average observable illustrating the choice of return time curves after GKLT implementation.

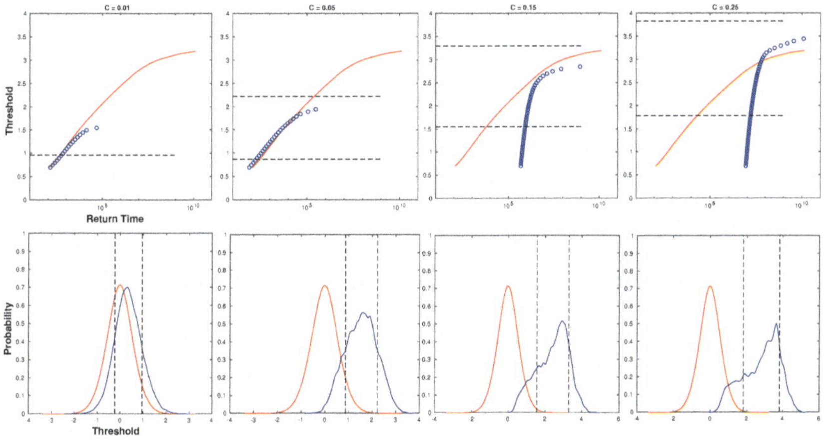

Fig. 8. Return time estimates for the Ornstein Uhlenbeck process time average observable illustrating the breakdown of the distributions for large values of C.

initial values may be taken at much shorter intervals. We choose one year intervals because this initial data was readily available from the long control run. We estimate the resampling time $\tau = 8$ days as the approximate time for autocorrelation to reach near zero. For each experiment, we use 100 years (100 initial values) run for 17 complete steps of the GKLT algorithm, or 136 days, to estimate anomaly recurrence times for the $T = 8$-day time average. We remark that the choice of T and τ here are the same, however this is not a requirement of the algorithm as illustrated in the Ornstein-Uhlenbeck system in Sect. 3.2. Results are compared to a 400 year (144,000 day) control run. Added noise to ensure

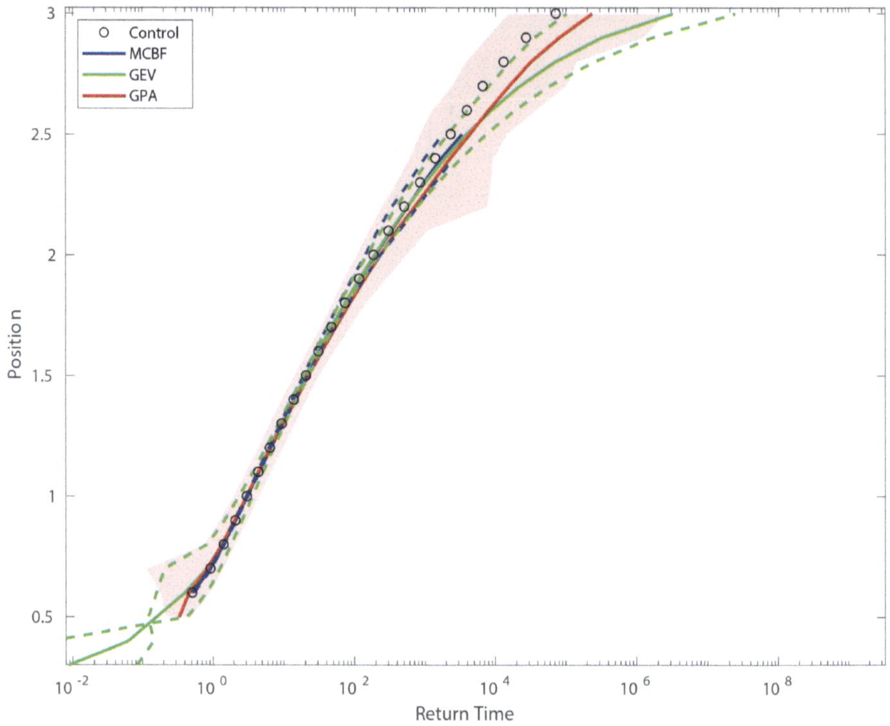

Fig. 9. Return time estimates for the Ornstein Uhlenbeck process using GPA for 3 different C values estimated over 10 experiments, GEV, and Monte Carlo brute forces methods with numerical cost $3 \cdot 10^3$. Relative error estimates for GEV and MC methods (dashed lines) and GPA (red region) are estimated from 10 experiments.

divergence of cloned trajectories is sampled uniformly from (preprogrammed noise) $[-\varepsilon\sqrt{2}, \varepsilon\sqrt{2}]$ where $\varepsilon = O(10^{-4})$.

Six experiments of the GKLT algorithm are performed on a starting ensemble of $N = 100$ trajectories with initial values taken as the starting value of the European spatial average at the beginning of each year. The values $C = [0.01, 0.05]$ (3 experiments per C value) are chosen to tilt the distribution of the spatial-time average at resampling times $\tau = 8$ days. We remark that constants $C = [0.1, 2]$ are also tested with less favorable results; however, these tests are not included in the total numerical cost of MC brute force and GEV methods. We choose the observable described by Eq. (15), with $\phi(x_n(t))$ taken as the European spatial average temperature, to estimate return time curves of the 8-day time average of European spatial averaged temperature.

We refer to Fig. 16 for this discussion. GEV and MC methods agree almost completely up to return times of $1 \cdot 10^6$ with the GEV continuing to provide estimates for longer return times. 95% confidence intervals for the GEV (green thin lines) are a result of the likelihood function. The return time curve for GKLT

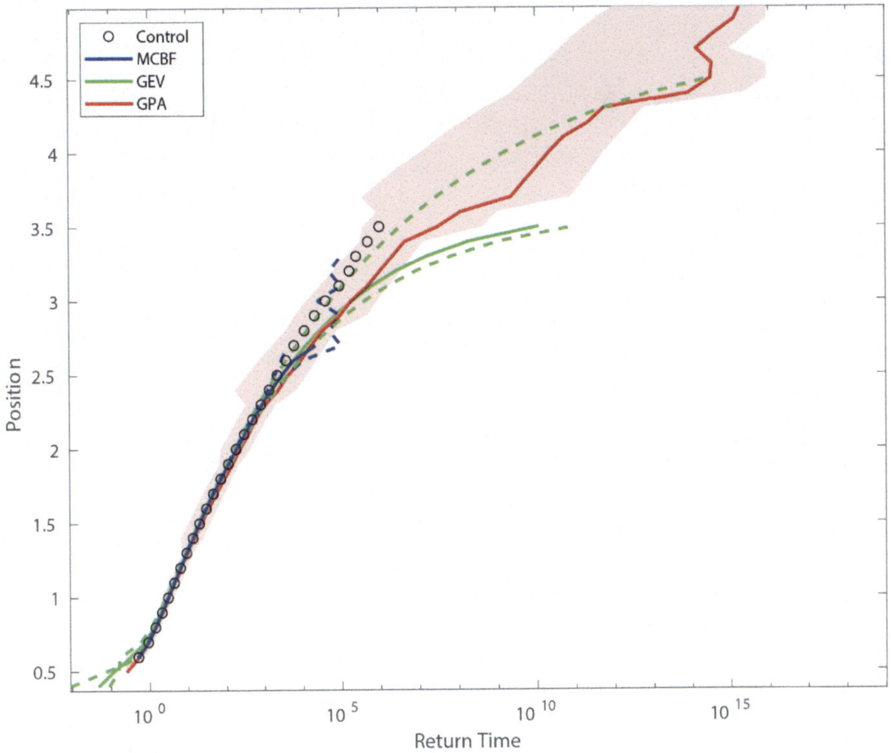

Fig. 10. Return time estimates for the Ornstein Uhlenbeck process using GPA for 10 different C values estimated over 30 experiments, GEV, and Monte Carlo brute forces methods with numerical cost $3 \cdot 10^3$. Relative error estimates for GEV amd MC methods (dashed lines) and GPA (red region) are estimated from 30 experiments.

is formed by the set of return time values from each of the 6 experiments that fall within 1/2 standard deviation of the mean of the shifted distribution. Hence, the deviation for GKLT (red region) is estimated by the minimum and maximum deviation of anywhere between 2 and 6 return time values for each threshold. Compared to that of the long control run, GKLT provides reliable estimates for return times up to $1 \cdot 10^4$, while GEV estimates remain near those of the long control run for return times up to $1 \cdot 10^6$. Deviation estimates for GKLT are smaller than the 95% confidence interval for the GEV for return times longer than $1 \cdot 10^3$; however, this may be the result of a low number of experiments. We also remark that the deviation estimate of the GKLT method for return times of the 8-day average anomaly near 1.5 K are much smaller compared to other thresholds. This reduction suggests that at least one of the C values chosen in GKLT is close to optimal for the 1.5 K threshold.

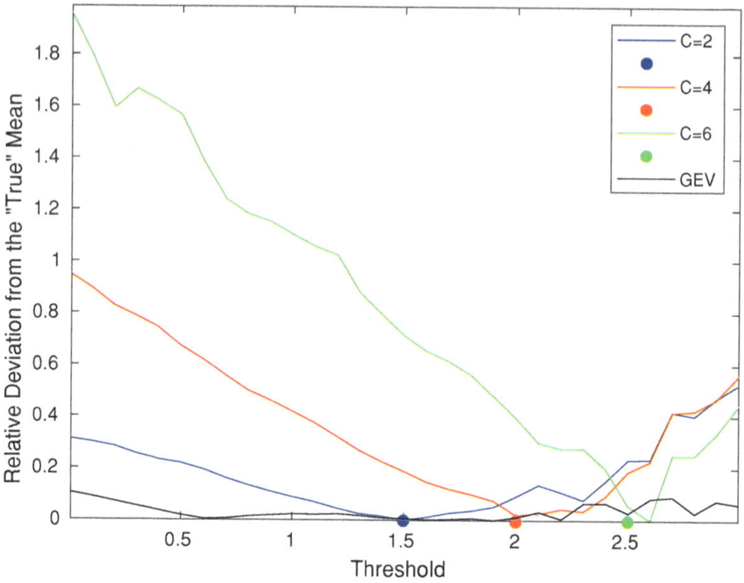

Fig. 11. Relative deviation for the OU process of the estimated mean $\mu(\hat{\gamma_A})$ from $K = 100$ runs of GPA with $N = 1000$ from the assumed, asymptotic mean γ. This deviation is only near zero for thresholds whose optimal tilting value C is chosen in the weight function (marked with a $\circ$). Relative deviation of the estimated mean from the GEV method is consistently near zero, suggesting that even though the deviation is larger, the estimate is more reliable.

4 Discussion

In this paper we have discussed two importance sampling (IS) methods: Genealogical particle analysis (GPA) which is used to estimate tail probabilities of a time series of observations at a fixed sampling rate, and GKLT which is used to estimate tail probabilities of a corresponding time average. Both methods work by tilting the distribution of observations in a reversible way so that the rare events corresponding to tail probabilities are sampled more often. We have illustrated the particular case when the observations of interest are distributed according to a roughly symmetric distribution and a rare event consists of an exceedance of a certain level where the natural choice of tilt corresponds to a shift towards the tail.

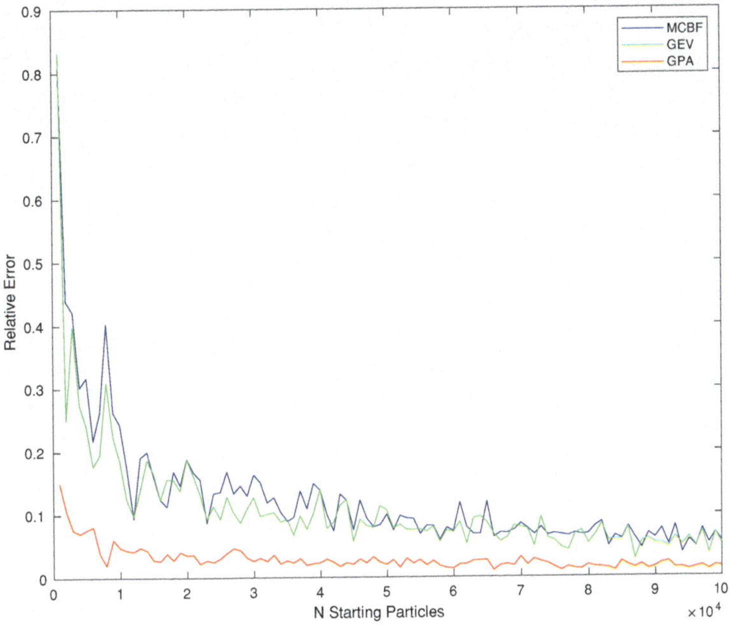

Fig. 12. Relative error for OU process of MC, GEV and GPA probability estimates of fixed threshold 2 and corresponding optimal tilting value $C = 4$.

We compare results of these two methods with classical statistics where rare event estimation is given by the Generalized Extreme Value (GEV) distribution. Under the goal of obtaining a return level curve, we have shown that the GEV outperforms both IS methods for all three systems used in this analysis by providing generally lower relative error and longer return time estimates. We have also illustrated a few disadvantages in IS methods including the strict dependence of the tilting value to initial conditions and requirement of multiple runs for return time curve and relative error estimation while demonstrating that classical GEV results only require a single run to estimate return time curves and follow standard brute force relative error growth. On the other hand, we have shown that our results do not conflict with previous literature and that both the GEV and IS methods outperform Monte Carlo brute force methods in estimating longer return times. In fact, following previous literature we have shown that IS methods can result in lower relative error than that of the GEV on subsets of tail probabilities (and hence, that of MC brute force) provided the optimal tilting value can be chosen.

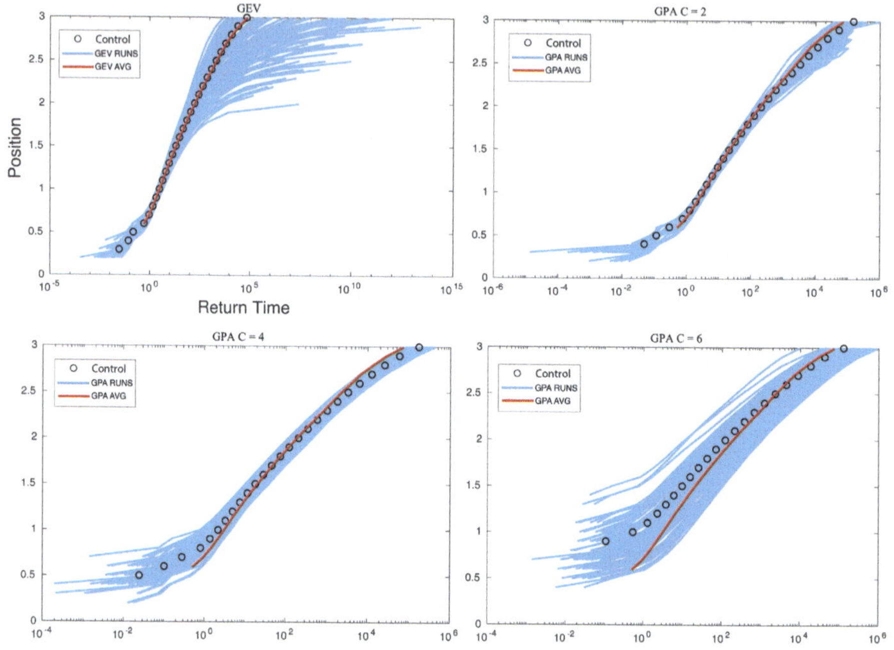

Fig. 13. Illustration for the OU process of the deviation of the return time curves from the control for GEV and 3 different tilting C values of GPA. Notice that the average return time curve (red) for the GEV fits the control (black ∘) for all long return times while accurate estimates for GPA only occur near the optimal threshold value.

In general, these results support the idea of using GEV methods over IS under the condition that optimal tilting values cannot be determined a priori and/or return time curves, rather than returns for a single level, are of interest. We emphasize that these results should not be taken to discount the value of importance sampling. The power of these methods can be seen in the decrease in relative error when optimal tilting values can be chosen. It would be interesting to see more theoretical work in estimating such values which, at the moment, requires an explicit formula of the (unknown) distribution of the observable. Other numerical work can also be completed using IS methods which does not involve tail probability estimation. One particular perspective we plan to explore is the algorithms' ability to provide the set of trajectories which most likely end in an extreme event.

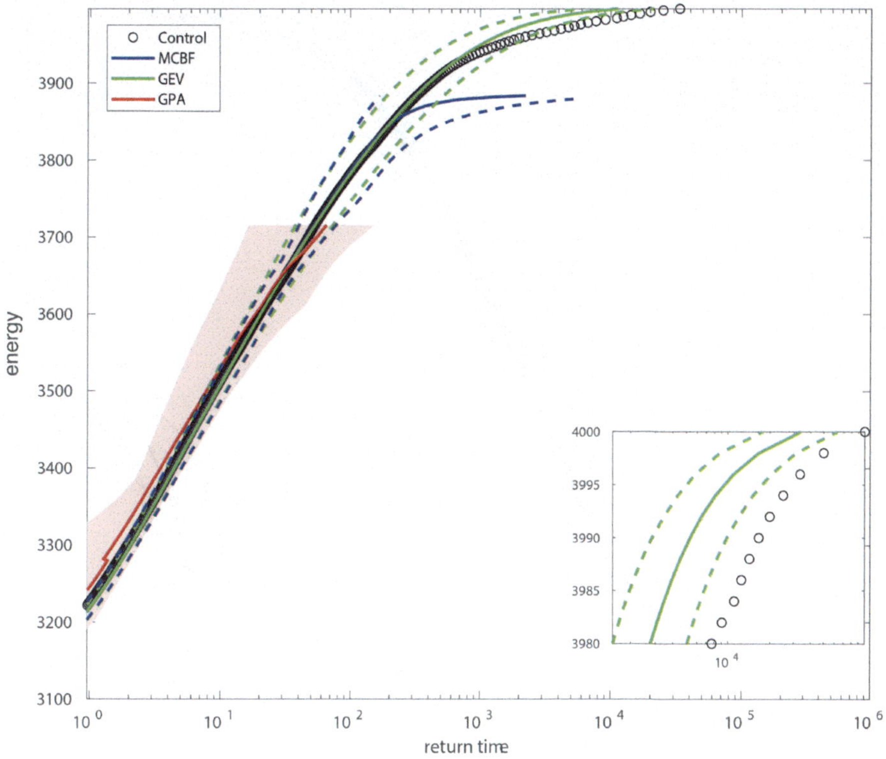

Fig. 14. Return time estimates for the Lorenz'96 process using GPA for $C = [3.1 \cdot 10^{-3}, 6.4 \cdot 10^{-3}]$ estimated over 10 experiments for $N = 2000$ starting particles, GEV, and Monte Carlo brute forces methods with numerical cost $4 \cdot 10^4$. Relative error estimates for GEV amd MC methods (dashed lines) and GPA (red region) are estimated from 10 experiments.

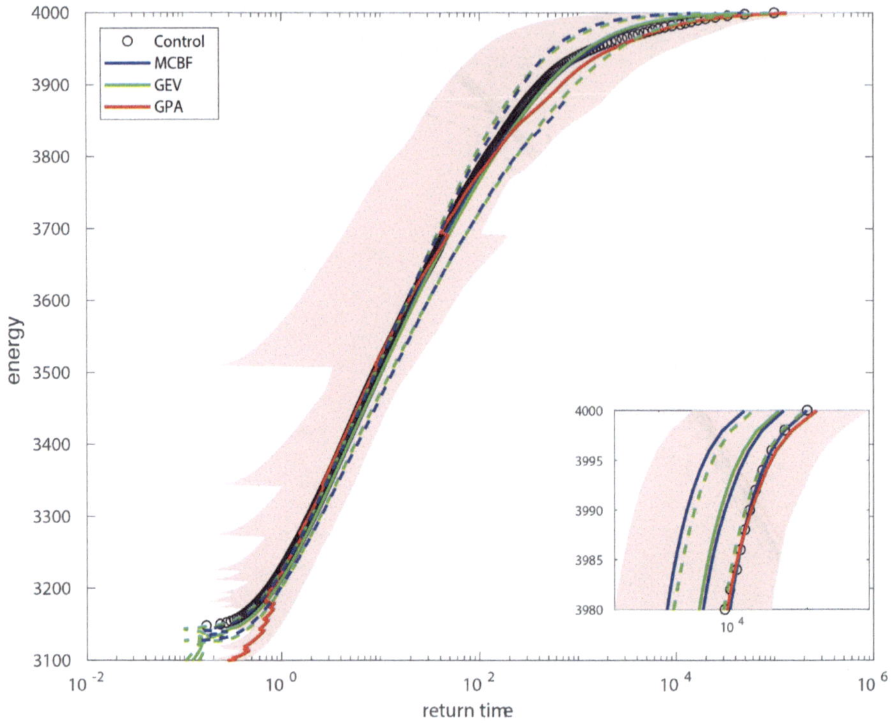

Fig. 15. Return time estimates for the Lorenz'96 process using GPA for $C = [3.1 \cdot 10^{-3}, 6.4 \cdot 10^{-3}]$ estimated over 10 experiments for $N = 5000$ starting particles, GEV, and Monte Carlo brute forces methods with numerical cost $1 \cdot 10^5$. Relative error estimates for GEV amd MC methods (dashed lines) and GPA (red region) are estimated from 10 experiments.

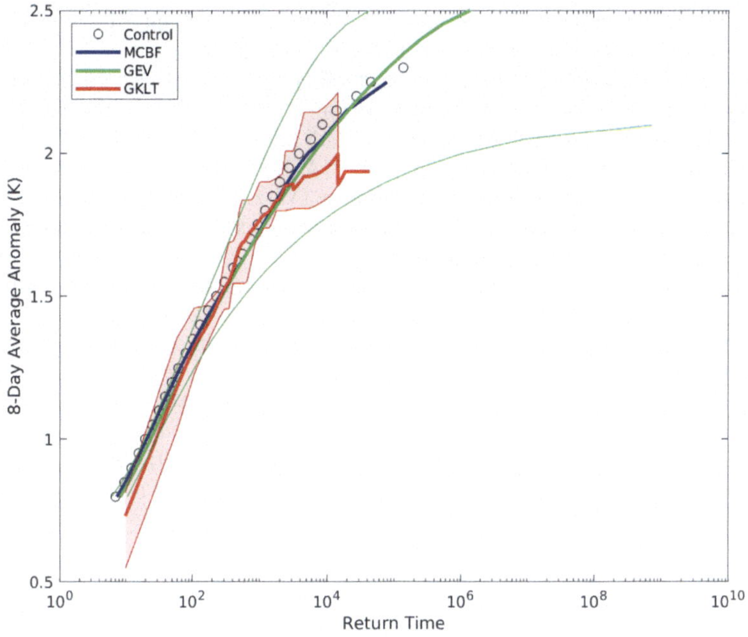

Fig. 16. Return time estimates for 8-day average temperature anomalies from PlaSim using GKLT for $C = 5 \cdot 10^{-2}$ for $N = 100$ over 136 days starting particles, GEV and Monte Carlo estimates are provided with numerical cost $6 \times 100 \times 136$ days. The control return time curve comes from a long brute-force run of 144,000 days. Green outer lines indicate the 95% confidence interval of the GEV. Red filled region indicates the deviation of the GKLT algorithm estimated over 6 runs.

Acknowledgements. We warmly thank Frank Lunkeit at Universität Hamburg for very helpful discussions and advice concerning PlaSim. MN was supported in part by NSF Grant DMS 1600780.

References

1. Carney, M., Azencott, R., Nicol, M.: Non-stationarity of summer temperature extremes in Texas. Int. J. Climatol. **40**(1), 620–640 (2020)
2. Lucarini, V., et al.: Extremes and Recurrence in Dynamical Systems, 312 pp. Wiley, Hoboken (2016)
3. Bucklew, J.: Introduction to Rare Event Simulation. Springer Series in Statistics. Springer, New York (2004)
4. Carney, M., Kantz, H.: Robust Regional clustering and modeling of nonstationary summer temperature extremes across Germany (preprint)
5. Coles, S.: An Introduction to Statistical Modeling of Extreme Values. Springer Series in Statistics, 4th edn. Springer, New York (2007)
6. Collet, P.: Statistics of closest return for some non-uniformly hyperbolic systems. Ergodic Theorem Dyn. Syst. **21**, 401–420 (2001)

7. Giardina, C., Kurchan, J., Lecomte, V., Tailleur, J.: Simulating rare events in dynamical processes. J. Stat. Phys. **145**, 787–811 (2011)
8. Gumbel, E.J.: Statistics of Extremes. Columbia University Press, New York (1958)
9. Hoskins, B., Simons, A.: A multi-layer spectral model and the semi-implicit method. Q. J. R Meteorol. Soc. **101**, 637–655 (1975)
10. Fraedrich, K., Kirk, E., Lunkeit, F.: PUMA Portable University Model of the Atmosphere. World Data Center for Climate (WDCC) at DKRZ (2009)
11. Freitas, J., Freitas, A., Todd, M.: Hitting times and extreme values. Probab. Theory Relat. Fields **147**(3), 675–710 (2010)
12. Freita, A.C., Freitas, J., Todd, M.: Speed of convergence for laws of rare events and escape rates. Stoch. Proc. App. **125**, 1653–1687 (2015)
13. Galambos, J.: The Asymptotic Theory of Extreme Order Statistics. Wiley, Hoboken (1978)
14. Galfi, V., Lucarini, V., Wouters, J.: A large deviation theory-based analysis of heat waves and cold spells in a simplified model of the general circulation of the atmosphere. J. Stat. Mech. Theory Exp. **3**(3), 033404 (2019). 39 pp
15. Gupta, C., Holland, M., Nicol, M.: Extreme value theory and return time statistics for dispersing billiard maps and flows, Lozi maps and Lorenz-like maps. Ergodic Theory Dyn. Syst. **31**(5), 1363–1390 (2011)
16. Wouters, J., Bouchet, F.: Rare event computation in deterministic chaotic systems using genealogical particle analysis. J. Phys. A: Math. Theor. **49**, 374002 (2016)
17. Del Moral, P., Garnier, J.: Genealogical particle analysis of rare events. Ann. App. Prob. **15**(4), 2496–2534 (2005)
18. Ragone, F., Wouters, J., Bouchet, F.: Computation of extreme heat waves in climate models using a large deviation algorithm. PNAS **115**(1), 24–29 (2018)
19. Giardina, C., Kurchan, J., Peliti, L.: Direct evaluation of large-deviation functions. Phys. Rev. Lett. **96**, 120603 (2006)
20. Tailleur, J., Kurchan, J.: Probing rare physical trajectories with Lyapunov weighted dynamics. Nat. Phys. **3**, 203–207 (2007)
21. Hall, P.: On the rate of convergence of normal extremes. J. Appl. Prob. **16**(2), 433–439 (1979)
22. Holland, M., Nicol, M.: Stochast. Dyn. **15**(4), 1550028 (2015). 23 pp
23. Leadbetter, M.R., Lindgren, G., Rootzén, H.: Extremes and Related Properties of Random Sequences and Processes. Springer, Heidelberg (1980)
24. Lorenz, E.N.: Predictability–a problem partly solved. In: Seminar on Predictability, vol. I, ECMWF (1996)
25. Lorenz, E.N.: Designing chaotic models. J. Atmos. Sci. **62**(5), 1574–1587 (2005)
26. Del Moral, P.: Feynman-Kac Formulae. Genealogical and Interacting Particle Systems with Applications. Probability and its Applications. Springer, New York (2004)
27. Ragone, F., Wouters, J., Bouchet, F.: Computation of extreme heat waves in climate models using a large deviation algorithm. Proc. Natl. Acad. Sci. U.S.A. **115**(1), 24–29 (2018)
28. Rubino, G., Tuffin, B.: Introduction to Rare Event Simulation. Rare Event Simulation Using Monte Carlo Methods, pp. 1-13. Wiley, Chichester (2009)
29. Wouters, J., Bouchet, F.: Rare event computation in deterministic chaotic systems using genealogical particle analysis. J. Phys. A **49**(37), 374002 (2016). 24 pp

Dynamical Systems Theory and Algorithms for NP-hard Problems

Tuhin Sahai(✉)

Raytheon Technologies Research Center, 2855 Telegraph Ave. Suite 410, Berkeley, CA 94705, USA
tuhin.sahai@gmail.com

Abstract. This article surveys the burgeoning area at the intersection of dynamical systems theory and algorithms for NP-hard problems. Traditionally, computational complexity and the analysis of non-deterministic polynomial-time (NP)-hard problems have fallen under the purview of computer science and discrete optimization. However, over the past few years, dynamical systems theory has increasingly been used to construct new algorithms and shed light on the hardness of problem instances. We survey a range of examples that illustrate the use of dynamical systems theory in the context of computational complexity analysis and novel algorithm construction. In particular, we summarize a) a novel approach for clustering graphs using the wave equation partial differential equation, b) invariant manifold computations for the traveling salesman problem, c) novel approaches for building quantum networks of Duffing oscillators to solve the MAX-CUT problem, d) applications of the Koopman operator for analyzing optimization algorithms, and e) the use of dynamical systems theory to analyze computational complexity.

Keywords: Computational complexity · Dynamical systems theory · NP-hardness · Heuristic algorithms · Combinatorial optimization

1 Introduction

Dynamical systems theory and computational complexity have, predominantly, been developed as independent areas of research over the last century with little interaction and mutual influence. Dynamical systems theory has its origins in the seminal work of Henri Poincaré [1] on celestial mechanics. Computational complexity theory, on the other hand, originated in the works of Alan Turing [2] and Alonzo Church [3] in the 1930s and has played an intimate role in the computing revolution of the twentieth century.

Eventually, dynamical systems theory (or nonlinear dynamics) found broad application beyond celestial mechanics. In particular, it has been used extensively to model and analyze engineering systems [4], physics of natural phenomena, biological [5] and chemical processes [6], fluid dynamics [7], and epidemiology [8] to

O. Junge et al. (Eds.): SON 2020, SSDC 304, pp. 183–206, 2020.
https://doi.org/10.1007/978-3-030-51264-4_8

name a few. Moreover, the analysis of dynamical systems is typically intimately tied to numerical methods [9,10] and scientific computation [11].

Links to the applications (outlined in the previous paragraph) have played a critical role in the theoretical development of the field. For example, they have influenced the development of various sub-areas within nonlinear dynamics such as ergodicity [12], chaos theory [13], and symbolic dynamics [14] to name a few. For a broad overview of the theoretical approaches to dynamical systems, we refer the reader to [15]. Although, dynamical systems theory has found wide application in engineering and the sciences, it has received scant attention from the computer science community.

Local continuous optimization techniques such as Nesterov's method [16] have recently been analyzed from a dynamical systems perspective [17]. Nesterov's method is an optimal gradient descent algorithm in terms of convergence rate. In [17], the authors derive a dynamical system by invoking a continuous time limit of the optimization step size. They then analyze the resulting ordinary differential equations (ODEs) to provide valuable insight into the algorithm and its associated optimality. Additionally, in [18] the authors use calculus of variations to gain additional insight into the convergence rates of accelerated gradient descent schemes. Although, this body of work does fall under the category of novel application of dynamical systems theory to optimization methods, we will not discuss it at length in this paper for two reasons (a) this work has sparked extensive follow-on work and consequently, various summary articles and presentations are already available, and (b) they appear to be restricted to accelerated gradient methods with no clear extension to the broader theory of computational complexity.

Non-deterministic polynomial-time (NP)-hard and -complete complexity classes can be traced to seminal work by Cook in 1971 [19]. The broad applicability of this work was outlined in a highly influential publication by Karp [20]. NP-hard problems such as the traveling salesman problem (TSP) [21] and lattice-based vector problems [23] arise in a wide variety of applications ranging from DNA sequencing and astronomy [22] to encryption [23]. In essence, the computation of optimal solutions for these problems quickly becomes intractable with the size of the instance (unlike problems that lie in the P complexity class). Note that some problems such as graph isomorphism [24] lie in the NP complexity class but are not expected to be NP-complete or NP-hard. Over the last few years, several efficient heuristic algorithms for approximating the solutions of NP-hard problems have been developed. For example, careful implementations of the Lin-Kernighan [25] and branch-and-bound [26] heuristics have been successful in computing optimal solutions of several large instances of the TSP. However, most NP-hard problems suffer from a lack of scalable approaches. Moreover, as long as $P \neq NP$ (where P is the complexity class of problems that can be solved in poylnomial time on a deterministic Turing machine), even efficient heuristics for some of these problems will remain elusive. See Fig. 1 for the hypothesized relationship between the most popular classes.

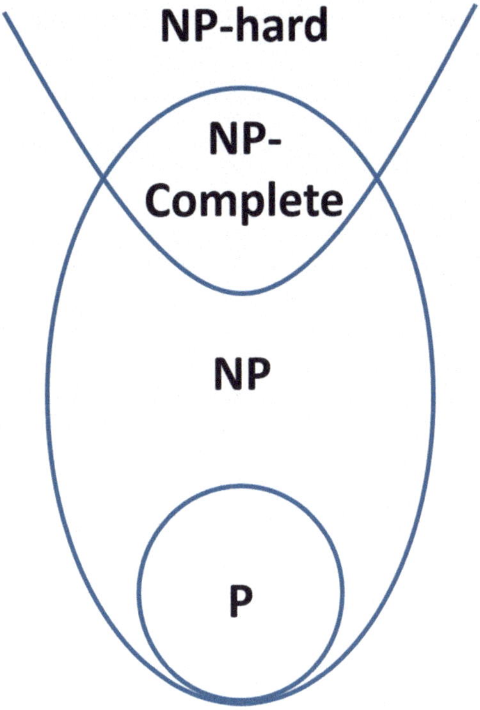

Fig. 1. The computational complexity map for the most common complexity classes.

In this work, we start by surveying the use of dynamical systems in the context of constructing state-of-the-art algorithms for NP-hard problems. In particular, we will cover the use of dynamical systems theory for constructing decentralized graph clustering algorithms [27,28], solutions for the TSP [29], and quantum-inspired networks of Duffing oscillators for solving the MAX-CUT problem [30]. We then switch to the use of dynamical systems theory for analysis of algorithms [31] and the underlying problems [32,33].

The goal of this survey paper is to highlight the potential application of dynamical systems theory for optimization of complex functions and analysis of computational complexity theory. This is a nascent field which presents the possibility of tremendous impact. Additionally, we expect this area to lead to new theoretical developments in nonlinear dynamics theory and novel algorithms for computationally intractable problems.

Ziessler, Surana, Speranzon, Klus, Dellnitz, and Banaszuk have all served as co-authors in my efforts in this area. However, my extensive discussions with Prof. Michael Dellnitz inspired me to delve deeper into the area of dynamical systems and the analysis of algorithms!

2 Novel Algorithm Construction: Decentralized Graph Clustering

Overview

Algorithms for graph analysis have a wide variety of applications such as routing, pattern recognition, database searches, network layout, and Internet PageRank to name a few [34]. Although some of the problems can be solved efficiently on present day computing devices, several graph analysis problems are computationally intractable [35]. For example, the problem of partitioning graphs into equal size sets while minimizing the weights of cut edges arises in a range of settings such as social anthropology, gene networks, protein sequences, sensor networks, computer graphics, and Internet routing algorithms [28]. To avoid unbalanced cuts, size restrictions are typically placed on the clusters; instead of minimizing inter-connection strength, if one minimizes the ratio of the inter-connection strength to the size of individual clusters, the problem becomes NP-complete [36,37].

In [27,28], a novel decentralized algorithm for clustering/partitioning graphs that exploits fundamental properties of a dynamically evolving networked system was constructed. In particular, by propagating waves in a graph, one can compute partitions or clusters in a completely decentralized setting. The method is orders of magnitude faster than existing approaches [39]. This is our first example of a dynamical systems theory based algorithm for a combinatorial optimization problem. We now discuss the details of the approach.

Let $\mathcal{G} = (V, E)$ be a graph with vertex set $V = \{1, \ldots, N\}$ and edge set $E \subseteq V \times V$, where a weight $\mathbf{W}_{ij} \geq 0$ is associated with each edge $(i, j) \in E$, and $\mathbf{W}$ is the $N \times N$ weighted adjacency matrix of $\mathcal{G}$. We assume that $\mathbf{W}_{ij} = 0$ if and only if $(i, j) \notin E$. The (normalized) graph Laplacian is defined as,

$$\mathbf{L}_{ij} = \begin{cases} 1 & \text{if } i = j \\ -\mathbf{W}_{ij} / \sum_{\ell=1}^{N} \mathbf{W}_{i\ell} & \text{if } (i,j) \in E \\ 0 & \text{otherwise}, \end{cases} \tag{1}$$

or equivalently, $\mathbf{L} = \mathbf{I} - \mathbf{D}^{-1}\mathbf{W}$ where $\mathbf{D}$ is the diagonal matrix with the row sums of $\mathbf{W}$.

Note that in [28], only undirected graphs were considered. The smallest eigenvalue of the Laplacian matrix is $\lambda_1 = 0$, with an associated eigenvector $\mathbf{v}^{(1)} = \mathbf{1} = [1, 1, \ldots, 1]^T$. Eigenvalues of $\mathbf{L}$ can be ordered as, $0 = \lambda_1 \leq \lambda_2 \leq \lambda_3 \leq \cdots \leq \lambda_N$ with associated eigenvectors $\mathbf{1}, \mathbf{v}^{(2)}, \mathbf{v}^{(3)} \cdots \mathbf{v}^{(N)}$ [36]. It is well known that the multiplicity of λ_1 is the number of connected components in the graph [36].

Given the Laplacian matrix $\mathbf{L}$, associated with a graph $\mathcal{G} = (V, E)$, spectral clustering divides $\mathcal{G}$ into two clusters by computing the signs of the N elements of the second eigenvector $\mathbf{v}^{(2)}$, or Fiedler vector. For further details about the computation of two or more clusters see [36].

There are many algorithms to compute eigenvectors, such as the Lanczos method or orthogonal iteration [38]. Although some of these methods are distributable, convergence is slow [38] and these algorithms do not consider/take advantage of the fact that the matrix for which the eigenvalues and eigenvectors need to be computed is the adjacency matrix of the underlying graph. In [39], the authors propose an algorithm to compute the first k largest eigenvectors (associated with the first k eigenvalues with greatest absolute value)[1] of a symmetric matrix. The algorithm in [39] emulates the behavior of orthogonal iteration using a decentralized process based on gossip algorithms or deterministic random walks on graphs. This approach can be slow as it converges after $O(\tau \log^2 N)$ iterations [39] where τ is the mixing time for the random walk on the graph and N is the number of nodes.

This procedure is equivalent to evolving the discretized heat equation on the graph and can be demonstrated as follows. The heat equation is given by,

$$\frac{\partial u}{\partial t} = \Delta u\,,$$

where u is a function of time and space, $\partial u/\partial t$ is the partial derivative of u with respect to time, and Δ is the Laplace operator [28].

When the above equation is discretized on a graph $\mathcal{G} = (V, E)$ one gets the following equation:

$$\mathbf{u}_i(t+1) = \mathbf{u}_i(t) - \sum_{j \in \mathcal{N}(i)} \mathbf{L}_{ij}\mathbf{u}_j(t)\,,$$

for $i, j \in V$. Here $\mathbf{u}_i(t)$ is the scalar value of u on node i at time t and $\mathcal{N}(i)$ are the neighbors of node i in the graph. The graph Laplacian $\mathbf{L} = [\mathbf{L}_{ij}]$ is the discrete counterpart of the Δ operator. The above iteration can be re-written, in matrix form, $\mathbf{u}(t+1) = (\mathbf{I} - \mathbf{L})\,\mathbf{u}(t)$ where $\mathbf{u}(t) = (\mathbf{u}_1(t), \ldots, \mathbf{u}_N(t))^T$. The solution of this iteration is,

$$\mathbf{u}(t) = C_0\mathbf{1} + C_1(1-\lambda_2)^t\mathbf{v}^{(2)} + \cdots + C_N(1-\lambda_N)^t\mathbf{v}^{(N)}\,, \tag{2}$$

where constants C_j depend on the initial condition $\mathbf{u}(0)$. It is interesting to note that in Eq. (2), the dependence of the solution on higher eigenvectors and eigenvalues of the Laplacian decays with increasing iteration count. Thus, it is difficult to devise a fast and distributed method for clustering graphs based on the heat equation.

In [27,28], a novel algorithm based on the idea of permanent excitation of the eigenvectors of $\mathbf{I} - \mathbf{L}$ using dynamical systems theory is constructed. In a theme similar to Mark Kac's question “Can one hear the shape of a drum?” [40], it was demonstrated that by evolving the wave equation in the graph, nodes can “hear”

[1] Note that in the case of spectral clustering we desire to compute the smallest k eigenvectors of $\mathbf{L}$. The algorithm is still applicable if we consider the matrix $\mathbf{I} - \mathbf{L}$.

the eigenvectors of the graph Laplacian using only local information. Moreover, it was shown, both theoretically and on examples, that the wave equation based algorithm is orders of magnitude faster than random walk based approaches for graphs with large mixing times. The overall idea of the wave equation based approach is to simulate, in a distributed fashion, the propagation of a wave through the graph and capture the frequencies at which the graph "resonates". In other words, it was shown that by using these frequencies one can compute the eigenvectors of the Laplacian, thus clustering the graph.

The wave equation based clustering approach can be described as follows. Analogous to the heat equation case (Eq. 2), the solution of the wave equation can be expanded in terms of the eigenvectors of the Laplacian. However, unlike the heat equation where the solution eventually converges to the first eigenvector of the Laplacian, in the wave equation all the eigenvectors remain eternally excited (a consequence of the second derivative of u with respect to time). This observation is used to develop a simple, yet powerful, distributed eigenvector computation algorithm. The algorithm involves evolving the wave equation on the graph and then computing the eigenvectors using local FFTs. The graph decomposition/partitioning algorithm based on the discretized wave equation on the graph is given by,

$$\mathbf{u}_i(t) = 2\mathbf{u}_i(t-1) - \mathbf{u}_i(t-2) - c^2 \sum_{j \in \mathcal{N}(i)} \mathbf{L}_{ij}\mathbf{u}_j(t-1), \tag{3}$$

where $\sum_{j \in \mathcal{N}(i)} \mathbf{L}_{ij}\mathbf{u}_j(t-1)$ originates from the discretization of the spatial derivatives in the wave equation. The rest of the terms originate from discretization of the $\partial^2 u/\partial t^2$ term in the wave equation. To update $\mathbf{u}_i$ using Eq. (3), one needs only the value of $\mathbf{u}_j$ at neighboring nodes and the connecting edge weights (along with previous values of $\mathbf{u}_i$).

The main steps of the algorithm are shown as Algorithm 1. Note that at each node (node i in the algorithm) one only needs nearest neighbor weights $\mathbf{L}_{ij}$ and the scalar quantities $\mathbf{u}_j(t-1)$ also at nearest neighbors. We emphasize, again, that $\mathbf{u}_i(t)$ is a scalar quantity and `Random`$([0,1])$ is a random initial condition on the interval $[0,1]$. The vector $\mathbf{v}_i^{(j)}$ is the i-th component of the j-th eigenvector, T_{max} is a positive integer derived in [27,28], `FrequencyPeak(Y,j)` returns the frequency at which the j-th peak occurs and `Coefficient`(ω_j) return the corresponding Fourier coefficient.

Proposition 1. *The clusters of graph $\mathcal{G}$, determined by the signs of the elements of the eigenvectors of* $\mathbf{L}$*, can be computed using the frequencies and coefficients obtained from the Fast Fourier Transform of* $(\mathbf{u}_i(1), \ldots, \mathbf{u}_i(T_{max}))$*, for all i and some $T_{max} > 0$. Here $\mathbf{u}_i$ is governed by the wave equation on the graph (shown in Eq. 3) with the initial condition $\mathbf{u}(-1) = \mathbf{u}(0)$ and $0 < c < \sqrt{2}$.*

Proof. For the proofs see [27,28]. □

Algorithm 1. Wave equation based eigenvector computation algorithm for node i. At node i one computes the sign of the i-th component of the first k eigenvectors. The cluster assignment is obtained by interpreting the vector of k signs as a binary number.

$\mathbf{u}_i(0) \leftarrow \texttt{Random}\,([0,1])$
$\mathbf{u}_i(-1) \leftarrow \mathbf{u}_i(0)$
$t \leftarrow 1$
while $t < T_{max}$ **do**
  $\mathbf{u}_i(t) \leftarrow 2\mathbf{u}_i(t-1) - \mathbf{u}_i(t-2) - c^2 \sum_{j \in \mathcal{N}(i)} \mathbf{L}_{ij} \mathbf{u}_j(t-1)$
  $t \leftarrow t+1$
end while
$Y \leftarrow \texttt{FFT}\,([\mathbf{u}_i(1), \ldots\ldots, \mathbf{u}_i(T_{max})])$
for $j \in \{1, \ldots, k\}$ **do**
  $\omega_j \leftarrow \texttt{FrequencyPeak}\,(Y, j)$
  $\mathbf{v}_i^{(j)} \leftarrow \texttt{Coefficient}(\omega_j)$
  if $\mathbf{v}_i^{(j)} > 0$ **then**
    $A_j \leftarrow 1$
  else
    $A_j \leftarrow 0$
  end if
end for
ClusterNumber $\leftarrow \sum_{j=1}^{k} A_j 2^{j-1}$

The above proof demonstrates that the approach is fundamentally *decentralized*. Moreover, it is shown in [27,28] that the convergence of the wave equation based eigenvector computation depends on the mixing time of the underlying Markov chain on the graph, and is given by,

$$T_{max} = O\left(\arccos\left(\frac{2 + c^2(e^{-1/\tau} - 1)}{2}\right)^{-1}\right) + O(N)\,, \tag{4}$$

where τ is the mixing time of the Markov chain. Thus, the wave equation based algorithm has better scaling with τ for graphs of any size (given by N, see Fig. 2).

The above work is an example of the construction of a state-of-the-art algorithm using dynamical systems theory. This work has also found application in distributed numerical computations [41] and uncertainty quantification [42]. We now present another example of constructing novel algorithms for NP-hard problems using the theory of nonlinear dynamics and invariant manifold computations.

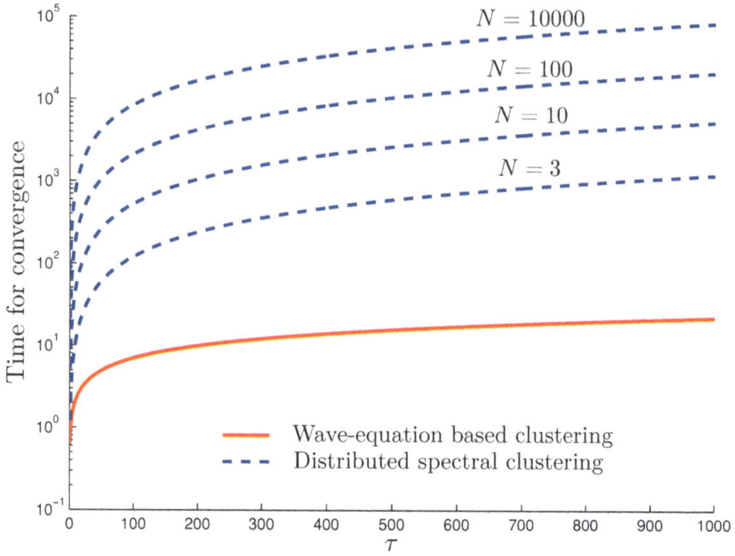

Fig. 2. Comparison of convergence rates between the distributed algorithm in [39] and our proposed wave equation algorithm for $c^2 = 1.99$. The wave equation based algorithm has better scaling with τ for graphs of any size (given by N). The plots are upper bounds on the convergence speed. For more details see [28].

3 Novel Algorithm Construction: Invariant Manifolds and the Traveling Salesman Problem

Overview

Recently, dynamical systems theory was used to construct novel algorithms for another iconic NP-hard problem [29]. The traveling salesman problem (TSP) has a long and rich history in the areas of computer science, optimization theory, and computational complexity, and has received decades of interest [19]. This combinatorial optimization problem arises in a wide variety of applications related to genome map construction, telescope management, and drilling circuit boards. The TSP also naturally occurs in applications related to target tracking [43], vehicle routing, and communication networks to name a few. We refer the reader to [19,29] for further details.

In its basic form, the statement of the TSP is exceedingly simple. The task is to find the shortest Hamiltonian circuit through a list of cities, given their pairwise distances. Despite its simplistic appearance, the underlying problem is NP-hard [20]. Several heuristics have been developed over the years to solve the problem [19] including ant colony optimization, cutting plane methods, Christofides heuristic algorithm, and the Lin–Kernighan heuristic.

In [29], inspired by dynamical systems theory, the authors construct novel orthogonal relaxation based approximations to the TSP. In particular, the constructed dynamical system captures the flow on the manifold of orthogonal matrices and ideally converges to a permutation matrix that minimizes the tour length. However, in general, the flow typically converges to local minima that are not competitive when compared to state-of-the-art heuristics. Inspired by this continuous relaxation, the authors compute the solution to a two-sided orthogonal Procrustes problem [44] that relaxes the TSP to the manifold of orthogonal matrices. They then combine the Procrustes approach with the Lin–Kernighan heuristic [25] for computing solutions of the TSP. Additionally, the authors use set-oriented methods to study the stability of optimal solutions and their stable manifolds, thereby providing insight into the associated basins of attraction and the resulting computational complexity of the problem.

Given a list of n cities $\{C_1, C_2, \ldots, C_n\}$ and the associated distances between cities C_i and C_j, denoted by d_{ij}, the TSP aims to find an ordering σ of $\{1, 2, \ldots, n\}$ such that the tour cost, given by

$$c = \sum_{i=1}^{n-1} d_{\sigma(i),\sigma(i+1)} + d_{\sigma(n),\sigma(1)}, \tag{5}$$

is minimized. For the Euclidean TSP, for instance, $d_{ij} = ||x_i - x_j||_2$, where $x_i \in \mathbf{R}^d$ is the position of C_i. In general, however, the distance matrix $D = (d_{ij})$ does not have to be symmetric (for example see [45]). The ordering σ can be represented as a unique permutation matrix P. Note, however, that due to the underlying cyclic symmetry, multiple orderings – corresponding to different permutation matrices – have the same cost.

There are several equivalent ways to define the cost function of the TSP. The authors restrict themselves to the trace[2] formulation. Let $\mathcal{P}_n$ denote the set of all $n \times n$ permutation matrices, then the TSP can be written as a combinatorial optimization problem of the form

$$\min_{P \in \mathcal{P}_n} \operatorname{tr}\left(A^T P^T B P\right), \tag{6}$$

where $A = D$ and $B = T$. Here, T is defined to be the adjacency matrix of the cycle graph of length n.

One uses the undirected cycle graph adjacency matrix for symmetric TSPs and the one corresponding to the directed cycle graphs for asymmetric TSPs.

---

[2] The trace of a matrix $A \in \mathbf{R}^{n \times n}$ is defined to be the sum of all diagonal entries, i.e., $\operatorname{tr}(A) = \sum_{i=1}^{n} a_{ii}$.

The matrices are defined as,

$$T_{\text{dir}} = \begin{pmatrix} 0 & 1 & & & \\ & 0 & 1 & & \\ & & \ddots & \ddots & \\ & & & 0 & 1 \\ 1 & & & & 0 \end{pmatrix} \quad \text{or} \quad T_{\text{undir}} = \begin{pmatrix} 0 & 1 & & & 1 \\ 1 & 0 & 1 & & \\ & \ddots & \ddots & \ddots & \\ & & 1 & 0 & 1 \\ 1 & & & 1 & 0 \end{pmatrix}.$$

The work in [29] focuses on the undirected version of the TSP. By relaxing the TSP problem to the manifold of orthogonal matrices (since permutation matrices are orthogonal matrices restricted to 0 or 1 entries), one can use the two sided Procrustes problem to solve the problem exactly, as outlined in the theorem below.

Theorem 1. *Given two symmetric matrices A and B, whose eigenvalues are distinct, let $A = V_A \Lambda_A V_A^T$ and $B = V_B \Lambda_B V_B^T$ be eigendecompositions, with $\Lambda_A = diag\left(\lambda_A^{(1)}, \ldots, \lambda_A^{(n)}\right)$, $\Lambda_B = diag\left(\lambda_B^{(1)}, \ldots, \lambda_B^{(n)}\right)$, and $\lambda_A^{(1)} \geq \cdots \geq \lambda_A^{(n)}$ as well as $\lambda_B^{(1)} \geq \cdots \geq \lambda_B^{(n)}$. Then every orthogonal matrix P^* which minimizes*

$$\min_{P \in \mathcal{O}_n} ||A - P^T B P||_F \tag{7}$$

has the form

$$P^* = V_B S V_A^T,$$

where $S = diag(\pm 1, \ldots, \pm 1)$.

A proof of this theorem can be found in [46]. If the eigenvalues of A and B are distinct, then there exist 2^n different solutions with the same cost. If one or both of the matrices possess repeated eigenvalues, then the eigenvectors in the matrices V_A and V_B are determined only up to basis rotations, which further increases the size of the solution space. The Procrustes problem is related to a dynamical system formulation of the TSP as outlined below.

The orthogonal relaxation of the combinatorial optimization problem (6), given by (7), can be solved using a steepest descent method on the manifold of orthogonal matrices. For more details about this formulation see [29]. One can pose the TSP as a constrained optimization problem of the form,

$$\min_{P \in \mathcal{O}_n} \operatorname{tr}\left(A^T P^T B P\right), \tag{8}$$

$$s.t. \;\; G(P) = 0. \tag{9}$$

This formulation gives rise to the following set of equations,

$$\begin{aligned} \dot{P} &= -P\left(\{P^T B P, A\} + \{P^T B^T P, A^T\}\right) - \lambda P\left((P \circ P)^T P - P^T (P \circ P)\right), \\ \dot{\lambda} &= \frac{1}{3} \operatorname{tr}\left(P^T\left(P - (P \circ P)\right)\right). \end{aligned} \tag{10}$$

The above set of equations are obtained by using gradient descent on the Lagrangian cost function.

Example 1. In order to illustrate the gradient flow approach, let us consider a simple TSP with 10 cities. Using (10), we obtain the results shown in Fig. 3. In this example, the dynamical system converges to the optimal tour.

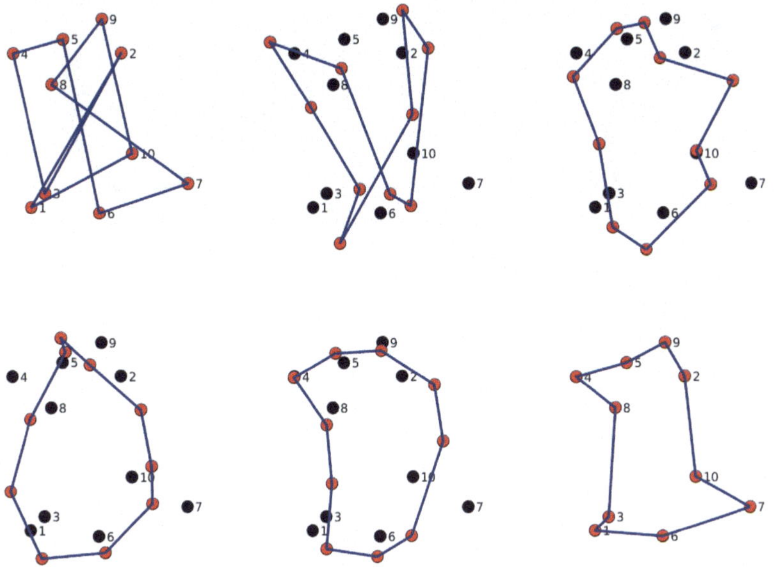

Fig. 3. Traveling salesman problem with 10 cities solved using the gradient flow (10). The original positions of the cities are shown in black, the positions transformed by the orthogonal matrix P in red. (a) Initial trivial tour given by $\sigma = (1, \ldots, 10)$. b–d) Intermediate solutions. (e) Convergence to an orthogonal matrix which is "close" to a permutation matrix with respect to any matrix norm. (f) Extraction of the corresponding permutation matrix. The initial tour was transformed into the optimal tour by the gradient flow.

The dynamical system without constraints converges to equilibria that are given by the Procrustes solutions. To shed light on the stability and local dynamics around the optimal TSP solutions one can approximate *subsets of the stable manifold* of the Procrustes solutions such that two permutation matrices are inside these sets. This numerical study enables the analysis of the robustness of Procrustes solutions under small perturbations of the initial permutation matrix and the assessment of the 'closeness' the Procrustes solution is to the optimal permutation matrix. In order to compute the sets of interest, set-oriented continuation techniques developed in [47] are used in [29]. An example computation is depicted in Fig. 4.

Moreover, one can also use set oriented methods to compute basins of attraction of optimal permutation matrices for small instances of the TSP. These basins (subsets of the stable manifold) are computed by perturbing the optimal solutions and integrating the flow backward in time [29]. The solutions are shown

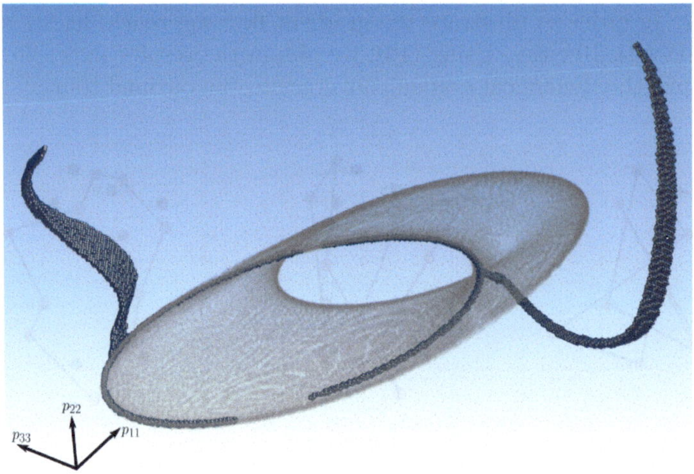

Fig. 4. Three-dimensional projection of two subsets of the stable manifold. The omega-limit sets of a small neighborhood of the permutation matrices P_1 and P_2 form a half circle on their corresponding Procrustes set.

in Fig. 5. These computations are interesting and capture the "hardness" of the problem. In particular, one can see that the solutions of relaxed versions of the problem (such as the relaxations to the manifolds of orthogonal matrices) do not, in general, lie in the basin of attraction of the optimal solutions of the original problem. Other such instances of analysis of relaxed solutions of the TSP using dynamical systems theory are outlined in [29].

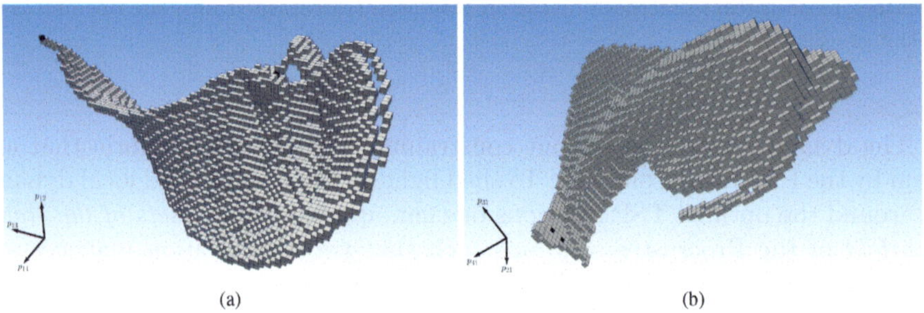

Fig. 5. (a)–(b) Three-dimensional projections of the basin of attraction of $(\widetilde{P}, \widetilde{\lambda})$. The dark cells depict the stationary solutions of the gradient flow (10) backward in time.

Although, dynamical systems theory demonstrates that the Procrustes solutions do not typically lie in the basin of attraction of the optimal solutions of the TSP, a new biasing scheme for the Lin–Kernighan heuristic is constructed using the aforementioned relaxation [29].

The Lin–Kernighan heuristic is a popular heuristic for the TSP [25]. Starting from an initial tour, the approach progresses by extracting edges from the tour and replacing them with new edges, while maintaining the Hamiltonian cycle constraint. If k edges in the tour are simultaneously replaced, this is known as the k-opt move [25]. To prune the search space, the algorithm relies on minimum spanning trees to identify edges that are more likely to be in the tour. This "importance" metric for edges is called α-nearness and described in [25,29]. The algorithm has found great success on large instances of the TSP, see [19] for more details.

In [29], the α-nearness metric is replaced with a new Procrustes solution–based metric that prunes/identifies important potential edges to include in the "candidate set list". This list is then used to generate the k-opt moves. The metric is captured in Fig. 6. The Procrustes solution tends to capture the longer edges that are important. To increase the inclusion of the short edges, the approach in [29] constructs a homotopy between the Procrustes (P-nearness) solution and the distance matrix. Using a graph Laplacian approach, the mixture of the two matrices is compared to the α-nearness approach on 22 well-known instances of the TSP. P-nearness based LKH converges to lower cost values in 18 of the instances when compared to α-nearness based LKH. Moreover, for 50 random TSP instances of size 1000 (cities) it is found that P-nearness has lower tour costs after a fixed number of k-opt moves in 31 of the instances, translating into an improvement for 62% of the instances.

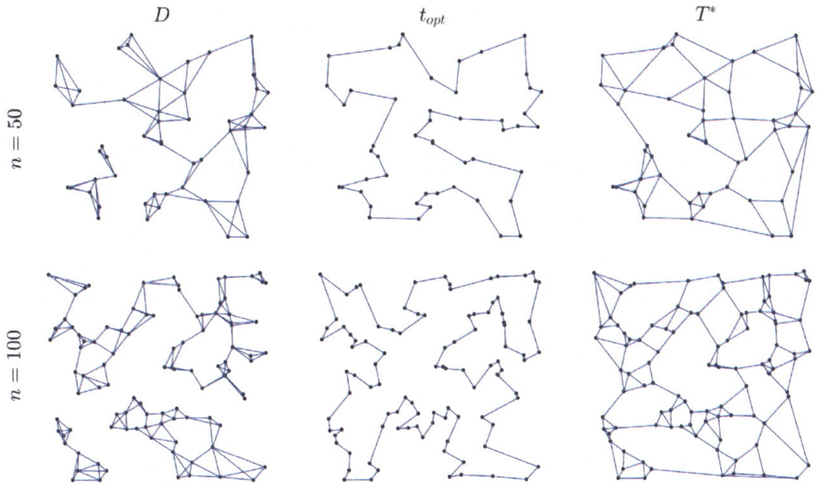

Fig. 6. Illustration of P-nearness for random TSP instances of size 50 and 100. The left column contains the edges with shortest distance, the center column has the optimal tour for the instances, and the right column contains the edges with the highest P-nearness values for each city. For each city, we plotted the three edges with the highest nearness values.

Thus, this is another example that demonstrates that dynamical systems theory can be used to analyze NP-hard problems and construct improved heuristics.

We now show how networks of Duffing oscillators can be used to construct a new algorithm for the iconic MAX-CUT problem.

4 Novel Algorithm Construction: Network of Duffing Oscillators for the MAX-CUT Problem

Overview

MAX-CUT [20] is a well-known NP-hard problem that arises in graph theory. Simply stated, the goal is to compute a subset S of the vertex set in a graph $\mathcal{G}$, such that the number of edges between S and the rest of the graph are maximized. The best known approximation ratio of 0.878 can be achieved in polynomial time using semi-definite programming [48]. The problem naturally arises in VLSI design and statistical physics, and has been extensively studied.

In [30], the authors construct an optimization algorithm by simulating the adiabatic evolution of Hamiltonian systems which can be used to approximate the MAX-CUT solution of an all-to-all connected graph. The approach is inspired by quantum adiabatic optimization for Ising systems [49] with the following energy,

$$E_{Ising}(s) = -\frac{1}{2}\sum_{i=1}^{N}\sum_{j=1}^{N} J_{ij} s_i s_j, \tag{11}$$

where s_i are the spins which can take values $\{-1,1\}$ and J_{ij} is the coupling coefficient. Finding the lowest energy state of the Ising system is computationally challenging (for a system with N spins, the potential number of states is 2^N). Note that one can map the Ising problem to the MAX-CUT problem by setting $J_{ij} = -w_{ij}$, where w_{ij} is the weight of the edge that connects nodes i and j. It is easy to show that minimizing the energy in Eq. (11) corresponds computing the solution of the MAX-CUT problem.

The approach outlined in [30] relies on the adiabatic evolution of a network of nonlinear oscillators. This system exhibits a bifurcation (called "simulated bifurcation") for each nonlinear oscillator. The two branches correspond to the -1 and $+1$ values for each spin. The authors exploit Graphical Processing Units (GPU) and Field Programmable Gate Array (FPGA) platforms to compute the solution of these Hamiltonian systems. This method is compared to existing methods and displays orders-of-magnitude improvement. The approach is demonstrated on an Ising system with 100,000 spins.

Consider the Hamiltonian that arises in Kerr-nonlinear parametric oscillators,

$$H(x,y,t) = \sum_{i=1}^{N}\left[\frac{K}{4}(x_i^2+y_i^2)^2 - \frac{p(t)}{2}(x_i^2-y_i^2) + \frac{\Delta_i}{2}(x_i^2+y_i^2)\right] - \frac{\xi_0}{2}\sum_{i=1}^{N}\sum_{j=1}^{N} J_{ij}(x_i x_j + y_i y_j). \tag{12}$$

Here x_i and y_i are position and momentum of the i-th oscillator respectively, K is the Kerr coefficient, $p(t)$ is the parametric pumping amplitude, and Δ_i is the detuning frequency between the natural frequency of the i-th oscillator and half the pumping frequency. Using the standard Hamiltonian formulation, one can derive equations of motion for each oscillator. Evolving these system of equations for x_i and y_i converges to low energy solutions of an Ising system (Eq. 11) with high probability. Thus, the sign of x_i at the end of the simulation determines the i-th spin. However, the above equations are computationally challenging to simulate from a numerical perspective.

Instead of using the equations that arise from the "full" Hamiltonian in Eq. (12), the authors (in [30]) construct a simplified Hamiltonian of the form,

$$\begin{aligned} H(x,y,t) &= \sum_{i=1}^{N}\frac{\Delta}{2}y_i^2 + V(x,t) \\ &= \sum_{i=1}^{N}\frac{\Delta}{2}y_i^2 + \left[\frac{K}{4}x_i^4 + \frac{\Delta - p(t)}{2}x_i^2\right] - \frac{\xi_0}{2}\sum_{i=1}^{N}\sum_{j=1}^{N} J_{ij}x_i x_j. \end{aligned} \tag{13}$$

The above Hamiltonian corresponds to the following system of equations,

$$\begin{aligned} \dot{x}_i &= \Delta y_i \\ \dot{y}_i &= -\left[Kx_i^2 - p(t) + \Delta\right]x_i + \xi_0 \sum_{i=1}^{N}\sum_{j=1}^{N} J_{ij}x_j. \end{aligned} \tag{14}$$

It is easy to see that the above system of equations are a network of Duffing oscillators [15]. The separability of the Hamiltonian (Eq. 13) makes the numerical integration of the system of equations easier. In particular, the authors use an explicit symplectic Euler scheme which makes it amenable for one to hard wire the resulting computational circuits on an FPGA platform. The computation proceeds as follows: all x and y variables are initially set to zero, $p(t)$ is then increased from 0 and the system in Eq. (14) is evolved. The sign of the final value of x_i serves as an approximation of the i-th spin of the associated Ising system.

The system in Eq. (14) has two branches of solutions as $p(t)$ is increased from zero. It is easy to see that these branches correspond to $\pm\sqrt{p - \Delta/K}$ for each oscillator and, consequently, leads to a 2^N solution space. If one varies $p(t)$ slowly, the adiabatic theorem ensures that if one converges to a low energy solution for $p(t)$ close to 0, the final solutions (for large $p(t)$) will also correspond to low energy.

This method is compared to state-of-the-art approaches for two instances of the MAX-CUT problem. In the first instance, an all-to-all 2000 node MAX-CUT problem is solved on an FPGA using the above approach. The authors demonstrate that the above framework successfully converges to the best known solutions [48] very quickly. Moreover, they test the approach on a 100,000 size problem (with 5×10^9 edges) and find that their approach converges to the answer 100–1000 times faster than existing software on GPU hardware. For more details of the work and associated results we refer the reader to [30].

Thus far, we have summarized three examples in which dynamical systems theory was used to construct novel algorithms for NP-hard problems. We now discuss approaches that exploit nonlinear dynamics theory to analyze optimization algorithms for NP-hard problems.

5 Analysis of Algorithms: Koopman Operators Based Analysis of Algorithms

Overview

Koopman operator theory is one of the most active and exciting sub-areas within dynamical systems theory [50–53]. The approach is based on the construction of an infinite dimensional linear operator that captures the evolution of the observables of the underlying nonlinear system. Consequently, the spectra and eigenfunctions of the operator, capture system dynamics. This methodology has been used used in a wide variety of settings, including system control and identification. An advantage of Koopman operator based methodologies is that the computations are typically based on time trace data of system evolution [54]. In recent work [31], Koopman operator theory was used to analyze algorithms. In particular, the authors consider optimization algorithms that evolve their state in the form of iterations. An assumption is made that the algorithm state spaces $X \subseteq \mathbb{R}^d$ are smooth k-dimensional Riemannian submanifolds in d-dimensional Euclidean spaces. A single iteration of the state x_n is represented as,

$$x_{n+1} = a(x_n), n \in \mathbb{N}, \tag{15}$$

where n is the iteration count. These iterative algorithms can sometimes be represented in continuous form (akin to the process used in [16]). In other words, one can (in the limit) represent the algorithm as a continuous vector field $v : X \rightarrow \mathbb{R}^k$. If one starts at an initial condition x_0, the continuous time representation is of the form [31],

$$\frac{dc(s)}{ds} = v(c(s)), \quad c(0) = x_0. \tag{16}$$

As shown in [31], this continuous form can be approximated using a Koopman operator framework. The authors then use this approach to study gradient descent and Newton-Raphson from a global dynamics perspective. Although the examples fall under the category of continuous optimization, the approach can certainly be used to study combinatorial optimization algorithms in the future.

Using the same nomenclature as in [31], consider a dynamical system of the form,

$$\frac{dS_t(x_0)}{dt} = v(S_t(x_0)), \tag{17}$$

then the family of Koopman operators $\mathscr{K}^t$ acts on the function space of observables $g : X \to \mathbb{C}$ as follows,

$$[\mathscr{K}^t g](x) = (g \circ S_t)(x). \tag{18}$$

An L^2 function space with an inner product is typically chosen for the space of observables. For more details, on the approach and choice of function space see [31]. Note that the Koopman operator is the adjoint of the Perron-Frobenius operator [55]. Letting $t = 1$, without loss of generality, The Koopman operator can be expanded in terms of its spectrum,

$$\mathscr{K} = \sum_k \lambda_k P_{\lambda_k} + \int_{\sigma_{ac}} \lambda dE(\lambda) \tag{19}$$

where λ_k lie in the discrete part and σ_{ac} is the continuous spectrum of the operator. P_λ and $dE(\lambda)$ are projection operators for their corresponding eigenspaces. The eigenfunctions of the Koopman operator are,

$$[\mathscr{K}^t \phi_\lambda](x) = (\phi_\lambda \circ S_t)(x) = \lambda^t \phi_\lambda(x). \tag{20}$$

Thus, one can predict the evolution of observables,

$$\mathscr{K}^t g = \sum_k c_k \lambda^t \phi_{\lambda,k}. \tag{21}$$

The operator can be numerically approximated using a data driven approach as outlined in [52,53]. The popular extended dynamic mode decomposition (EDMD) methodology introduced in [52] is used to analyze algorithms using the Koopman operator lens [31].

The EDMD approach approximates the action of the infinite dimensional operator using a finite set of real-valued functions (also called "dictionary"). In particular, given a smooth manifold $M \subset \mathbb{R}^d$ sampled by a finite set of points $X = \{x_i \in M\}$, the EDMD approach computes the action of the Koopman operator on the dictionary of points in X. The operator itself is approximated using a least squares approach [52] as outlined below. Given a dictionary of N_D observables $D = \{d_i : M \to \mathbb{R} | \ i = 1, \ldots, N_D\}$ one can define a matrix of the form $G = [d_1(X), d_2(X), \ldots, d_{N_D}(X)]$. Then the Koopman operator $\mathscr{K}^t$ can be approximated as,

$$K = \frac{1}{N_X^2}(G^T G)^\dagger (A^T A), \tag{22}$$

where N_X is the size of the dataset and $A = [\mathscr{K}^t d_1(X), \mathscr{K}^t d_2(X), \ldots, \mathscr{K}^t d_{N_D}(X)]$. For more details see [52].

The operator gives a local approximation of the underlying algorithm applied to a specific instance of a problem. In particular, one can use the data of a short burst of computation to compute a local approximation of the dynamics of the algorithm to *accelerate* its convergence. The eigenvalues, vectors, and modes are computed using EDMD. This approximation is used as a data-driven surrogate for the system to accelerate optimization. In [31], the authors use the following cost function,

$$f(x_1, x_2) = (x_1^2 + x_2 - 11)^2 + (x_1 + x_2^2 - 7)^2, \tag{23}$$

to demonstrate utility of the Koopman approach. The function has one local maximum and four local minima [31]. The authors study the the popular gradient descent algorithm using the Koopman operator framework. In particular, they use radial basis functions to form a dictionary and compute 503 eigenvalues and eigenfunctions. They show that one can construct an ergodic decomposition of the state space, thereby separating the different basins of attraction [31]. The approach is able to capture the global dynamics of the algorithm in this setting, providing valuable insight regarding the performance and limitations of the algorithm.

Additionally, the authors demonstrate the use of Koopman operators for studying global dynamics of algorithms in high-dimensional spaces. They take the example of a 100-dimensional problem and show that the dynamics quickly contracts to a low dimensional subset. They demonstrate that one can accelerate the prediction of trajectories of gradient descent with high accuracy. The work concludes with the illustration of the utility of Koopman operators for analyzing the iconic Newton-Raphson method for root finding [31]. For a complex polynomial of degree two, they show that the eigenfunction diverges at the roots. They also show that in cases of chaotic behavior of Newton-Raphson, one can use approximations [56] of the continuous spectrum of the Koopman operator to study statistical properties of the emergent chaos.

Although the above Koopman methodology was demonstrated on problems of continuous optimization, it provides a new technique with which one can study combinatorial optimization problems. We anticipate that the Koopman operator approach will be a new tool with which to study algorithms for NP-hard problems and improve their performance.

6 Analysis of Algorithms: Chaos and Dynamical Systems for Analyzing the Satisfiability (SAT) Problem

Overview

The satisfiability problem is another iconic problem that frequently arises in the study of computational complexity theory. The challenge here is to find a satisfying assignment for a logical formula. In particular, a k-SAT Boolean formula $\phi(x)$ of N Boolean variables and m clauses, $\phi : \{0,1\}^N \rightarrow \{0,1\}$, is written in the conjunctive normal form (CNF) [59] as follows,

$$\phi(x) = \bigwedge_{i=1}^{m} C_i = \bigwedge_{i=1}^{m} (x_{i_1} \vee x_{i_2} \vee \ldots \vee x_{i_k}), \tag{24}$$

where x_{i_l} is the lth literal in clause C_i. A SAT formula is said to be *satisfiable* if there exists an assignment for the binary variables $\mathbf{x}$ such that $\phi(\mathbf{x}) = 1$ (true). It is well known that the satisfiability problem is NP-complete [19]. A critical parameter associated with the satisfiability problem is the clause density $\alpha = m/N$ [58]. In particular, the probability that a random k-SAT instance is satisfiable undergoes a phase transition as a function of α $(N \rightarrow \infty)$ [58]. The MAX-SAT problem (and the corresponding weighted version) [59] requires one to find that assignment (or assignments) that maximize the number (or the cumulative weights) of satisfied clauses. Consider a SAT formula ϕ, then every assignment x can be mapped to an "energy" $\Phi(x)$ such that,

$$\Phi(\mathbf{x}) = \sum_{i=1}^{m} C_i, \tag{25}$$

where $C_i = 1$, if the i-th clause evaluates to true. In other words, the goal under the MAX-SAT problem is to find the assignment for $\mathbf{x}$ such that the number of satisfied clauses (or energy) is maximized. The MAX-SAT problem is harder (from a computational standpoint) than the SAT problem. In particular, it is known to be strongly NP-hard (there are no polynomial time approximation schemes). The problem of computing density of states (DOS) encompasses the SAT and MAX-SAT problems. Classical and quantum algorithms for estimating the DOS of logical formulae were constructed in [60].

In a series of seminal papers [32,33,57], the authors construct a dynamical systems approach to study satisfiability problems. They construct a dynamical system that computes the solutions of SAT instances. Here, the equilibria of the dynamical system correspond to literal values for which the SAT formula in Eq. (24) evaluates to `true`. They prove that the dynamical system admits no *false* equilibria or limit cycles. Additionally, they relate the emergence of transient chaos and fractal boundaries with optimization hardness of the problem instance [32], pointing to a deep connection between dynamical systems theory and computational complexity.

As mentioned in the overview, in [32], the authors embed SAT equations into a system of ordinary differential equations using the the following mapping: let $s_{x_i} = [-1, 1]$, i.e. s_{x_i} can take values between -1 and 1 such that,

$$s_{x_i} = \begin{cases} -1, & \text{if } x_i = \text{FALSE} \\ 1, & \text{if } x_i = \text{TRUE}. \end{cases}$$

Generalizing the dynamical system for satisfiability problems constructed in [32], one can define c_{mi} and K_m as follows,

$$c_{mi} = \begin{cases} -1, & \text{if } s_{x_i} \text{ appears in negated form in m-th clause} \\ 1, & \text{if } s_{x_i} \text{ appears in direct form in m-th clause} \\ 0, & \text{otherwise,} \end{cases}$$

$$K_m(s) = 2^{-k} \prod_{j=0}^{k-1} \prod_{i=1}^{N} (1 - c_{mi} s_{x_i}) \quad \forall m = 1, 2, \ldots, M.$$

Note that $K_m(s) = 0$, if and only if m-th clause is satisfied i.e. $c_{mi} s_{x_i} = 1$ for at least one variable x_i that appears in clause m. In [32], the authors define an energy function of the form $V(s) = \sum_{m=1}^{M} a_m K_m(s)^2$ such that $V(s^*) = 0$ only at a solution s^* of the satisfiability problem. The *auxiliary* variables $a_m \in (0, \infty)$ prevent the non-solution attractors from trapping the search dynamics (for more information see [32]).

In [32], the authors find that as the constraint density of the k-SAT problem increases, the trajectories of the dynamical system display intermittent chaos with fractal basin boundaries [32]. Note that, in this work, the existence of chaos is associated with the emergence of positive finite size Lyapunov exponents (FSLE) [15] and the emergence of chaos corresponds to the well known phase transitions in the k-SAT problem [32].

In [33], the authors further exploit the above system of equations to study the k-SAT problem with increasing constraint density. They find that hardness appears as a second order phase transition and discover that the resulting transient chaos displays a novel exponential-algebraic scaling. In [57], the authors exploit the above framework to construct novel solvers for the MAX-SAT problem. This body of work demonstrates that dynamical systems theory can, in fact, be used to simultaneously study computational complexity theory and construct novel algorithms for NP-hard problems.

7 Conclusion

Combinatorial optimization is a wide and important area of research with numerous applications. For decades, computer scientists have developed novel algorithms for addressing these problems. Some problems are amenable to algorithms and theory developed thus far (examples include graph routing and sorting), while others, in general, remain intractable from a computational standpoint (such as the traveling salesman problem and MAX-SAT) despite significant efforts. The classification of problems into different classes (such as NP, NP-hard, and PSPACE) and associated analyses has given rise to the field of computational complexity theory.

Nonlinear dynamics, on the other hand, arises in a multitude of engineering and scientific settings. The theory has been used to explain system behavior in a diverse set of fields such as fluidics, structural mechanics, population dynamics, epidemiology, optics, and aerospace propulsion. However, the application of the theory of dynamical systems to combinatorial optimization and computational complexity remains limited.

In this survey article, we summarize five recent examples of using dynamical systems theory for constructing and analyzing combinatorial optimization problems. In particular, we cover (a) a novel approach for clustering graphs using the wave equation partial differential equation (PDE), (b) invariant manifold computations for the traveling salesman problem, (c) novel approaches for building quantum networks of Duffing oscillators to solve the MAX-CUT problem, (d) applications of the Koopman operator for analyzing optimization algorithms, and (e) the use of dynamical systems theory to analyze computational complexity of the SAT problem.

We note that the above set of examples are not comprehensive and there are several examples that have been omitted in this survey. However, the goal of this article is not to provide a complete list of all such examples but to demonstrate that dynamical systems theory can be exploited to construct algorithms and approaches for optimization problems. Even more importantly, we hope to inspire others to extend existing dynamical systems approaches to construct the next-generation of techniques and insights for combinatorial optimization.

Acknowledgements. This material is based upon work supported by the Defense Advanced Research Projects Agency (DARPA) and Space and Naval Warfare Systems Center, Pacific (SSC Pacific) under Contract No. N6600118C4031.

References

1. Poincaré, H.: Sur le problème des trois corps et les équations de la dynamique. Acta Mathe. **13**, A3–A270 (1890)
2. Turing, A.: On computable problems with an application to the entscheidungsproblem. Proc. Lond. Math. Soc. **2**, 42 (1936)
3. Church, A.: A set of postulates for the foundation of logic. Ann. Math. **33**, 346–366 (1932)
4. Strogatz, S.H.: Nonlinear Dynamics and Chaos: With Applications to Physics, Biology, Chemistry, and Engineering. CRC Press, Boca Raton (2018)
5. Walleczek, J. (ed.): Self-Organized Biological Dynamics and Nonlinear Control: Toward Understanding Complexity, Chaos and Emergent Function in Living Systems. Cambridge University Press (2006)

6. Kevrekidis, I.G., Schmidt, L.D., Aris, R.: On the dynamics of periodically forced chemical reactors. Chem. Eng. Commun. **30**(6), 323–330 (1984)
7. Holmes, P., Lumley, J.L., Berkooz, G., Rowley, C.W.: Turbulence, Coherent Structures, Dynamical Systems and Symmetry. Cambridge University Press, Cambridge (2012)
8. Liu, W.M., Hethcote, H.W., Levin, S.A.: Dynamical behavior of epidemiological models with nonlinear incidence rates. J. Math. Biol. **25**(4), 359–380 (1987)
9. Dellnitz, M., Junge, O.: Set oriented numerical methods for dynamical systems. Handb. Dyn. Syst. **2**, 221–264 (2002)
10. Mezić, I.: Analysis of fluid flows via spectral properties of the Koopman operator. Annu. Rev. Fluid Mech. **45**, 357–378 (2013)
11. Hummer, G., Kevrekidis, I.G.: Coarse molecular dynamics of a peptide fragment: free energy, kinetics, and long-time dynamics computations. J. Chem. Phys. **118**(23), 10762–10773 (2003)
12. Hasselblatt, B., Katok, A. (eds.) Handbook of Dynamical Systems. Elsevier, Amsterdam (2002)
13. Lorenz, E.N.: Deterministic nonperiodic flow. J. Atmos. Sci. **20**(2), 130–141 (1963)
14. Morse, M., Hedlund, G.A.: Symbolic dynamics. Am. J. Math. **60**(4), 815–866 (1938)
15. Guckenheimer, J., Holmes, P.: Nonlinear Oscillations, Dynamical Systems, and Bifurcations of Vector Fields, vol. 42. Springer, New York (2013)
16. Nesterov, Y.: A method for unconstrained convex minimization problem with the rate of convergence $O(1/k^2)$. In: Doklady an Ussr, vol. 269, pp. 543–547 (1983)
17. Su, W., Boyd, S., Candes, E.: A differential equation for modeling Nesterov's accelerated gradient method: theory and insights. In: Advances in Neural Information Processing Systems, pp. 2510–2518 (2014)
18. Wibisono, A., Wilson, A.C., Jordan, M.I.: A variational perspective on accelerated methods in optimization. Proc. Natl. Acad. Sci. **113**(47), E7351–E7358 (2016)
19. Cook, S.A.: The complexity of theorem-proving procedures. In: Proceedings of the Third Annual ACM Symposium on Theory of Computing, pp. 151–158 (1971)
20. Karp, R.M.: Reducibility among combinatorial problems. In: Complexity of Computer Computations, pp. 85–103. Springer, Boston (1972)
21. Lin, S.: Computer solutions of the traveling salesman problem. Bell Syst. Tech. J. **44**(10), 2245–2269 (1965)
22. Cook, W.J.: In Pursuit of the Traveling Salesman: Mathematics at the Limits of Computation. Princeton University Press, Princeton (2011)
23. Peikert, C.: A decade of lattice cryptography. Found. Trends® Theoret. Comput. Sci. **10**(4), 283–424 (2016)
24. Klus, S., Sahai, T.: A spectral assignment approach for the graph isomorphism problem. Inf. Inference J. IMA **7**(4), 689–706 (2018)
25. Helsgaun, K.: An effective implementation of the Lin-Kernighan traveling salesman heuristic. Eur. J. Oper. Res. **126**(1), 106–130 (2000)
26. Applegate, D., Bixby, R., Chvatal, V., Cook, W. Concorde TSP Solver (2006)
27. Sahai, T., Speranzon, A., Banaszuk, A.: Wave equation based algorithm for distributed eigenvector computation. In: 49th IEEE Conference on Decision and Control (CDC), pp. 7308–7315, December 2010
28. Sahai, T., Speranzon, A., Banaszuk, A.: Hearing the clusters of a graph: a distributed algorithm. Automatica **48**(1), 15–24 (2012)
29. Sahai, T., Ziessler, A., Klus, S., Dellnitz, M.: Continuous relaxations for the traveling salesman problem. Nonlinear Dyn. **97**(4), 2003–2022 (2019)

30. Goto, H., Tatsumura, K., Dixon, A.R.: Combinatorial optimization by simulating adiabatic bifurcations in nonlinear Hamiltonian systems. Sci. Adv. **5**(4), eaav2372 (2019)
31. Dietrich, F., Thiem, T.N., Kevrekidis, I.G.: On the Koopman operator of algorithms. arXiv preprint arXiv:1907.10807 (2019)
32. Ercsey-Ravasz, M., Toroczkai, Z.: Optimization hardness as transient chaos in an analog approach to constraint satisfaction. Nat. Phys. **7**(12), 966–970 (2011)
33. Varga, M., Sumi, R., Toroczkai, Z., Ercsey-Ravasz, M.: Order-to-chaos transition in the hardness of random Boolean satisfiability problems. Phys. Rev. E **93**(5), 052211 (2016)
34. Even, S.: Graph Algorithms. Cambridge University Press, UK (2011)
35. Cormen, T.H., Leiserson, C.E., Rivest, R.L., Stein, C.: Introduction to Algorithms. MIT Press (2009)
36. Von Luxburg, U.: A tutorial on spectral clustering. Stat. Comput. **17**(4), 395–416 (2007)
37. Wagner, D., Wagner, F.: Between min cut and graph bisection. In: International Symposium on Mathematical Foundations of Computer Science, pp. 744–750. Springer, Berlin (1993)
38. Golub, G.H., Van Loan, C.F.: Matrix Computations. Johns Hopkins University Press, Baltimore and London (1996)
39. Kempe, D., McSherry, F.: A decentralized algorithm for spectral analysis. J. Comput. Syst. Sci. **74**(1), 70–83 (2008)
40. Kac, M.: Can one hear the shape of a drum? Am. Math. Monthly **73**(4P2), 1–23 (1966)
41. Klus, S., Sahai, T., Liu, C., Dellnitz, M.: An efficient algorithm for the parallel solution of high-dimensional differential equations. J. Comput. Appl. Math. **235**(9), 3053–3062 (2011)
42. Surana, A., Sahai, T., Banaszuk, A.: Iterative methods for scalable uncertainty quantification in complex networks. Int. J. Uncertain. Quantific. **2**(4) (2012)
43. Englot, B., Sahai, T., Cohen, I.: Efficient tracking and pursuit of moving targets by heuristic solution of the traveling salesman problem. In: 52nd IEEE Conference on Decision and Control, pp. 3433–3438. IEEE, December 2013
44. Gower, J.C., Dijksterhuis, G.B.: Procrustes Problems, vol. 3. Oxford University Press, New York (2004)
45. Sahai, T., Klus, S., Dellnitz, M.: A Traveling Salesman Learns Bayesian Networks. arXiv preprint arXiv:1211.4888 (2012)
46. Schönemann, P.H.: On two-sided orthogonal Procrustes problems. Psychometrika **33**(1), 19–33 (1968)
47. Dellnitz, M., Hohmann, A.: The computation of unstable manifolds using subdivision and continuation. In: Nonlinear Dynamical Systems and Chaos, pp. 449–459. Birkhäuser, Basel (1996)
48. Goemans, M.X., Williamson, D.P.: Improved approximation algorithms for maximum cut and satisfiability problems using semidefinite programming. J. ACM **42**(6), 1115–1145 (1995)
49. Glauber, R.J.: Time-dependent statistics of the Ising model. J. Math. Phys. **4**(2), 294–307 (1963)
50. Budisić, M., Mohr, R., Mezić, I.: Applied Koopmanism. Chaos Interdiscipl. J. Nonlinear Sci. **22**(4), 047510 (2012)
51. Klus, S., Koltai, P., Schütte, C.: On the numerical approximation of the Perron-Frobenius and Koopman operator. arXiv preprint arXiv:1512.05997 (2015)

52. Williams, M.O., Kevrekidis, I.G., Rowley, C.W.: A data-driven approximation of the Koopman operator: extending dynamic mode decomposition. J. Nonlinear Sci. **25**(6), 1307–1346 (2015)
53. Lusch, B., Kutz, J.N., Brunton, S.L.: Deep learning for universal linear embeddings of nonlinear dynamics. Nat. Commun. **9**(1), 1–10 (2018)
54. Kutz, J.N., Brunton, S.L., Brunton, B.W., Proctor, J.L.: Dynamic Mode Decomposition: Data-driven Modeling of Complex Systems. Society for Industrial and Applied Mathematics (2016)
55. Dellnitz, M., Hohmann, A., Junge, O., Rumpf, M.: Exploring invariant sets and invariant measures. Chaos Interdiscipl. J. Nonlinear Sci. **7**(2), 221–228 (1997)
56. Korda, M., Putinar, M., Mezić, I.: Data-driven spectral analysis of the Koopman operator. Appl. Comput. Harmonic Anal. (2018)
57. Molnár, B., Molnár, F., Varga, M., Toroczkai, Z., Ercsey-Ravasz, M.: A continuous-time MaxSAT solver with high analog performance. Nat. Commun. **9**(1), 1–12 (2018)
58. Biere, A., Heule, M., van Maaren, H. (eds.) Handbook of Satisfiability, vol. 185. IOS Press, Amsterdam (2009)
59. Krentel, M.W.: The complexity of optimization functions. J. Comput. Syst. Sci. **36**, 490–509 (1988)
60. Sahai, T., Mishra, A., Pasini, J.M., Jha, S.: Estimating the density of states of Boolean satisfiability problems on classical and quantum computing platforms. In: 34th AAAI Conference on Artificial Intelligence (AAAI) (2020)

Optimal Control

Symmetry in Optimal Control: A Multiobjective Model Predictive Control Approach

Kathrin Flaßkamp[1(✉)], Sina Ober-Blöbaum[2], and Sebastian Peitz[2]

[1] Systems Modeling and Simulation, Saarland University, Saarbrücken, Germany
kathrin.flasskamp@uni-saarland.de
[2] Department of Mathematics, Paderborn University, Paderborn, Germany
{sinaober,speitz}@math.upb.de

Abstract. Many dynamical systems possess symmetries, e.g. rotational and translational invariances of mechanical systems. These can be beneficially exploited in the design of numerical optimal control methods. We present a model predictive control scheme which is based on a library of precomputed motion primitives. The primitives are equivalence classes w.r.t. the symmetry of the optimal control problems. Trim primitives as relative equilibria w.r.t. this symmetry, play a crucial role in the algorithm. The approach is illustrated using an academic mobile robot example.

Keywords: Dynamical systems with symmetry · Model predictive control · Motion planning · Motion primitives · Multiobjective optimization · Optimal control · Relative equilibria · Scalarization methods

1 Introduction

Symmetry and dynamical systems are closely intertwined [52]. Besides discrete symmetries such as reflections or rotations (by fixed angles), dynamical systems typically possess continuous symmetries. Colloquially, an object is called symmetric if its reflection looks the same. For dynamical systems, symmetric trajectories are indistinguishable, i.e., the dynamic behavior is the same and thus, it is called *invariant* with respect to the symmetry action. Symmetry as a mathematical concept is well studied in the literature, cf. the classical textbooks [33,48], for instance. Besides revealing and analyzing symmetry structures in dynamical systems, the existence of symmetry can be used to simplify (i.e., reduce) the formal description. For mechanical systems moving in $\mathbb{R}^2$ or $\mathbb{R}^3$, symmetries are often given by translational or rotational invariances. Throughout this chapter, we focus on autonomous systems, such that translation in time provides another symmetry indeed.

O. Junge et al. (Eds.): SON 2020, SSDC 304, pp. 209–237, 2020.
https://doi.org/10.1007/978-3-030-51264-4_9

The concept of symmetry in dynamical systems takes over to control systems [7,8,13]. From an engineering point of view, a control system can be steered via inputs such that a given control task is performed. It is thus natural to search for a control input that performs best w.r.t. given performance criteria. Studying optimal control of dynamical systems also has a long history, dating back to the famous Brachistochrone problem [64]. Improved computational means over the past 50 years today allow for the numerical treatment of increasingly challenging optimal control problems, including the simultaneous consideration of multiple conflicting criteria. In this case, one cannot hope for single (local) optima. Instead, the set of optimal compromises – the *Pareto set* – has to be computed.

In very early works, symmetries of optimal control problems have been considered based on symmetries of the optimal control Hamiltonian (see e.g. [67]) and used to construct decompositions of optimal feedback laws [31]. Symmetric optimal control problems have also been studied in [5,16,65,66] where a Noether theorem for optimal control problems is proven leading to generalized conserved quantities along the solutions. However, so far, symmetry in optimal control has been less studied in motion planning and model predictive control applications even though it may be advantageously exploited for computational efficiency.

Optimal control and symmetries in dynamical systems both become crucial when addressing up-to-date problems raising in the design and control of modern technical systems. Since intelligent technical systems shall autonomously adapt to current situations and environmental conditions, as well as to tasks and requirements, multiobjective optimal control problems arise continuously during operation, and technical systems have to be equipped with highly efficient on-board solution methods.

Having these challenges in mind, this contribution is concerned with model predictive control (MPC) of symmetric dynamical systems subject to multiple objectives. We study symmetry properties in the optimal control problems which have to be solved continuously in MPC, and we provide a symmetry exploiting planning method for trajectory optimization. Similar methods have been presented in [57] as well as in [36] for noisy systems. Here, we extend these ideas by additionally taking into account the concept of trim primitives in motion planning [24,25,28], where trims and maneuvers appear alternatingly. For illustration, we use an academic example of a mobile robot system, resembling a box-shaped robot moving on the ground via hovercraft-type propulsion (as introduced e.g. in [21,55]).

Example 1 (Mobile robot). The simplified dynamics of a mechanical system moving on a plane (cf. Fig. 1) is defined by the Lagrangian function

$$L(q, \dot{q}) = \frac{1}{2}m(\dot{x}_1^2 + \dot{x}_2^2) + \frac{1}{2}\Theta\dot{x}_3,$$

where $q = (x_1, x_2, x_3)$ is the robot's configuration comprised of the two positions in the plane, as well as the orientation with respect to the horizontal axis, and m and Θ are the robot's mass and inertia. Considering two control inputs as depicted in Fig. 1, the dynamics can be described by the following second order differential equation:

$$\ddot{q} = \begin{pmatrix} \frac{1}{m}\left(\cos(x_3)u_1 - \sin(x_3)u_2\right) \\ \frac{1}{m}\left(\sin(x_3)u_1 + \cos(x_3)u_2\right) \\ -\frac{ru_2}{\Theta} \end{pmatrix}. \tag{1}$$

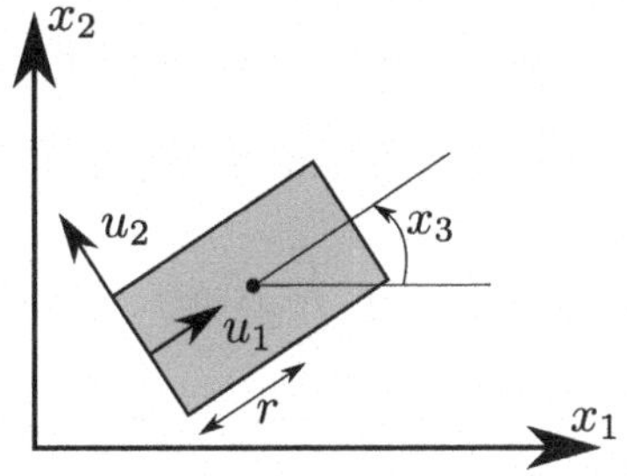

Fig. 1. Simple model of an mobile robot in the plane with second order dynamics.

The remainder of this chapter is organized as follows: We collect mathematical preliminaries in Sect. 2. Symmetry in multiobjective optimal control is studied in Sect. 3. We present as symmetry exploiting MPC scheme in Sect. 4. In Sect. 5, numerical results are presented, before we conclude in Sect. 6.

2 Preliminaries

In this section, we summarize the basic concepts in the fields of multiobjective optimal control, model predictive control (MPC) and symmetry that will be required throughout the chapter.

2.1 Multiobjective Optimal Control

In multiobjective optimal control one is interested in minimizing multiple conflicting objectives while taking the system dynamics into account which are here described by an ordinary differential equation (see also [57]):

$$\min_{x\in\mathcal{X},u\in\mathcal{U}} J(x,u) = \begin{pmatrix} \int_{t_0}^{t_e} C_1(x(t),u(t))\,dt + \Phi_1(x(t_e)) \\ \vdots \\ \int_{t_0}^{t_e} C_k(x(t),u(t))\,dt + \Phi_k(x(t_e)) \end{pmatrix} \tag{2a}$$

$$\text{s.t.} \qquad \dot{x}(t) = f(x(t),u(t)), \qquad t \in (t_0,t_e], \tag{2b}$$

$$x(t_0) = x^0, \tag{2c}$$

$$g_i(x(t),u(t)) \leq 0, \quad i = 1,\ldots,l, \qquad t \in (t_0,t_e], \tag{2d}$$

$$h_j(x(t),u(t)) = 0, \quad j = 1,\ldots,m, \qquad t \in (t_0,t_e], \tag{2e}$$

with system state $x(t) \in \mathbb{R}^{n_x}$ and control $u(t) \in U \subset \mathbb{R}^{n_u}$ with U being closed and convex and $\mathcal{X} = W^{1,\infty}([t_0, t_e], \mathbb{R}^{n_x})$ and $\mathcal{U} = L^\infty([t_0, t_e], U)$ being the corresponding function spaces for the curves x and u respectively. The cost function $J : \mathcal{X} \times \mathcal{U} \to \mathbb{R}^k$ involves k conflicting objectives and the functions $C_i : \mathbb{R}^{n_x} \times U \to \mathbb{R}$, $\Phi_i : \mathbb{R}^{n_x} \to \mathbb{R}$ are continuously differentiable. $f : \mathbb{R}^{n_x} \times U \to \mathbb{R}^{n_x}$ is Lipschitz continuous, and $g : \mathbb{R}^{n_x} \times U \to \mathbb{R}^l$, $g = (g_1, \dots, g_l)^\top$ and $h : \mathbb{R}^{n_x} \times U \to \mathbb{R}^m$, $h = (h_1, \dots, h_m)^\top$, are continuously differentiable inequality and equality constraint functions, respectively.

(x, u) is a *feasible pair* if it satisfies the constraints (2b)–(2e). The space of the control trajectories $\mathcal{U}$ is the *decision space* and the image of all feasible pairs forms the *objective space.*

By introducing the flow of the dynamical system

$$\varphi_u(x^0, t) = x^0 + \int_{t_0}^{t} f(x(t), u(t))\, dt, \tag{3}$$

the explicit dependency of J, g and h on x can be removed. This leads to the following simplified multiobjective optimal control problem:

$$\begin{aligned} &\min_{u \in \mathcal{U}} \hat{J}(x^0, u) \\ &\hat{g}_i(x^0, \mathfrak{u}) \leq 0, \quad i = 1, \dots, l, \qquad t \in (t_0, t_e], \\ &\hat{h}_j(x^0, \mathfrak{u}) = 0, \quad j = 1, \dots, m, \qquad t \in (t_0, t_e], \end{aligned} \tag{MOCP}$$

with

$$\hat{J}_i(x^0, u) = \int_{t_0}^{t_e} \hat{C}_i(x^0, \mathfrak{u})\, dt + \hat{\Phi}_i(x^0, \mathfrak{u})$$

with $\hat{C}_i(x^0, \mathfrak{u}) := C_i(\varphi_u(x^0, t), u(t))$ and $\hat{\Phi}_i(x^0, \mathfrak{u}) := \Phi_i(\varphi_u(x^0, t_e))$ for $i = 1, \dots, k$. We use the notation $\mathfrak{u} := u|_{[t_0, t]}$ to represent the control function up to time t which is assumed to uniquely define the state $x(t)$ at time t. The constraints $\hat{g}(x^0, \mathfrak{u})$ and $\hat{h}(x^0, \mathfrak{u})$ are defined accordingly. u is a *feasible curve* if it satisfies the equality and inequality constraints $\hat{g}_i, i = 1, \dots, l$, and $\hat{h}_j, j = 1, \dots, m$.

Since there exists no total order of the objective function values in $\mathbb{R}^k$ with $k \geq 2$, we introduce the following partial order:

Definition 1. Let $v, w \in \mathbb{R}^k$. The vector v is *less than* w (denoted by $v < w$), if $v_i < w_i$ for all $i \in \{1, \dots, k\}$. The relation $\leq$ is defined in an analogous way.

In general, we cannot expect to find isolated optimal curves for (MOCP). Instead, we look for a *set of optimal compromises* (also called the *Pareto set* or *set of non-dominated curves*) which is defined as follows.

Definition 2. Consider problem (MOCP). Then

1. a feasible curve u^* *dominates* a curve u, if $\hat{J}(x^0, u^*) \leq \hat{J}(x^0, u)$ and $\hat{J}(x^0, u^*) \neq \hat{J}(x^0, u)$.

2. a feasible curve u^* is called *globally Pareto optimal* if there exists no feasible curve $u \in \mathcal{U}$ dominating u^*. The image $\hat{J}(x^0, u^*)$ of a globally Pareto optimal curve u^* is called a *globally Pareto optimal value*. If this property holds in a neighborhood $U(u^*) \subset \mathcal{U}$, then u^* is called *locally Pareto optimal*.
3. the set of non-dominated feasible curves is called the *Pareto set* $\mathcal{P}$, its image the *Pareto front* $\mathcal{P}_F$.

An example of Pareto optimality is illustrated in Fig. 2 for a finite-dimensional problem, i.e. curves reduce to points in the Euclidean space. By varying a point in the Pareto set (red line in Fig. 2 (a)), we can only improve one objective by accepting a trade-off in at least one other objective, as shown by the red line in Fig. 2 (b).

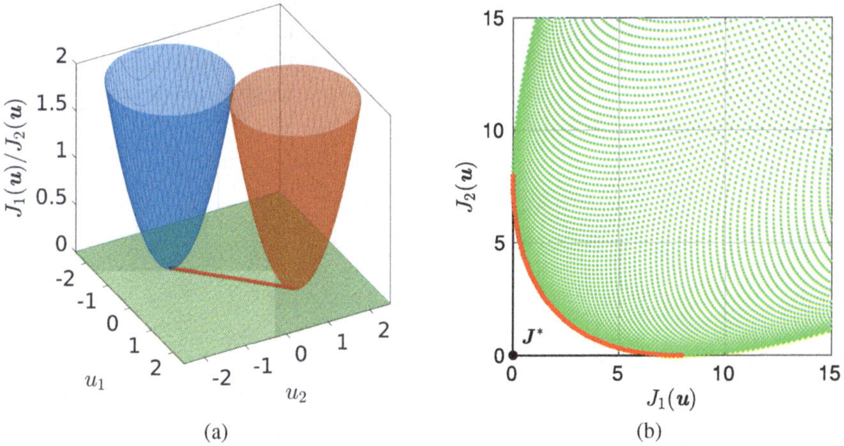

Fig. 2. Red lines: Pareto set $\mathcal{P}$ (a) and Pareto front $\mathcal{P}_F$ (b) of an example problem (two paraboloids) of the form $\min_{u \in \mathbb{R}^2} J(u)$, $J : \mathbb{R}^2 \to \mathbb{R}^2$. The point $J^* = (0, 0)^\top$ is called the utopian point.

For the numerical solution of optimal control problems, there exist many fundamentally different approaches which can be categorized into *direct or indirect solution methods*. Direct methods rely on a discretization of the objective functions, the dynamics and the constraints to transform the problem into a finite-dimensional multiobjective optimization problem (MOP) – also denoted as "first discretize, then optimize", see e.g., [56]). On the other hand, indirect methods derive necessary optimality conditions based on *Pontryagin's Maximum Principle* (also denoted as "first optimize, then discretize"). For an overview of different methods, see e.g. [6,45]).

Solution methods for MOCPs typically rely on a direct approach such that one is faced with finite-dimensional MOPs [46,58,62]. Well-established methods for solving such MOPs are *scalarization techniques*, *continuation methods*, *evolutionary algorithms* [9], *set-oriented methods* [15,62], *or cell-to-cell mapping techniques* [35,63]. For example, continuation methods make use of the fact that under certain conditions, the Pareto set is a smooth manifold of dimension $k-1$ that can be approximated using *predictor-corrector* schemes [37]. More recent continuation methods are, for example, the *Pareto Tracer* for multiobjective problems and the *Pareto Explorer* for many objective optimization problems [50].

In scalarization, solution approaches for single objective optimization problems are extended to the multiobjective problem by transforming the MOCP into a sequence of scalar-valued problems[1]. In this way, the Pareto set is approximated by a finite set of Pareto optimal curves. There exists a large variety of scalarization approaches in the literature such as the *weighted-sum method*, the *ϵ-constraint method*, *normal boundary intersection*, or *reference point methods* [18]. In this work, we focus on the latter to solve MOCPs, which is illustrated in Fig. 3. In the reference point method, the minimization of the vector valued objective function in (MOCP) is replaced by the minimization of a scalar function which is the euclidean distance between a feasible point $J(u^{(i)})$ and an infeasible *target* $T^{(i)} < J(u^{(i)})$:

$$\begin{aligned} &\min_{u^{(i)}\in\mathcal{U}} \left\| \hat{J}(x^0, u^{(i)}) - T^{(i)} \right\|_2^2 \\ &\hat{g}_i(x^0, \mathrm{u}^{(i)}) \leq 0, \quad i = 1, \ldots, l, \quad t \in (t_0, t_e], \\ &\hat{h}_j(x^0, \mathrm{u}^{(i)}) = 0, \quad j = 1, \ldots, m, \quad t \in (t_0, t_e], \end{aligned} \tag{RP}$$

The solution of problem (RP) for one $T^{(i)}$ yields one Pareto optimal curve and the corresponding point on the Pareto front. Once two points of the Pareto front are known, these are used to approximate the tangent space of the front and to construct the target $T^{(i+1)}$ for the next scalar problem, cf. Fig. 3 (b). This is done by shifting the point first parallel (i.e., along the vector $\hat{J}(x^0, u^{(i)}) - \hat{J}(x^0, u^{(i-1)})$) and then orthogonal (parallel to the direction $T^{(i)} - \hat{J}(x^0, u^{(i)})$) to the Pareto

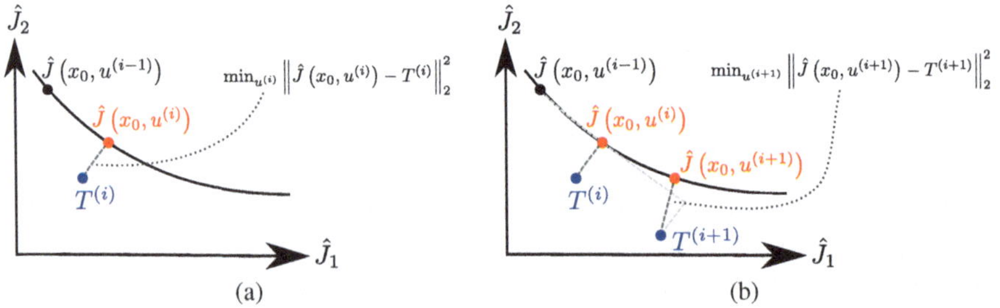

Fig. 3. Reference point method. (a) Solution of the ith scalar problem. (b) Construction of the next target $T^{(i+1)}$, prediction step $u^{(p,i+1)}$, and solution of the next scalar control problem.

[1] In this situation, indirect approaches can be used as well, cf. e.g. [61].

front. To accelerate the solution process for the next scalar problem (RP), linear extrapolation is used to compute a predictor $u^{(p,i+1)}$ based on the two points $u^{(i)}$ and $u^{(i-1)}$. This approach ensures an almost equidistant covering of the front. For a more detailed description see, e.g., [59, pp. 24–26].

2.2 Model Predictive Control

For taking unforeseen events, disturbances, or model inaccuracies during the control design into account, MPC methods (also referred to as *moving horizon control* or *receding horizon control*) have become very popular in recent years [32]. In MPC, open-loop problems are solved repeatedly on a finite *prediction horizon* of length $t_p = ph$, where $p \in \mathbb{N}^{>0}$ and $h \in \mathbb{R}^{>0}$ is the sample time (cf. Fig. 4). More concretely, problem (MOCP) is solved on a moving horizon $(s = 0, 1, \dots)$ with time constants and initial conditions given as

$$
\begin{aligned}
t_0 &= t_s, \\
t_e &= t_s + hp = t_{s+p}, \\
x^0 &= x(t_s).
\end{aligned}
$$

After each finite horizon optimization, a fraction of the control is applied to the system over the *control horizon* $t_c = qh \leq ph = t_p$ while the optimization is repeated with the prediction horizon moving forward by the sample time h.

This results in a closed-loop control which is more flexible to react to system disturbances and unforeseen events. Furthermore, the solution of finite horizon MOCPs is computationally less expensive because shorter horizons are considered compared to the full problem. Nevertheless, for complex problems, it is still hard to meet real-time requirements (note that the optimal control problem has to be solved within the control horizon t_c). This is already challenging for scalar-valued MPC problems, and considering multiple objectives is clearly infeasible without taking further measures such as scalarization methods (see

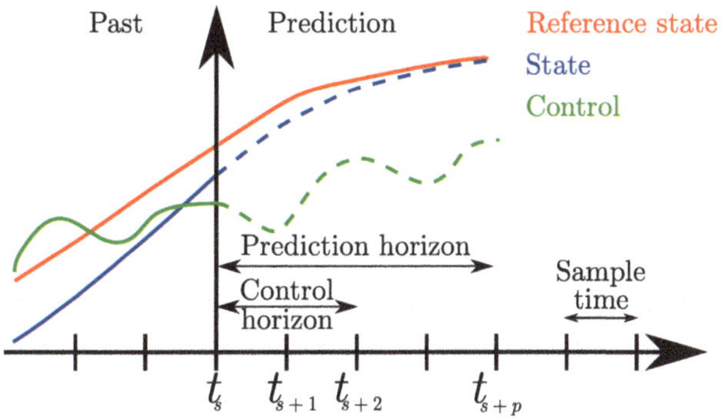

Fig. 4. Sketch of the MPC method. Due to the real-time constraints, the optimization problem has to be solved within the control horizon $t_c = qh$.

e.g. [4,69,70]), crude approximation of the entire front [29,44,54] or offline computation of control parameters [34,43]. For a survey of feedback control with multiple objectives we refer to [60]. Another way to meet real-time requirements is *explicit MPC* [3], see also [1] for a survey. Here, the MPC problem is reformulated as a multiparametric optimization problem which can be solved in an offline phase and the solutions are stored in a library. During the MPC loop, the computation of an optimal solution is then replaced by extracting the optimal input from the library. The approach presented in this contribution can be categorized as an explicit MPC method since we pre-compute Pareto optimal controls in an offline phase. In addition, symmetries of the underlying system are exploited to reduce the numerical effort. During operation, the best compromise solution is picked online based on the current situation and the user's preference.

2.3 Symmetry

Many dynamical systems show symmetry properties. Roughly speaking, a symmetric system possess a finite or infinite number of solutions "which look the same". In engineering systems, symmetry comes by design, e.g. by modularized system structures or by reducing the system's complexity. Here, we focus on symmetries which can be formally described by actions of Lie groups. We introduce motion primitives as equivalence classes w.r.t. the symmetry and identify trim primitives as motions with special properties. The presentation follows the lines of [25,57] and [24].

2.3.1 Equivariance and Equivalence

We formally describe symmetries by a finite-dimensional Lie group G and its group action ψ which are defined in the following way (see also [25]). A Lie group is a group $(G, \circ)$, which is also a smooth manifold, for which the group operations $(g, h) \mapsto g \circ h$ and $g \mapsto g^{-1}$ are smooth. If, in addition, a smooth manifold M is given, we call a map $\psi : G \times M \to M$ a left-action of G on M if and only if the following properties hold:

- $\psi(e, \mathbf{x}) = \mathbf{x}$ for all $\mathbf{x} \in M$ where e denotes the neutral element of $(G, \circ)$.
- $\psi(g, \psi(h, \mathbf{x})) = \psi(g \circ h, \mathbf{x})$ for all $g, h \in G$ and $\mathbf{x} \in M$.

For convenience, we define $\psi_g : M \to M$ with $\psi_g(x) := \psi(g, x)$ for $g \in G$ and $x \in M$.

Definition 3 (Symmetry (cf. [28])). Let $(G, \circ)$ a Lie-group and ψ a left-action of G on $\mathbb{R}^{n_x}$. Then a dynamical control system described by

$$\dot{x}(t) = f(x(t), u(t)) \quad \forall t \in [t_0, t_e], \quad x(t_0) = x^0, \tag{4}$$

is *invariant under the group action* ψ, or equivalently, *G is a symmetry group for the system* (4), if for all $g \in G$, $x^0 \in \mathbb{R}^{n_x}$, $t \in [t_0, t_e]$ and $u \in \mathcal{U}$ it holds

$$\psi_g(\varphi_u(x^0, t)) = \varphi_u(\psi_g(x^0), t). \tag{5}$$

This means that the flow commutes with the group action. Invariance under a group action implies equivalence of trajectories in the following sense (see also [24,28,57]).

Definition 4 (Equivalence of trajectories (cf. [28])). Two trajectories $\pi_1 : t \in [t_{0,1}, t_{e,1}] \mapsto (x_1(t), u_1(t))$ and $\pi_2 : t \in [t_{0,2}, t_{e,2}] \mapsto (x_2(t), u_2(t))$ of Eq. (4) are *equivalent*, if it holds that

(i) $t_{e,1} - t_{0,1} = t_{e,2} - t_{0,2}$ and
(ii) there exist a $g \in G$ and an $T \in \mathbb{R}$ such that $x_1(t) = \psi_g(x_2(t - T))$ and $u_1(t) = u_2(t - \mathrm{T})\,\forall t \in [t_{0,1}, t_{e,1}]$.

Thus, two trajectories are equivalent if one trajectory can be represented solely by a time translation and the group action applied to the other trajectory. All equivalent trajectories can be summed up in an equivalence class. By abuse of notation, we call the equivalence class as well as its representative a *motion primitive* (cf. [28]).

Remark 1 (Equivariance of vector fields (cf. [25,57]*)).* Symmetry of a dynamical control system can be described by the equivariance of the underlying vector field, i.e., by the condition

$$f(\psi_g(x), u) = D_x\psi_g(f(x, u)) \quad \forall x \in \mathbb{R}^{n_x}, g \in G, \tag{6}$$

where $D_x\psi_g : \mathbb{R}^{n_x} \to \mathbb{R}^{n_x}$ is the tangent lift of ψ_g which acts on $v = \dot{x}$ as $D_x\psi_g(v) = \frac{d}{dx}\psi_g(x) \cdot v$.

Remark 2 (Invariance of Lagrangian systems). Symmetries of mechanical systems which can be derived by a Lagrangian $L(q, \dot{q})$ and corresponding Euler-Lagrange equations $\frac{d}{dt}\frac{\partial L}{\partial \dot{q}}(q, \dot{q}) - \frac{\partial L}{\partial q}(q, \dot{q}) = 0$ can be described by the invariance of the Lagrangian function which is defined by the condition

$$L(\psi_g(q), D_q\psi_g(\dot{q})) = L(q, \dot{q}) \quad \forall (q, \dot{q}) \in \mathbb{R}^n \times \mathbb{R}^n, g \in G,$$

where $D_q\psi_g$ is the tangent lift of the group action ψ_g as defined in Remark 1.

Typical symmetries of mechanical systems are translational ($G = \mathbb{R}^n$), rotational ($G = SO(n)$) and combined ($G = SE(n) \approx SO(n) \times \mathbb{R}^n$) symmetries. The corresponding action and its lift are given by $\psi_g(q) = Rq + \Delta q$ and $D_q\psi_g(v) = Rv$ with $R \in SO(n)$ and $\Delta q \in \mathbb{R}^n$. $SO(n)$ is the special orthogonal group, which can be represented by the set of matrices $\{R \in \mathbb{R}^{n,n} \mid R^\top R = I, \det(R) = 1\}$. The dimension of a Lie group is given by the number of elements required to represent a Lie group element $g \in G$. For the examples above, we have $\dim(\mathbb{R}^n) = n$, $\dim(SO(n)) = n(n-1)/2$ and $\dim(SE(n)) = n(n-1)/2 + n$.

2.3.2 Symmetry Exploiting Motion Planning

It was first shown by Frazzoli et al. [26,28] how to formally take advantage of symmetry in motion planning. Following the idea of quantization (see also [24,27,40]), such so-called *motion primitives* are generated by solving optimal control problems for intermediate problems which can be combined into various sequences. The problem is thus reduced to searching for the optimal sequence out of all admissible sequences in a library of motion primitives which can be realized using global search methods. Extensions towards hybrid systems have been considered in [22,23].

Definition 5 (Trim primitive (cf. [28])). A solution (x, u) to system dynamics (4) on $[t_0, t_e]$ with initial value x^0, that can be written as

$$x(t) = \psi(\exp(t\xi), x^0),\ \ u(t) \equiv \bar{u} = \text{const.}\, \forall t \in [t_0, t_e] \tag{7}$$

with $\xi \in \mathfrak{g}$, the Lie algebra which corresponds to symmetry group G and $\exp : \mathfrak{g} \to G, \xi \mapsto \exp(t\xi) \in G$ the exponential map, is called a *trim primitive*, *trim* for short.

Note that the time parametrization defines via $\exp(t\xi)$ a one-parameter group orbit in G. If this generates a solution to the system dynamics via the symmetry lift, we call this trajectory a trim.

Remark 3 (Trim primitives and relative equilibria in Lagrangian/Hamiltonian systems). In dynamical system theory, trim primitives are known as relative equilibria. Typically, uncontrolled autonomous systems with symmetry are considered. Here, a relative equilibrium is a solution which is generated by the symmetry action and thus, an equilibrium in the non-symmetric part of the system, i.e. the *shape space*. Lagrangian or Hamiltonian dynamics allow nice characterizations of relative equilibria: Classically, Routh studied the case of cyclic variables (see e.g. [7]), i.e. Lagrangian systems that are independent of some of the configuration variables (θ). The dynamics can then be reduced to the remaining variables $(q, \dot{q})$,

$$R^\mu(q, \dot{q}) = \left[L(q, \dot{q}, \dot{\theta}) - \mu\dot{\theta}\right]_{p=\mu} \quad \Rightarrow \quad \frac{\partial}{\partial q} R^\mu(q, \dot{q}) - \frac{d}{dt}\frac{\partial}{\partial \dot{q}} R^\mu(q, \dot{q}) = 0,$$

and equilibria of the reduced dynamics correspond to relative equilibria of the original system. In [49], the generalized version of Lagrangian reduction has been studied and applied to a double spherical pendulum. In the general Lagrangian case, a relative equilibrium can be characterized as the critical point of the amended potential

$$V_\mu = V + \frac{1}{2}\left\langle \mu, \mathbb{I}^{-1}\mu \right\rangle$$

where μ is the value of the conserved momentum, $\mathbb{I}$ the locked inertia tensor, and $\langle \cdot \rangle$ denotes the vector-covector pairing (see [47,49] for details). This can then be extended to controlled Lagrangian systems: Assuming control that acts as a potential force, a trim primitive can be characterized as a critical point of the constantly controlled amended potential,

$$V_\mu^u = V - \nu + \frac{1}{2}\left\langle \mu, \mathbb{I}^{-1}\mu \right\rangle \qquad \text{with } \frac{\partial}{\partial q}\nu(q) = u,$$

see [21,24] for details.

Trims are uniquely defined by their initial value x^0, their control $\bar{u}$ and the Lie algebra element ξ, which makes them easy to store and handle in a library of motion primitives. More concretely, only the *shape space coordinates* of the trim have to be fixed, since any trim, as it is a motion primitive, can be shifted by the symmetry action. A second benefit of trims is that they are simply parametrized by time, i.e. their duration need not be fixed in advance, but can be adjusted during the sequencing (cf. [24,28] for details).

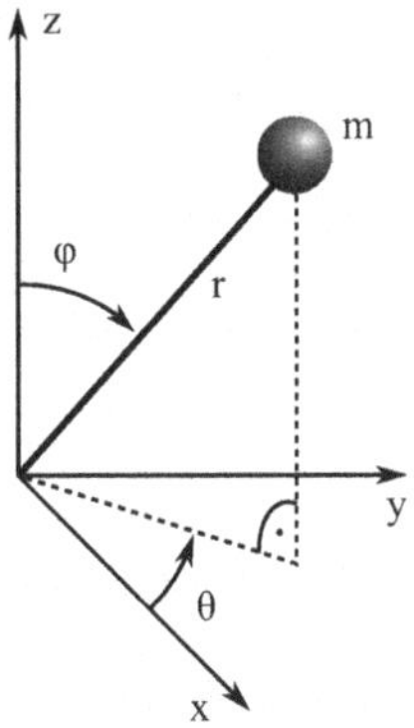

Fig. 5. The spherical pendulum.

Example 2 (Spherical Pendulum (cf. [21]*))* Consider a mathematical pendulum that can move with two degrees of freedom, i.e. on a sphere as depicted in Fig. 5. Its dynamics (apart from the north/south pole) are described by

$$\ddot{\theta} = -2\frac{\cos(\varphi)}{\sin(\varphi)}\dot{\varphi}\dot{\theta}$$
$$\ddot{\varphi} = \sin(\varphi)\cos(\varphi)\dot{\theta}^2 + \frac{g}{r}\sin(\varphi).$$

The system is symmetric w.r.t. rotations about the vertical axis and the symmetry group is $G = S^1$, acting by addition only in the horizontal coordinate.

Relative equilibria fulfill $\dot{\theta}^2 = -\frac{g}{r\cdot\cos(\varphi)}$, i.e. they are purely horizontal rotations ($\dot{\varphi} = 0$) in the lower hemisphere (cf. Fig. 6(a)). For every angle φ in the lower hemisphere, there is exactly one (up to the sign) rotational velocity which generates a relative equilibrium, i.e. a trim with zero control. If we allow control in φ-direction, the rotational velocity and the height of a trim can be chosen arbitrarily with $u_\varphi = -mgr\sin(\varphi) - mr^2\sin(\varphi)\cos(\varphi)\dot{\theta}^2$ (cf. Fig. 6(b)).

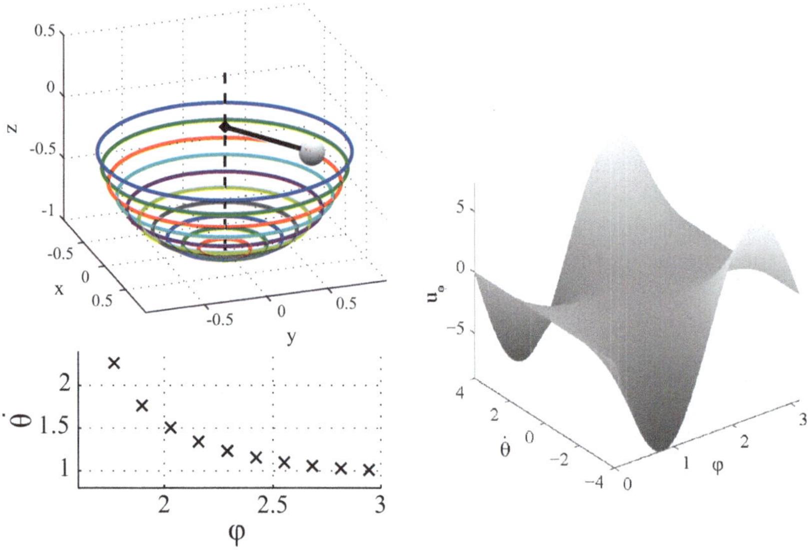

(a) (b)

Fig. 6. Trim primitives for the spherical pendulum system: (a) trims without control, i.e. relative equilibria exist in the lower hemisphere for correct rotational velocity. (b) If φ and $\dot{\theta}$ shall be chosen independently of one another, a constant control value u_φ has to be chosen according to the control manifold, see [21] for a detailed discussion.

Maneuver Automaton: Typically, it is not possible to continuously switch a control system from one trim to another (kinematic mechanical systems being an exception). Thus, control maneuvers have to be provided in order to smoothly concatenate motion primitives. A finite set of trims and maneuvers form the maneuver automaton which can be used for motion planning (Fig. 7). The design procedure is as follows. For details and formal definition in terms of automata theory, we refer to [28].

0. Assume the system dynamics (4), the symmetry (G, ψ) and the corresponding Lie algebra $\mathfrak{g}$ to be known.
1. Choose a finite set of trim primitives $\mathcal{T}$, e.g. by gridding the Lie algebra. Store their generating triples $(x^0, \bar{u}, \xi)$.
2. Compute a finite set of maneuvers $\mathcal{M}$ that start and end on trims $t_i, t_e \in \mathcal{T}$. This can be done by optimal control methods, for instance [24,40]. Maneuvers have to be stored as time-discretized control and state trajectories.

3. Set up the graph structure of the maneuver automaton. Trim primitives form the vertices and maneuvers the edges. A maneuver $m_{ie} \in \mathcal{M}$ which starts at trim t_i and end in trim t_e is a directed edge between these two nodes.

Note that every maneuver and every trim of $\mathcal{T} \cup \mathcal{M}$ is a motion primitive, i.e. it can be shifted by the symmetry action in order to build trajectory sequences. For given initial and final states on corresponding trims, a sequence can be found by planning methods, i.e. sampling-based roadmap algorithms (see [28,40]) or a modified version of the A* or the D* planning method[2].

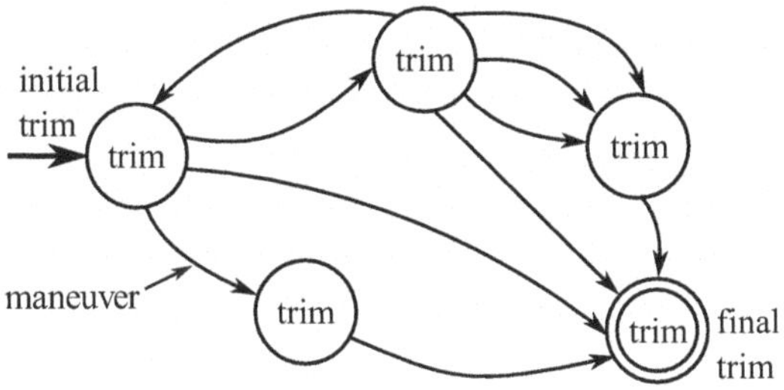

Fig. 7. Maneuver automaton adapted from [28]. Trim primitives form the nodes, maneuvers are the edges. A solution from initial to final trim consists of an alternating sequence of trims and maneuvers. This resembles the execution of a hybrid automaton, [22].

Remark 4 (Specially structured maneuvers). The definition of a maneuver is quite broad; the only restriction is that it is compatible with the dynamics and it has to start and end on a trim primitive of the automaton. This allows to consider specially structured maneuvers: In [21,24], maneuvers on stable and unstable manifolds of the uncontrolled dynamics have been integrated into the motion primitive concept. This formalizes the idea of energy efficient trajectory generation in astrodynamics (see [10,12,14,30,41,42,51]) by exploiting inherent system structures, namely stable and unstable manifolds of invariant objects in the uncontrolled dynamics.

3 Symmetries in Multiobjective Optimal Control

Symmetries do not only play a major role in dynamical control systems but also in optimal control problems. In this section, we will introduce the notion of symmetry in multiobjective optimal control and discuss how it can be exploited for the solution of multiobjective optimal control problems. We mainly follow

[2] See e.g. D* code for planning with primitives provided by Marin Kobilarov, Autonomous Systems, Control and Optimization (ASCO) Laboratory, Johns Hopkins University at https://github.com/jhu-asco/dsl.

the definitions and explanation in [57]. Considering symmetries in multiobjective optimal control problems, we are interested in Pareto optimal solutions of (MOCP) that remain Pareto optimal when the initial conditions are transformed by the symmetry group action such that

$$\arg\min_u \hat{J}(x^0, u) = \arg\min_u \hat{J}(\psi_g(x^0), u) \quad \forall g \in G. \tag{8}$$

This means that we require the Pareto set to be invariant under group actions on the initial conditions. Symmetries in single-objective, linear-quadratic explicit MPC have been studied in [11] and relations to the approach above and related methods are discussed in [57].

The following theorem provides conditions under which Pareto sets are invariant under group actions, i.e., that satisfy Eq. (8).

Theorem 1 (Symmetry of (MOCP) **(cf.** [57]**)).** *Let* $\mathcal{X} = W^{1,\infty}([t_0, t_e], \mathbb{R}^{n_x})$ *and* $\mathcal{U} = L^\infty([t_0, t_e], \mathbb{R}^{n_u})$. *If*

1. *the dynamics are invariant under the Lie Group action* ψ, *i.e. Eq.* (5) *holds for* $t \in [t_0, t_e]$;
2. *there exist* $\alpha, \beta, \delta \in \mathbb{R}$, $\alpha \neq 0$, *such that the cost functions* C_i *and the Mayer terms* Φ_i, $i = 1, \ldots, k$, *are invariant under the Lie Group action* ψ *up to linear transformations, i.e.,*

$$C_i(\psi_g(x), u) = \alpha C_i(x, u) + \beta \tag{9}$$

and

$$\Phi_i(\psi_g(x_e)) = \alpha \Phi_i(x_e) + \delta \qquad \textit{for } i = 1, \ldots, k; \tag{10}$$

3. *the constraints* g_i, $i = 1, \ldots, l$ *and* h_j, $j = 1, \ldots, m$, *are invariant under the Lie Group action* ψ, *i.e.,*

$$g_i(\psi_g(x), u) = g_i(x, u) \qquad \textit{for } i = 1, \ldots, l, \tag{11}$$

$$h_j(\psi_g(x), u) = h_j(x, u) \qquad \textit{for } j = 1, \ldots, m, \tag{12}$$

then we have

$$\arg\min_u \hat{J}(\psi_g(x^0), u) = \arg\min_u \hat{J}(x^0, u) \quad \forall g \in G. \tag{8}$$

We say that problem (MOCP) *is invariant under the Lie group action* ψ_g, *or equivalently,* G *is a symmetry group for problem* (MOCP).

A proof can be found in [57].

In principle, Theorem 1 states that if the objective function and the constraints are invariant (up to linear transformations) under the same group action as the dynamical control system (Conditions 2 and 3), then all trajectories contained in an equivalence class defined by (5) (Condition 1) will also be contained in an equivalence class defined by (8). However, the latter class may contain more solutions since we do not explicitly pose restrictions on the state but only require the solutions of (MOCP) (i.e., the control u) to be identical. This is demonstrated with the following example which involves no dynamical control system and consequently, no invariance condition on the flow has to be imposed (see also Corollary 1 in [57]).

Example 3 (Invariance of the arg min*)* Let us consider the parameter dependent objective function $J(u,\gamma)$ from [68, Example 3.12] with $J : \mathbb{R}^2 \times \mathbb{R} \to \mathbb{R}^2$:

$$\min_{u\in\mathbb{R}^2} \begin{pmatrix} \frac{1}{2}\left(\sqrt{1+(u_1+u_2)^2}+\sqrt{1+(u_1-u_2)^2}\right)+u_1-u_2\right)+\gamma e^{-(u_1-u_2)^2} \\ \frac{1}{2}\left(\sqrt{1+(u_1+u_2)^2}+\sqrt{1+(u_1-u_2)^2}\right)-u_1+u_2\right)+\gamma e^{-(u_1-u_2)^2} \end{pmatrix}. \tag{13}$$

Here, no dynamical control system is considered as extra constraint, which yields a finite-dimensional MOP. The Pareto sets and fronts for varying values of γ are shown in Fig. 8. Clearly, $\mathcal{P}$ is invariant under translations in γ (i.e., the arg min of (13) is invariant under translations in γ).

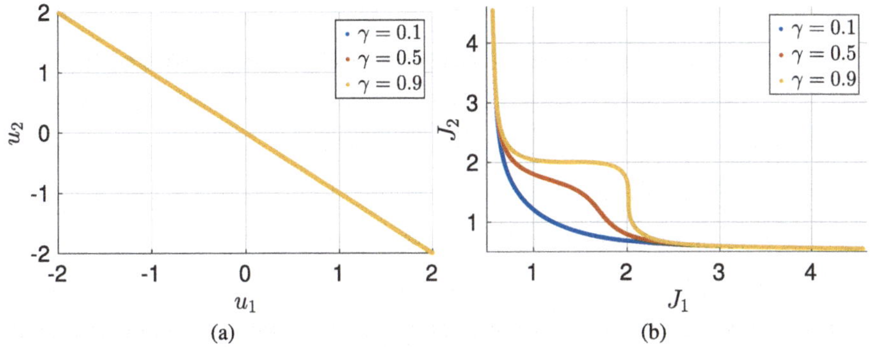

Fig. 8. Pareto set (a) and Pareto front (b) of Problem (13) for varying values of γ. Although the fronts vary, $\mathcal{P}$ is invariant under translations in γ.

Remark 5 (Group actions on controls and parameters, cf. [57]*).* One can extend the notion of symmetry for optimal control problems by introducing the Lie group actions χ_h on the control trajectories u and ξ_l on extra parameters $\gamma \in \mathbb{R}^{n_\gamma}$ of the optimal control problem (with h and l being elements of the Lie groups H and L, respectively). More concretely, we consider the group action triple $(\psi_g, \chi_h, \xi_l) : \mathbb{R}^{n_x} \times \mathbb{R}^{n_u} \times \mathbb{R}^{n_\gamma} \to \mathbb{R}^{n_x} \times \mathbb{R}^{n_u} \times \mathbb{R}^{n_\gamma}$. For a parameter dependent flow $\varphi_u(x^0, t; \gamma)$ the invariance condition (5) for the dynamical control system is then replaced by

$$\psi_g(\varphi_u(x^0, t; \gamma)) = \varphi_{\chi_h u}(\psi_g(x^0), t; \xi_l(\gamma)). \tag{14}$$

and the invariance condition for a parameter dependent optimal control problem (MOCP) reads

$$\arg\min_{u} \hat{J}(x^0, u, \gamma) = \arg\min_{u} \hat{J}(\psi_g(x^0), \chi_h(u), \xi_l(\gamma)) \quad \forall g \in G, h \in H, l \in L. \tag{15}$$

By Theorem 1, (15) is satisfied if dynamics, cost functions, Mayer terms and constraints are all invariant under the Lie group action (ψ_g, χ_h, ξ_l).

Example 4 (Parameter dependent problems, cf. [57]*)* An example for parameter dependent problems are tracking problems with cost functions in the following form

$$C(x, \gamma) = \|x - \gamma\|_2^2,$$

where x is the state and γ some reference to be tracked. For translation and rotation invariant dynamical control systems, i.e., $\psi_g(x) = R \cdot x + \Delta x$, invariance of the cost function is ensured by applying the same Lie group action ψ_g to γ, i.e., $C(\psi_g(x), \psi_g(\gamma)) = \|\psi_g(x) - \psi_g(\gamma)\|_2^2 = \|R \cdot x + \Delta x - (R \cdot \gamma + \Delta x)\|_2^2 = C(x, \gamma)$ where the last equality follows from the orthogonality of R. Another parameter dependent optimal control problem is considered in Sect. 5.

4 Symmetry Exploiting Model Predictive Control

In this work, we exploit all the above mentioned concepts, i.e.,

- multiobjective optimization,
- model predictive control,
- motion planning with motion primitives, and
- symmetry exploitation in optimal control

in order to develop a real-time capable feedback algorithm for nonlinear systems with the ability to adjust the prioritization of conflicting objectives online. A similar algorithm was proposed in [57] and extended to uncertainties in [36]. The main difference here is that we additionally use the concept of motion planning with trim primitives, i.e., two maneuvers are connected by a trim primitive. In contrast to classical motion planning with precomputed primitives, we do not store entire maneuvers and trims in a library, but only the control inputs resulting in the desired behavior. The dynamics themselves are then "created" when the plant is running and the optimal control is applied. In general, this can be done by numerical simulation. For trim primitives, in particular, the defining expression via Lie algebra, exponential map, and symmetry action can be used in order to avoid integration errors and improve numerical efficiency.

Remark 6 (Relation to existing methods). For single-objective, linear-quadratic optimal control, explicit MPC was introduced by Bemporad et al. [3], where an offline-online decomposition was introduced such that during operation, one only has to select the precomputed optimal control from a library. For the construction of the library, the control problem is reformulated as a parametric

Algorithm 1. Offline phase

Require: Bounds $x^0_{\mathsf{min}}, x^0_{\mathsf{max}} \in \mathbb{R}^{n_x}$, $\gamma_{\mathsf{min}}, \gamma_{\mathsf{max}} \in \mathbb{R}^{n_\gamma}$, distance between grid points $\delta \in \mathbb{R}^{n_x+n_\gamma}$.

1: Dimension reduction by exploiting symmetries: Decrease dimension of parameter $(x^0, \gamma) \in \mathbb{R}^{n_x+n_\gamma}$ to $(\widetilde{x}^0, \widetilde{\gamma}) \in \mathbb{R}^{\widetilde{n}_x+\widetilde{n}_\gamma}$ by exploiting the symmetry groups G and L (cf. Section 3).
2: Construction of library: Create an $(\widetilde{n}_x + \widetilde{n}_\gamma)$-dimensional grid $\mathcal{L}$ for the parameter $(\widetilde{x}^0, \widetilde{\gamma})$ between $(\widetilde{x}^0_{\mathsf{min}}, \widetilde{\gamma}_{\mathsf{min}})$ and $(\widetilde{x}^0_{\mathsf{max}}, \widetilde{\gamma}_{\mathsf{max}})$ with distance δ_i in the ith direction.
3: Compute the Pareto sets $\mathcal{P}_{(\widetilde{x}^0, \widetilde{\gamma})}$ for all $(\widetilde{x}^0, \widetilde{\gamma}) \in \mathcal{L}$ using the reference point method.

optimization problem with the initial condition x^0 as the parameter. For linear-quadratic problems, only a finite number of problems has to be solved in the offline phase to obtain the exact solution for all parameters, since the optimal control is an affine function of the initial condition x^0. The explicit MPC concept was later extended to account for symmetries in [11]. This way, the numerical effort of the offline phase could be significantly reduced. For nonlinear problems, it is no longer possible to obtain an exact offline solution for all parameter values. Instead, interpolation between existing library entries has to be performed [2,17,38,39].

The algorithm consists of an offline phase and an online phase. In the **offline phase**, the parameter dependent problem (MOCP) is solved for many different parameter values, where (x^0, γ) can be composed of both the initial condition and external parameters. The parametrization is discretized on an equidistant grid with dimension $n_x + n_\gamma = \dim(x^0, \gamma)$. This means that the numerical effort of the offline phase grows exponentially with the parameter dimension. In order to reduce the parameter dimension and thus the cost of the offline phase, symmetries in the dynamical control system are exploited. Then, one problem (MOCP) is solved for each entry in the resulting symmetry-reduced library and stored to the library $\mathcal{L}$. The entire procedure is summarized in Algorithm 1.

The **online phase** (see Algorithm 2) is strongly related to classical explicit MPC, but extended by the two steps

a) selection of a compromise solution according to a user-defined weight ρ (line 4 in Algorithm 2), and
b) usage of trim primitives (with $u = 0$) in case they dominate the solution from the library with respect to the global objectives (lines 7–10 in Algorithm 2).

Algorithm 2. Online phase

Require: Weight $\rho \in \mathbb{R}^k$ with $\sum_{i=1}^{k} \rho_i = 1$ and $\rho \geq 0$.

1: **for** $t = t_0, t_1, t_2, \ldots$ **do**

2: Obtain the current initial condition $\widetilde{x}^0 = \widetilde{x}(t)$ and the parameter value $\widetilde{\gamma}$ from the plant.

3: Identify the $2(\widetilde{n}_x + \widetilde{n}_\gamma)$ neighboring grid points of $(\widetilde{x}^0, \widetilde{\gamma})$ in $\mathcal{L}$ (i.e., closest below and above in each component of $(\widetilde{x}^0, \widetilde{\gamma})$). These points are collected in the index set $\mathcal{I}$.

4: From each of the corresponding Pareto sets $\mathcal{P}_{(\widetilde{x}^0, \widetilde{\gamma})_i}$, $i \in \mathcal{I}$, select a Pareto optimal control u_i according to the weight ρ.

5: Compute the distances d_i between the entries of the library and $(\widetilde{x}^0, \widetilde{\gamma})$:

$$d_i = \|(\widetilde{x}^0, \widetilde{\gamma})_i - (\widetilde{x}^0, \widetilde{\gamma})\|_2.$$

6: Interpolate between the entries in $\mathcal{I}$:

$$u = \frac{\sum_{i=1}^{|\mathcal{I}|} \frac{1}{d_i} u_i}{\sum_{i=1}^{|\mathcal{I}|} \frac{1}{d_i}}.$$

7: Compute the resulting objective values corresponding to the zero control trim $u = 0$.

8: **if** the zero control trim dominates the control selected from the library **then**

9: $u = 0$

10: **end if**

11: Apply u to the plant for the control horizon length t_c.

12: **end for**

5 Results for the Mobile Robot

Consider the mobile robot dynamics given in (1). Let's first transform the second order system (1) into a first order system:

$$\begin{pmatrix} \dot{x}_1 \\ x_2 \\ x_3 \\ v_1 \\ v_2 \\ v_3 \end{pmatrix} = \begin{pmatrix} v_1 \\ v_2 \\ v_3 \\ \frac{1}{m}(\cos(x_3)u_1 - \sin(x_3)u_2) \\ \frac{1}{m}(\sin(x_3)u_1 + \cos(x_3)u_2) \\ -\frac{ru_2}{\Theta} \end{pmatrix} = f(x, u), \tag{16}$$

with $x = (x_1, x_2, x_3, v_1, v_2, v_3)$ and $u = (u_1, u_2)$. Let us, furthermore, introduce the variables $q = (x_1, x_2, x_3)$, $\dot{q} = (v_1, v_2, v_3)$ such that $x = (q, \dot{q})$ and $w = (\dot{q}, \ddot{q})$.

The aim is to control the robot in such a way that a desired destination x^d is reached both as fast and as control-efficient as possible:

$$\begin{aligned} \min_{x\in\mathcal{X},u\in\mathcal{U}} & \begin{pmatrix} T \\ \int_0^T \|u(t)\|_2^2\,dt \end{pmatrix} \\ \text{s.t.} \quad & \dot{x}(t) = f(x(t),u(t)), \quad x(t_0) = x^0, \qquad\qquad t \in (0,T], \\ & \|(x_1(T),x_2(T))^\top - x^d\| \leq r^d \end{aligned} \tag{17}$$

where T is the time for the entire trajectory, and the destination area is defined by a circle[3] with fixed radius r^d and center x^d. Introducing the flow notation (3), a reduced parameter dependent problem of form (MOCP) can be derived with eight parameters. Of these, six stem from the initial condition of the robot, and two from the destination:

$$(x^0,\gamma) = (x_1^0, x_2^0, x_3^0, v_1^0, v_2^0, v_3^0, x_1^d, x_2^d),$$

see also Fig. 9.

Now we identify the symmetries of the MOCP (17). Consider the Lie group action $\psi_g : \mathbb{R}^3 \to \mathbb{R}^3$ with

$$\psi_g(q) = Q \cdot q + \Delta q \tag{18}$$

with $Q = \begin{pmatrix} R_{g_3} & 0 \\ 0 & 1 \end{pmatrix}$, $R_{g_3} = \begin{pmatrix} \cos g_3 & -\sin g_3 \\ \sin g_3 & \cos g_3 \end{pmatrix}$ and $\Delta q = (g_1, g_2, g_3)^\top$. This action represents translations of the robot's center of mass (q_1, q_2) and simultaneous translation in the robot's orientation x_3 and rotation about its center of mass. The corresponding symmetry group is thus given by $G = L = SE(2)$ with

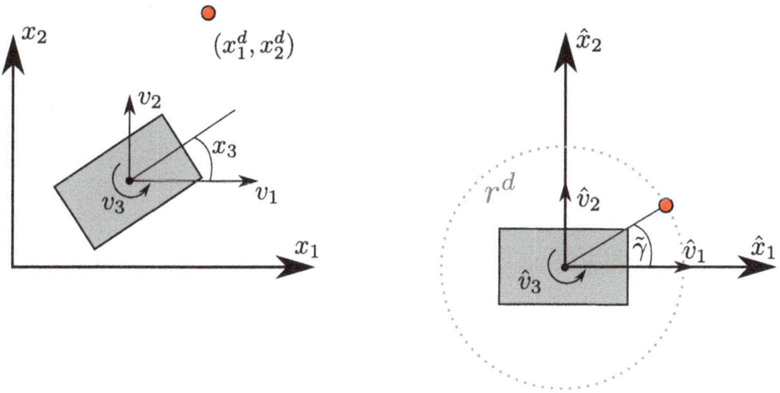

Fig. 9. Left: Problem with eight degrees of freedom, from which six are the robot's degrees of freedom and two are the position x^d of the destination. Right: Symmetry-reduced problem with four degrees of freedom.

[3] The reason for specifying the target as a circular area is that this way, the parameter dimension can be further reduced by one, as we will see below.

$\dim(G) = \dim(L) = 3$. The corresponding action $D\psi_g : \mathbb{R}^6 \to \mathbb{R}^6$ acting on $x = (q, \dot{q})$ is given by

$$D\psi_g(x) = \begin{pmatrix} Q & \mathbf{0}_{3\times3} \\ \mathbf{0}_{3\times3} & Q \end{pmatrix} \cdot x + \Delta x \tag{19}$$

with $\Delta x = (g_1, g_2, g_3, \mathbf{0}_{3,1})^\top$. Similarly, we can compute $D^2\psi_g : \mathbb{R}^6 \to \mathbb{R}^6$ acting on $w = (\dot{q}, \ddot{q})$ as

$$D^2\psi_g(w) = \begin{pmatrix} Q & \mathbf{0}_{3\times3} \\ \mathbf{0}_{3\times3} & Q \end{pmatrix} \cdot w \tag{20}$$

We are now ready to prove the following result.

Proposition 1. *Problem* (17) *with* $q = (x_1, x_2, x_3)$, $\dot{q} = (v_1, v_2, v_3)$, $x = (q, \dot{q})$ *and a destination area given by* $\gamma = x^d$ *is invariant under the group action* (ψ_g, ξ_l) *,where* ψ_g *is given in* (18) *and*

$$\xi_l(\gamma) = R_{g_3} \cdot \gamma + \Delta\gamma \tag{21}$$

with $R_{g_3} = \begin{pmatrix} \cos g_3 & -\sin g_3 \\ \sin g_3 & \cos g_3 \end{pmatrix}$ *and* $\Delta\gamma = (g_1, g_2)^\top$.

Proof Equivariance of vector field: Using the first order system formulation (16), we have

$$\begin{aligned} f(D\psi_g(x), u) &= \begin{pmatrix} Q \cdot \dot{q} \\ \frac{1}{m}(\cos(x_3 + g_3)u_1 - \sin(x_3 + g_3)u_2) \\ \frac{1}{m}(\sin(x_3 + g_3)u_1 + \cos(x_3 + g_3)u_2) \\ -\frac{ru_2}{\Theta} \end{pmatrix} \\ &= \begin{pmatrix} Q \cdot \dot{q} \\ \begin{pmatrix} \cos(g_3) & -\sin(g_3) \\ \sin g_3 & \cos(g_3) \end{pmatrix} \begin{pmatrix} \frac{1}{m}(\cos(x_3)u_1 - \sin(x_3)u_2) \\ \frac{1}{m}(\sin(x_3)u_1 + \cos(x_3)u_2) \end{pmatrix} \\ -\frac{ru_2}{\Theta} \end{pmatrix} \\ &= \begin{pmatrix} Q & \mathbf{0}_{3\times3} \\ \mathbf{0}_{3\times3} & Q \end{pmatrix} \cdot f(x, u) = D^2\psi_g(f(x, u)). \end{aligned}$$

Invariance of Cost Functions and Constraints: Since the cost function is independent from the state x, it is invariant under the group action. The invariance of the constraint $\|(x_1(T), x_2(T))^\top - x^d\| \leq r^d$ follows with the orthogonality of R.

The statement follows with Theorem 1. □

Remark 7. To ensure symmetry of the MOCP, also the destination point x^d has to be translated by the same group action as the center of mass (x_1, x_2).

Next, we compute the trim primitives based on the Lie algebra of the symmetry group. The Lie algebra of the group of rotation matrices R_{g_3} is the group of skew symmetric matrices

$$\begin{pmatrix} 0 & -\xi_3 \\ \xi_3 & 0 \end{pmatrix}$$

and the Lie algebra of the group of 2-dimensional translation is $\mathbb{R}^2$.

Proposition 2 (Trim primitives). *The robot with dynamics* (16) *and symmetry action* $D\psi_g$ *as defined in* (19) *has two types of trim primitives:*

a) Straight translation: *Going straight with constant velocity*

$$\begin{pmatrix} x_1 \\ x_2 \\ x_3 \\ v_1 \\ v_2 \\ v_3 \end{pmatrix}(t) = \begin{pmatrix} x_1^0 + \xi_1 \cdot t \\ x_2^0 + \xi_2 \cdot t \\ x_3^0 \\ v_1^0 \\ v_2^0 \\ 0 \end{pmatrix} \quad \textit{with} \quad \begin{aligned} \xi_1 &= v_1^0, \\ \xi_2 &= v_2^0, \\ \xi_3 &= 0, \\ u_1 &= 0, \\ u_2 &= 0. \end{aligned}$$

b) Circular motion:

$$\begin{pmatrix} x_1 \\ x_2 \\ x_3 \\ v_1 \\ v_2 \\ v_3 \end{pmatrix}(t) = \begin{pmatrix} \cos(\xi_3 t)x_1^0 - \sin(\xi_3 t)x_2^0 + \xi_1/\xi_3 \sin(\xi_3 t) - (1 - \cos(\xi_3 t))\xi_2/\xi_3 \\ \sin(\xi_3 t)x_1^0 + \cos(\xi_3 t)x_2^0 + \xi_1/\xi_3(1 - \cos(\xi_3 t)) + \sin(\xi_3)\xi_2/\xi_3 \\ \xi_3 t \\ \cos(\xi_3 t)v_1^0 - \sin(\xi_3 t)v_2^0 \\ \sin(\xi_3 t)v_1^0 + \cos(\xi_3 t)v_2^0 \\ v_3^0 \end{pmatrix}$$

with

$$\begin{aligned} \xi_1 &= v_1^0 + \xi_3 x_2^0, \\ \xi_2 &= v_2^0 - \xi_3 x_1^0, \\ \xi_3 &= v_3^0, \\ u_2 &= 0, \\ \cos(\xi_3 t)v_1^0\xi_3 - \sin(\xi_3 t)v_2^0\xi_3 &= 1/m \sin(x_3^0 + \xi_3 t)u_1, \\ -\sin(\xi_3 t)v_1^0\xi_3 - \cos(\xi_3 t)v_2^0\xi_3 &= 1/m \cos(x_3^0 + \xi_3 t))u_1. \end{aligned}$$

Proof. When defining the Lie algebra as a subgroup of $\mathfrak{se}(2)$ and using the Rodriguez formula (cf. e.g. [53]) for the exponential map, the trim representation follows from direct computation. Constraints on the Lie algebra elements ξ_1, ξ_2, ξ_3, the initial value $(x_1^0, x_2^0, x_3^0, v_1^0, v_2^0, v_3^0)$ and the constant control values (u_1, u_2) are derived from comparison with the dynamics (16), since a trim has to be a valid motion. □

In our numerical example, we will make use of trims of type a), i.e. straight motions with constant velocity and zero control.

As an eight-dimensional parameter results in a prohibitively expensive offline phase, we exploit the translational symmetries in x_1, x_2, and x_3. As long as the target point x^d is translated and rotated accordingly, this leaves the solution of Problem (17) unchanged and results in a parameter of dimension five:

$$(x^0, \gamma)' = (v_1^0, v_2^0, v_3^0, \delta_1, \delta_2),$$

where $\delta = R_{g_3} x^d$ now denotes the distance between the robot and the destination in the rotated frame of reference. In order to further reduce the dimension by one, we project the destination point onto the circle with radius r^d such that it can be specified exclusively by the angle $\tilde{\gamma} = \arctan(\delta_2/\delta_1)$, cf. Fig. 9 for an illustration. This results in the four-dimensional parameter

$$(\tilde{x}^0, \tilde{\gamma}) = (v_1^0, v_2^0, v_3^0, \tilde{\gamma}),$$

for which we can now introduce a discretization according to Table 1. Note that we can restrict $\tilde{\gamma}$ to positive values by exploiting an additional discrete reflection symmetry, which reduces the computational effort by another 50%. This results in an offline phase with 14 784 problems of the following form:

$$\begin{aligned} \min_{\hat{x}\in\mathcal{X},u\in\mathcal{U}} &= J((\tilde{x}^0,\tilde{\gamma}),u) = \min_{\hat{x}\in\mathcal{X},u\in\mathcal{U}} \begin{pmatrix} t_e \\ \int_{t_0}^{t_e} \|u(t)\|_2^2\, dt \end{pmatrix} \\ \text{s.t.} & \qquad (16), \\ \hat{x}(t_0) &= \left(0,0,0,v_1^0,v_2^0,v_3^0\right)^\top, \quad x(t_e) = \begin{pmatrix} r^d\cos(\tilde{\gamma}) \\ r^d\sin(\tilde{\gamma}) \end{pmatrix}, \end{aligned} \tag{22}$$

where the $\hat{x}$ notation indicates the rotated and translated frame of reference.

For the numerical solution, we use a direct method. We discretize both the state and the control input on an 11-dimensional time grid with a constant time step $\delta t = \kappa h$ with $h = 0.1$ and approximate the dynamics by a fourth-order Runge-Kutta scheme. The variable κ is a scaling factor for the time step which we use to allow for a variable end time t_e in order to address the first objective. This way, we obtain a 89-dimensional nonlinear MOP ($6 \cdot 11$ state variables, $2 \cdot 11$ control variables and κ) for each parameter $(\tilde{x}^0, \tilde{\gamma})$ that we solve using the

Table 1. Parameters for the library $\mathcal{L}$.

Variable	Minimal value	Maximal value	Step size	Number of grid points
v_1	-7	7	2	8
v_2	-7	7	2	8
v_3	-5	5	1	11
$\tilde{\gamma}$	0	π	$\pi/20$	21

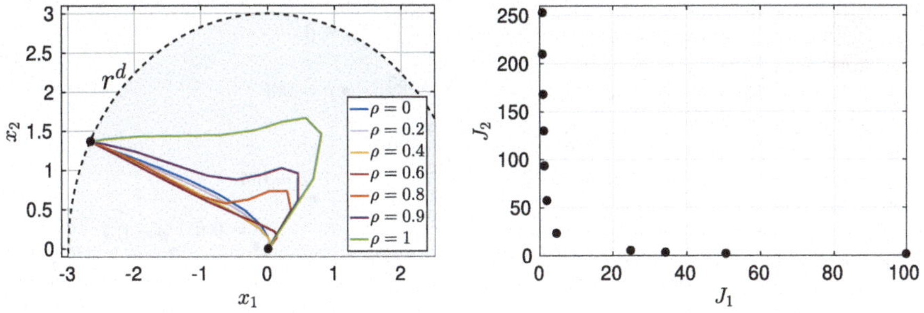

Fig. 10. Visualization of the offline phase. The figures represent one entry of the library $\mathcal{L}$ at $(\tilde{x}^0, \tilde{\gamma}) = ((1,1,0), 0.85\pi)$. Left: Pareto optimal trajectories for Problem (22). As the discretization of the Pareto set consists of 11 entries, the respective solutions are identified via weights $\rho \in \{0, 0.1, \ldots, 1\}$. Right: The corresponding Pareto front.

reference point method (RP). For more details on the automated solution of a large number of MOPs, see [57].

Figure 10 depicts an exemplary solution corresponding to one entry of the library $\mathcal{L}$. On the left, the trajectories resulting from the Pareto optimal inputs u are shown. Here, the weight ρ indicates to which extent the second objective is priorized, i.e., which of the entries from the Pareto set $\mathcal{P}_{(\tilde{x}^0,\tilde{\gamma})}$ is selected and applied to the plant. Consequently, $\rho = 0$ results in fast and $\rho = 1$ in control-efficient driving. The figure on the right side then shows the corresponding Pareto front.

In Fig. 11, we see several solutions of the online phase (Algorithm 2) with the initial condition $(x^0) = (0,0,0,1,0,0)$ and a control horizon of length $t_c = 0.3$ for different weights ρ. In the left plot, several trajectories are shown from the initial point $(0,0)$ to the target area around $x^d = (-40, 20)$ with $r^d = 3$. One can easily identify the parts of the trajectories where the zero control trim dominates, as these result in straight lines. This becomes also apparent in Fig. 12, where the corresponding controls that are applied to the plant are shown. Consequently, the MPC behavior resembles the classical motion planning concept, where trims and maneuvers are used alternatingly. The Pareto front of the corresponding *global objectives* (i.e., the objectives evaluated over the entire trajectory) is shown on the right, and we see that the desired trade-off behavior is achieved by varying the weight. In addition, online adaptations of ρ (for instance, in order to react to changing priorization) are easily performed. Two examples are also visualized in Fig. 11 as dashed lines, where the weight is adapted towards a faster driving style after 5 and 15 s, respectively.

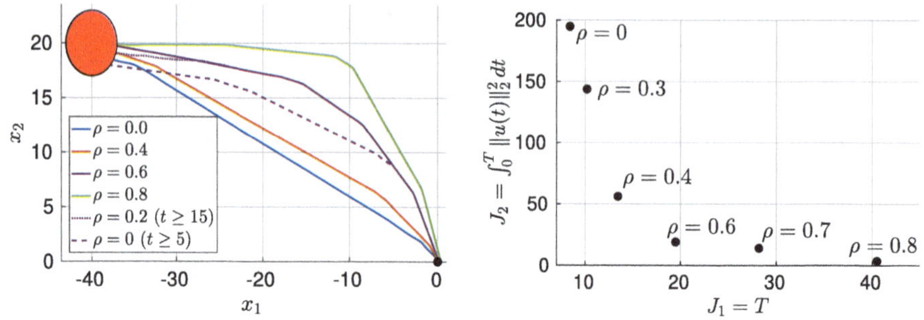

Fig. 11. Left: Several robot trajectories for different weights ρ, generated by the MOMPC Algorithm 2 based on Problem (17). The straight parts correspond to zero control trims (cf. Fig. 12). The two dashed trajectories represent cases where the weights are changed during operation. Right: The corresponding Pareto front for the *global objectives*.

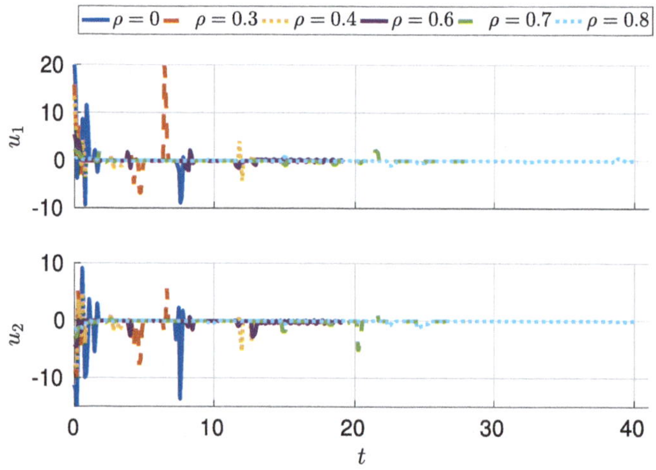

Fig. 12. The control input generated by the MOMPC Algorithm 2. The large parts where $u = 0$ correspond to trim primitives.

Remark 8. (Numerical challenges). It should be noted that the offline-online multiobjective MPC procedure presented here faces several difficulties from a numerical perspective:

a) Interpolation needs to be performed for parameter values that are not contained in the library $\mathcal{L}$.
b) The performance can be sensitive to the choice of the prediction horizon length t_c. In particular, a good choice may also depend on the weight ρ, as optimal controls with a high priority on control costs often result in uncontrolled (straight ahead) driving on the first part of the prediction horizon. As

a consequence, a repeated application of this first part may result in failure to reach the destination.

These issues will be subject of further research. In particular, an additional – much cheaper – online phase appears to be promising [36].

6 Concluding Remarks

We have presented an MPC algorithm for nonlinear dynamical systems with multiple objectives which exploits symmetries in dynamical systems to reduce the computational effort. Based on Lie group symmetries, Pareto optimal motion primitives can be identified and stored in a motion planning library in an offline phase. In the online phase – which is closely related to Explicit MPC – a Pareto optimal motion primitive is selected based on the decision maker's preference and applied to the plant over the control horizon length. In contrast to classical approaches from motion planning with motion primitives, only Pareto optimal controls instead of controls and state trajectories have to be stored due to our definition of symmetry in optimal control problems. Furthermore, as an extension to [57], we take trim primitives into account which describe motions generated by the symmetry action and thus, are very easy to store. For the mobile robot example, the MPC algorithm is applied using trims corresponding to straight motions with constant velocity. Future work can extend this to all possible types of trim primitives. However, this requires a shortest path search in the maneuver automaton. A formal proof of stability can be based on [25]. Moreover, revealing the relation between motion primitives and turnpikes (cf. [19,20]) within MPC is promising. Finally, increased robustness (cf. Remark 8) can be achieved by introducing an online adaptive library. In this case, the library could grow (by solving additional MOCPs online) if a required entry is missing.

Acknowledgments. We would like to acknowledge Michael Dellnitz, who has been an inspiration to us from our initial PhD time until today in many areas of research that jointly resulted in works on multiobjective optimization, dynamical systems, optimal control and symmetries.

References

1. Alessio, A., Bemporad, A.: A survey on explicit model predictive control. In: Magni, L., Raimondo, D.M., Allgöwer, F. (eds.) Nonlinear Model Predictive Control: Towards New Challenging Applications. volume 384, pp. 345–369. Springer, Heidelberg (2009)
2. Bemporad, A., Filippi, C.: An algorithm for approximate multiparametric convex programming. Comput. Optim. Appl. **35**(1), 87–108 (2006)
3. Bemporad, A., Morari, M., Dua, V., Pistikopoulos, E.N.: The explicit linear quadratic regulator for constrained systems. Automatica **38**(1), 3–20 (2002)
4. Bemporad, A., de la Peña, D.M.: Multiobjective model predictive control. Automatica **45**(12), 2823–2830 (2009)

5. Betsch, P., Becker, C.: Conservation of generalized momentum maps in mechanical optimal control problems with symmetry. Int. J. Numer. Meth. Eng. **111**(2), 144–175 (2017)
6. Betts, J.T.: Survey of numerical methods for trajectory optimization. AIAA J. Guidance Control Dyn. **21**(2), 193–207 (1998)
7. Bloch, A.M.: Nonholonomic Mechanics and Control. Springer (2003)
8. Bullo, F., Lewis, A.D.: Geometric Control of Mechanical Systems. Texts in Applied Mathematics, vol. 49. Springer (2004)
9. Coello Coello, C.A., Lamont, G.B., Van Veldhuizen, D.A.: Evolutionary Algorithms for Solving Multi-Objective Problems, vol. 2. Springer (2007)
10. Conley, C.: Low energy transit orbits in the restricted three-body problem. SIAM J. Appl. Math. **16**(4), 732–746 (1968)
11. Danielson, C., Borrelli, F.: Symmetric Explicit Model Predictive Control, vol. 4. IFAC (2012)
12. Dellnitz, M., Junge, O., Post, M., Thiere, B.: On target for Venus - set oriented computation of energy efficient low thrust trajectories. Celestial Mech. Dyn. Astron. **95**, 357–370 (2006)
13. Dellnitz, M., Klus, S.: Sensing and control in symmetric networks. Dyn. Syst. **32**(1), 61–79 (2017)
14. Dellnitz, M., Ober-Blöbaum, S., Post, M., Schütze, O., Thiere, B.: A multi-objective approach to the design of low thrust space trajectories using optimal control. Celestial Mech. Dyn. Astron. **105**, 33–59 (2009)
15. Dellnitz, M., Schütze, O., Hestermeyer, T.: Covering Pareto sets by multilevel subdivision techniques. J. Optim. Theory Appl. **124**(1), 113–136 (2005)
16. Djukić, D.S.: Noether's theorem for optimum control systems. Int. J. Control **18**(3), 667–672 (1973)
17. Dua, V., Pistikopoulos, E.N.: Algorithms for the solution of multiparametric mixed-integer nonlinear optimization problems. Ind. Eng. Chem. Res. **38**(10), 3976–3987 (1999)
18. Ehrgott, M.: Multicriteria Optimization, 2nd edn. Springer, Heidelberg (2005)
19. Faulwasser, T., Flaßkamp, K., Ober-Blöbaum, S., Worthmann, K.: Towards velocity turnpikes in optimal control of mechanical systems. IFAC-PapersOnLine **52**(16), 490–495 (2019)
20. Faulwasser, T., Flaßkamp, K., Ober-Blöbaum, S., Worthmann, K.: A dissipativity characterization of velocity turnpikes in optimal control problems for mechanical systems. In: Accepted for 24th International Symposium on Mathematical Theory of Networks and Systems 2021. arXiv:2002.04388 (2020)
21. Flaßkamp, K.: On the optimal control of mechanical systems – hybrid control strategies and hybrid dynamics. Ph.D. thesis, University of Paderborn (2013)
22. Flaßkamp, K., Hage-Packhäuser, S., Ober-Blöbaum, S.: Symmetry exploiting control of hybrid mechanical systems. J. Comput. Dyn. **2**(1), 25–50 (2015)
23. Flaßkamp, K., Ober-Blöbaum, S.: Motion planning for mechanical systems with hybrid dynamics. In: Fontes, M., Günther, M., Marheineke, N. (eds.) Progress in Industrial Mathematics at ECMI 2012. The European Consortium for Mathematics in Industry, vol. 19. Springer, Heidelberg (2014)
24. Flaßkamp, K., Ober-Blöbaum, S., Kobilarov, M.: Solving optimal control problems by exploiting inherent dynamical systems structures. J. Nonlinear Sci. **22**(4), 599–629 (2012)
25. Flaßkamp, K., Ober-Blöbaum, S., Worthmann, K.: Symmetry and motion primitives in model predictive control. Math. Control Signals Syst. **31**, 445–485 (2019)

26. Frazzoli, E.: Robust hybrid control for autonomous vehicle motion planning. Ph.D. thesis, Massachusetts Institute of Technology (2001)
27. Frazzoli, E., Bullo, F.: On quantization and optimal control of dynamical systems with symmetries. In: Proceedings of the 41st IEEE Conference on Decision and Control, pp. 817–823 (2002)
28. Frazzoli, E., Dahleh, M.A., Feron, E.: Maneuver-based motion planning for nonlinear systems with symmetries. IEEE Trans. Rob. **21**(6), 1077–1091 (2005)
29. García, J.J.V., Garay, V.G., Gordo, E.I., Fano, F.A., Sukia, M.L.: Intelligent Multi-Objective Nonlinear Model Predictive Control (iMO-NMPC): towards the 'on-line' optimization of highly complex control problems. Expert Syst. Appl. **39**, 6527–6540 (2012)
30. Gómez, G., Koon, W.S., Lo, M.W., Marsden, J.E., Masdemont, J., Ross, S.D.: Connecting orbits and invariant manifolds in the spatial three-body problem. Nonlinearity **17**, 1571–1606 (2004)
31. Grizzle, J.W., Marcus, S.I.: Optimization of systems possessing symmetries. In: Bensoussan, A., Lions, J.L. (eds.) Analysis and Optimization of Systems, pp. 513–524. Springer, Heidelberg (1984)
32. Grüne, L., Pannek, J.: Nonlinear Model Predictive Control, 2 edn. Springer (2017)
33. Guckenheimer, J., Holmes, P.: Nonlinear Oscillations, Dynamical Systems, and Bifurcations of Vector Fields. Applied Mathematical Sciences, vol. 42. Springer (1983)
34. Hernández, C., Naranjani, Y., Sardahi, Y., Liang, W., Schütze, O., Sun, J.Q.: Simple cell mapping method for multi-objective optimal feedback control design. Int. J. Dyn. Control **1**(3), 231–238 (2013)
35. Hernández, C., Schütze, O., Sun, J.-Q.: Global multi-objective optimization by means of cell mapping techniques. In: EVOLVE–A Bridge between Probability, Set Oriented Numerics and Evolutionary Computation VII, pp. 25–56. Springer (2017)
36. Hernández, C., Ober-Blöbaum, S., Peitz, S.: Explicit multi-objective model predictive control for nonlinear systems under uncertainty. arXiv:2002.06006 (2020)
37. Hillermeier, C.: Nonlinear Multiobjective Optimization: A Generalized Homotopy Approach. Birkhäuser (2001)
38. Johansen, T.A.: On multi-parametric nonlinear programming and explicit nonlinear model predictive control. In: 41st IEEE Conference on Decision and Control, vol. 3, pp. 2768–2773 (2002)
39. Johansen, T.A.: Approximate explicit receding horizon control of constrained nonlinear systems. Automatica **40**, 293–300 (2004)
40. Kobilarov, M.: Discrete geometric motion control of autonomous vehicles. Ph.D. thesis, University of Southern California (2008)
41. Koon, W.S., Lo, M.W., Marsden, J.E., Ross, S.D.: Shoot the Moon. Spaceflight Mech. **105**(2), 1017–1030 (2000)
42. Koon, W.S., Lo, M.W., Marsden, J.E., Ross, S.D.: Low energy transfer to the Moon. Celestial Mech. Dyn. Astron. **81**(1–2), 63–73 (2001)
43. Krüger, M., Witting, K., Trächtler, A., Dellnitz, M.: Parametric model-order reduction in hierarchical multiobjective optimization of mechatronic systems. In: Proceedings of the 18th IFAC World Congress 2011, Milano, Italy, vol. 18, pp. 12611–12619. Elsevier, Oxford (2011)
44. Laabidi, K., Bouani, F., Ksouri, M.: Multi-criteria optimization in nonlinear predictive control. Math. Comput. Simul. **76**(5–6), 363–374 (2008)
45. Liberzon, D.: Calculus of Variations and Optimal Control Theory: A Concise Introduction. Princeton University Press (2012)

46. Logist, F., Sager, S., Kirches, C., Van Impe, J.F.: Efficient multiple objective optimal control of dynamic systems with integer controls. J. Process Control **20**(7), 810–822 (2010)
47. Marsden, J.E.: Lectures on Mechanics. Number 174 in London Mathematical Society Lecture Note Series. Cambridge University Press (1993)
48. Marsden, J.E., Ratiu, T.S.: Introduction to Mechanics and Symmetry. Texts in Applied Mathematics, 2nd edn., vol. 17. Springer (1999)
49. Marsden, J.E., Scheurle, J.: Lagrangian reduction and the double spherical pendulum. Z. Math. Phys. (ZAMP) **44**, 17–43 (1993)
50. Martín, A., Schütze, O.: Pareto Tracer: a predictor-corrector method for multi-objective optimization problems. Eng. Optim. **50**(3), 516–536 (2018)
51. McGehee, R.: Some homoclinic orbits for the restricted three-body problem. Ph.D. thesis, University of Wisconsin (1969)
52. Melbourne, I., Dellnitz, M., Golubitsky, M.: The structure of symmetric attractors. Arch. Ration. Mech. Anal. **123**(1), 75–98 (1993)
53. Richard, M., Murray, S., Shankar, S., Li, Z.: A Mathematical Introduction to Robotic Manipulation. CRC Press Inc., USA (1994)
54. Núñez, A., Cortés, C.E., Sáez, D., De Schutter, B., Gendreau, M.: Multiobjective model predictive control for dynamic pickup and delivery problems. Control Eng. Pract. **32**, 73–86 (2014)
55. Ober-Blöbaum, S.: Discrete Mechanics and Optimal Control. Ph.D. thesis, University of Paderborn (2008)
56. Ober-Blöbaum, S., Junge, O., Marsden, J.E.: Discrete mechanics and optimal control: an analysis. Control Optim. Calc. Var. **17**(2), 322–352 (2011)
57. Ober-Blöbaum, S., Peitz, S.: Explicit multiobjective model predictive control for nonlinear systems with symmetries. arXiv:1809.06238 (2018)
58. Ober-Blöbaum, S., Ringkamp, M., zum Felde, G.: Solving multiobjective optimal control problems in space mission design using discrete mechanics and reference point techniques. In: 51st IEEE International Conference on Decision and Control, pp. 5711–5716, Maui, HI, USA, 10–13 December 2012
59. Peitz, S.: Exploiting structure in multiobjective optimization and optimal control. Ph.D. thesis Paderborn University (2017)
60. Peitz, S., Dellnitz, M.: A survey of recent trends in multiobjective optimal control - surrogate models, feedback control and objective reduction. Math. Comput. Appl. **23**(2), 30 (2018)
61. Peitz, S., Ober-Blöbaum, S., Dellnitz, M.: Multiobjective optimal control methods for the Navier-Stokes equations using reduced order modeling. Acta Applicandae Mathematicae **161**(1), 171–199 (2019)
62. Schütze, O., Witting, K., Ober-Blöbaum, S., Dellnitz, M.: Set oriented methods for the numerical treatment of multiobjective optimization problems. In: Tantar, E., Tantar, A.-A., Bouvry, P., Del Moral, P., Legrand, P., Coello Coello, C.A., Schütze, O. (eds.) EVOLVE - A Bridge between Probability, Set Oriented Numerics and Evolutionary Computation. Studies in Computational Intelligence, vol. 447, pp. 187–219. Springer, Heidelberg (2013)
63. Sun, J.-Q., Xiong, F.-R., Schütze, O., Hernández, C.: Cell Mapping Methods–algorithmic Approaches and Applications. Nonlinear Systems and Complexity, vol. 99 (2019)
64. Sussmann, H.J., Willems, J.C.: 300 years of optimal control: from the Brachystochrone to the maximum principle. IEEE Control Syst. **17**(3), 32–44 (1997)

65. Torres, D.F.M.: Conservation laws in optimal control. In: Colonius, F., Grüne, L. (eds.) Dynamics, Bifurcations, and Control, pp. 287–296. Springer, Heidelberg (2002)
66. Torres, D.F.M.: On the Noether theorem for optimal control. Eur. J. Control **8**(1), 56–63 (2002)
67. van der Schaft, A.J.: Symmetries in optimal control. SIAM J. Control Optim. **25**(2), 245–259 (1987)
68. Witting, K.: Numerical algorithms for the treatment of parametric multiobjective optimization problems and applications. Ph.D. thesis, University of Paderborn (2012)
69. Zavala, V.M.: A multiobjective optimization perspective on the stability of economic MPC. IFAC-PapersOnLine **48**(8), 974–980 (2015)
70. Zavala, V.M., Flores-Tlacuahuac, A.: Stability of multiobjective predictive control: a utopia-tracking approach. Automatica **48**(10), 2627–2632 (2012)

POD-Based Mixed-Integer Optimal Control of Evolution Systems

Christian Jäkle and Stefan Volkwein(✉)

Department of Mathematics and Statistics, University of Konstanz,
Universitätsstraße 10, 78457 Konstanz, Germany
{Christian.Jaekle,Stefan.Volkwein}@uni-konstanz.de

Abstract. In this chapter the authors consider the numerical treatment of a mixed-integer optimal control problem governed by linear convection-diffusion equations and binary control variables. Using relaxation techniques (introduced by [31] for ordinary differential equations) the original mixed-integer optimal control problem is transferred into a relaxed optimal control problem with no integrality constraints. After an optimal solution to the relaxed problem has been computed, binary admissible controls are constructed by a sum-up rounding technique. This allows us to construct – in an iterative process – binary admissible controls such that the corresponding optimal state and the optimal cost value approximate the original ones with arbitrary accuracy. However, using finite element (FE) methods to discretize the state and adjoint equations often yield to extensive systems which make the frequently calculations time-consuming. Therefore, a model-order reduction based on the proper orthogonal decomposition (POD) method is applied. Compared to the FE case, the POD approach yields to a significant acceleration of the CPU times while the error stays sufficiently small.

Keywords: Mixed-integer optimal control · Integer programming · Relaxation methods · Evolution problems · Proper orthogonal decomposition

1 Introduction

A simplified optimal control problem is considered which is motivated by energy efficient building operation. The goal is to reach a certain desired temperature distribution in a room while choosing an optimal (underfloor) heating strategy. The temperature is governed by a heat equation with convection which extends our results in [3,4], where no convection was involved in the modeling of the heat transfer. Since the heating is described by a time-depending discrete control, the optimization problem involves continuous and discrete variables. These kinds of problems are considered in [10,11,22,32]. For partial differential equations (PDEs) we refer to the note [24]. In particular, mixed-integer problems for

O. Junge et al. (Eds.): SON 2020, SSDC 304, pp. 238–264, 2020.
https://doi.org/10.1007/978-3-030-51264-4_10

hyperbolic PDEs are considered, e.g., for problems in gas transportation systems [15], electric transmission lines [12] and traffic flow [16,17].

Frequently, integer problems are solved with the branch-and-bound method (see, e.g., [5]) to guarantee global optimality. Especially for finite-dimensional linear integer programming, the branch-and-bound method is the method of choice. However, this is often not possible or very expensive for optimal control problems, where infinite-dimensional control spaces are involved. Therefore, methods are used to approximate an optimal integer solution by sufficiently accurate solutions, which can be computed by techniques from infinite-dimensional optimization. In this work we apply relaxation methods which can be found in [31] for the case of ordinary differential equations and in [16,17] for the case of PDEs. To solve the relaxed optimal control problems, we rely on techniques from PDE-constrained optimization ([25,36]). Utilizing sum-up-rounding strategies (introduced by Sager in [31,33]) we construct discrete controls from the continuous ones; see also [26,27].

To speed-up the numerical solution of the relaxed optimal control problems we apply reduced-order modeling; cf. [1,34], for instance. In this work the relaxed optimal control problems are solved by POD Galerkin projection methods; cf. [14,19]. The POD method is known to be very efficient for dynamical systems. An POD a-posteriori error analysis – developed in [35] for optimal control problems – is extended in such a way that the error of the computed suboptimal POD solution can be controlled. This leads to an efficient and a certified optimization method which is also analytically based on theoretical results in [17].

Let us mention that there are other reduced-order approaches available, e.g., the reduced basis methods; cf. [13]. Especially for non-linear problems, the proper orthogonal decomposition (POD) method is a popular and widely used method. Here, predefined points in time are considered by a previously released dynamic system to build up the so-called snapshot space. The leading eigenfunctions of a singular value decomposition are then chosen as the basis for the reduced space, see for example [14]. It has been shown, that this method has good properties in the context of optimal control problems, especially thanks to an available a-posteriori estimate, see [23,35].

The chapter is organized as follows: In Sect. 2 the mixed-integer optimal control problem is introduced. Its relaxation is explained in Sect. 3. The numerical solution approach is described in Sect. 4 and Sect. 5 is devoted to present numerical results. Finally, we draw some conclusions in Sect. 6.

2 Problem Formulation

Let $\Omega \subset \mathbb{R}^n$, $n \in \{1,2,3\}$, be a bounded domain with Lipschitz-continuous boundary $\Gamma = \partial\Omega$. For $T > 0$ we set $Q = (0,T) \times \Omega$ and $\Sigma = (0,T) \times \Gamma$. Moreover, let H and V denote the standard real and separable Hilbert spaces $L^2(\Omega)$ and $H^1(\Omega)$, respectively, endowed with the usual inner products

$$\langle \varphi, \psi \rangle_H = \int_\Omega \varphi\psi \,\mathrm{d}\boldsymbol{x}, \quad \langle \varphi, \psi \rangle_V = \int_\Omega \varphi\psi + \nabla\varphi \cdot \nabla\psi \,\mathrm{d}\boldsymbol{x}$$

and associated induced norms. For more details on Lebesgue and Sobolev spaces we refer to [9]. Recall the Hilbert space $W(0,T) = \{\varphi \in L^2(0,T;V) \,|\, \varphi_t \in L^2(0,T;V')\}$ endowed with the common inner product [8, pp. 472–479]. It is well-known that $W(0,T)$ is continuously embedded into $C([0,T];H)$, the space of continuous functions from $[0,T]$ to H. When t is fixed, the expression $\varphi(t)$ stands for the function $\varphi(t,\cdot)$ considered as a function in Ω only.

In this work we consider the following mixed-integer optimal control problem:

$$\min_{(y,u)} J(y,u) = \frac{1}{2}\int_0^T \int_\Omega |y(t,\boldsymbol{x}) - y^d(t,\boldsymbol{x})|^2 \,\mathrm{d}\boldsymbol{x}\mathrm{d}t + \frac{\gamma}{2}\sum_{i=1}^m \int_0^T |u_i(t)|^2 \,\mathrm{d}t \tag{1a}$$

subject to a convection-diffusion equation

$$y_t(t,\boldsymbol{x}) - \Delta y(t,\boldsymbol{x}) + v(\boldsymbol{x})\cdot\nabla y(t,\boldsymbol{x}) = f(t,\boldsymbol{x}) + \sum_{i=1}^m u_i(t)b_i(\boldsymbol{x}), \qquad (t,\boldsymbol{x}) \in Q, \tag{1b}$$

$$\frac{\partial y}{\partial n}(t,\boldsymbol{s}) + q(\boldsymbol{s})y(t,\boldsymbol{s}) = g(t,\boldsymbol{s}), \qquad (t,\boldsymbol{s}) \in \Sigma, \tag{1c}$$

$$y(0,\boldsymbol{x}) = y_\circ(\boldsymbol{x}), \qquad \boldsymbol{x} \in \Omega \tag{1d}$$

and binary control constraints

$$u(t) \in \{0,1\}^m = \{u^i\}_{i=1}^N \quad \text{in } [0,T] \text{ a.e. (almost everywhere)}, \tag{1e}$$

where the u^i's are 0–1-vectors in $\mathbb{R}^m$ and $N = 2^m$ holds.

The desired temperature fulfills $y^d \in L^\infty(Q)$. For the regularization parameter we have $\gamma > 0$. The convection field v is supposed to be in $L^\infty(\Omega;\mathbb{R}^n)$. The heat source function satisfied $f \in C(\overline{Q})$. For $m \in \mathbb{N}$ we assume that the control shape functions fulfill $b_1,\dots,b_m \in C(\overline{\Omega})$ and $b_i \geq 0$ on Ω a.e., but at least for one $i \in \{1,\dots,m\}$ it holds $b_i > 0$ on Ω a.e. The isolation function satisfies $q \in L^\infty(\Gamma)$ with $q \geq 0$ on Γ a.e. The outer temperature is described by g and belongs to $C(\overline{\Sigma})$. Finally, for the initial temperature distribution we have $y_\circ \in C(\overline{\Omega})$.

Since we are interested in weak solutions to the state equation (1b)–(1d), we recall this solution concept for our case: A solution $y \in W(0,T)$ to (1b)–(1d) is understood as a weak solution, i.e., y belongs to $W(0,T)$ and satisfies

$$\frac{\mathrm{d}}{\mathrm{d}t}\langle y(t),\varphi\rangle_H + a(y(t),\varphi) = \langle \mathcal{F}(t,u(t)),\varphi\rangle_{V',V} \quad \text{for all } \varphi \in V \text{ in } (0,T], \tag{2a}$$

$$\langle y(0),\varphi\rangle_H = \langle y_\circ,\varphi\rangle_H \quad \text{for all } \varphi \in V, \tag{2b}$$

where the bilinear form $a : V \times V \to \mathbb{R}$ is defined as

$$a(\varphi,\phi) = \int_\Omega \nabla\varphi\cdot\nabla\phi \,\mathrm{d}\boldsymbol{x} + \int_\Omega (v\cdot\nabla\varphi)\phi \,\mathrm{d}\boldsymbol{x} + \int_\Gamma q\varphi\phi \,\mathrm{d}\boldsymbol{s} \quad \text{for } \varphi,\phi \in V$$

and the inhomogeinity $\mathcal{F} : [0,T] \times \mathbb{R}^m \to V'$ is given by

$$\langle \mathcal{F}(t,u),\varphi\rangle_{V',V} = \int_\Omega \Big(f(t) + \sum_{i=1}^m u_ib_i\Big)\varphi \,\mathrm{d}\boldsymbol{x} + \int_\Gamma g(t)\varphi \,\mathrm{d}\boldsymbol{s}$$

for $(t, u) \in [0, T] \times \mathbb{R}^m$, $u = (u_i)_{1 \le i \le m}$ and $\varphi \in V$. Note that the mapping $\mathcal{F}(\cdot, u)$ is continuous for every $u \in \mathbb{R}^m$. The next proposition follows from the results in [20, Chapter 5].

Proposition 1. *Under the above assumptions on the data the following properties hold:*

1) *The bilinear form $a(\cdot, \cdot)$ is continuous and coercive, i.e., there are constants $\eta \ge 0$, $\eta_1 > 0$ and $\eta_2 \ge 0$ satisfying*

$$|a(\varphi, \phi)| \le \eta \, \|\varphi\|_V \|\phi\|_V \qquad \text{for all } \varphi, \phi \in V,$$
$$|a(\varphi, \varphi)| \ge \eta_1 \, \|\varphi\|_V^2 - \eta_2 \, \|\varphi\|_H^2 \qquad \text{for every } \varphi \in V.$$

2) *For any $u \in U = L^2(0, T; \mathbb{R}^m)$ there exists a unique solution $y \in W(0, T)$ to (2) that satisfies*

$$\|y\|_{W(0,T)} \le C\big(\|\mathcal{F}(\cdot, u(\cdot))\|_{C([0,T];V')} + \|y_\circ\|_H\big)$$

for a constant $C > 0$.

Remark 1. The bilinear form $a(\cdot, \cdot)$ defines a bounded linear operator $\mathcal{A} : V \to V'$ by

$$\langle \mathcal{A}\varphi, \phi \rangle_{V',V} = a(\varphi, \phi) \quad \text{for } \varphi, \phi \in V.$$

Furthermore, the operator $\mathcal{A}$ can also be considered as an unbounded operator on H with domain $\mathscr{D}(\mathcal{A}) = H^2(\Omega) \cap V \cap C(\overline{\Omega})$ which is dense in $C(\overline{\Omega})$. The operator $-\mathcal{A}$ generates a C_0-semigroup on $C(\overline{\Omega})$ and the solution y to (2) belongs to $W(0, T) \cap C(\overline{Q})$; cf. [29, Chapter 5]. Utilizing the continuity assumptions for f, $b_1, \ldots, b_m$, g and $y_\circ$ we can write (2) as the Cauchy problem

$$\dot{y}(t) = -\mathcal{A}y(t) + \mathcal{F}(t, u(t)) \text{ for } t \in (0, T], \quad y(0) = y_\circ$$

posed in $C(\overline{\Omega})$. It is proved in [28, Theorem 4.3] that $-\mathcal{A}$ also generates a holomorphic semigroup on $C(\overline{\Omega})$. $\Diamond$

Throughout this work the binary problem (1a)–(1e) is called **(BN)**. Its cost value at an admissible solution is denoted by J^{BN}. Furthermore, we introduce a relaxed problem, where (1e) is replaced by the relaxation

$$u(t) \in [0, 1]^m \text{ in } [0, T] \text{ a.e.,} \tag{1e'}$$

Problem (1a)–(1d) together with (1e') is denoted by **(RN)**. We write J^{RN} for the objective value obtained by an admissible solution for **(RN)**. Let us mention that (1a)–(1d) together with (1e') does not involve any integrality constraints. Thus, solution methods from continuous optimization can be applied.

3 Relaxation Method

Commonly, mixed-integer problems are solved with the branch-and-bound method (see e.g. [5]) to guarantee global optimality. However, for optimal control problems this is often computationally too expensive. In order to get an optimal control problem without any integer restrictions we apply therefore the approach in [17] which leads to convexified relaxed problems that can be solved by available techniques from PDE-constrained optimization; see [18,35], for instance.

3.1 Convexification

Using (1e) we introduce the following representation of the control variable

$$\beta(t) \in \{0,1\}^N, \ \sum_{i=1}^N \beta_i(t) = 1 \text{ and } u(t) = \sum_{i=1}^N \beta_i(t) u^i \quad \text{for } t \in [0,T].$$

To solve our mixed-integer optimal control problem **(BN)** we consider the following convexification (cf. [17, Section 2])

$$\min_{(y_\beta,\beta)} \frac{1}{2} \int_0^T \int_\Omega |y_\beta(t,\boldsymbol{x}) - y^d(t,\boldsymbol{x})|^2 \,\mathrm{d}\boldsymbol{x}\mathrm{d}t + \frac{\gamma}{2} \sum_{i=1}^N \|u^i\|_{\mathbb{R}^m}^2 \int_0^T \beta_i(t)\,\mathrm{d}t \tag{3a}$$

subject to

$$\frac{\mathrm{d}}{\mathrm{d}t} \langle y_\beta(t), \varphi\rangle_H + a(y_\beta(t),\varphi) = \sum_{i=1}^N \beta_i(t) \, \langle \mathcal{F}(t,u^i),\varphi\rangle_{V',V} \quad \forall \varphi \in V \text{ in } (0,T], \tag{3b}$$

$$\langle y_\beta(0),\varphi\rangle_H = \langle y_\circ,\varphi\rangle_H \quad \forall\varphi \in V, \tag{3c}$$

$$\beta(t) = \big(\beta_i(t)\big)_{1\le i\le N} \in \{0,1\}^N \quad \text{in } [0,T], \tag{3d}$$

$$\sum_{i=1}^N \beta_i(t) = 1 \quad \text{in } [0,T]. \tag{3e}$$

The convexification (3) of **(BN)** is called **(BL)** and we write J^{BL} for the objective value obtained by an admissible solution. Of course, **(BL)** still contains the integrality constraint (3d). Therefore, we introduce its relaxation – that we call **(RL)** – by

$$\min_{(y_\alpha,\alpha)} \frac{1}{2} \int_0^T \int_\Omega |y_\alpha(t,\boldsymbol{x}) - y^d(t,\boldsymbol{x})|^2 \,\mathrm{d}\boldsymbol{x}\mathrm{d}t + \frac{\gamma}{2} \sum_{i=1}^N \|u^i\|_{\mathbb{R}^m}^2 \int_0^T \alpha_i(t)\,\mathrm{d}t \tag{4a}$$

subject to

$$\frac{\mathrm{d}}{\mathrm{d}t} \langle y_\alpha(t), \varphi\rangle_H + a(y_\alpha(t),\varphi) = \sum_{i=1}^N \alpha_i(t) \langle \mathcal{F}(t,u^i),\varphi\rangle_{V',V} \quad \forall \varphi \in V \text{ in } (0,T], \tag{4b}$$

$$\langle y_\alpha(0),\varphi\rangle_H = \langle y_\circ,\varphi\rangle_H \quad \forall\varphi \in V, \tag{4c}$$

$$\alpha(t) = \big(\alpha_i(t)\big)_{1\le i\le N} \in [0,1]^N \quad \text{in } [0,T], \tag{4d}$$

$$\sum_{i=1}^N \alpha_i(t) = 1 \quad \text{in } [0,T]. \tag{4e}$$

We write J^{RL} for the objective value obtained by an admissible solution.

In the following theorem we show that the convexification does not change the optimal values of the original problem **(BN)** and the convexified problem **(BL)**. The proof is similar to the one in [31, Theorem 4.6] and therefore adapted from there.

Theorem 1. *If the convexified binary optimal control problem* **(BL)** *has an optimal solution* (y_β^*, β^*) *with objective value* J^{BL}*, then there exists an m-dimensional control function* u^* *such that* (y^*, u^*) *is an optimal solution of the binary optimal control problem* **(BN)** *with objective value* J^{BN} *satisfying* $J^{BL} = J^{BN}$*. The converse holds true as well.*

Proof. Assume that (y_β^*, β^*) is a minimizer of **(BL)**. Since it is feasible we have the special order set property (3e) with $\beta_i^*(\cdot) \in \{0, 1\}$ for all $i = 1, \ldots, N$. Thus, there exists one index $1 \leq j(t) \leq N$ for almost all (f.a.a.) $t \in [0, T]$ such that

$$\beta_{j(t)}^* = 1 \text{ and } \beta_i^* = 0, \quad i \neq j(t).$$

The binary control function

$$u^*(t) = u^{j(t)}, \quad t \in [0, T] \text{ a.e.}$$

is therefore well-defined and yields an identical right-hand side function value

$$\mathcal{F}(t, u^*(t)) = \mathcal{F}(t, u^{j(t)}) = \beta_{j(t)}^* \mathcal{F}(t, u^{j(t)}) = \sum_{i=1}^{N} \beta_i^*(t) \mathcal{F}(t, u^i), \quad t \in [0, T] \text{ a.e.}$$

and identical objective function

$$J(y^*, u^*) = J(y^*, u^{j(\cdot)}) = \beta_{j(\cdot)}^* J(y_\beta^*, u^{j(\cdot)}) = \sum_{i=1}^{N} \beta_i^*(\cdot) J(y_\beta^*, u^i)$$

compared to the feasible and optimal solution (y_β^*, β^*) of **(BL)**. Therefore (y^*, u^*) is a feasible solution of **(BN)** with objective value J^{BL}. Next we show – by contradiction – that there exists no admissible solution to **(BN)** with a smaller cost value than J^{BL}. Hence, we assume that a feasible solution $(\hat{y}, \hat{u})$ of **(BN)** exists with objective value $\hat{J}^{BN} < J^{BL}$. Since the set $\{u^1, \ldots, u^N\}$ contains all feasible assignments of $\hat{u}$, there exists again an index function $\hat{j}(\cdot)$ such that $\hat{u}$ can be written as

$$\hat{u}(t) = u^{\hat{j}(t)}, \quad t \in [0, T] \text{ a.e.}$$

With the same arguments above, β is defined as

$$\beta_i(t) = \begin{Bmatrix} 1 & \text{if } i = \hat{j}(t), \\ 0 & \text{otherwise} \end{Bmatrix} \quad \text{for } i = 1, \ldots, 2^N \text{ and } t \in [0, T] \text{ a.e..}$$

Consequently, β is feasible for **(BL)** with objective function value $\hat{J}^{BN} < J^{BL}$ which contradicts the optimality assumption of problem **(BN)**. Thus (y^*, u^*) is an optimal solution of problem **(BN)**.
The converse is proven with the same argumentation starting from the optimal solution **(BN)**.

In the following we want to apply [17, Theorem 1]. For that purpose we additionally suppose that $f(\cdot, \boldsymbol{x})$ and $g(\cdot\,, \boldsymbol{s})$ belong to $W^{1,\infty}(0,T)$ for almost all $\boldsymbol{x} \in \Omega$ and $\boldsymbol{s} \in \Gamma$, respectively. Next we verify assumptions (H0) to (H3) of [17, Theorem 1]:

- (H0): In [20, Theorem 5.13] is proved that **(RL)** has a unique optimal solution.
- (H1): Due to our regularity assumptions for the inhomogeneities f and g, the objective J and $\mathcal{F}$ are (locally) Lipschitz-continuous.
- (H2): Utilizing the $W^{1,\infty}$-regularity for the inhomogeneities f and g again, we notice that $\mathcal{F}(\cdot\,, u^i)$ belongs also $W^{1,\infty}(0,T)$ for any $i = 1, \ldots, N$. Now, (H2) follows from [17, Proposition 1], i.e., there exists a constant $C > 0$ with

$$\left\| \frac{\mathrm{d}}{\mathrm{d}\tau} \left(e^{-\mathcal{A}(t-\tau)} \mathcal{F}(\tau, u^i) \right) \right\|_{C(\overline{\Omega})} \leq C \quad \text{for } 0 < \tau < t < T \text{ a.e. and } 1 \leq i \leq N.$$

- (H3): It follows also that the mapping $t \mapsto \mathcal{F}(t, u^i)$ is essentially bounded in $C(\overline{\Omega})$ for any $i \in \{1, \ldots, N\}$.

Summarizing we have the following result [17, Theorem 1 and Corollary 1]:

Proposition 2. *Let the regularity conditions for the data stated in Sect. 2 hold. Moreover, $f(\cdot\,, \boldsymbol{x})$ and $g(\cdot\,, \boldsymbol{s})$ belong to $W^{1,\infty}(0,T)$ for almost all $\boldsymbol{x} \in \Omega$ and $\boldsymbol{s} \in \Gamma$, respectively. Suppose that (y_α^*, α^*) is the solution to the relaxed problem* **(RL)** *with objective value J^{RL}. Choose an arbitrary $\varepsilon > 0$. Then there exists a feasible solution $(y_\varepsilon^*, u_\varepsilon^*)$ of problem* **(BN)** *satisfying*

$$J^{BN} \leq J^{RL} + \varepsilon.$$

Remark 2. Notice that a feasible solution $(y_\varepsilon^*, u_\varepsilon^*)$ of problem ***(BN)*** can be constructed from (y_α^*, α^*) by sum-up rounding; cf. Sect. 4.2 and [17, Algorithm 1]. ◊

4 Numerical Solution Method

In the following we describe in detail how to apply the theoretical results in a numerical realization. We utilize Algorithm 1 which is based on the approach described in [17]. To guarantee convergence in a finite number of steps, the sequences of non-negative accuracies $\{\varepsilon_k\}_{k\in\mathbb{N}}$ and the time discretizations should be chosen such that $\varepsilon_k \to 0$ and $\Delta t_k = \max_{i=1,\ldots,\nu^k}\{t_i^k - t_{i-1}^k\} \to 0$, according to Theorem 1 in [17].

4.1 Solution of the Relaxed Problem

Let us introduce two particular solutions for the state and dual equations:

Algorithm 1. Relaxation Method for the FEM Model

Choose a time discretization $\mathcal{G}^0 = \{0 = t_0^0 < t_1^0 \ldots < t_{\nu^0}^0 = T\}$, a sequence of non-negative accuracies $\{\varepsilon_k\}_{k\in\mathbb{N}}$ and some fixed tolerance $\varepsilon > 0$.
for $k = 0, 1, \ldots$ **do**
    Find an optimal control α^k of **(RL)** which stops with a tolerance of ε_k.
    Set $J^k_{\mathsf{rel}} = \hat{J}(\alpha^k)$.
    if α^k is binary admissible **then**
        break
    end if
    Using $\mathcal{G}^k$ and α^k to define a piecewise constant function β^k as described in 4.3.
    Determine $J^k = \hat{J}(\beta^k)$.
    if $|J^k_{\mathsf{rel}} - J^k| \le \varepsilon/2$ and $0 < \varepsilon_k \le \frac{\varepsilon}{2}$ **then**
        break
    end if
    Choose $\mathcal{G}^{k+1} = \{0 = t_0^{k+1} < t_1^{k+1} \ldots < t_{\nu^{k+1}}^{k+1} = T\}$ such that $\mathcal{G}^k \subset \mathcal{G}^{k+1}$.
end for
Set $y^*_{\mathsf{bin}} = S\beta^k + \hat{y}$, $\beta^* = \beta^k$, $y^* = S\alpha^k + \hat{y}$ and $u^*(t) = \sum_{j=1}^N u^j \beta_j^k(t)$.

- $\hat{y} \in W(0,T)$ is the weak solution to

$$\begin{aligned} \hat{y}_t(t,\boldsymbol{x}) - \Delta\hat{y}(t,\boldsymbol{x}) + v(\boldsymbol{x}) \cdot \nabla\hat{y}(t,\boldsymbol{x}) &= f(t,\boldsymbol{x}) && \text{in } Q \text{ a.e.,} \\ \frac{\partial\hat{y}}{\partial n}(t,\boldsymbol{s}) + q(\boldsymbol{s})\hat{y}(t,\boldsymbol{s}) &= g(t,\boldsymbol{s}) && \text{on } \Sigma \text{ a.e.,} \\ \hat{y}(0,\boldsymbol{x}) &= y_\circ(\boldsymbol{x}) && \text{in } \Omega \text{ a.e.} \end{aligned}$$

- Further, $\hat{p} \in W(0,T)$ is the weak solution to

$$\begin{aligned} -\hat{p}_t(t,\boldsymbol{x}) - \Delta\hat{p}(t,\boldsymbol{x}) - \nabla \cdot \big(v(\boldsymbol{x})\hat{p}(t,\boldsymbol{x})\big) &= y^d(t,\boldsymbol{x}) - \hat{y}(t,\boldsymbol{x}) && \text{in } Q \text{ a.e.,} \\ \frac{\partial\hat{p}}{\partial n}(t,\boldsymbol{s}) + \big(q(\boldsymbol{s}) + v(\boldsymbol{s}) \cdot \boldsymbol{n}(\boldsymbol{s})\big)\hat{p}(t,\boldsymbol{s}) &= 0 && \text{on } \Sigma \text{ a.e.,} \\ \hat{p}(T,\boldsymbol{x}) &= 0 && \text{in } \Omega \text{ a.e.,} \end{aligned}$$

  where $\boldsymbol{n}$ denotes the outwart normal vector.

The first step in Algorithm 1 is to solve **(RL)**. To do so we use a first-order augmented Lagrange method. Thus, we consider for $c \ge 0$

$$\min \hat{J}(\alpha) + \frac{c}{2}\int_0^T \left|\sum_{j=1}^N \alpha(t) - 1\right|^2 \mathrm{d}t \quad \text{s.t.} \quad \alpha \in \mathcal{A}_{\mathsf{ad}} \text{ and } \sum_{j=1}^N \alpha_j(t) = 1, \tag{5}$$

where the penalty term

$$\frac{c}{2}\int_0^T \left|\sum_{j=1}^N \alpha(t) - 1\right|^2 \mathrm{d}t$$

is the augmentation term, $\hat{J}(\alpha) = J(y_\alpha, \alpha)$ is the reduced cost functional and y_α solves (4b)–(4c). The set $\mathcal{A}_{\mathsf{ad}}$ is defined by

$$\mathcal{A}_{\mathsf{ad}} = \big\{\alpha \in \mathcal{A} \,\big|\, \alpha_j(t) \in [0,1] \text{ on } [0,T] \text{ a.e. for } j = 1,\ldots,N\big\}$$

and $\mathcal{A} = L^2(0,T;\mathbb{R}^N)$. For $c > 0$ the augmented Lagrangian is given as

$$\mathcal{L}_c(\alpha,\lambda) = \hat{J}(\alpha) + \left\langle \sum_{j=1}^N \alpha_j(\cdot) - 1, \lambda \right\rangle_{L^2(0,T)} + \frac{c}{2}\left\| \sum_{j=1}^N \alpha_j(\cdot) - 1 \right\|_{L^2(0,T)}^2.$$

For the inner optimization (i.e., the minimization of $\mathcal{L}_c(\cdot\,\lambda)$ with respect to the primal variable α) we choose a multiplier $\lambda^0 \in \mathcal{A}$ and set $k = 0$. Then, for $k = 0, 1, \ldots$, we solve for $c_k > 0$

$$\min \mathcal{L}_{c_k}(\alpha, \lambda^k) \quad \text{s.t.} \quad \alpha \in \mathcal{A}_{\mathsf{ad}} \tag{$\mathbf{P}_c^k$}$$

and set

$$\lambda^{k+1} = \lambda^k + c_k \left(\sum_{j=1}^N \alpha_j(\cdot) - 1 \right).$$

For more details about Lagrangian methods see, e.g., [6, Chapter 3 and 4]. We repeat this process until we have

$$\left\| \sum_{j=1}^N \alpha_j(\cdot) - 1 \right\|_{L^2(0,T)}^2 \leq \varepsilon$$

for a given tolerance $\varepsilon > 0$. The optimality conditions are given as

$$\partial_\alpha \mathcal{L}(\bar{\alpha},\bar{\lambda})(\alpha - \bar{\alpha}) = \langle \nabla_\alpha \mathcal{L}_c(\bar{\alpha},\bar{\lambda}), \alpha - \bar{\alpha} \rangle_{\mathcal{A}} \geq 0 \quad \text{for all } \alpha \in \mathcal{A}_{\mathsf{ad}},$$

where $\partial_\alpha \mathcal{L}(\bar{\alpha},\bar{\lambda}) : \mathcal{A} \to \mathbb{R}$ stands for the partial derivative with respect to α, $\mathcal{L}_c(\bar{\alpha},\bar{\lambda}) \in \mathcal{A}$ is the gradient with respect to α. Moreover, $\bar{\alpha}$ is a local optimal solution to (P_c^k), and we have

$$\begin{aligned}
\partial_\alpha \mathcal{L}(\bar{\alpha},\bar{\lambda})\alpha^\delta &= \langle \nabla_\alpha \mathcal{L}_c(\bar{\alpha},\bar{\lambda}), \alpha^\delta \rangle_{\mathcal{A}} \\
&= \hat{J}'(\bar{\alpha})\alpha^\delta + \sum_{j=1}^N \langle \bar{\alpha}_j^\delta, \bar{\lambda} \rangle_{L^2(0,T)} + c \sum_{j=1}^N \sum_{l=1}^N \langle \bar{\alpha}_j - 1, \alpha_l^\delta \rangle_{L^2(0,T)}
\end{aligned}$$

for all directions $\alpha^\delta \in \mathcal{A}$. For a given point $\alpha \in \mathcal{A}_{\mathsf{ad}}$ and a direction $\alpha^\delta \in \mathcal{A}$ the directional derivative $\hat{J}'(\alpha)\alpha^\delta$ can be computed as follows:

1) Compute for a given $\alpha = (\alpha_i)_{1 \leq i \leq N} \in \mathcal{A}_{\mathsf{ad}}$ the state y_α solving

$$\begin{aligned}
(y_\alpha)_t(t,\boldsymbol{x}) - \Delta y_\alpha(t,\boldsymbol{x}) + v(\boldsymbol{x}) \cdot \nabla y_\alpha(t,\boldsymbol{x}) &= \sum_{j=1}^N \left(\sum_{i=1}^m b_i(\boldsymbol{x}) u_i^j \right) \alpha_j && \text{in } Q \text{ a.e.},\\
\frac{\partial y_\alpha}{\partial n}(t,\boldsymbol{s}) + q(\boldsymbol{s}) y_\alpha(t,\boldsymbol{s}) &= 0 && \text{on } \Sigma \text{ a.e.},\\
y_\alpha(0,\boldsymbol{x}) &= 0 && \text{in } \Omega \text{ a.e.}
\end{aligned}$$

and set $y = \hat{y} + y_\alpha$.

2) Solve the adjoint equation

$$\begin{aligned} -(p_\alpha)_t(t,\boldsymbol{x}) - \Delta p_\alpha(t,\boldsymbol{x}) - \nabla\cdot\big(v(\boldsymbol{x})p_\alpha(t,\boldsymbol{x})\big) &= -y_\alpha(t,\boldsymbol{x}) && \text{in } Q \text{ a.e.,}\\ \frac{\partial p_\alpha}{\partial n}(t,\boldsymbol{s}) + \big(q(\boldsymbol{s}) + v(\boldsymbol{s})\cdot\boldsymbol{n}(\boldsymbol{s})\big)p_\alpha(t,\boldsymbol{s}) &= 0 && \text{on } \Sigma \text{ a.e.,}\\ p_\alpha(T,\boldsymbol{x}) &= 0 && \text{in } \Omega \text{ a.e.} \end{aligned}$$

and set $p = \hat{p} + p_\alpha$.

3) Set for $\alpha^\delta \in \mathcal{A}$

$$\begin{aligned} \hat{J}'(\alpha)\alpha^\delta &= \frac{\gamma}{2}\sum_{j=1}^N\Big(\int_0^T\sum_{i=1}^m u_i^j\alpha_j^\delta(t)\,\mathrm{d}t\Big) - \int_0^T\int_\Omega\sum_{j=1}^N\Big(\sum_{i=1}^m b_i(\boldsymbol{x})u_i^j\Big)\alpha_j^\delta(t)p\,\mathrm{d}\boldsymbol{x}\mathrm{d}t\\ &= \sum_{j=1}^N\int_0^T\Big(\frac{\gamma}{2}\sum_{i=1}^m\big(u_i^j\alpha_j^\delta(t)\big) - \alpha_j^\delta(t)\int_\Omega p(t,\boldsymbol{x})\sum_{i=1}^m\big(b_i(\boldsymbol{x})u_i^j\big)\,\mathrm{d}\boldsymbol{x}\Big)\,\mathrm{d}t\\ &= \sum_{j=1}^N\int_0^T\Big(\frac{\gamma}{2}\sum_{i=1}^m u_i^j - \int_\Omega p(t,\boldsymbol{x})\sum_{i=1}^m\big(b_i(\boldsymbol{x})u_i^j\big)\,\mathrm{d}\boldsymbol{x}\Big)\alpha_j^\delta(t)\,\mathrm{d}t\\ &= \Big\langle\Big(\frac{\gamma}{2}\sum_{i=1}^m u_i^j - \int_\Omega p(\cdot,\boldsymbol{x})\sum_{i=1}^m\big(b_i(\boldsymbol{x})u_i^j\big)\,\mathrm{d}\boldsymbol{x}\Big)_{1\le j\le N},\alpha^\delta\Big\rangle_{L^2(0,T)}. \end{aligned}$$

Therefore we can introduce the Riesz representant of the linear, bounded functional $\hat{J}'(\alpha) : \mathcal{A} \to \mathbb{R}$ by

$$\Big(\frac{\gamma}{2}\sum_{i=1}^m u_i^j - \int_\Omega p(\cdot,\boldsymbol{x})\sum_{i=1}^m\big(b_i(\boldsymbol{x})u_i^j\big)\,\mathrm{d}\boldsymbol{x}\Big)_{1\le j\le N} =: \nabla\hat{J}(\alpha) \in \mathcal{A}.$$

In particular, $\hat{J}'(\alpha) = \langle\nabla\hat{J}(\alpha),\cdot\rangle_{L^2(0,T)}$ holds true.

Remark 3. The first order optimality conditions for problem (P_c^k) are given by the variational inequality

$$\begin{aligned} &\Big\langle\Big(\frac{\gamma}{2}\sum_{i=1}^m u_i^j - \int_\Omega p(\cdot,\boldsymbol{x})\sum_{i=1}^m\big(b_i(\boldsymbol{x})u_i^j\big)\,\mathrm{d}\boldsymbol{x} + \bar{\lambda}\Big)_{1\le j\le N},\alpha-\bar{\alpha}\Big\rangle_{\mathcal{A}}\\ &\qquad + c\sum_{j=1}^N\sum_{l=1}^N\langle\bar{\alpha}_j - 1,\alpha_l - \bar{\alpha}_l\rangle_{L^2(0,T)} \ge 0 \end{aligned}$$

for all $\alpha = (\alpha_i)_{1\le i\le N} \in \mathcal{A}_{\mathsf{ad}}$, where $\bar{\alpha}$ is the optimal solution to (5). For more details see [20, Chapter 5]. ◊

4.2 Sum-Up Rounding

Assume that we have found an optimal solution α to **(RL)**. The next step in Algorithm 1 is to construct a binary admissible control function β for **(BL)**.

There we need to guarantee that we will not lose the special ordered set property (SOS-1). Therefore we construct β by using the following so-called sum up rounding strategy (compare to [17] and [31, Section 5.1]) as follows.

Let $\mathcal{G} = \{t_0, t_1, \dots, t_\nu\}$ be a time grid with $0 = t_0 < t_1 < \dots < t_\nu = T$. Define $\beta = (\beta_1, \dots, \beta_N) : [0, T] \to \{0, 1\}^N$ by

$$\beta_i(t) = p_{i,j}, \quad t \in [t_j, t_{j+1}), \quad i = 1, \dots, N,\ j = 0, \dots, \nu - 1,$$

where for all $i = 1, \dots, N$, $j = 0, \dots, \nu - 1$

$$p_{i,j} = \begin{cases} 1 & \text{if } (\hat{p}_{i,j} \geq \hat{p}_{l,j}\, \forall l \in \{1, \dots, N\} \setminus \{i\}) \text{ and} \\ & (i < l : \forall l \in \{1, \dots, N\} \setminus \{i\} : \hat{p}_{i,j} = \hat{p}_{l,j}) \\ 0 & \text{else} \end{cases}$$

$$\hat{p_{j,i}} = \int_0^{t_{j+1}} \alpha_i(\tau) d\tau - \sum_{l=0}^{j-1} p_{i,l}(t_{l+1} - t_l).$$

Finally, to define a binary admissible control function for **(BN)** $u : [0, T] \to \{0, 1\}^m$ we set

$$u(t) = \sum_{j=1}^{N} u^j \beta_j(t).$$

Thanks to Theorem 1 we get then $J^{BL} = J^{BN}$, where J^{BL} depends on β and J^{BN} depends on u.

4.3 Redefine the Time Discretization

If the values between the cost functions of **(BN)** and **(RL)** are not small enough we need to redefine the time grid to get a better solution. To do so, there are several strategies given in [31, Section 5.3]. However, the simplest way would be just to define the grid in an equidistant way by double it and to use the old solution as a warmstart. In our numerical experiments this is the way we have done.

4.4 The POD Method

The most expensive part in Algorithm 1 is to find an optimal control α of **(RL)**. If one use here, e.g., finite elements to discretize the state and adjoint equations these leads to huge computational time. Therefore, to reduce the cost of the numerical solution method we apply a POD-method. To apply POD to the convexified problem assume that we have already computed a POD basis of rank ℓ and the corresponding POD space $V^\ell = \text{span}\{\psi_1, \dots, \psi_\ell\} \subset V$ is given. Moreover, we assume that we have computed the inhomogeneous part $\hat{y}$ of the solution to the state equation. With the same notations as in Sect. 4.1 we introduce the weak formulation of the homogeneous part of the reduced-order state equation

$$\frac{\mathrm{d}}{\mathrm{d}t}\langle y_\alpha^\ell(t),\psi\rangle_H + a(y_\alpha^\ell(t),\psi) = \Big\langle \sum_{i=1}^m u_i b_i, \psi \Big\rangle_H \qquad \forall \psi \in V^\ell \text{ in } (0,T], \tag{6}$$

$$\langle y_\alpha^\ell(0),\psi\rangle_H = 0 \qquad \forall \psi \in V^\ell. \tag{7}$$

and set $y^\ell = \hat{y} + y_\alpha^\ell$. Moreover we define the POD approximated reduced cost function by

$$\hat{J}^\ell(\alpha) := \frac{1}{2}\left\|y^\ell - y_d\right\|^2_{L^2(Q)} + \frac{\gamma}{2}\sum_{j=1}^N \|u^j\|^2_{\mathbb{R}^m}\int_0^T \alpha_j(t)\,\mathrm{d}t.$$

With the definition of the POD approximated reduced cost we can define the POD approximated reduced problem for the inner optimization of the Lagrange method, more precisely for (P_c^k). For the inner optimization we consider therefore the POD approximated reduced problem

$$\min \mathcal{L}_{c_k}^\ell(\alpha,\lambda^k) \quad \text{s.t.} \quad \alpha \in \mathcal{A}_{\mathsf{ad}} \tag{$\mathbf{P}_c^{\ell,k}$}$$

for a given c_k and λ^k, where we have for $c \geq 0$

$$\mathcal{L}_c^\ell(\alpha,\lambda^k) = \hat{J}^\ell(\alpha) + \Big\langle \sum_{j=1}^N \alpha_j - 1, \lambda^k \Big\rangle_{L^2(0,T)} + \frac{c}{2}\Big\|\sum_{j=1}^N \alpha_j - 1\Big\|^2_{L^2(0,T)}.$$

Then the gradient of $\hat{J}^\ell$ is given by

$$\nabla \hat{J}^\ell(\alpha) = \left(\frac{\gamma}{2}\sum_{i=1}^m u_i^j - \int_\Omega p^\ell(\cdot,\boldsymbol{x})\sum_{i=1}^m \big(b_i(\boldsymbol{x})u_i^j\big)\,\mathrm{d}\boldsymbol{x}\right)_{1\leq j\leq N} \in \mathcal{A},$$

where $p^\ell = \hat{p} + p_\alpha^\ell$ holds true und p_α^ℓ solves the adjoint problem

$$-\frac{\mathrm{d}}{\mathrm{d}t}\langle p_\alpha^\ell(t),\psi\rangle_H + a(\psi,p_\alpha^\ell(t)) = -\langle y_\alpha^\ell(t),\psi\rangle_H \qquad \forall \psi \in V^\ell \text{ in } (0,T], \tag{8}$$

$$\langle p_\alpha^\ell(T),\psi\rangle_H = 0 \qquad \forall \psi \in V^\ell. \tag{9}$$

In [20, Theorem 5.38] is the following a-priori convergence result given.

Theorem 2. *Suppose assumptions from Sect. 2 hold. Let the linear, bounded operator* $\mathcal{B} : \mathcal{A} \to L^2(0,T;(V^\ell)')$ *be given as*

$$\langle \mathcal{B}\alpha)(t),\psi\rangle_{(V^\ell)',V^\ell} = \sum_{j=1}^N \alpha_j(t)\sum_{i=1}^m u_i^j\langle b_i,\psi\rangle_H \quad \textit{for } \alpha \in \mathcal{A} \textit{ in } [0,T] \textit{ a.e.}$$

We assume that $\mathcal{B}$ *is injective. For arbitrarily given* $\alpha \in \mathcal{A}$ *we suppose that the solutions* y_α *and* p_α *to (6) and (9), respectively, belong to* $H^1(0,T;V)\setminus\{0\}$.

1) *If we compute the POD space* V^ℓ *by solving*

$$\min \sum_{j=1}^{4} \int_0^T \Big\| y^j(t) - \sum_{i=1}^{\ell} \langle y^j(t), \psi_i \rangle_V \psi_i \Big\|_V^2 \,\mathrm{d}t \;\; s.t. \;\; \{\psi_i\}_{i=1}^{\ell} \subset V,\; \langle \psi_i, \psi_j \rangle_V = \delta_{ij}$$

using the snapshots $y^1 = y_\alpha$, $y^2 = (y_\alpha)_t$, $y^3 = p_\alpha$ *and* $y^4 = (p_\alpha)_t$, *then the optimal solution* $\bar\alpha$ *of* (P_c^k) *and the optimal solution* $\bar\alpha^\ell$ *to the reduced problem* $(\mathrm{P}_c^{\ell,k})$ *satisfy*

$$\lim_{\ell\to\infty} \|\bar\alpha^\ell - \bar\alpha\|_{\mathcal{A}} = 0.$$

2) *If an optimal POD basis of rank* ℓ *is computed by choosing the snapshots* $y^1 = y_{\bar\alpha}$, $y^2 = (y_{\bar\alpha})_t$, $y^3 = p_{\bar\alpha}$ *and* $y^4 = (p_{\bar\alpha})_t$, *then we have*

$$\lim_{\ell\to\infty} \|\bar\alpha^\ell - \bar\alpha\|_{\mathcal{A}} \le C \sum_{i=\ell+1}^{\infty} \mu_i,$$

where $\{\mu\}_{i\in\mathbb{N}}$ *are the eigenvalues of the corresponding POD problem satisfying the error formula*

$$\sum_{j=1}^{4} \int_0^T \Big\| y^j(t) - \sum_{i=1}^{\ell} \langle y^j(t), \psi_i \rangle_V \psi_i \Big\|_V^2 \,\mathrm{d}t = \sum_{i=\ell+1}^{\infty} \mu_i;$$

cf. [14], *for instance.*

The following a-posteriori error estimate guarantees that the error stays small in the numerical solution method. A proof can be found in [20, Theorem 5.39].

Theorem 3. *Let all assumptions of Theorem 2 hold. For arbitrarily given* $\alpha \in \mathcal{A}$ *we choose the snapshots* $y^1 = y_\alpha$, $y^2 = (y_\alpha)_t$, $y^3 = p_\alpha$ *and* $y^4 = (p_\alpha)_t$. *Define the function* $\zeta^\ell \in \mathcal{A}$ *by*

$$\zeta_i^\ell(t) = \begin{cases} -\min(0, \xi_i^\ell(t)) & \text{a.e. in } \mathfrak{A}_{0,i}^\ell = \{t \in [0,T] \,|\, \bar\alpha_i^l(t) = 0\}, \\ \;\;\max(0, \xi_i^\ell(t)) & \text{a.e. in } \mathfrak{A}_{1,i}^\ell = \{t \in [0,T] \,|\, \bar\alpha_i^l(t) = 1\}, \\ -\xi_i^\ell(t) & \text{a.e. in } [0,T] \setminus (\mathfrak{A}_{0,i}^\ell \cup \mathfrak{A}_{1,i}^\ell), \end{cases}$$

where $\xi^\ell = \nabla_\alpha \mathcal{L}_c(\alpha^\ell, \lambda)$ *in* $\mathcal{A}$. *Then, for* $c > 0$ *we get the a-posteriori error estimate*

$$\|\bar\alpha - \bar\alpha^\ell\|_{\mathcal{A}}^2 \le \frac{1}{c} \|\zeta^\ell\|_{\mathcal{A}}, \tag{10}$$

and in particular, $\lim_{\ell\to\infty} \|\zeta^\ell\|_{\mathcal{A}} = 0$.

With these results, we can solve **(RL)** with the POD method and control the error with the a-posteriori error estimate in our numerical solution approach.

Table 1. Parameter and function values for the numerical experiments

Symbol	Value	Description
T	1	Final time
Ω	$(0,1)^2$	Spatial domain
$v(x)$	$(1,1)^T$ for all $x \in \Omega$	Convection term
$q(x)$	0.1 on $\partial\Omega$	Isolation of the room
$f(t,x)$	0	No influence from the source term f
$g(t,x)$	see Fig. 1	Outside temperature modeled by a polynomial with degree 3

5 Numerical Experiments

In this section we investigate the mixed-integer optimal control problem numerically by the method introduced in Sect. 4. To recall, we consider the following mixed-integer optimal control problem:

$$\min_{y,u} J(y,u) = \frac{1}{2}\int_0^T \int_\Omega |y(t,\boldsymbol{x}) - y^d(t,\boldsymbol{x})|^2 \mathrm{d}\boldsymbol{x}\mathrm{d}t + \frac{\gamma}{2}\sum_{i=1}^m \int_0^T |u_i(t)|^2 \mathrm{d}t \tag{11a}$$

subject to a convection-diffusion equation

$$2y_t - \Delta y + v \cdot \nabla y = f + \sum_{i=1}^m u_i b_i \qquad \text{in } \Omega \tag{11b}$$

$$\frac{\partial y}{\partial n} + qy = g \qquad \text{on } \Sigma \tag{11c}$$

$$y(0) = y_0 \qquad \text{in } \Omega \tag{11d}$$

and binary admissibility of $u(\cdot)$

$$u(t) \in \{0,1\}^m \text{ in } [0,T] \text{ a.e. (almost everywhere).} \tag{11e}$$

We will use the parameters and function values which are given in Table 1. Moreover, for the desired temperature we first want a decrease in the temperature to 10° until $t = 0.25$, an increase to 18° until $t = 0.75$ and then again a decrease to 10°. With this we want to avoid simple solutions which are like $u(t) = 1$ or $u(t) = 0$ for all $t \in [0,T]$. Therefore we set for all $x \in \Omega$

$$y^d(t,x) = \begin{cases} 10 & \text{if } t < 0.25 \text{ or } t > 0.75, \\ 18 & \text{otherwise.} \end{cases}$$

In the following we will use all notations from the previous sections. The implementations are done in Python where we use the packages NumPy, SciPy and Matplotlib, see [21] as well as FEniCS, see [2,30]. All computations are done on a standard laptop (Acer Aspire 5, Intel(R) Core(TM) i5-8250U, 1.6 GHz (up to 3.4 GHz), 8 GB DDR4-RAM).

For solving the relaxed optimal control problems **(RL)** we use for the inner optimization in each iteration of the augmented Lagrange method (e.g. solving (P_c^k)) the L-BFGS-B method from SciPy, [7]. The maximum number of iterations for the L-BFGS-B method is set up to 100. For the first time grid we utilize a tolerance of $\varepsilon_0 = 10^{-4}$. After each modification of the time grid we divided ε_0 by 10 until $\varepsilon_k = 10^{-7}$ which give us a sequence of non-negative accuracies $\{\varepsilon_k\}$. To solve the integer problem we use as a tolerance $\varepsilon = 10^{-5}$. The first time discretization is given by an equidistant time grid $\mathcal{G}^0 \subset [0, T]$ with 50 time steps. The first optimal control problem **(RL)** is solved with a (pseudo) random initial control function. After that one we reuse the old solution to get faster convergence results. We stop the algorithm either if

$$\left|J^{BN} - J^{RL}\right| \leq \varepsilon$$

or if the size of the time grid is bigger than 800 grid points. We redefine the time grid in an equidistant way by double it.

In the tests we consider a tow dimensional control and impose a floor heating in the subdomains $\Omega_{b_1} = (0, 0.25) \times (0, 0.5) \subset \Omega$ and $\Omega_{b_2} = (0.25, 0.5) \times (0, 0.5) \subset \Omega$. Therefore we set $b_1(\boldsymbol{x}) = 1$ for all $\boldsymbol{x} \in \Omega_{b_1}$ and $b_2(\boldsymbol{x}) = 1$ for all $\boldsymbol{x} \in \Omega_{b_2}$. Set $u_1^1 = 0$, $u_1^2 = 1$, $u_2^1 = 0$ and $u_2^2 = 1$ the convexification and the relaxation leads to **(RL)** which has the form

$$\min_{y,\alpha} J(y, \alpha) = \frac{1}{2} \int_0^T \int_\Omega |y(t, \boldsymbol{x}) - y^d(t, \boldsymbol{x})|^2 \mathrm{d}\boldsymbol{x}\mathrm{d}t + \frac{\gamma}{2} \int_0^T (\alpha_2(t) + \alpha_3(t) + 2\alpha_4(t))\mathrm{d}t$$

subject to

$$\begin{aligned} y_t - \Delta y + v \cdot \nabla y &= f + b_1\alpha_2 + b_2\alpha_3 + (b_1 + b_2)\alpha_4, && \text{in } \Omega, \\ \frac{\partial y}{\partial n} + qy &= g, && \text{on } \Sigma, \\ y(0) &= y_0, && \text{in } \Omega, \\ \alpha(t) &\in [0, 1]^4, && \text{in } [0, T] \text{ a.e.}, \\ \sum_{i=1}^4 \alpha_i(t) &= 1, && \text{in } [0, T] \text{ a.e..} \end{aligned}$$

After finding an optimal control α we use the sum-up rounding strategy as described in Sect. 4.2 to get a binary admissible β and set

$$u(t) = \begin{pmatrix} 0 \\ 0 \end{pmatrix} \beta_1(t) + \begin{pmatrix} 1 \\ 0 \end{pmatrix} \beta_2(t) + \begin{pmatrix} 0 \\ 1 \end{pmatrix} \beta_3(t) + \begin{pmatrix} 1 \\ 1 \end{pmatrix} \beta_4(t)$$

for all $t \in [0, T]$.

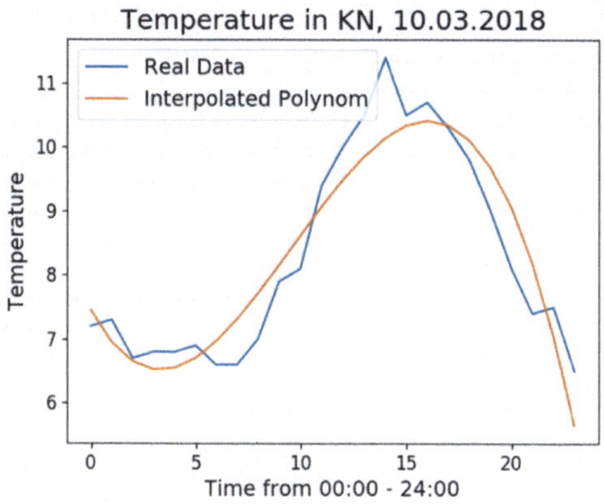

Fig. 1. Temperature outside given by real data and the interpolated polynom.

5.1 Full Finite Element Method Model

In our numerical experiments we compare the behavior of the algorithm with different regularization parameters γ. Notice, that for $\gamma = 0$ the problem leads to a bang-bang solution. Therefore, as bigger the regularization parameter gets, as more complicated is the integer problem. But on the other hand, big regularization parameters should make the relaxed problem easier to handle.

5.1.1 Case $\gamma = 0.01$

The first test is done with $\gamma = 0.01$. Here we use the initial condition that represents a constant temperature of 16° in the whole room, i.e. $y_0(\boldsymbol{x}) = 16$ for all $\boldsymbol{x} \in \Omega$. After four times redefining the grid, the algorithm has found a solution with a difference between $J^{BN} = J^{BL}$ and J^{RL} around $6 \cdot 10^{-5}$. The convergence behavior is given in the left subfigure of Fig. 2. Notice, that in the same figure we see also the difference between $J^{BN} = J^{BL}$ and J^{RL} (blue line) as well as the difference between J^{BN} and J^{RN} (orange line) which is close to the other one, caused of the small regularization parameter γ. Notice as well that we always have that $J^{RL} < J^{RN}$ which we could expect from the theoretical results. In Fig. 3 are the optimal control functions u_1 and u_2 and the corresponding binary functions β_1, β_2, β_3 and β_4. Notice that all control functions are close to bang-bang. Notice as well that to guarantee the SOS-1 property for the binary control functions β_i we need in β_2 an additional 1 at time step $t \approx 0.08$ which we don't see in the relaxed control α_2.

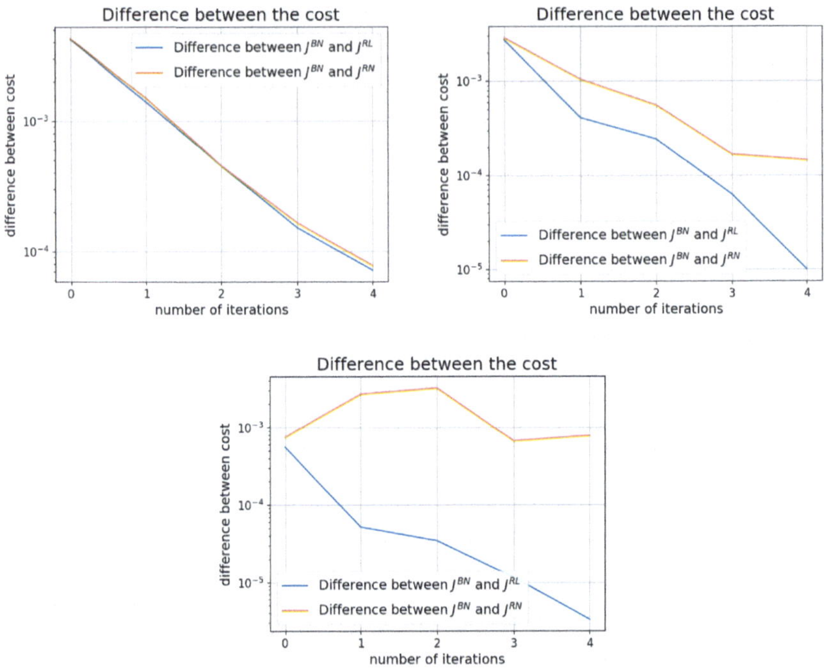

Fig. 2. Difference between J^{BN} and J^{RL} as well as between J^{BN} and J^{RN}. Case $\gamma = 0.01$ on the top left, $\gamma = 0.1$ on the top right and $\gamma = 1$ on the bottom middle.

5.1.2 Case $\gamma = 0.1$

Next we increase the regularization parameter and set therefore $\gamma = 0.1$. The algorithm modifies the time grid four times to reach a tolerance around 10^{-6} for the difference of $J^{BN} = J^{BL}$ and J^{RL}. The convergence behavior is given in the right subfigure of Fig. 2. Notice that the difference of $J^{BN} = J^{BL}$ and J^{RN} is bigger compared to the case $\gamma = 0.01$ and would not converge to a solution which is as close as that one we have found. Or in other words, just relax the integer problem leads to big duality gaps which do not close by just redefining the time grid.

5.1.3 Case $\gamma = 1$

Finally, we do the same test with $\gamma = 1$. To avoid solutions which are zero everywhere, we set $y_0(\boldsymbol{x}) = 14$ for all $\boldsymbol{x} \in \Omega$. The algorithm is done after four times redefinition of the time grid and a difference between $J^{BN} = J^{BL}$ and J^{RL} which is $\approx 10^{-6}$. The convergence behavior is given in the bottom subfigure of Fig. 2. Notice that there is a big difference between J^{BN} and J^{RN}. Here we see again the nice benefit of the convexification. Without that, it would be impossible to get such a small difference as for the solution, and we could say nothing how good our solution would be. Again we see that the duality gab

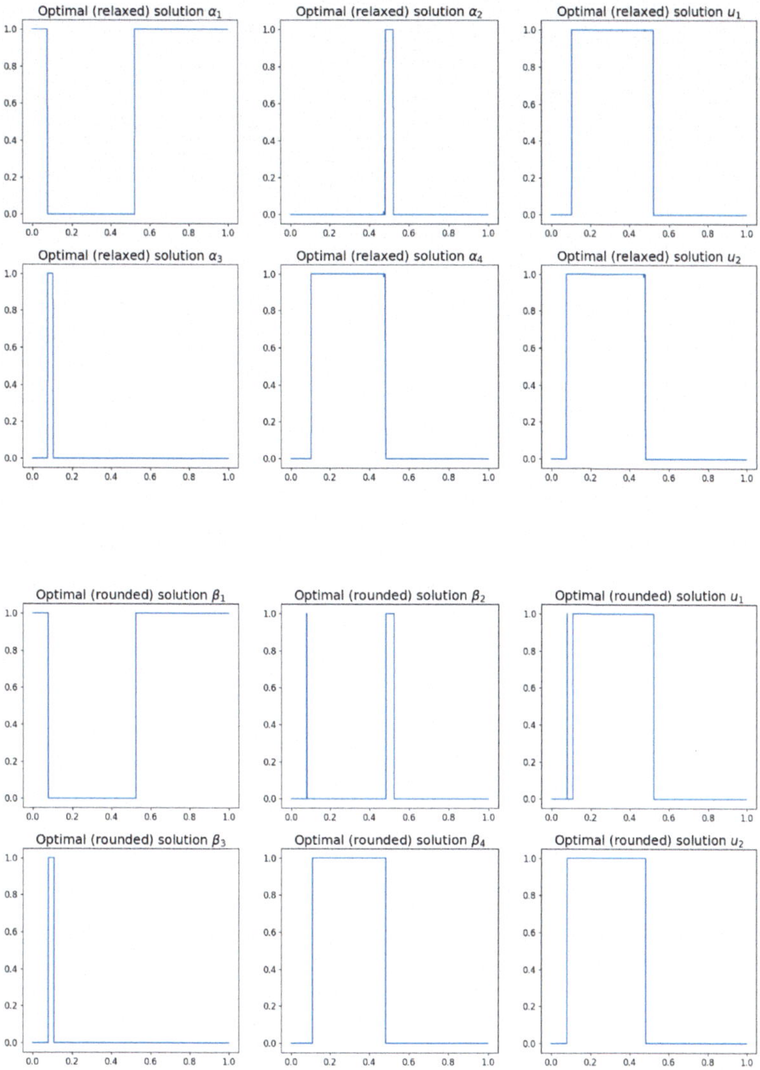

Fig. 3. Optimal control functions (relaxed and rounded) and corresponding optimal binary functions β_i for $\gamma = 0.01$.

between the integer solution and the relaxed solution without convexification is huge compared to the gap of the confexified problem.

If we have a look on Fig. 4 we can see the influence of the convection term in the control functions. The second control function is on the right side in the domain and from the convection therefore more expansive in the costs, although we do not weight our control functions. Therefore we get a zero control function for the second control. For the first one we have a little (but less than in the

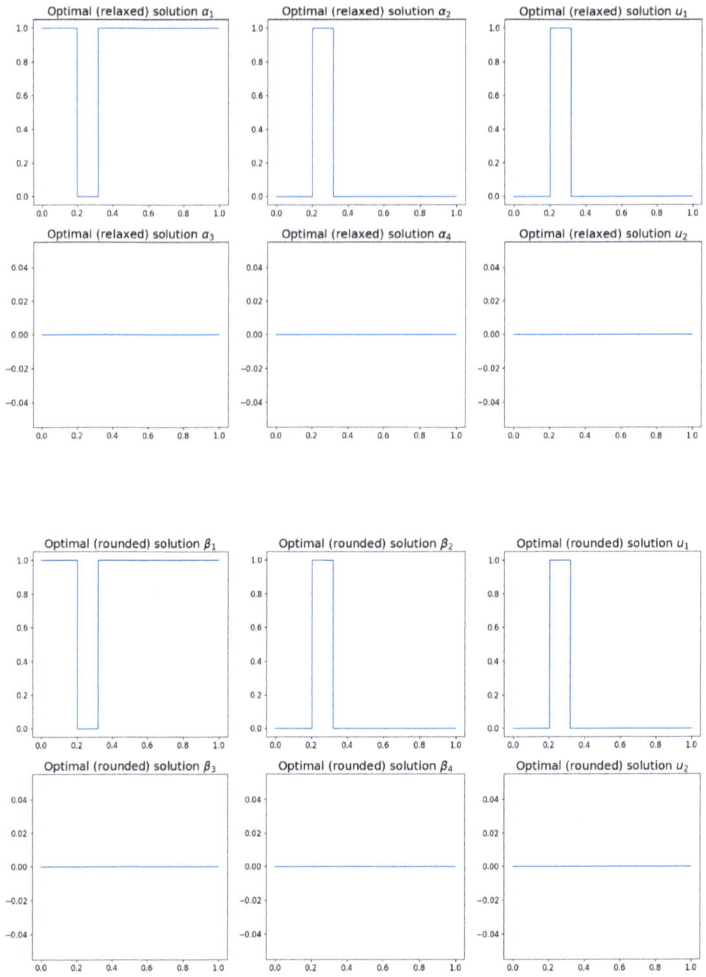

Fig. 4. Optimal control functions (relaxed and rounded) and corresponding optimal binary functions β_i for $\gamma = 1$.

case without convexification) chattering behavior. Summarized we see for a two-dimensional problem that the convexification leads to good solutions and the gap between integer and relaxed closes nicely. We also see, that the estimate between J^{BN} and J^{RL} is sharp, in contrast to J^{RN} which we could expect from the theoretical results. Moreover, we see a nice linear convergence behavior for the solutions by redefining the time grid in an equidistant way by doubling it.

5.2 The Reduced POD Model

In the following we do the same tests as in Sect. 5.1 but apply the POD method on the corresponding relaxed problems. We investigated the quality of the POD approximated solutions with the a-posteriori error estimate from Theorem 3 and compare the computational time. Since we are solving an integer problem we expect similar solutions as before and compare therefore the values of the cost functions corresponding to **(RL)** as well as **(RN)**.

To generate a POD basis of rank l we use the snapshots $y^1 = \mathcal{S}\alpha_0$ and $y^2 = \mathcal{A}\alpha_0$, where α_0 is a (pseudo) random initial control function which we use for both problems as initial one e.g. we start both optimal control problems with this one. We compute the POD basis as describes in [20, Section 4] using trapezoidal weights and solve the eigenvalue problem corresponding to the POD problem with the SVD method. As weighting matrix we use the mass matrix and we use $l = 10$ snapshots. The offline phase, which mean calculating the POD ansatz functions need in all three cases around 0.025 s.

5.2.1 Case $\gamma = 0.01$

For the FEM method, the algorithm needs to redefine the grid four times and found a solution after 1994.2 s. The algorithm spends the most time (around 1400 s) in the last time grid. This could come from a bad initial condition since in all other cases, the reuse of the old solution works pretty well. The difference between $J^{BN} = J^{BL}$ and J^{RL} is $\approx 7 * 10^{-5}$.

The POD method needs 154.2 s and is therefore more than 12 times faster, although we have use the same initial condition. In this test for u_2, at $t \approx 0.5$ the POD method has found a solution where $u_2^{POD}(t) = 1$ and $u_2^{FEM}(t) = 0$. The rest is equal. The convergence behavior of the full problem and the POD problem is given in Fig. 5 as well as the difference between J^{RL}_{FEM} and J^{RL}_{FEM}. In Table 2 we have given all values of the cost in the different time grids for the full problem as well as for the POD problem. Notice the interesting behavior of the a-posteriori error functions ζ.

Finally we have a look of the difference of the controls in the final grid. Here we have

$$\left\|u^{Rel}_{FEM} - u^{Rel}_{POD}\right\|_{A} = 0.02171, \quad \left\|u^{Int}_{FEM} - u^{Int}_{POD}\right\|_{A} = 0.02.$$

Therefore we can conclude that the POD method founds a similar solution and is much faster than the full method. Moreover, we can see that the POD method has found a solution in each time grid which is equal or slightly bigger than that one from the full model.

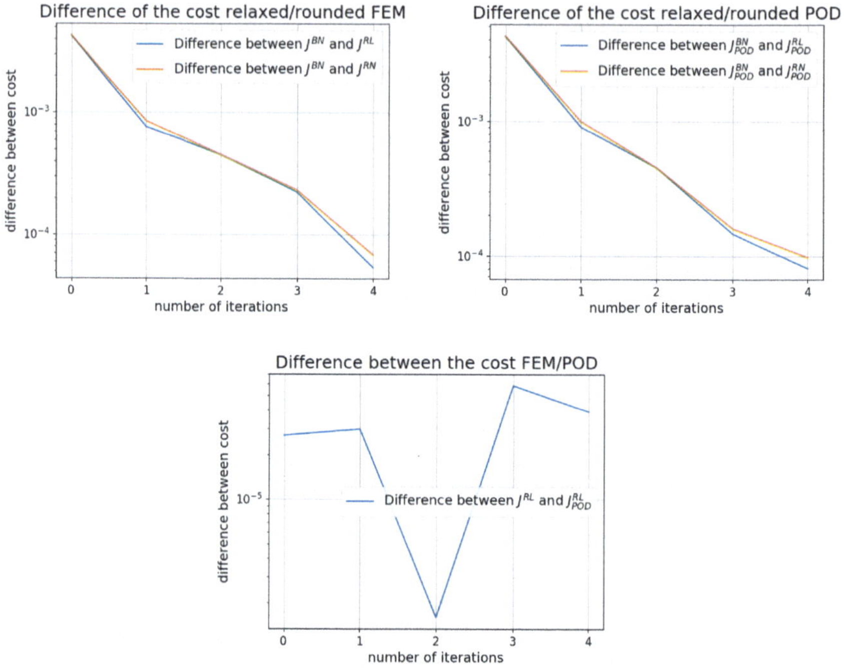

Fig. 5. Convergence behavior of the cost for $\gamma = 0.01$. On the top left, difference between the relaxed and the binary cost for the FEM. On the top right the difference for the POD method and on the bottom middle the difference between the convexified relaxed cost for the FEM and the POD method.

Table 2. Summarized values in the different time grids for $\gamma = 0.01$.

Time instances	50	100	200	400	800
$\frac{1}{c}\lVert\zeta\rVert$	$6 \cdot 10^{-8}$	0.00011	10^{-8}	$6 \cdot 10^{-5}$	$8 \cdot 10^{-6}$
$J^{BN} = J^{BL}$	11.5186	11.4813	11.4924	11.5009	11.5027
J^{RN}	11.5143	11.4805	11.4919	11.5006	11.5026
J^{RL}	11.5144	11.4806	11.4919	11.5006	11.5026
$J^{BN}_{POD} = J^{BL}_{POD}$	11.5186	11.4815	11.4924	11.5009	11.5028
J^{RN}_{POD}	11.5143	11.4805	11.4919	11.5007	11.5027
J^{RL}_{POD}	11.5144	11.4806	11.4919	11.5007	11.5027

5.2.2 Case $\gamma = 0.1$

Like before both algorithms need to redefine the time grid four times. For the full model the algorithm needs 1789.4 s. Again, the augmented Lagrange method in the final time grid needs the most time of the whole process. The difference between $J^{BN} = J^{BL}$ and J^{RL} is $\approx 10^{-6}$ and therefore reached our accuracy.

For the reduced POD model the algorithm needs 144.9 s and is therefore again more than 12 times faster. But the difference between $J^{BN}_{POD} = J^{BL}_{POD}$ and J^{RL}_{POD} is $\approx 10^{-5}$ and therefore a bit worse than for the full model. The convergence behavior of the costs for the full model and the POD model is given in Fig. 6. Table 3 shows again the different values of the cost and the behavior of the error functions ζ.

Again we have a look at the difference of the control in the final grid. This time we have

$$\left\|u^{Rel}_{FEM} - u^{Rel}_{POD}\right\|_A = 0.039, \quad \left\|u^{Int}_{FEM} - u^{Int}_{POD}\right\|_A = 0.141.$$

The discrete control u_1 of the full model and the POD model differs for

$$t \in \{0.13625, 0.48875, 0.49, 0.49125\}$$

and the discrete control u_2 differs for

$$t \in \{0.13375, 0.135, 0.41, 0.4125, 0.49125\}$$

which causes the difference of the binary control functions. Notice, again the value of the cost function J^{BN}_{POD} is equal at every time grid except in the case where the grid is of the size 200. Here the value of J^{BN}_{POD} is slightly bigger.

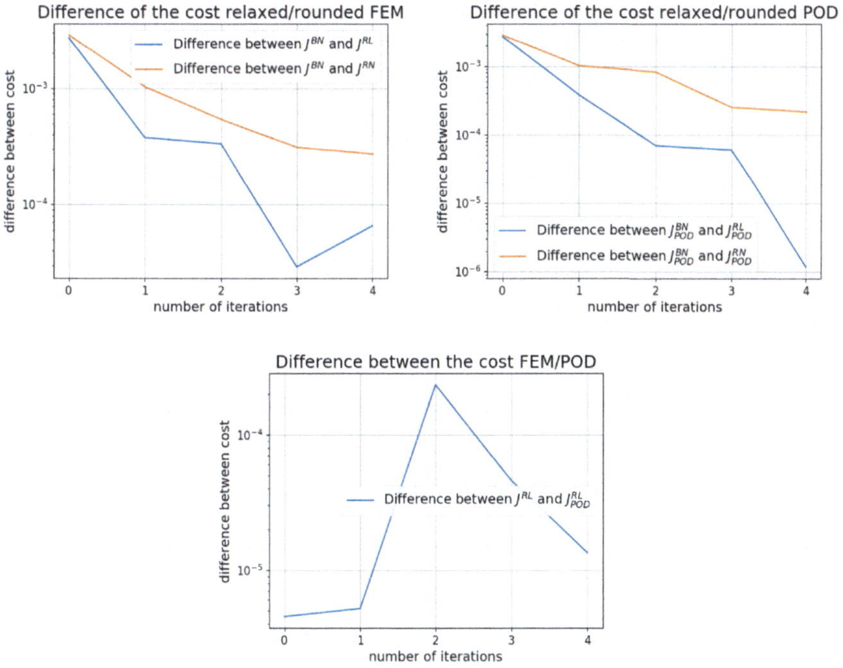

Fig. 6. Convergence behavior of the cost for $\gamma = 0.1$. On the top left, difference between the relaxed and the binary cost for the FEM. On the top right the difference for the POD method and on the bottom middle the difference between the convexified relaxed cost for the FEM and the POD method.

Table 3. Summarized values in the different time grids for $\gamma = 0.1$.

Time instances	50	100	200	400	800
$\frac{1}{c}\|\zeta\|$	$1 \cdot 10^{-9}$	$8 \cdot 10^{-5}$	$6 \cdot 10^{-5}$	$5 \cdot 10^{-7}$	$4 \cdot 10^{-6}$
$J^{BN} = J^{BL}$	11.5494	11.5126	11.5246	11.5338	11.5355
J^{RN}	11.5465	11.5115	11.5241	11.5335	11.5353
J^{RL}	11.5467	11.5122	11.5243	11.5337	11.5355
$J^{BN}_{POD} = J^{BL}_{POD}$	11.5494	11.5126	11.5247	11.5338	11.5355
J^{RN}_{POD}	11.5465	11.5116	11.5238	11.5335	11.5353
J^{RL}_{POD}	11.5466	11.5122	11.5246	11.5337	11.5352

5.2.3 Case $\gamma = 1$

The full model redefines the time grid three times and needed 277.1 s. The difference between $J^{BN} = J^{BL}$ and J^{RL} is $\approx 10^{-6}$, so we reach our tolerance what is the reason why the algorithm needs much less time and one time grid less than in the cases of $\gamma = 0.01$ and $\gamma = 0.1$. This could come from a good initial control u_0. Notice in the test of Subsect. 5.1 the algorithm needed one time grid more to reach this tolerance. Using the reduced POD model the algorithm needs 42.8 s and redefined the time grids three times. The plots of the convergence behavior are given in Fig. 7 and in Table 4 we have summarized our findings. Notice this time the value of J^{BL}_{POD} is a bit smaller or equal than the cost J^{BL} for the full model. The difference between $J^{BN}_{POD} = J^{BL}_{POD}$ and J^{RL}_{POD} is $\approx 10^{-6}$.

Having a look at the difference of the control functions in the final time grid we get

$$\left\|u^{Rel}_{FEM} - u^{Rel}_{POD}\right\|_A = 0.0073,$$
$$\left\|u^{Int}_{FEM} - u^{Int}_{POD}\right\|_A = 0.02.$$

The only difference in the binary control functions is for u_1 at $t = 0.205$.

Table 4. Summarized values in the different time grids for $\gamma = 1$.

Time instances	50	100	200	400
$\frac{1}{c}\|\zeta\|$	$2 \cdot 10^{-9}$	$1 \cdot 10^{-6}$	$6 \cdot 10^{-6}$	$1 \cdot 10^{-6}$
$J^{BN} = J^{BL}$	12.2703	12.2492	12.2561	12.2936
J^{RN}	12.2697	12.2466	12.2529	12.2903
J^{RL}	12.2697	12.2491	12.2566	12.2936
$J^{BN}_{POD} = J^{BL}_{POD}$	12.2703	12.2492	12.2561	12.2936
J^{RN}_{POD}	12.2697	12.2466	12.2579	12.2903
J^{RL}_{POD}	12.2697	12.2491	12.2561	12.2936

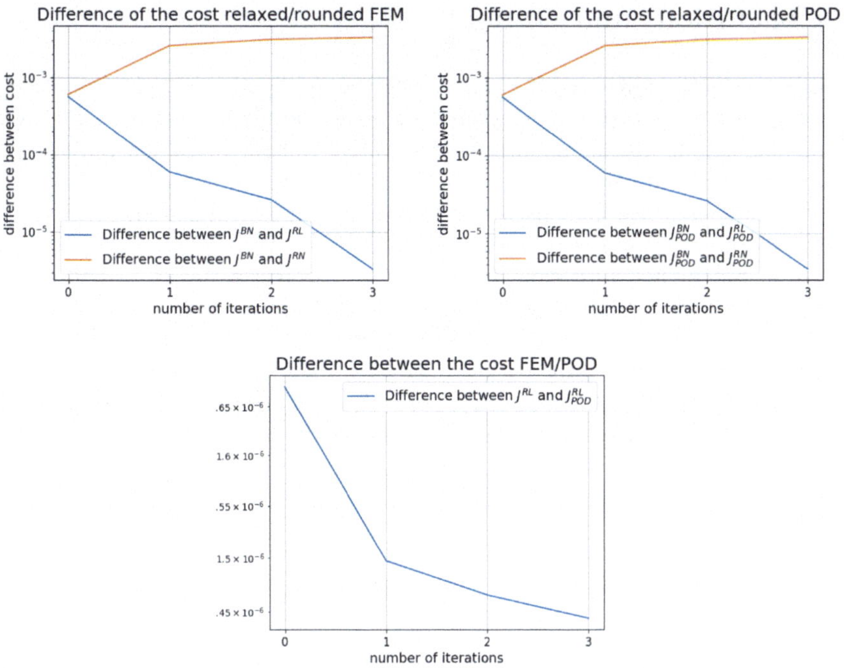

Fig. 7. Convergence behavior of the cost for $\gamma = 1$. On the top left, difference between the relaxed and the binary cost for the FEM. On the top right the difference for the POD method and on the bottom middle the difference between the convexified relaxed cost for the FEM and the POD method.

Table 5. Different computational times for the FEM method and the POD method for different regularization parameter.

Computational time			
			FEM/POD
$\gamma = 0.01$	FEM	1994.2 s	12, 9
	POD	154.2 s	
$\gamma = 0.1$	FEM	1789.4 s	12.3
	POD	144.9 s	
$\gamma = 1$	FEM	277.1 s	6.5
	POD	42.8 s	

Summarized we can definitely say that working with a reduced POD model instead of the full model gives a huge improvement of the computational time (see Table 5 for summarizing the speed up). Moreover, we get with this approach similar solutions which can lead incidental to smaller values of the cost function, therefore even better solutions. We have also seen, that a very small number l of ansatz functions for the POD basis are enough to reach this good solutions.

6 Conclusions and Outlook

In this chapter the authors have dealt with the application of relaxation methods combined with proper orthogonal decomposition (POD) methods for model order reduction to solve mixed-integer optimal control problems governed by linear convection-diffusion equations. After adopting the algorithm of [17] and verifying that this problem satisfies the assumptions of Theorem 1 in [17] to guarantee convergence a detailed description of the numerical solution method was given. Since the finite element method to discretize the state and adjoint equations from the optimization procedure leads to huge systems which have to be solved frequently, the POD method was introduced. This reduced the time-consuming optimization process and leads to a significant acceleration of the CPU times while the error remains small. The functionality of the algorithm and this behavior was verified by numerical experiments.

References

1. Antoulas, A.C.: Approximation of Large-Scale Dynamical Systems. SIAM, Philadelphia (2005)
2. Alnæs, M., Blechta, J., Hake, J., Johansson, A., Kehlet, B., Logg, A., Richardson, C., Ring, J., Rognes, M., Wells, G.: The FEniCS project version 1.5. Arch. Numer. Softw. **3**, 9–23 (2015)
3. Bachmann, F.: A branch-and-bound approach to mixed-integer optimal control using POD. Master's thesis, Department of Mathematics and Statistics, University of Konstanz (2017). http://nbn-resolving.de/urn:nbn:de:bsz:352-0-408645
4. Bachmann, F., Beermann, B., Lu, J., Volkwein, S.: POD-based mixed-integer optimal control of the heat equation. J. Sci. Comput. **81**, 48–75 (2019)
5. Belotti, P., Kirches, Ch., Leyffer, S., Linderoth, J., Luedtke, J., Mahajan, A.: Mixed-integer nonlinear optimization. Acta Numerica **22**, 1–131 (2013)
6. Bertsekas, D.: Nonlinear Programming. Athena Scientific Optimization and Computation Series. Athena Scientific, Belmont (1999)
7. Byrd, R., Lu, P., Nocedal, J., Zhu, C.: A limited memory algorithm for bound constrained optimization. SIAM J. Sci. Comput. **16**, 1190–1208 (1995)
8. Dautray, R., Lions, J.-L.: Mathematical Analysis and Numerical Methods for Science and Technology. Volume 5: Evolution Problems I. Springer, Berlin (2000)
9. Evans, L.C.: Partial Differential Equations. American Mathematical Society, Providence (2002)
10. Fügenschuh, A., Geißler, B., Martin, A., Morsi, A.: The transport PDE and mixed-integer linear programming. In: Models and Algorithms for Optimization in Logistics, 21–26 June 2009 (2009)
11. Gerdts, M.: Optimal Control of ODEs and DAEs. De Gruyter, Berlin (2011)
12. Göttlich, S., Hante, F., Potschka, A., Schewe, L.: Penalty alternating direction methods for mixed-integer optimal control with combinatorial constraints. Preprint (2019). arXiv:1905.13554v2
13. Grepl, M.A., Maday, Y., Nguyen, N.C., Patera, A.T.: Efficient reduced-basis treatment of nonaffine and nonlinear partial differential equations. ESAIM: Math. Model. Numer. Anal. **41**, 575–605 (2007)

14. Gubisch, M., Volkwein, S.: Proper orthogonal decomposition for linear-quadratic optimal control (Chap. 1). In: Model Reduction and Approximation. Computational Science and Engineering, pp. 3–63 (2017)
15. Gugat, M., Leugering, G., Martin, A., Schmidt, M., Sirvent, M., Wintergerst, D.: MIP-based instantaneous control of mixed-integer PDE-constrained gas transport problems (2017)
16. Hante, F.: Relaxation methods for hyperbolic PDE mixed-integer optimal control problems. Optim. Control Appl. Methods **38**, 1103–1110 (2017)
17. Hante, F., Sager, S.: Relaxation methods for mixed-integer optimal control of partial differential equations. Comput. Optim. Appl. **55**, 197–225 (2013)
18. Hinze, M., Pinnau, R., Ulbrich, M., Ulbrich, S.: Optimization with PDE Constraints. Mathematical Modelling: Theory and Applications, vol. 23. Springer, Berlin (2009)
19. Holmes, P., Lumley, J., Berkooz, G., Rowley, C.: Turbulence, Coherent Structures, Dynamical Systems and Symmetry, Cambridge Monographs on Mechanics, 2nd edn. Cambridge University Press, Cambridge (2012)
20. Jäkle, C.: POD-based mixed-integer optimal control of convection-diffusion equations. Master thesis, Department of Mathematics and Statistics, University of Konstanz (2019). http://nbn-resolving.de/urn:nbn:de:bsz:352-2-qgy7rxfhsbp89
21. Jones, E., Oliphant, T., Peterson, P., et al.: SciPy: Open source scientific tools for Python (2001)
22. Kirches, C.: Fast Numerical Methods for Mixed-Integer Nonlinear Model-Predictive Control. Advances in Numerical Mathematics. Vieweg+Teubner Verlag (2011)
23. Kunisch, K., Volkwein, S.: Optimal snapshot location for computing POD basis functions. ESAIM: Math. Model. Numer. Anal. **44**, 503–529 (2010)
24. Leyffer, S., Cay, P., Kouri, D., van Bloemen Waanders, B.: Mixed-integer PDE-constrained optimization. Technical report, Oberwolfach Report (2015)
25. Lions, J.L.: Optimal Control of Systems Governed by Partial Differential Equations. Springer-Verlag, Grundlehren der mathematischen Wissenschaften (1971)
26. Manns, P., Bestehorn, F., Hansknecht, C., Kirches, C., Lenders, F.: Approximation properties of sum-up rounding and consequences for mixed-integer PDE-constrained optimization. In: Liberti, L., Sager, S., Wiegele, A. (eds.) Mixed-Integer Nonlinear Optimization: A Hatchery for Modern Mathematics, Oberwolfach Reports **26**, 40-42 (2019)
27. Manns, P., Kirches, C., Lenders, F.: A linear bound on the integrality gap for Sum-Up Rounding in the presence of vanishing constraints. Submitted (2018). http://www.optimization-online.org/DB_FILE/2018/04/6580.pdf
28. Nittka, R.: Regularity of solutions of linear second order elliptic and parabolic boundary value problems on Lipschitz domains. J. Differ. Equ. **251**, 860–880 (2011)
29. Nittka, R.: Elliptic and parabolic problems with robin boundary conditions on Lipschitz domains. Ph.D. thesis, Faculty of Mathematics and Economics, Ulm University (2010)
30. Ølgaard, K., Logg, A., Wells, G.: Automated code generation for discontinuous Galerkin methods. SIAM J. Sci. Comput. **31**, 849–864 (2008)
31. Sager, S.: Numerical Methods for Mixed-Integer Optimal Control Problems. Der andere Verlag, Tönning (2005). https://mathopt.de/PUBLICATIONS/Sager2005.pdf
32. Sager, S., Bock, H., Diehl, M.: The integer approximation error in mixed-integer optimal control. Math. Program. **133**, 1–23 (2012)

33. Sager, S., Reinelt, G., Bock, H.G.: Direct methods with maximal lower bound for mixed-integer optimal control problems. Math. Program. **18**, 109–149 (2009)
34. Schilders, W., van der Vorst, H., Rommes, J.: Model Order Reduction: Theory. The European Consortium for Mathematics in Industry. Research Aspects and Applications. Springer, Heidelberg (2008)
35. Tröltzsch, F., Volkwein, S.: POD a-posteriori error estimates for linear-quadratic optimal control. Comput. Optim. Appl. **44**, 83–115 (2009)
36. Tröltzsch, F.: Optimal Control of Partial Differential Equations: Theory, Methods, and Applications. American Mathematical Society, Providence (2010)

From Bellman to Dijkstra: Set-Oriented Construction of Globally Optimal Controllers

Lars Grüne[1] and Oliver Junge[2(✉)]

[1] Mathematical Institute, University of Bayreuth, Bayreuth, Germany
lars.gruene@uni-bayreuth.de

[2] Department of Mathematics, Technical University of Munich, Munich, Germany
oliver.junge@tum.de

> An optimal policy has the property that whatever the initial state and initial decision are, the remaining decisions must constitute an optimal policy with regard to the state resulting from the first decision.
>
> Richard Bellman, 1957

Abstract. We review an approach for discretizing Bellman's optimality principle based on piecewise constant functions. By applying this ansatz to a suitable dynamic game, a discrete feedback can be constructed which robustly stabilizes a given nonlinear control system. Hybrid, event and quantized systems can be naturally handled by this construction.

1 Introduction

Whenever the state of some dynamical system can be influenced be repeatedly applying some control ("decision") to the system, the question might arise how the sequence of controls – the policy – can be chosen in such a way that some given objective is met. For example, one might be interested in steering the system to an equilibrium point, i.e. to *stabilize* the otherwise unstable point. In many contexts, the application of some control comes at some cost (fuel, money, time, ...) which then is accumulated over time. Typically, one is interested in meeting the given objective at minimal accumulated cost. This is the context of Richard Bellman's famous quote which already hints at how to solve the problem: One can recursively construct an optimal sequence of controls backwards in time by starting at the/some final state. It just so happens that this is also the idea of Edsger Dijkstra's celebrated algorithm for finding shortest paths in weighted directed graphs.

At the core, this procedure requires one to store the minimal accumulated cost at each state, the *value function*. According to the recursive construction

O. Junge et al. (Eds.): SON 2020, SSDC 304, pp. 265–294, 2020.
https://doi.org/10.1007/978-3-030-51264-4_11

of the sequence of optimal controls, the value function satisfies a recursion, i.e. a fixed point equation, the *Bellman equation*. From the value function at some state, the optimal control associated to that state can be recovered by solving a static optimization problem. This assignment defines a function on (a subset of) the states into the set of all possible control values and so the state can be *fed back* into the system, yielding a dynamical system without any external input. By construction, the accumulated cost along some trajectory of this *closed loop system* will be minimal.

In the case of a finite state space (with a reasonable number of states), storing the value function is easy. In many applications from, e.g., the engineering sciences, however, the state space is a subset of Euclidean space and thus the value function a function defined on a continuum of states. In this case, the value function typically cannot be represented in a closed form. Rather, some approximation scheme has to be decided upon and the value function (and thus the feedback) has to be approximated numerically.

In this chapter, we review contributions by the authors developing an approach for approximating the value function and the associated feedback by *piecewise constant functions*. This may seem like a bad idea at first, since in general one would prefer approximation spaces of higher order. However, it turns out that this ansatz enables an elegant solution of the discretized problem by standard shortest path algorithms (i.e. Dijkstra's algorithm). What is more, it also enables a unified treatment of system classes which otherwise would require specialized algorithms, like hybrid systems, event systems or systems with quantized state spaces.

As is common for some discretization, the discrete value function does not inherit a crucial property of the true one: In general, it does not decrease monotonically along trajectories of the closed loop system. In other words, it does not constitute a *Lyapunov function* of the closed loop system. As a consequence, the associated feedback may fail to stabilize some initial state. This deficiency can be cured by considering a more general problem class, namely a system which can be influenced by two independent controls – a *dynamic game*. In particular, if the second input is interpreted as some perturbation induced by the discretization, a discrete feedback results which retains the Lyapunov function property.

On the other hand, as any construction based on the Bellman equation, or more generally as any computational scheme which requires to represent a function with domain in some Euclidean space, our construction is prone to the *curse of dimension* (a term already coined by Bellman): In general, i.e. unless some specialized approximation space is employed, the computational cost for storing the value function grows exponentially in the dimension of state space. That is, in practice, our approach is limited to systems with a low dimensional state space (i.e. of dimension ≤ 4, say).

2 Problem Formulation

We are given a control system in discrete time

$$x_{k+1} = f(x_k, u_k, w_k), \quad k = 0, 1, \ldots, \tag{1}$$

where $x_k \in X$ is the *state* of the system, $u_k \in U$ is the *control input* and $w_k \in W$ is some external *perturbation*. We are further given an *instantaneous cost* function g which assigns the cost

$$g(x_k, u_k) \geq 0$$

to any transition $x_k \mapsto f(x_k, u_k, w)$, $w \in W$.

Our task is to globally and optimally stabilize a given *target set* $T \subset X$ by constructing a *feedback* $u : S \to U$, $S \subset X$, such that T is an asymptotically stable set for the *closed loop system*

$$x_{k+1} = f(x_k, u(x_k), w_k), \quad k = 0, 1, \ldots \tag{2}$$

with $x_0 \in S$ for *any* sequence $(w_k)_k$ of perturbations and such that the *accumulated cost*

$$\sum_{k=0}^{\infty} g(x_k, u(x_k)) \tag{3}$$

is minimal.

System Classes. Depending on the choice of the spaces X, U and W and the form of the map f, a quite large class of systems can be modelled by (1). Most generally, X, U and W have to be compact metric spaces – in particular, they may be discrete. Common examples which will also be considered later, include

- sampled-data systems: X, U and W are compact subsets of Euclidean space, f is the time-T-map of the control flow of some underlying continuous time control system and g typically integrates terms along the continuous time solution over one sampling interval;
- hybrid systems: $X = Y \times D$, where $Y \subset \mathbb{R}^n$ compact and D is finite, U and W may be continuous (compact) sets or finite (cf. Sect. 8);
- discrete event systems: f may be chosen as a (generalized) Poincaré map (cf. Sect. 8).
- quantized systems: The feedback may receive only quantized information on the state x, i.e. x is projected onto a finite subset of X before u is evaluated on this quantized state.

3 The Optimality Principle

The construction of the feedback law u will be based on a discretized version of the optimality principle. In order to convey the basic idea more clearly, we start by considering problem (1) without perturbations, i.e.

$$x_{k+1} = f(x_k, u_k), \quad k = 0, 1, \ldots \tag{4}$$

and assume that $X \subset \mathbb{R}^d$ and $U \subset \mathbb{R}^m$ are compact, $0 \in X$ and $0 \in U$. We further assume that $0 \in X$ is a fixed point of $f(\cdot, 0)$, i.e. $f(0, 0) = 0$, constituting our target set $T := \{0\}$, that $f : X \times U \to X$ and $g : X \times U \to [0, \infty)$ are continuous, that $g(0, 0) = 0$ and $\inf_{u \in U} g(x, u) > 0$ for all $x \neq 0$.

For a given initial state $x_0 \in X$ and a given sequence $\boldsymbol{u} = (u_0, u_1, \dots) \in U^{\mathbb{N}}$ of controls, there is a unique trajectory $\boldsymbol{x}(x_0, \boldsymbol{u}) = (x_k(x_0, \boldsymbol{u}))_{k \in \mathbb{N}}$ of (4). For $x \in X$, let

$$\mathcal{U}(x) = \{\boldsymbol{u} \in U^{\mathbb{N}} : x_k(x, \boldsymbol{u}) \to 0 \text{ as } k \to \infty\}$$

denote the set of *stabilizing control sequences* and

$$S = \{x \in X : \mathcal{U}(x) \neq \emptyset\}$$

the *stabilizable subset* of X. The accumulated cost along some trajectory $\boldsymbol{x}(x_0, \boldsymbol{u})$ is given by

$$J(x_0, \boldsymbol{u}) = \sum_{k=0}^{\infty} g(x_k(x_0, \boldsymbol{u}), u_k). \tag{5}$$

Note that this series might not converge for some $(x_0, \boldsymbol{u})$. The least possible value of the accumulated cost over all stabilizing control sequences defines the *(optimal) value function* $V : X \to [0, \infty]$,

$$V(x) = \inf_{\boldsymbol{u} \in \mathcal{U}(x)} J(x, \boldsymbol{u}) \tag{6}$$

of the problem. Let $S_0 := \{x \in X : V(x) < \infty\}$ be the set of states in which the value function is finite. Clearly, $S_0 \subset S$. On S_0, the value function satisfies the *optimality principle* [2]

$$V(x) = \inf_{u \in U} \{g(x, u) + V(f(x, u))\}. \tag{7}$$

The right hand side

$$L[v](x) := \inf_{u \in U} \{g(x, u) + v(f(x, u))\}$$

of (7) defines the *Bellman operator* L on real valued functions on X. The value function V is the unique fixed point of L satisfying the boundary condition $V(0) = 0$.

Using the value function V, one can construct the feedback $u : S_0 \to U$,

$$u(x) := \operatorname*{argmin}_{u \in U} \{g(x, u) + V(f(x, u))\}, \tag{8}$$

whenever this minimum exists. Obviously, V then satisfies

$$V(x) \geq g(x, u(x)) + V(f(x, u(x))), \tag{9}$$

for $x \in S_0$, i.e. the optimal value function is a Lyapunov function for the closed loop system on S_0 (provided that V is continuous at $T = \{0\}$[1]) – and this guarantees asymptotic stability of $T = \{0\}$ for the closed loop system. By construction, this feedback u is also optimal in the sense that the accumulated cost J is minimized along any trajectory of the closed loop system.

[1] This property can be ensured by suitable asymptotic controllability properties and bounds on g.

4 A Discrete Optimality Principle

In general, the value function (resp. the associated feedback) cannot be determined exactly and some numerical approximation has to be sought. Here, we are going to approximate V by functions which are piecewise constant on some partition of X. This approach is motivated by the fact that the resulting discrete problem can be solved efficiently and that, via a generalization of the framework to perturbed systems in Sect. 5 the feedback is also piecewise constant and can be computed offline.

Let $\mathcal{P}$ be a finite partition of the state space X, i.e. a finite collection of pairwise disjoint subsets of X whose union covers X. For $x \in X$, let $\pi(x) \in \mathcal{P}$ denote the partition element that contains x. In what follows, we identify any subset $\{P_1, \ldots, P_k\}$ of $\mathcal{P}$ with the corresponding subset $\bigcup_{i=1,\ldots k} P_i$ of X.

Let $\mathbb{R}^{\mathcal{P}} \subset \mathbb{R}^X = \{v : X \to \mathbb{R}\}$ be the subspace of real valued functions on X which are piecewise constant on the elements of $\mathcal{P}$. Using the projection

$$\psi[v](x) := \inf_{x' \in \pi(x)} v(x'), \tag{10}$$

from $\mathbb{R}^X$ onto $\mathbb{R}^{\mathcal{P}}$, we define the *discretized Bellman operator*

$$L_{\mathcal{P}} := \psi \circ L.$$

Again, this operator has a unique fixed point $V_{\mathcal{P}}$ satisfying the boundary condition $V_{\mathcal{P}}(0) = 0$, which will serve as an approximation to the exact value function V.

Explicitely, the discretized operator reads

$$L_{\mathcal{P}}[v](x) = \inf_{x' \in \pi(x)} \left\{ \inf_{u \in U} \{g(x', u) + v(f(x', u))\} \right\}.$$

and $V_{\mathcal{P}}$ satisfies the optimality principle

$$V_{\mathcal{P}}(x) = \inf_{x' \in \pi(x), u \in U} \{g(x', u) + V_{\mathcal{P}}(f(x', u))\}. \tag{11}$$

Recalling that $V_{\mathcal{P}}$ is constant on each element P of the partition $\mathcal{P}$, we write $V_{\mathcal{P}}(P)$ in order to denote the value $V_{\mathcal{P}}(x)$ for some $x \in P$. We can rewrite (11) as

$$V_{\mathcal{P}}(x) = \min_{P} \inf_{(x', u)} \{g(x', u) + V_{\mathcal{P}}(P)\} \tag{12}$$

where the min is taken over all $P \in \mathcal{P}$ for which $P \cap f(\pi(x), U) \neq \emptyset$ and the inf over all pairs $x' \in \pi(x)$, $u \in U$ such that $f(x', u) \in P$. Now define the multivalued map $\mathcal{F} : \mathcal{P} \rightrightarrows \mathcal{P}$,

$$\mathcal{F}(P) = \{P' \in \mathcal{P} : P' \cap f(P, U) \neq \emptyset\} \tag{13}$$

and the cost function $\mathcal{G} : \mathcal{P} \times \mathcal{P} \to [0, \infty)$,

$$\mathcal{G}(P, P') = \inf_{u \in U} \{g(x, u) \mid x \in P, f(x, u) \in P'\}. \tag{14}$$

Equation (12) can then be rewritten as

$$V_{\mathcal{P}}(P) = \min_{P' \in \mathcal{F}(P)} \{\mathcal{G}(P, P') + V_{\mathcal{P}}(P')\}.$$

Graph Interpretation. It is useful to think of this reformulation of the discrete optimality principle in terms of a directed weighted graph $G_{\mathcal{P}} = (\mathcal{P}, E_{\mathcal{P}})$. The nodes of the graph are given by the elements of the partition $\mathcal{P}$, the edges are defined by the map $\mathcal{F}$: there is an edge $(P, P') \in E_{\mathcal{P}}$ whenever $P' \in \mathcal{F}(P)$ and the edge $e = (P, P')$ is weighted by $\mathcal{G}(e) := \mathcal{G}(P, P')$, cf. Fig. 1. In fact, the value $V_{\mathcal{P}}(P)$ is the length $\mathcal{G}(p) := \sum_{k=1}^{m} \mathcal{G}(e_k)$ of the shortest path $p = (e_1, \ldots, e_m)$ from P to the element $\pi(0)$ containing 0 in this graph. As such, it can be computed by (e.g.) the following algorithm with complexity $\mathcal{O}(|\mathcal{P}| \log(|\mathcal{P}|) + |E|)$:

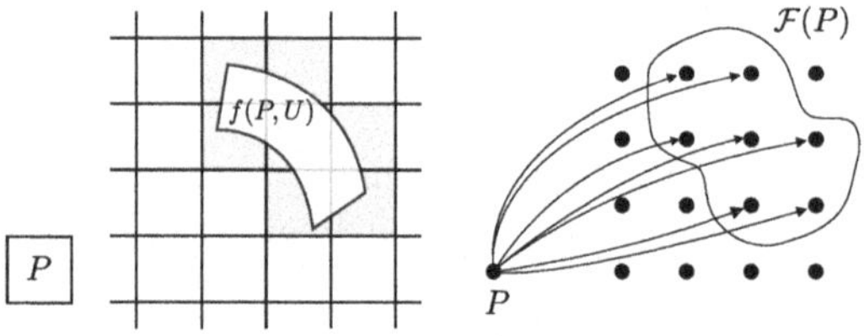

Fig. 1. Partition of phase space, image of an element (left) and corresponding edges in the induced graph (right).

Algorithm DIJKSTRA [5]

for each $P \in \mathcal{P}$: $V(P) := \infty$; $V(\pi(0)) := 0$; $\mathcal{Q} := \mathcal{P}$
while $\mathcal{Q} \neq \emptyset$
  $P := \text{argmin}_{P' \in \mathcal{Q}} V(P')$
  $\mathcal{Q} := \mathcal{Q} \backslash \{P\}$
  for each $Q \in \mathcal{P}$ with $(Q, P) \in E_{\mathcal{P}}$
    if $V(Q) > \mathcal{G}(Q, P) + V(P)$ then
      $V(Q) := \mathcal{G}(Q, P) + V(P)$

□

The time complexity of this algorithm depends on the data structure which is used in order to store the set $\mathcal{Q}$. In our implementation we use a binary heap which leads to a complexity of $\mathcal{O}((|\mathcal{P}| + |E|) \log |\mathcal{P}|)$. This can be improved to $\mathcal{O}(|\mathcal{P}| \log |\mathcal{P}| + |E|)$ by employing a Fibonacci heap.

A similar idea is at the core of *fast marching methods* [16,18] and *ordered upwind methods* [17].

Implementation. We use the approach from [3,4] as implemented in GAIO in order to construct a cubical partition of X, stored in binary tree. For the construction of the edges and their weights, we use a finite set of sample points $\tilde{U} \subset U$ and $\tilde{P} \subset P$ for each $P \in \mathcal{P}$ and compute the approximate image

$$\tilde{\mathcal{F}}(P) = \{P' \in \mathcal{P} : P' \cap f(\tilde{P}, \tilde{U}) \neq \emptyset\}, \tag{15}$$

so that the set of edges is approximately given by all pairs (P, P') for which $P' \in \tilde{\mathcal{F}}(P)$. Correspondingly, the weight of the edge (P, P') is approximated by

$$\tilde{\mathcal{G}}(P, P') = \min_{(x,u) \in \tilde{P} \times \tilde{U}} \{g(x, u) \mid f(x, u) \in P'\}.$$

This construction of the graph via the mapping of sample points indeed constitutes the main computational effort in computing the discrete value function. It might be particularly expensive if the control system f is given by the control flow of a continuous time system. Note, however, that a sampling of the system will be required in any method that computes the value function. In fact, in standard methods like value iteration, the same point might be sampled multiple times (in contrast to the approach described here).

Certainly, this approximation of the box images introduces some error, i.e. one always has that $\tilde{\mathcal{F}}(P) \subset \mathcal{F}(P)$, but typically $\mathcal{F}(P) \subsetneq \tilde{\mathcal{F}}(P)$. In experiments, one often increases the number of sample points until the result of the computation stabilizes. Alternatively, in the case that one is interested in a *rigorous* computation, either techniques based on Lipschitz estimates [13] or interval arithmetic [19] can be employed.

Example 1 ***(A simple 1D system).*** Consider the system

$$x_{k+1} = x_k + (1 - a)u_k x_k, \quad k = 0, 1, \ldots, \tag{16}$$

where $x_k \in X = [0, 1], u_k \in U = [-1, 1]$ and $a \in (0, 1)$ is a fixed parameter. Let

$$g(x, u) = (1 - a)x,$$

such that the optimal control policy is to steer to the origin as fast as possible, i.e. for every x, the optimal sequence of controls is $(-1, -1, \ldots)$. This yields $V(x) = x$ as the value function.

For the experiment, we consider $a = 0.8$ and use partitions of equally sized subintervals of $[0, 1]$. The edge weights (14) are approximated by minimizing over 100 equally spaced sample points in each subinterval and 10 equally spaced points in U. Figure 2 shows the exact and two discrete value functions, resulting from running the code in Fig. 3 in Matlab (requires the GAIO toolbox[2]).

[2] Available at http://www.github.com/gaioguy/gaio.

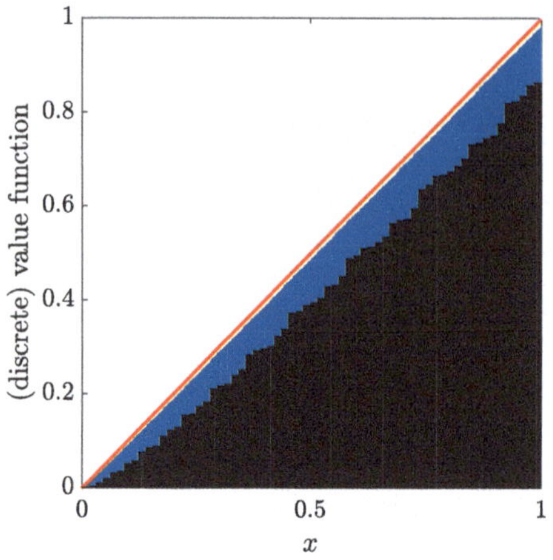

Fig. 2. Exact (red) and discrete value functions for the simple example on partitions of 64 (black) and 1024 (blue) intervals.

```
a = 0.8;
f = @(x,u) [x + (1-a).*x.*u, (1-a)*x];        % control system
X = linspace(-1,1,100)';                      % state samples
U = linspace(-1,1,10)';                       % control samples

depth = 6; c = 0.5; r = 0.5;
tree = Tree(c, r);                            % construct full tree
subdivide(tree, depth);                       % construct partition
A = dpgraph(tree, f, X, U, depth);            % compute graph
D = tree.search(0, depth);                    % find destination box
[V,~] = dijkstra(A', D);                      % compute value function

n = 2^depth; dx = 1/n; x = linspace(dx/2,1-dx/2,n);
clf;  plot(0:1,0:1,'r'); hold on; bar(x,V,1);
axis tight; axis square;
xlabel('$x$'); ylabel('(discrete) value function');
```

Fig. 3. Code: value function for a simple 1d system.

4.1 The Discrete Value Function

Proposition 1 [14]. *For every partition* $\mathcal{P}$ *of* X, $V_{\mathcal{P}}(x) \le V(x)$ *for all* $x \in X$.

Proof. The statement obviously holds for $x \in X$ with $V(x) = \infty$. So let $x \in S_0$, i.e. $V(x) < \infty$. For arbitrary $\varepsilon > 0$, let $\boldsymbol{u} = (u_0, u_1, \ldots) \in \mathcal{U}(x)$ be a control sequence such that $J(x, \boldsymbol{u}) < V(x) + \varepsilon$ and $(x_k(x, \boldsymbol{u}))_k$ the associated trajectory of (4). Consider the path

$$(e_1, \ldots, e_m), \quad e_k = (\pi(x_{k-1}), \pi(x_k)), \quad k = 1, \ldots, m,$$

where $x = x_0$ and and m is minimal with $x_m \in \pi(0)$. The length of this path is

$$\begin{aligned}\sum_{k=1}^{m} \mathcal{G}(e_k) &= \sum_{k=1}^{m} \inf_{u\in U} \{g(x,u) \mid x \in \pi(x_{k-1}), f(x,u) \in \pi(x_k)\} \\ &\leq \sum_{k=1}^{m} g(x_{k-1}, u_{k-1}) \leq \sum_{k=1}^{\infty} g(x_{k-1}, u_{k-1}) = J(x, \boldsymbol{u}),\end{aligned}$$

yielding the claim. □

This property immediately yields an efficient aposteriori error estimate for $V_{\mathcal{P}}$: For $x \in S_0$ consider

$$e(x) = \inf_{u\in U} \{g(x,u) + V_{\mathcal{P}}(f(x,u))\} - V_{\mathcal{P}}(x). \tag{17}$$

Note that $e(x) \geq 0$. Since

$$\begin{aligned}V(x) - V_{\mathcal{P}}(x) &= \inf_{u\in U} \{g(x,u) + V(f(x,u))\} - V_{\mathcal{P}}(x) \\ &\geq \inf_{u\in U} \{g(x,u) + V_{\mathcal{P}}(f(x,u))\} - V_{\mathcal{P}}(x) = e(x),\end{aligned}$$

we obtain

Proposition 2. *The function $e : S_0 \to [0, \infty)$ is a lower bound on the error between the true value function V and its approximation $V_{\mathcal{P}}$:*

$$e(x) \leq V(x) - V_{\mathcal{P}}(x), \quad x \in S_0.$$

Now consider a sequence $(\mathcal{P}^{(\ell)})_{\ell\in\mathbb{N}}$ of partitions of X which is nested in the sense that for all ℓ and every $P \in \mathcal{P}^{(\ell+1)}$ there is a $P' \in \mathcal{P}^{(\ell)}$ such that $P \subset P'$. For the next proposition recall that $S \subset X$ is the set of initial conditions that can be asymptotically controlled to 0.

Proposition 3 [14]. *For fixed $x \in S$, the sequence $(V_{\mathcal{P}^{(\ell)}}(x))_{\ell\in\mathbb{N}}$ is monotonically increasing.*

Proof. For $x \in S$, the value $V_{\mathcal{P}^{(\ell)}}(x)$ is the length of a shortest path $p = (e_1, \ldots, e_m)$, $e_k \in E_{\mathcal{P}^{(\ell)}}$, connecting $\pi(x)$ to $\pi(0)$ in $\mathcal{P}^{(\ell)}$. Suppose that the claim was not true, i.e. for some ℓ there are shortest paths p in $G_{\mathcal{P}^{(\ell)}}$ and p' in $G_{\mathcal{P}^{(\ell+1)}}$ such that $\mathcal{G}(p') < \mathcal{G}(p)$. Using p', we are going to construct a path $\tilde{p}$ in $G_{\mathcal{P}^{(\ell)}}$ with $\mathcal{G}(\tilde{p}) < \mathcal{G}(p)$, contradicting the minimality of p: Let $p' = (e'_1, \ldots, e'_{m'})$, with $e'_k = (P'_{k-1}, P'_k) \in E_{\mathcal{P}^{(\ell+1)}}$. Hence, $f(P'_{k-1}, U) \cap P'_k \neq \emptyset$, for $k = 1, \ldots, m'$. Since the partitions $\mathcal{P}^{(\ell)}$ are nested, there are sets $\tilde{P}_k \in \mathcal{P}^{(\ell)}$ such that $P'_k \subset \tilde{P}_k$ for $k = 0, \ldots, m'$. Thus, $f(\tilde{P}_{k-1}, U) \cap \tilde{P}_k \neq \emptyset$, i.e. $\tilde{e}_k = (\tilde{P}_{k-1}, \tilde{P}_k)$ is an edge in $E_{\mathcal{P}^{(\ell)}}$ and $\tilde{p} = (\tilde{e}_1, \ldots, \tilde{e}_{m'})$ is a path in $G_{\mathcal{P}^{(\ell)}}$. Furthermore, for $k = 1, \ldots, m'$,

$$\begin{aligned}\mathcal{G}(\tilde{e}_k) &= \inf_{u\in U} \{g(x,u) \mid x \in \tilde{P}_{k-1}, f(x,u) \in \tilde{P}_k\} \\ &\leq \inf_{u\in U} \{g(x,u) \mid x \in P'_{k-1}, f(x,u) \in P'_k\} = \mathcal{G}(e'_k).\end{aligned}$$

This yields $\mathcal{G}(\tilde{p}) \leq \mathcal{G}(p') < \mathcal{G}(p)$, contradicting the minimality of p. □

So far we have shown that for every $x \in S$ we have a monotonically increasing sequence $(V_{\mathcal{P}^{(\ell)}}(x))_{\ell\in\mathbb{N}}$, which is bounded by $V(x)$ due to Proposition 1. The following theorem states that for points $x \in S$ the limit is indeed $V(x)$ if the maximal diameter of the partition elements goes to 0. For some finite partition $\mathcal{P}$ of X, let $\operatorname{diam}(\mathcal{P}) := \max_i \operatorname{diam}(P_i)$ be the *diameter of the partition* $\mathcal{P}$.

Theorem 1 [14]. *If* $\operatorname{diam}(\mathcal{P}^{(\ell)}) \to 0$ *then* $V_{\mathcal{P}^{(\ell)}}(x) \to V(x)$ *as* $\ell \to \infty$ *for all* $x \in S$.

4.2 The Discrete Feedback

Recall that an optimally stabilizing feedback can be constructed using the (exact) value function for the problem (cf. (8)). We will use this idea, replacing V by its approximation $V_{\mathcal{P}}$: using $\tilde{U}$ from (15)[3], for $x \in S$ we define

$$u_{\mathcal{P}}(x) := \operatorname*{argmin}_{u\in\tilde{U}}\{g(x,u) + V_{\mathcal{P}}(f(x,u))\} \tag{18}$$

(the minimum exists because $\tilde{U}$ is a finite set) and consider the closed loop system

$$x_{k+1} = f(x_k, u_{\mathcal{P}}(x_k)), \quad k = 0, 1, \ldots. \tag{19}$$

The following theorems state in which sense this feedback is stabilizing and approximately optimal. Let again $(\mathcal{P}^{(\ell)})_{\ell\in\mathbb{N}}$ be a nested sequence of partitions of X and $D \subseteq S$, $0 \in D$, an open set with the property that for each $\varepsilon > 0$ there exists $\ell_0(\varepsilon) > 0$ such that

$$\max_{x\in D} |V(x) - V_{\mathcal{P}^{(\ell)}}(x)| \le \varepsilon, \quad \text{for } \ell \ge \ell_0(\varepsilon).$$

Let further $c > 0$ be the largest value such that

$$V_{\mathcal{P}^{(1)}}^{-1}([0,c]) \subset D.$$

Note that by Proposition 3 this implies that $V_{\mathcal{P}^{(\ell)}}^{-1}([0,c]) \subset D$ for all $\ell \in \mathbb{N}$.

Theorem 2 [7]. *Under the assumptions above, there exists* $\varepsilon_0 > 0$ *and a function* $\delta : \mathbb{R} \to \mathbb{R}$ *with* $\lim_{\alpha\to 0}\delta(\alpha) = 0$, *such that for all* $\varepsilon \in (0, \varepsilon_0]$, *all* $\ell \ge \ell_0(\varepsilon/2)$, *all* $\eta \in (0,1)$ *and all* $x_0 \in V_{\mathcal{P}^{(\ell)}}^{-1}([0,c])$ *the trajectory* $(x_k)_k$ *generated by the closed loop system* (19) *with feedback* $u_{\mathcal{P}^{(\ell)}}$ *satisfies*

$$V(x_k) \le \max\left\{ V(x_0) - (1-\eta)\sum_{j=0}^{k-1} g(x_j, u_{\mathcal{P}^{(\ell)}}(x_j)),\ \delta(\varepsilon/\eta) + \varepsilon \right\}.$$

[3] The subsequent statements remain true if we replace $\tilde{U}$ by any set $\hat{U} \subset U$ with $\tilde{U} \subset \hat{U}$ for which the argmin in (18) exists.

This apriori estimate shows in which sense the feedback $u_{\mathcal{P}}$ approximately yields optimal performance. However, the theorem does not give information about the partition $\mathcal{P}$ which is needed in order to achieve a desired level of accuracy. This can be achieved by employing the error function e from above.

Consider some partition $\mathcal{P}$ of X. Let $g_0(x) := \inf_{u \in U} g(x,u)$ and $C_\varepsilon(\mathcal{P}) := \{x \in V_{\mathcal{P}}^{-1}([0,c] \,|\, g_0(x) \le \varepsilon\}$ and define $\delta(\varepsilon) := \sup_{x \in C_\varepsilon} V(x)$. Note that if V is continuous at $T = \{0\}$ then $\delta(\varepsilon) \to 0$ as $\varepsilon \to 0$ because $C_\varepsilon(\mathcal{P})$ shrinks down to 0 since g and thus g_0 are continuous.

Theorem 3 [7]. *Assume that for some $\varepsilon > 0$ and some $\eta \in (0,1)$, the error function e satisfies*

$$e(x) \le \max\{\eta g_0(x), \varepsilon\} \quad \textit{for all } x \in V_{\mathcal{P}}^{-1}([0,c]). \tag{20}$$

Then, for each $x_0 \in V_{\mathcal{P}}^{-1}([0,c]$, the trajectory $(x_k)_k$ generated by the closed loop system (19) satisfies

$$V_{\mathcal{P}}(x_k) \le \max\left\{ V_{\mathcal{P}}(x_0) - (1-\eta)\sum_{j=0}^{k-1} g(x_j, u_{\mathcal{P}}(x_j)),\ \ \delta(\varepsilon/\eta) + \varepsilon \right\}. \tag{21}$$

Example 2 ***(An inverted pendulum).*** We consider a model for an inverted pendulum on a cart, cf. [7,14]. We ignore the dynamics of the cart, and so we only have one degree of freedom, namely the angle $\varphi \in [0, 2\pi]$ between the pendulum and the upright vertical. The origin $(\varphi, \dot\varphi) = (0,0)$ is an unstable equilibrium (with the pendulum pointing upright) which we would like to stabilize. The model reads

$$\left(\tfrac{4}{3} - m_r \cos^2\varphi\right)\ddot\varphi + \tfrac{m_r}{2}\dot\varphi^2 \sin 2\varphi - \tfrac{g}{\ell}\sin\varphi = -u\,\tfrac{m_r}{m\ell}\cos\varphi, \tag{22}$$

where $m = 2$ is the mass of the pendulum, $M = 8$ the mass of the cart, $m_r = m/(m+M)$, $\ell = 0.5$ the length of the pendulum and $g = 9.8$ the gravitational constant. We consider the discrete time control system (4) with $f(x,u) = \Phi^t(x,u)$, $x = (\varphi, \dot\varphi)$, for $t = 0.1$, where $\Phi^t(x,u)$ denotes the controlled flow of (22) with constant control input $u(\tau) = u$ for $\tau \in [0,t]$. For the instantaneous cost function we choose

$$g(x,u) = \int_0^t q(\Phi^\tau(x,u), u)\, d\tau,$$

with the quadratic cost $q(x,u) = \frac{1}{2}\left(0.1\varphi^2 + 0.05\dot\varphi^2 + 0.01u^2\right)$.

We use the classical Runge-Kutta scheme of order 4 with step size 0.02 in order to approximate Φ^t, choose $X = [-8,8] \times [-10,10]$ as state space for $x = (\varphi, \dot\varphi)$, which we partition into $2^9 \times 2^9$ boxes of equal size, and $U = [-64, 64]$ as the control space. In approximating the graph's edges and their weights, we map an equidistant grid of 3×3 points on each partition box, choosing from 17 equally spaced values in U.

Figure 4 shows the discrete value function as well as the trajectory generated by the discrete feedback for the initial value $(3.1, 0.1)$, as computed by the GAIO code in Fig. 6. As shown on the right of this figure, the discrete value function does not decrease monotonically along the feedback trajectory, indicating that the assumptions of Theorem 3 are not satisfied. And indeed, as shown in Fig. 5, this trajectory repeatedly moves through regions in state space where the error function e is not smaller than g_0. In fact, on a coarser partition ($2^7 \times 2^7$ boxes), the discrete feedback (18) is not even stabilizing this initial condition any more. We will adress this deficiency in the next sections.

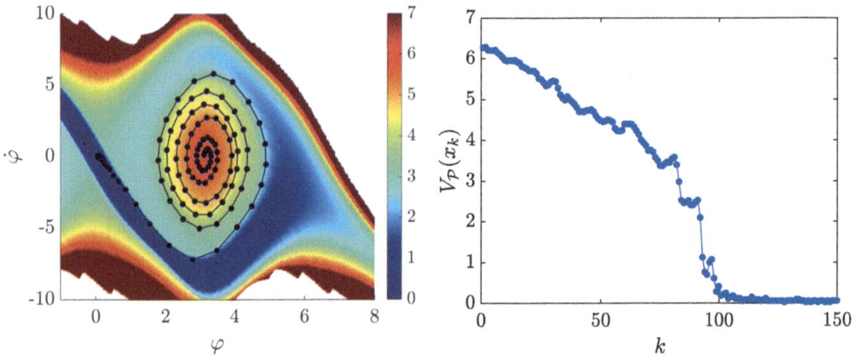

Fig. 4. Left: Discrete value function and feedback trajectory for the inverted pendulum. Right: Behaviour of the discrete value function along the feedback trajectory.

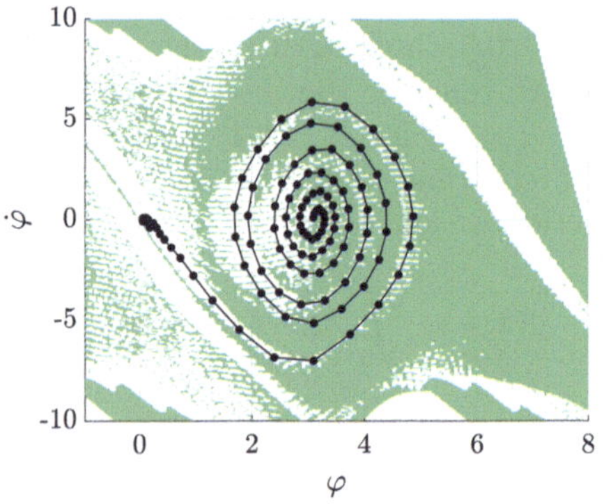

Fig. 5. Inverted pendulum: region where $e(x) < g_0(x)$ (green) and feedback trajectory.

```
m = 2; M = 8; m_r = m/(m+M); l = 0.5; g = 9.8;
q1 = 0.1; q2 = 0.05; r0 = 0.01;
v = @(x,u) [ x(:,2), ...                              % vector field & cost
             (g/l*sin(x(:,1)) - 0.5*m_r*x(:,2).^2.*sin(2*x(:,1)) - ...
             m_r/(m*l)*u.*cos(x(:,1)))./(4.0/3.0 - m_r*cos(x(:,1)).^2), ...
             0.5*( q1*(x(:,1).^2) + q2*(x(:,2).^2) + r0*u.^2 )];
h = 0.02; steps = 5;                                  % step size, # of steps
f = @(x,u) rk4u(v,[x zeros(size(x,1),1)],u,h,steps);% control system
n = 2; x1 = linspace(-1,1,n)';
[XX,YY] = meshgrid(x1,x1); X = [ XX(:) YY(:) ];       % sample points
U = linspace(-64,64,17)';                             % control samples

depth = 18; c = [0 0]; r = [8 10];
tree = Tree(c, r);
subdivide(trre, depth);                               % construct partition
G = dpgraph(tree, f, X, U, depth);                    % compute graph
dest = tree.search([0;0], depth);                     % target set
[V,~] = dijkstra(G', dest);                           % discrete value function

V(find(V == Inf)) = NaN; V(find(V > 7)) = 7;
clf; boxplot2(tree, 'density', V);                    % plot value function
colorbar; colormap jet; shading flat; axis('tight')
xlabel('$\varphi$'); ylabel('$\dot\varphi$');

%% discrete feedback, trajectory of the closed loop system
x(1,:) = [3.1, 0.1];
for k = 1:200
  fxc = f(ones(size(U))*x(k,:), U)';       % map current point under all controls
  bn = tree.search(fxc(1:2,:),depth)';     % determine corresp. boxes
  [v, k_min] = min(fxc(3,:) + V(bn));      % determine minimizing control
  V_fb(k) = V(bn(k_min));                  % value at next point
  x(k+1,:) = fxc(1:2,k_min);               % next iterate
end
hold on; plot(x(:,1),x(:,2),'k.-','linewidth',1,'markersize',22);
axis([-0.1 6 -8 8]);
figure(2); plot(V_fb,'.-','linewidth',1,'markersize',22)
xlabel('$k$'); ylabel('$V_\mathcal{P}(x_k)$');
```

Fig. 6. Code: discrete value function for the inverted pendulum

5 The Optimality Principle for Perturbed Systems

Let us now return to the full problem from Sect. 2 of optimally stabilizing the discrete time perturbed control system

$$x_{k+1} = f(x_k, u_k, w_k), \quad k = 0, 1, \ldots. \tag{23}$$

subject to an instantaneous cost $g(x_k, u_k)$. For the convergence statements later, we assume $f : X \times U \times W \to X$ and $g : X \times U \to [0, \infty)$ to be continuous and $X \subset \mathbb{R}^d, U \subset \mathbb{R}^m$ and $W \subset \mathbb{R}^\ell$ to be compact. More general spaces will be discussed in Sect. 8. For a given initial state $x_0 \in X$, a control sequence $\boldsymbol{u} = (u_k)_{k\in\mathbb{N}} \in U^{\mathbb{N}}$ and a perturbation sequence $\boldsymbol{w} = (w_k)_{k\in\mathbb{N}} \in W^{\mathbb{N}}$, we obtain the *trajectory* $(x_k(x, \boldsymbol{u}, \boldsymbol{w}))_{k\in\mathbb{N}}$ satisfying (23) while the associated accumulated cost is given by

$$J(x, \boldsymbol{u}, \boldsymbol{w}) = \sum_{k=0}^{\infty} g(x_k(x, \boldsymbol{u}, \boldsymbol{w}), u_k).$$

Recall that our goal is to derive a *feedback* $u : S \to U$, $S \subset X$, that *stabilizes* the closed loop system

$$x_{k+1} = f(x_k, u(x_k), w_k), \quad k = 0, 1, 2, \ldots \tag{24}$$

for any perturbation sequence $(w_k)_k$, i.e. for every trajectory $(x_k(x_0, \boldsymbol{w}))_k$ of (24) with $x_0 \in S$ and $\boldsymbol{w} \in W^{\mathbb{N}}$ arbitrary, we have $x_k \to T$ as $k \to \infty$, where $T \subset S$ is a given *target set*, and the *accumulated cost* $\sum_{k=0}^{\infty} g(x_k, u(x_k))$ is minimized.

The problem formulation can be interpreted as describing a *dynamic game* (see e.g. [6]), where at each step of the iteration (23) two *players* choose a control u_k and a perturbation w_k, respectively. The goal of the controlling player is to minimize J, while the perturbing player wants to maximize it. We assume that the controlling player chooses u_k first and that the perturbing player knows u_k when choosing w_k. We further assume that the perturbing player cannot foresee future choices of the controlling player. This can be formalized by restricting the possible $\boldsymbol{w}$ to

$$\boldsymbol{w} = \beta(\boldsymbol{u}),$$

where $\beta : U^{\mathbb{N}} \to W^{\mathbb{N}}$ is a *nonanticipating strategy*, i.e. a strategy satisfying

$$u_k = u'_k \quad \forall k \leq K \quad \Rightarrow \quad \beta_k(\boldsymbol{u}) = \beta_k(\boldsymbol{u}') \quad \forall k \leq K$$

for any $\boldsymbol{u} = (u_k)_k, \boldsymbol{u}' = (u'_k)_k \in U^{\mathbb{N}}$. We denote by $\mathcal{B}$ the set of all nonanticipating strategies $\beta : U^{\mathbb{N}} \to W^{\mathbb{N}}$.

The control task is finished once we are in T, we therefore assume that T is compact and robustly forward invariant, i.e. for all $x \in T$ there is a control $u \in U$ such that $f(x, u, w) \subset T$ for all $w \in W$, that $g(x, u) = 0$ for all $x \in T$, $u \in U$ and $g(x, u) > 0$ for all $x \notin T$, $u \in U$.

Our construction of the feedback $u : S \to U$ will be based on the *upper value function* $V : X \to [0, \infty]$,

$$V(x) = \sup_{\beta \in \mathcal{B}} \inf_{\boldsymbol{u} \in U^{\mathbb{N}}} J(x, \boldsymbol{u}, \beta(\boldsymbol{u})), \tag{25}$$

of the game (23), which is finite on the set $S_0 := \{x \in X \mid V(x) < \infty\}$. The upper value function satisfies the *optimality principle* [9]

$$V(x) = \inf_{u \in U} \left[g(x, u) + \sup_{w \in W} V(f(x, u, w)) \right], \quad x \in S_0. \tag{26}$$

The right hand side $L[v](x) = \inf_{u \in U} [g(x, u) + \sup_{w \in W} v(f(x, u, w))]$ of this fixed point equation again defines a *dynamic programming operator* $L : \mathbb{R}^X \to \mathbb{R}^X$. The upper value function is the unique fixed point of L satisying the boundary condition $V(x) = 0$, $x \in T$. Like in the unperturbed case, using the upper value function V, one can construct the feedback $u : S_0 \to U$,

$$u(x) := \operatorname*{argmin}_{u \in U} \left[g(x, u) + \sup_{w \in W} V(f(x, u, w)) \right], \tag{27}$$

whenever this minimum exists.

6 A Discrete Optimality Principle for Perturbed Systems

Analogously to the discretization in Sect. 4 we now derive a discrete version of (26), cf. [9]. Again, to this end, we will approximate the upper value function by a function which is piecewise constant on the elements of some partition of X. This approach will lead to a directed weighted hypergraph instead of the ordinary directed graph in Sect. 4 and, again, the approximate upper value function can be computed by an associated shortest path algorithm.

Let $\mathcal{P}$ be a finite partition of X. Using the projection (10), the *discretized dynamic game operator* $L_{\mathcal{P}} : \mathbb{R}^{\mathcal{P}} \to \mathbb{R}^{\mathcal{P}}$ is defined by

$$L_{\mathcal{P}} := \psi \circ L.$$

Again, this operator has a unique fixed point $V_{\mathcal{P}}$ satisfying the boundary condition $V_{\mathcal{P}}(x) = 0$, $x \in T$, which will serve as an approximation to the exact value function V.

Explicitely, the discretized operator reads

$$L_{\mathcal{P}}[v](x) = \inf_{x' \in \pi(x)} \left(\inf_{u \in U} \left[g(x', u) + \sup_{w \in W} v(f(x', u, w)) \right] \right)$$

and $V_{\mathcal{P}}$ satisfies the optimality principle

$$V_{\mathcal{P}}(x) = \inf_{x' \in \pi(x), u \in U} \left[g(x', u) + \sup_{w \in W} V_{\mathcal{P}}(f(x', u, w)) \right]. \tag{28}$$

Note that since $V_{\mathcal{P}}$ is constant on each partition element, we can rewrite this as

$$V_{\mathcal{P}}(x) = \inf_{x' \in \pi(x), u \in U} \left[g(x', u) + \sup_{P' \in \mathcal{F}(x', u)} V_{\mathcal{P}}(P') \right],$$

where

$$\mathcal{F}(x', u) = \{P \in \mathcal{P} \mid f(x', u, w) \in P \text{ for some } w \in W\}.$$

Since the partition $\mathcal{P}$ is finite, there are only finitely many possible sets $\mathcal{F}(x', u)$ and we can further rewrite (28) as

$$V_{\mathcal{P}}(x) = \min_{\mathcal{N}} \inf_{(x', u)} \left[g(x', u) + \sup_{P' \in \mathcal{N}} V_{\mathcal{P}}(P') \right],$$

where the min is taken over all collections $\mathcal{N} \in \{\mathcal{F}(x', u) \mid x' \in \pi(x), u \in U\}$ and the inf over all (x', u) such that $\mathcal{F}(x', u) = \mathcal{N}$. Now define the multivalued map $\mathcal{F} : \mathcal{P} \rightrightarrows 2^{\mathcal{P}}$,

$$\mathcal{F}(P) = \{\mathcal{F}(x, u) : (x, u) \in P \times U\},$$

and the cost function

$$\mathcal{G}(P, \mathcal{N}) = \inf_{u \in U} \{g(x, u) : x \in P, \mathcal{F}(x, u) = \mathcal{N}\}.$$

Equation (28) can then be rewritten as

$$V_{\mathcal{P}}(P) = \min_{\mathcal{N} \in \mathcal{F}(P)} \left[\mathcal{G}(P, \mathcal{N}) + \sup_{P' \in \mathcal{N}} V_{\mathcal{P}}(P') \right],$$

Graph Interpretation. Like in the unperturbed case, we can think of this reformulation of the optimality principle in terms of a graph. More precisely, we have a directed hypergraph $(\mathcal{P}, E_{\mathcal{P}})$ with the set $E \subset \mathcal{P} \times 2^{\mathcal{P}}$ of directed hyperedges given by

$$E_{\mathcal{P}} = \{(P, \mathcal{N}) \mid \mathcal{N} = \mathcal{F}(x, u) \text{ for some } (x, u) \in P \times U\},$$

and each edge $(P, \mathcal{N})$ is weighted by $\mathcal{G}(P, \mathcal{N})$, c.f. Fig. 7. The discrete upper value function $V_{\mathcal{P}}(P)$ is the length of a shortest path from P to some element P' which has a nonempty intersection with the target set T (and, thus, by the boundary condition, $V_{\mathcal{P}}(P') = 0$).

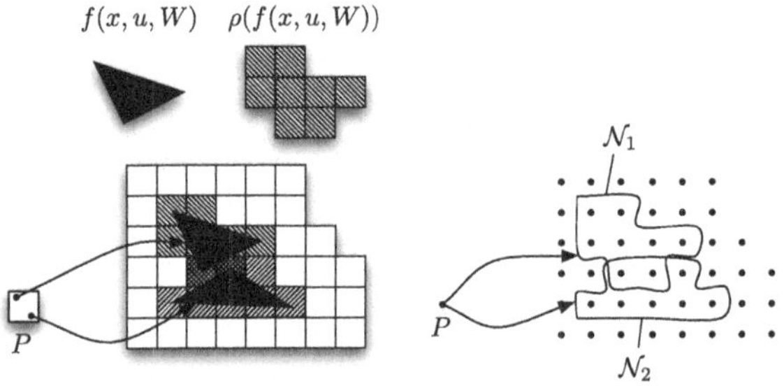

Fig. 7. Illustration of the construction of the hypergraph.

Shortest Paths in Hypergraphs. Algorithm 1 can be generalized to the hypergraph case, cf. [9,20]. To this end, we modify lines 5–7 such that the maximization over the perturbations is taken into account:

for each $(Q, \mathcal{N}) \in E_{\mathcal{P}}$ with $P \in \mathcal{N}$
    if $V(Q) > \mathcal{G}(Q, \mathcal{N}) + \max_{N \in \mathcal{N}} V(N)$ then
        $V(Q) := \mathcal{G}(Q, \mathcal{N}) + \max_{N \in \mathcal{N}} V(N)$

Note that during the while-loop of Algorithm 1,

$$V(P) \geq V(P') \quad \text{for all } P' \in \mathcal{P} \backslash \mathcal{Q}.$$

Thus, if $\mathcal{N} \subset \mathcal{P} \backslash \mathcal{Q}$, then $\max_{N \in \mathcal{N}} V(N) = V(P)$, and the value of the node Q will never be decreased again. On the other hand, if $\mathcal{N} \not\subset \mathcal{P} \backslash \mathcal{Q}$, then the value of Q will be further decreased at a later time – and thus we can save on changing it in the current iteration of the while-loop. We can therefore save on the explicit maximization and replace lines 5–7 by

for each $(Q, \mathcal{N}) \in E_{\mathcal{P}}$ with $P \in \mathcal{N}$
    if $\mathcal{N} \subset \mathcal{P} \backslash \mathcal{Q}$ then
        if $V(Q) > \mathcal{G}(Q, \mathcal{N}) + V(P)$ then
            $V(Q) := \mathcal{G}(Q, \mathcal{N}) + V(P)$

The overall algorithm for the hypergraph case is as follows. Here, $\mathcal{T} := \{P \in \mathcal{P} \mid P \cap T \neq \emptyset\}$ is the set of target nodes.

Algorithm MINMAX-DIJKSTRA

for each $P \in \mathcal{P}$: $V(P) := \infty$; for each $P \in \mathcal{T}$: $V(P) := 0$; $\mathcal{Q} := \mathcal{P}$
while $\mathcal{Q} \neq \emptyset$
    $P := \text{argmin}_{P' \in \mathcal{Q}} V(P')$
    $\mathcal{Q} := \mathcal{Q} \backslash \{P\}$
    for each $(Q, \mathcal{N}) \in E_{\mathcal{P}}$ with $P \in \mathcal{N}$
        if $\mathcal{N} \subset \mathcal{P} \backslash \mathcal{Q}$ then
            If $V(Q) > \mathcal{G}(Q, \mathcal{N}) + V(P)$ then
                $V(Q) := \mathcal{G}(Q, \mathcal{N}) + V(P)$

□

Time Complexity. In line 5, each hyperedge is considered at most N times, with N being a bound on the cardinality of the hypernodes $\mathcal{N}$. Additionally, we need to perform the check in line 6, which has linear complexity in N. Thus, the overall complexity of the minmax-Dijkstra algorithm is $\mathcal{O}(|\mathcal{P}| \log |\mathcal{P}| + |E| N (N + \log |\mathcal{P}|))$ (when using a binary heap for storing $\mathcal{Q}$), cf. [20].

Space Complexity. The storage requirement grows linearly with $|\mathcal{P}|$. This number, however, grows exponentially with the dimension of state space (if the entire state space is covered and under the assumption of uniformly large elements). The number of hyperedges is determined by the Lipschitz constant of f, the size of the hypernodes $\mathcal{N}$ will be given by the magnitude of the perturbation.

Implementation. We use the same approach as in the unperturbed case: A cubical partition is constructed hierarchically and stored in a binary tree. In order to approximate the set $E_{\mathcal{P}} \subset \mathcal{P} \times 2^{\mathcal{P}}$ of hyperedges, we choose finite sets $\tilde{P} \subset P$, $\tilde{U} \subset U$ and $\tilde{W} \subset W$ of sample points, set

$$\tilde{\mathcal{F}}(x, u) = \{P \in \mathcal{P} \mid f(x, u, w) \in P \text{ for some } w \in \tilde{W}\}$$

and compute

$$\tilde{\mathcal{F}}(P) := \{\tilde{\mathcal{F}}(x, u) : (x, u) \in \tilde{P} \times \tilde{U}\} \subset 2^{\mathcal{P}}$$

as an approximation to $\mathcal{F}(P)$. Correspondingly, the weight on the hyperedge $(P, \mathcal{N})$ is approximated by

$$\tilde{\mathcal{G}}(P, \mathcal{N}) = \min\{g(x, u) : (x, u) \in \tilde{P} \times \tilde{U}, \tilde{\mathcal{F}}(x, u) = \mathcal{N}\}.$$

Example: A simple 1D System. We reconsider system (16), adding a small perturbation at each time step:

$$x_{k+1} = x_k + (1 - a) u_k x_k + w_k, \quad k = 0, 1, \ldots,$$

with $x_k \in [0,1]$, $u_k \in [-1,1]$, $w_k \in [-\varepsilon, \varepsilon]$ for some $\varepsilon > 0$ and the fixed parameter $a \in (0,1)$. The cost function is still $g(x,u) = (1-a)x$ so that the optimal control policy is again $u_k = -1$ for all k, independently of the perturbation sequence. The optimal strategy for the perturbing player is to slow down the dynamics as much as possible, corresponding to $w_k = \varepsilon$ for all k. The dynamical system resulting from inserting the optimal strategies is

$$x_{k+1} = ax_k + \varepsilon, \quad k = 0, 1, \ldots.$$

This map has a fixed point at $x = \varepsilon/(1-a)$. In the worst case, i.e. $w_k = \varepsilon$ for all k, it is not possible to get closer than $\alpha_0 := \varepsilon/(1-a)$ to the origin. We therefore define $T = [0, \alpha]$ with $\alpha > \alpha_0$ as the target set. With

$$k(x) = \left\lceil \frac{\log \frac{\alpha - \alpha_0}{x - \alpha_0}}{\log a} \right\rceil + 1,$$

the exact optimal value function is

$$V(x) = (x - \alpha_0)\left(1 - a^{k(x)}\right) + \varepsilon k(x),$$

as shown in Fig. 8 for $a = 0.8$, $\varepsilon = 0.01$ and $\alpha = 1.1\alpha_0$. In that figure, we also show the approximate optimal value functions on partitions of 64, 256 and 1024 intervals, respectively. In the construction of the hypergraph, we used an equidistant grid of ten points in each partition interval, in the control space and in the perturbation space.

6.1 Convergence

It is natural to ask whether the approximate value function converges to the true one when the element diameter of the underlying partition goes to zero. This has been proven pointwise on the stabilizable set S in the unperturbed case [14], as well as in an L^1-sense on S and an L^∞ sense on the domain of continuity in the perturbed case, assuming continuity of V on the boundary of the target set T [9]. The same reference also provides an analysis for state constrained problems. Here an additional robustness condition is needed, namely that the optimal value function changes continuously with respect to the L^p-norm for some $p \in \{1, \ldots, \infty\}$ if the state constraints are tightened. If this condition holds, then the convergence statement remains valid under state constraints, with L^∞ replaced by L^p.

Due to the construction of the discretization, the approximation $V_{\mathcal{P}}$ of the optimal value function is always less or equal than the true optimal value function. This is not necessarily a good property. For instance, for proving stability of the system controlled by the numerical feedback law it would be convenient if $V_{\mathcal{P}}$ was a Lyapunov function. Lyapunov functions, however, are supersolutions to the dynamic programming equation, rather than subsolutions as our $V_{\mathcal{P}}$. In order to overcome this disadvantage, in the next section we present a particular construction of a dynamic game in which the discretization error is treated as a perturbation.

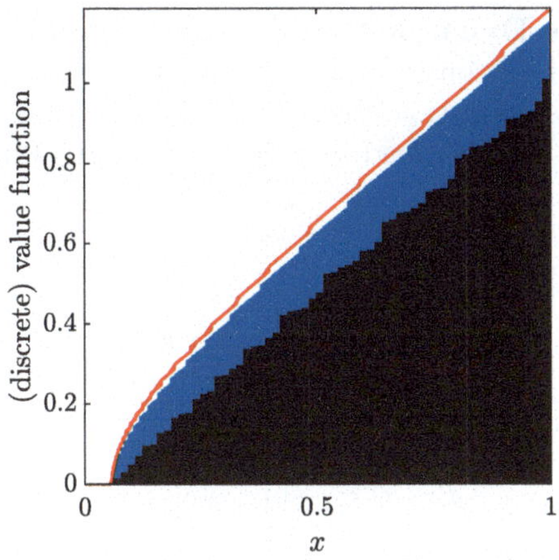

Fig. 8. Exact (red) and discrete upper value functions for the perturbed simple example on partitions of 64 (black) and 1024 (blue) intervals.

```
a = 0.8; ep = 0.01;
alpha0 = ep/(1-a); alpha = 1.1*alpha0;
f = @(x,u,w) [x + (1-a).*x.*u + w, (1-a)*x];       % control system
X = linspace(-1,1,5)';                              % state samples
U = linspace(-1,1,2)';                              % control samples
W = linspace(-ep,ep,3)';                            % perturbation samples

depth = 6; c = [0.5]; r = [0.5];
tree = Tree(c, r);                                  % construct full tree
subdivide(tree, depth);                             % construct partition

G = compute_hypergraph(tree, f, X, U, W, depth);    % construct hypergraph
Gt = trnsp_hgraph(G);                               % transpose hypergraph
T = tree.search_box([0],[alpha],depth);             % target boxes
V_P = minmax_dijkstra(G, Gt, T);                    % upper value function

n = 2^depth; dx = 1/n; x = linspace(dx/2,1-dx/2,n);
k = @(x) floor(log((alpha-alpha0)./(x-alpha0))/log(a))+1;
V = @(x) (x>alpha).*((x-alpha0).*(1-a.^k(x))+ep*k(x));% exact value function
bar(x,V_P,1,'k'); hold on; plot(x,V(x),'r');
axis tight; axis square; xlabel('$x$');
ylabel('(discrete) value function');
```

Fig. 9. Code: upper value function for the perturbed simple 1d system.

7 The Discretization as a Perturbation

As shown in Theorems 2 and 3, the discrete feedback (18) will practically stabilize the closed loop system (19) under suitable conditions. Our numerical experiment in Example 2, however, revealed that a rather fine partition might be needed in order to achieve stability. More generally, as we have seen in Fig. 4 (right), the discrete value function is not a Lyapunov function of the closed loop system in every case.

Construction of the Dynamic Game. In order to cope with this problem we are going to use the ideas on treating perturbed systems in Sect. 5 and 6. The idea is to view the discretization error as a perturbation of the original system. Under the discretization described in Sect. 4, the original map $(x, u) \mapsto f(x, u)$ is perturbed to

$$(x, u) \mapsto \hat{f}(x, u, w) := f(x + w, u), \quad x + w \in \pi(x).$$

Note that this constitutes a generalization of the setting in Sects. 5 and 6 since the perturbation space W here depends on the state, $W = W(x)$. Correspondingly, the associated cost function is

$$\hat{g}(x, u) = \sup_{x' \in \pi(x)} g(x', u). \tag{29}$$

Theorem 4 [8]. *Let V denote the value function* (6) *of the control system* (f, g)*, $\hat{V}$ the value function* (25) *of the associated game* $(\hat{f}, \hat{g})$ *and $V_{\mathcal{P}}$ the discrete value function* (28) *of* $(\hat{f}, \hat{g})$ *on a given partition $\mathcal{P}$ with numerical target set $T_{\mathcal{P}} \subset \mathcal{P}$, $T = \{0\} \subset T_{\mathcal{P}}$. Then $V_{\mathcal{P}}(x) = \hat{V}(x)$ and*

$$V(x) - \max_{y \in T_{\mathcal{P}}} V(y) \leq V_{\mathcal{P}}(x), \tag{30}$$

i.e. $V_{\mathcal{P}}$ is an upper bound for $V - \max V|_{T_{\mathcal{P}}}$. Furthermore, $V_{\mathcal{P}}$ satisfies

$$V_{\mathcal{P}}(x) \geq \min_{u \in U} \{g(x, u) + V_{\mathcal{P}}(f(x, u))\} \tag{31}$$

for all $x \in X \setminus T_{\mathcal{P}}$.

Proof. We first note that $\hat{V}$ is constant on the elements of $\mathcal{P}$: On $T_{\mathcal{P}}$, this is true since $T_{\mathcal{P}}$ is a union of partition elements by assumption. Outside of $T_{\mathcal{P}}$, by definition of the game $(\hat{f}, \hat{g})$ we have

$$\hat{V}(x) = \inf_{u \in U} \left\{ \sup_{x' \in \pi(x)} g(x', u) + \sup_{x' \in f(\pi(x), u)} \hat{V}(x') \right\},$$

so that $\inf_{x' \in \pi(x)} \hat{V}(x') = \hat{V}(x)$. On the other hand, according to [9, Proposition 7.1] we have $V_{\mathcal{P}}(x) = \inf_{x' \in \pi(x)} \hat{V}(x')$, so that $V_{\mathcal{P}} = \hat{V}$.

Now for $x \notin T_{\mathcal{P}}$, Eq. (26) yields

$$\begin{aligned} \hat{V}(x) &= \inf_{u \in U} \sup_{x' \in \pi(x)} \left\{ g(x', u) + \hat{V}(f(x', u)) \right\} \\ &\geq \min_{u \in U} \left\{ g(x, u) + \hat{V}(f(x, u)) \right\} \end{aligned} \tag{32}$$

which shows (31).

In order to prove (30), we order the elements $P_1, P_2, \ldots \in \mathcal{P}$ such that $i \geq j$ implies $V_{\mathcal{P}}(P_i) \geq V_{\mathcal{P}}(P_j)$. Since $\inf_{u \in U} g(x, u) > 0$ for $x \neq 0$ by assumption,

$V_{\mathcal{P}}(P_i) = 0$ is equivalent to $P_i \subseteq T_{\mathcal{P}}$. By the ordering of the elements this implies that there exists $i^* \geq 1$ such that $P_i \subseteq T_{\mathcal{P}} \Leftrightarrow i \in \{1, \dots, i^*\}$ and thus (30) holds for $x \in P_1, \dots, P_{i^*}$. We now use induction: fix some $i \in \mathbb{N}$, assume (30) holds for $x \in P_1, \dots, P_{i-1}$ and consider $x \in P_i$. If $V_{\mathcal{P}}(P_i) = \infty$ there is nothing to show. Otherwise, since V satisfies the dynamic programming principle, using (32) we obtain

$$\begin{aligned} V(x) - \hat{V}(x) &\leq \inf_{u \in U} \{g(x,u) + V(f(x,u))\} - \min_{u \in U} \left\{ g(x,u) + \hat{V}(f(x,u)) \right\} \\ &\leq V(f(x,u^*)) - \hat{V}(f(x,u^*)), \end{aligned}$$

where $u^* \in U$ realizes the minimum in (32). Now, since $g(x,u^*) > 0$, we have $\hat{V}(f(x,u^*)) < \hat{V}(x)$ implying $f(x,u^*) \in P_j$ for some $j < i$. Since by the induction assumption the inequality in (30) holds on P_j, this implies that it also holds on P_i which finishes the induction step. □

The Feedback Is the Shortest Path. As usual, we construct the discrete feedback by

$$u_{\mathcal{P}}(x) := \operatorname*{argmin}_{u \in U} \left[\hat{g}(x,u) + \sup_{x' \in f(\pi(x),u)} V_{\mathcal{P}}(x') \right].$$

By construction, this feedback is constant on each partition element. Moreover, we can directly extract $u_{\mathcal{P}}$ from the minmax-Dijkstra algorithm: We associate the minimizing control value $\underline{u}(P, \mathcal{N})$ to each hyperedge $(P, \mathcal{N})$,

$$\underline{u}(P, \mathcal{N}) = \operatorname*{argmin}_{u \in U, \mathcal{F}(P) = \mathcal{N}} \left[\sup_{x \in P} g(x,u) \right]. \tag{33}$$

The feedback is then immediately given by

$$u_{\mathcal{P}}(x) = \underline{u}(\pi(x), \underline{\mathcal{N}}(\pi(x))), \tag{34}$$

where

$$\underline{\mathcal{N}}(P) = \operatorname*{argmin}_{\mathcal{N} \in \mathcal{F}(P)} \left\{ \mathcal{G}(P, \mathcal{N}) + \sup_{N \in \mathcal{N}} V_{\mathcal{P}}(N) \right\}$$

is defining the hypernode of minimal value adjacent to some node P in the hypergraph. The computation of $\underline{\mathcal{N}}(P)$ can be done on the fly within the minmax-Dijkstra Algorithm 2:

Algorithm. MINMAX-DIJKSTRA WITH FEEDBACK

for each $P \in \mathcal{P}$: $V(P) := \infty$, $\underline{\mathcal{N}}(P) := \emptyset$; for each $P \in \mathcal{T}$: $V(P) := 0$; $\mathcal{Q} := \mathcal{P}$
while $\mathcal{Q} \neq \emptyset$
  $P := \text{argmin}_{P' \in \mathcal{Q}} V(P')$
  $\mathcal{Q} := \mathcal{Q} \backslash \{P\}$
  for each $(Q, \mathcal{N}) \in E_{\mathcal{P}}$ with $P \in \mathcal{N}$
    if $\mathcal{N} \subset \mathcal{P} \backslash \mathcal{Q}$ then
      if $V(Q) > \mathcal{G}(Q, \mathcal{N}) + V(P)$ then
        $V(Q) := \mathcal{G}(Q, \mathcal{N}) + V(P)$
        $\underline{\mathcal{N}}(Q) := \mathcal{N}$

□

Consequently, the discrete feedback $\underline{u}$ can be computed offline. Once $\underline{u}(P, \underline{\mathcal{N}}(P))$ has been computed for every partition element P, the only remaining online computation is the determination of $\pi(x_k)$ for each state x_k on the feedback trajectory. In our case, this can be done efficiently, since we store the partition in a binary tree. Note, however, that the fast online evaluation of the feedback law is enabled by a comparatively large offline computation, namely the construction of the hypergraph.

Behaviour of the Closed Loop System

Theorem 5 [8]**.** *Under the assumptions of Theorem 4, if $(x_k)_k$ denotes the trajectory of the closed loop system (19) with feedback (34) and if $V_{\mathcal{P}}(x_0) < \infty$, then there exists $k^* \in \mathbb{N}$ such that $x_{k^*} \in T$ and*

$$V_{\mathcal{P}}(x_k) \geq g(x_k, u_{\mathcal{P}}(x_k)) + V_{\mathcal{P}}(x_{k+1}), \quad k = 0, \ldots, k^* - 1.$$

Proof. From the construction of $u_{\mathcal{P}}$ we immediately obtain the inequality

$$V_{\mathcal{P}}(x_k) \geq g(x_k, u_{\mathcal{P}}(x_k)) + V_{\mathcal{P}}(x_{k+1}) \tag{35}$$

for all $k \in \mathbb{N}_0$ with $x_k \in X \setminus T_{\mathcal{P}}$. This implies the existence of k^* such that the first two properties hold since $g(x_k, u_{\mathcal{P}}(x_k)) > 0$ for $x_k \notin T_{\mathcal{P}}$, $V_{\mathcal{P}}$ is piecewise constant and equals zero only on $T_{\mathcal{P}}$. □

Theorem 5 implies that the closed-loop solution reaches the target $T_{\mathcal{P}}$ at time step k^* and that the optimal value function decreases monotonically until the target is reached, i.e., it behaves like a Lyapunov function. While it is in principle possible that the closed-loop solution leaves the target after time k^*, this Lyapunov function property implies that after such excursions it will return to $T_{\mathcal{P}}$.

If the system (4) is asymptotically controllable to the origin and V is continuous, then we can use the same arguments as in [9] in order to show that on increasingly finer partitions $\mathcal{P}_\ell$ and for targets $T_{\mathcal{P}_\ell}$ shrinking down to $\{0\}$ we obtain $V_{\mathcal{P}_\ell} \to V$. This can also be used to conclude that the distance of possible excursions from the target $T_{\mathcal{P}_\ell}$ become smaller and smaller as $\mathcal{P}_\ell$ becomes finer.

We note that the Lyapunov function property of $V_{\mathcal{P}}$ outside $T_{\mathcal{P}}$ holds regardless of the size of the partition elements. However, if the partition is too coarse then $V_{\mathcal{P}} = \infty$ will hold on large parts of X, which makes the Lyapunov function property useless. In case that large partition elements are desired—for instance, because they correspond to a quantization of the state space representing, e.g., the resolution of certain sensors—infinite values can be avoided by choosing the control value not only depending on one partition element but on two (or more) consecutive elements. The price to pay for this modification is that the construction of the hypergraph becomes significantly more expensive, but the benefit is that stabilization with much coarser discretization or quantization is possible. For details we refer to [10,11].

Example 3 ***(The inverted pendulum reconsidered.).*** We reconsider Example 2 and apply the construction from this section. Figure 10, which results from running the code shown in Fig. 11 as well as lines 25ff. from the code in Fig. 6, shows the discrete upper value function on a partition of 2^{16} boxes with target set $T = [-0.1, 0.1]^2$ as well as the trajectory generated by the discrete feedback (33) for the initial value (3.1, 0.1). As expected, the approximate value function is decreasing monotonically along this trajectory. Furthermore, this trajectory is clearly closer to the optimal one because it converges to the origin much faster.

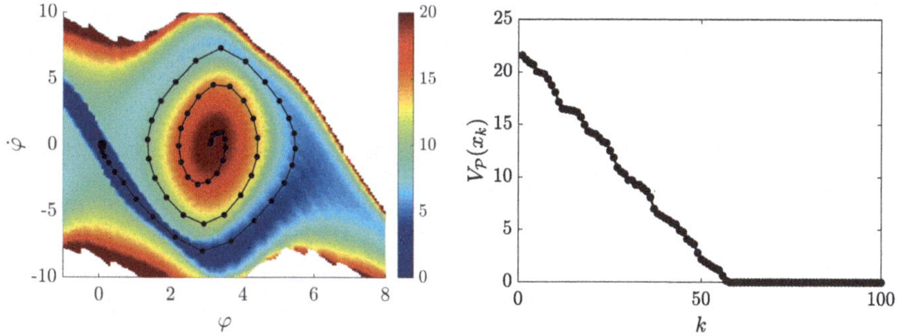

Fig. 10. Inverted pendulum: Discrete upper value function and robust feedback trajectory (left); decrease of the discrete value function along the feedback trajectory.

8 Hybrid, Event and Quantized Systems

Hybrid Systems. The discretization of the optimality principle described in Sects. 4–7 can be used in order to deal with *hybrid systems* in a natural way. Hybrid systems can often be modeled by a discrete time control system of the form

$$\begin{aligned} x_{k+1} &= f_c(x_k, y_k, u_k) \\ y_{k+1} &= f_d(x_k, y_k, u_k) \end{aligned} \qquad k = 0, 1, \ldots, \tag{36}$$

```
m = 2; M = 8; m_r = m/(m+M); l = 0.5; g = 9.8;
q1 = 0.1; q2 = 0.05; r0 = 0.01;
v = @(x,u) [ x(:,2), ...                                    % vector field & cost
             (g/l*sin(x(:,1)) - 0.5*m_r*x(:,2).^2.*sin(2*x(:,1)) - ...
             m_r/(m*l)*u.*cos(x(:,1)))./(4.0/3.0 - m_r*cos(x(:,1)).^2), ...
             0.5*( q1*(x(:,1).^2) + q2*(x(:,2).^2) + r0*u.^2 )];
h = 0.02; steps = 5;                                         % step size, # of steps
f = @(x,u) rk4u(v,[x zeros(size(x,1),1)],u,h,steps);% control system
n = 3; x1 = linspace(-1,1,n)';
[XX,YY] = meshgrid(x1,x1); X = [ XX(:) YY(:) ];              % sample points
U = linspace(-64,64,17)';                                    % control samples

depth = 16; c = [0 0]; r = [8 10];
tree = Tree(c, r);  sd = 8;                                  % construct full tree
subdivide(tree, depth),

G = dphgraph2(tree, f, X, U, depth);                         % construct hypergraph
Gt = trnsp_hgraph(G);                                        % transpose hypergraph
T = tree.search_box([0;0], [0.1;0.1], depth);                % target boxes
[V, u] = minmax_dijkstra(G, Gt, T);                          % value function,feedback

V(find(V == Inf)) = NaN;                                     % unstabilizable set
figure(1); clf; boxplot2(tree, 'density', V);                % plot value function
colorbar; shading flat; axis('tight')
xlabel('$\varphi$'); ylabel('$\dot\varphi$');
```

Fig. 11. Code: discrete upper value function and robust feedback for the inverted pendulum

with two maps $f_c : X \times Y \times U \to X \subset \mathbb{R}^n$ and $f_d : X \times Y \times U \to Y$. The set U of control inputs can be discrete or continuous, the (compact) set $X \subset \mathbb{R}^n$ is the continuous part of state space and the set Y of discrete states (or *modes*) is a finite set. The class of hybrid systems described by (36) is quite general: It comprises

- models with purely continuous state space (i.e. $Y = \{0\}$, $f_c(x, y, u) = f_c(x, u)$, $f_d \equiv 0$), but discrete or finite control space U;
- models in which the continuous part f_c is controlled by the mode y and only the discrete part f_d of the map is controlled by the input ($f_c(x, y, u) = f_c(x, y)$ and $f_d(x, y, u) = f_d(y, u)$ may be given by an automaton);
- models with state dependent switching: Here we have a general map f_c and $f_d(x, y, u) = f_d(x)$.

As in the previous chapters, we denote the solutions of (36) for initial values $x_0 = x$, $y_0 = y$ and some control sequence $\boldsymbol{u} = (u_0, u_1, \ldots) \in U^{\mathbb{N}}$ by $x_k(x, y, \boldsymbol{u})$ and $y_k(x, y, \boldsymbol{u})$, respectively. We assume that for each k, the map $x_k(\cdot, y, \boldsymbol{u})$ is continuous for each $y \in Y$ and each $\boldsymbol{u} \in U^{\mathbb{N}}$. We prescribe a target set $T \subset X$ (i.e. a subset of the continuous part of state space) and our aim is to find a control sequence $\boldsymbol{u} = (u_k)_{k \in \mathbb{N}}$ such that $x_k(x, y, \boldsymbol{u}) \to T$ as $k \to \infty$ for initial values x, y in some stabilizable set $S \subset X \times Y$, while minimizing the accumulated cost $\sum_{k=0}^{\infty} g(x_k, y_k, u_k)$, where $g : X \times Y \times U \to [0, \infty)$ is a given instantaneous cost with $g(x, y, u) > 0$ for all $x \notin T$, $y \in Y$ and $u \in U$. To this end, we would like to construct an approximately optimal feedback $u : S \to U$ such that a suitable asymptotic stability property for the resulting closed loop system holds. Again, the construction will be based on a discrete value function. For an appropriate choice of g this function is continuous in x at least in a neighborhood of T [12].

Computational Approach. Let $\mathcal{Q}$ be a partition of the continuous part X of state space. Then the sets

$$\mathcal{P} := \{Q_i \times \{y\} \mid Q_i \in \mathcal{Q},\, y \in Y\} \tag{37}$$

form a partition of the product state space $Z = X \times Y$. On $\mathcal{P}$ the approaches from Sects. 4–7 can be applied literally.

Example 4 ***(Example: A switched voltage controller).*** This is an example taken from [15]: Within a device for DC to DC conversion, a semiconductor is switching the polarity of a voltage source $V_{\rm in}$ in order to keep the ouput voltage x_1 as constant as possible close to a prescribed value $V_{\rm ref}$, cf. Fig. 12, while the load is varying and thus the output current $I_{\rm load}$ changes. The model is

$$\begin{aligned}\dot{x}_1 &= \frac{1}{C}(x_2 - I_{\rm load})\\ \dot{x}_2 &= -\frac{1}{L}x_1 - \frac{R}{L}x_2 + \frac{1}{L}uV_{\rm in}\\ \dot{x}_3 &= V_{\rm ref} - x_1,\end{aligned} \tag{38}$$

where $u \in \{-1, 1\}$ is the control input. The corresponding discrete time system is given by the time-t-map Φ^t ($t = 0.1$ in our case) of (38), with the control input held constant during this sampling period. We use the quadratic instantaneous cost function

$$g(x, u) = q_P(\Phi_1^t(x) - V_{\rm ref})^2 + q_D(\Phi_2^t(x) - I_{load})^2 + q_I\Phi_3^t(x)^3.$$

The third component in (38) is only used in order to penalize a large L^1-error of the output voltage. We slightly simplify the problem (over its original formulation in [15]) by using $x_3 = 0$ as initial value in each evaluation of the discrete map. Correspondingly, the map reduces to a two-dimensional one on the x_1, x_2-plane.

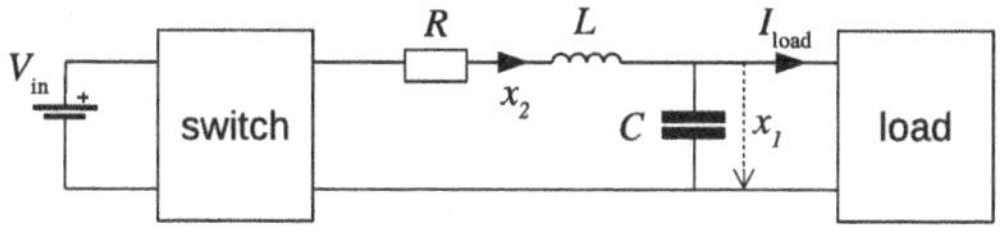

Fig. 12. A switched DC/DC converter (cf. [15]).

In the following numerical experiment we use the same parameter values as given in [15], i.e. $V_{\rm in} = 1V$, $V_{\rm ref} = 0.5$, $R = 1\Omega$, $L = 0.1H$, $C = 4F$, $I_{\rm load} = 0.3\,A$, $q_P = 1$, $q_D = 0.3$ and $q_I = 1$. Confining our domain of interest to the rectangle $X = [0, 1] \times [-1, 1]$, our target set is given by $T = \{V_{\rm ref}\} \times [-1, 1]$. For the construction of the finite graph, we employ a partition of X into 64×64 equally sized boxes. We use 4 test points in each box, namely their vertices, in order to construct the edges of the graph.

Using the resulting discrete value function (associated to a nominal $I_{load} = 0.3$ A) and the associated feedback, we repeated the stabilization experiment from [15], where the load current is changed after every 100 iterations. Figure 13 shows the result of this simulation, proving that our controller stabilizes the system as requested.

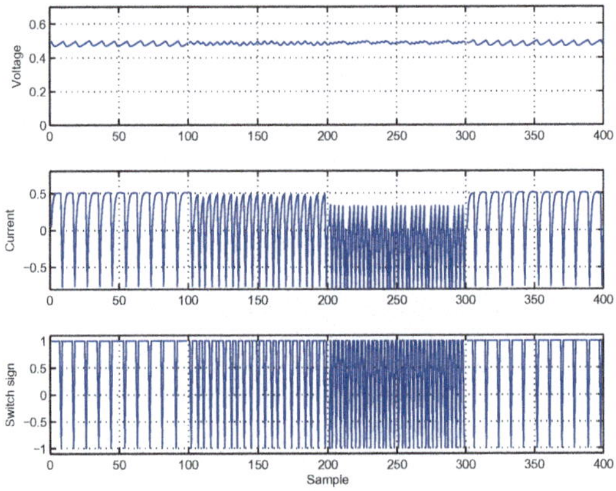

Fig. 13. Simulation of the controlled switched power converter.

Event Systems. In many cases, the discrete-time system (1) is given by time-sampling an underlying continuous time control system (an ordinary differential equation with inputs u and w), i.e. by the time-t-map of the flow of the continuous time system. In some cases, instead of fixing the time step t in each evaluation of f, it might be more appropriate to chosen t in dependence of the dynamics. Formally, based on the discrete time model (1) of the plant, we are dealing with the discrete time system

$$x_{\ell+1} = \tilde{f}(x_\ell, u_\ell), \quad \ell = 0, 1, \ldots, \tag{39}$$

where

$$\tilde{f}(x, u) = f^{r(x,u)}(x, u), \tag{40}$$

$r : X \times U \to \mathbb{N}_0$ is a given *event function* and the iterate f^r is defined by $f^0(x, u) = x$ and $f^r(x, u) = f(f^{r-1}(x, u), u)$, cf. [10]. The associated instantaneous cost $\tilde{g} : X \times U \to [0, \infty)$ is given by

$$\tilde{g}(x, u) = \sum_{k=0}^{r(x,u)-1} g(f^k(x, u), u). \tag{41}$$

The time k of the underlying system (1) can be recovered from the event time ℓ through

$$k_{\ell+1} = k_\ell + r(x_\ell, u_\ell).$$

Note that this model comprises an *event-triggered* scenario where the event function is constructed from a comparison of the state of (1) with the state of (39), as well as the scenario of *self-triggered control* (cf. [1]) where the event function is computed from the state of (1) alone.

Quantized Systems. The approach for discretizing the optimality principle described in Sects. 4–6 is based on a discretization of state space in form of a finite partition. While in general the geometry of the partition elements is arbitrary (except from reasonable regularity assumptions), in many cases (e.g. in our implementation in GAIO) cubical partitions are a convenient choice. In this case, the discretization can be interpreted as a *quantization* of (1), where the quantized system is given by the finite state system

$$P_{k+1} = F(P_k, u_k, \gamma_k), \quad k = 0, 1, \ldots, \tag{42}$$

with

$$F(P, u, \gamma) = \pi(f(\gamma(P), u)), \qquad P \in \mathcal{P}, u \in U,$$

where $\gamma : \mathcal{P} \to X$ is a function which chooses a point x from some partition element $P \in \mathcal{P}$, i.e. it satisfies $\pi(\gamma(P)) = P$ for all $\mathcal{P} \in P$ [10]. The choice function models the fact that it is unknown to the controller from which exact state x_k the system transits to the next cell P_{k+1}. It may be viewed as a perturbation which might prevent us from reaching the target set – in this sense, (42) constitutes a *dynamic game* in the sense of Sect. 6.

9 Lazy Feedbacks

In some applications, e.g. when data needs to be transmitted between the system and the controller over a channel with limited bandwidth, it might be desirable to minimize the amount of transmitted data. More specifically, the question might be how to minimize the number of times that a new control value has to be transmitted from the controller to the system. In this section, we show how this can be achieved in an optimization based feedback construction by defining a suitable instantaneous cost function.

In order to detect a change in the control value we need to be able to compare its current value to the one in the previous time step. Based on the setting from Sect. 2, we consider the discrete-time control system

$$z_{k+1} = \bar{f}(z_k, u_k), \quad k = 0, 1, 2, \ldots \tag{43}$$

with $z_k = (x_k, w_k) \in Z := X \times U$, $u_k \in U$ and

$$\bar{f}(z, u) = \bar{f}((x, w), u) := \begin{bmatrix} f(x, u) \\ u \end{bmatrix}.$$

Given some target set $T \subset X$, we define $\bar{T} := T \times U$ as the target set in the extended state space Z. The instantaneous cost function $\bar{g} : Z \times U \to [0, \infty)$, which penalizes control value changes is given by

$$\bar{g}_\lambda(z, u) = \bar{g}_\lambda((x, w), u) := (1 - \lambda)g(x, u) + \lambda\delta(u - w) \tag{44}$$

with

$$\delta(u) = \begin{cases} 0 & \text{if } u = 0, \\ 1 & \text{else.} \end{cases} \tag{45}$$

Here, $\lambda \in [0, 1)$ (in particular, $\lambda < 1$ in order to guarantee that $\bar{g}(z, u) = 0$ iff $z \in \bar{T}$).

In order to apply the construction from Sect. 7, we choose a finite partition $\mathcal{P}$ of X. Let $\hat{V}_\mathcal{P}$ denote the associated discrete upper value function, $\hat{S} = \{x \in X : \hat{V}_\mathcal{P}(x) < \infty\}$ the stabilizable set, and $\hat{u}_\mathcal{P}$ the associated feedback for the original system (f, g). For simplicity, we assume that U is finite and use $\mathcal{P} \times U$ as the partition of the extended state space Z. We denote the discrete upper value function of $(\bar{f}, \bar{g}_\lambda)$ by $\bar{V}_\lambda : Z \to [0, \infty]$, the stabilizable subset by $\bar{S}_\lambda := \{z \in Z : \bar{V}_\lambda(z) < \infty\}$ and the associated feedback by $\bar{u}_\lambda : \bar{S}_\lambda \to U$.

For some arbitrary feedback $u_\lambda : \bar{S}_\lambda \to U$, consider the closed loop system

$$z_{k+1} = \bar{f}(z_k, u_\lambda(z_k)), \quad k = 0, 1, 2, \ldots. \tag{46}$$

We will show that for any sufficiently large $\lambda < 1$ the closed loop system with $u_\lambda = \bar{u}_\lambda$ is asymptotically stable on $\bar{S}_\lambda$, more precisely that for $z_0 \in \bar{S}_\lambda$ the trajectory of (46) enters $\bar{T}$ in finitely many steps and that the number of control value changes along this trajectory is minimal.

To this end, for some initial state $z_0 \in \bar{S}_\lambda$, let $(z_k)_k \in Z^\mathbb{N}$, $z_k = (x_k, w_k)$, be the trajectory of (46). Let $\kappa(z_0, u_\lambda) = \min\{k \geq 0 : z_k \in \bar{T}\}$ be the minimal number of time steps until the trajectory reaches the target set $\bar{T}$,

$$E(z_0, u_\lambda) = \sum_{k=0}^{\kappa(z_0,u_\lambda)} \delta\big(u_\lambda(z_k) - w_k\big)$$

the number of control value changes along the corresponding trajectory as well as

$$J(z_0, u_\lambda) = \sum_{k=0}^{\kappa(z_0,u_\lambda)} g(x_k, u(z_k)), \quad \text{resp.} \quad \bar{J}(z_0, u_\lambda) = \sum_{k=0}^{\kappa(z_0,u_\lambda)} \bar{g}(z_k, u(z_k))$$

the associated accumulated costs. Note that

$$\bar{J}(z_0, u_\lambda) = (1 - \lambda)J(z_0, u_\lambda) + \lambda E(z_0, u_\lambda).$$

Theorem 6. *For all $\lambda \in [0, 1)$, $\hat{S} \times U \subset \bar{S}_\lambda$. Using the optimal feedback $\bar{u}_\lambda$ in (46) and for $z_0 \in \bar{S}_\lambda$, $z_k \to \bar{T}$ as $k \to \infty$. Further, there exists $\lambda < 1$ such that for any feedback $u_\lambda : \bar{S}_\lambda \to U$ and $z_0 \in \bar{S}_\lambda$ with $\kappa(z_0, u_\lambda) < K$ for some arbitrary $K \in \mathbb{N}$, we have $E(z_0, u_\lambda) \geq E(z_0, \bar{u}_\lambda)$.*

Proof. By construction, the system (43, 44) fulfills the assumptions of Theorem 5, so we have asymptotic stability of the closed loop system (46) with $u_\lambda = \bar{u}_\lambda$ for all $z_0 \in \bar{S}_\lambda$.

In order to show that $\hat{S} \times U \subset \bar{S}_\lambda$ for all $\lambda \in [0, 1)$, choose $\lambda \in [0, 1)$ and some initial value $z_0 = (x_0, u_0) \in \hat{S} \times U$. Consider the feedback

$$u(z) = u((x, u)) := \hat{u}_{\mathcal{P}}(x)$$

for system (43). This leads to a trajectory $(x_k, u_k)_k$ of the extended system with $(x_k)_k$ being a trajectory of the the closed loop system for f with feedback $\hat{u}_{\mathcal{P}}$. Since $x_0 \in \hat{S}$, $\hat{V}_{\mathcal{P}}(x_0)$ is finite and the accumulated cost $\bar{J}(z_0, u)$ for this trajectory does not exceed $(1-\lambda)\hat{V}_{\mathcal{P}}(x_0) + \lambda\kappa(z_0, u)$ which is finite. According to the optimality of V_λ,

$$V_\lambda(z_0) \leq (1-\lambda)\hat{V}_{\mathcal{P}}(x_0) + \lambda\kappa(z_0, u) < \infty$$

follows, i.e. $z_0 \in \bar{S}_\lambda$.

To show the optimality of $\bar{u}_\lambda$ with respect to the functional E, assume there exists a feedback $u_\lambda : \bar{S}_\lambda \to U$ with $E(z_0, u_\lambda) \leq E(z_0, \bar{u}_\lambda) - 1$ for some $z_0 \in \bar{S}_\lambda$. Since $\bar{u}_\lambda$ is optimal, the following inequality holds:

$$\begin{aligned}
(1-\lambda)J(z_0, u_\lambda) + \lambda E(z_0, u_\lambda) &= \bar{J}(z_0, u_\lambda) \\
&\geq \bar{J}(z_0, \bar{u}_\lambda) \\
&= (1-\lambda)J(z_0, \bar{u}_\lambda) + \lambda E(z_0, \bar{u}_\lambda) \\
&\geq (1-\lambda)J(z_0, \bar{u}_\lambda) + \lambda E(z_0, u_\lambda) + \lambda
\end{aligned}$$

and thus

$$(1-\lambda)J(z_0, u_\lambda) \geq (1-\lambda)J(z_0, \bar{u}_\lambda) + \lambda. \tag{47}$$

Let $C(u_\lambda) = \max_{z_0}\{J(z_0, u_\lambda) \mid \kappa(z_0, u_\lambda) < K\}$ which is finite. From (47) we get

$$(1-\lambda)C(u_\lambda) \geq (1-\lambda)C(\bar{u}_\lambda) + \lambda. \tag{48}$$

so that $\lambda \to 1$ leads to a contradiction. □

Acknowledgements. OJ thanks Michael Dellnitz for being his mentor, colleague and friend since more than 25 years. OJ and LG gratefully acknowledge the support through the Priority Programme SPP 1305 Control Theory of Digitally Networked Dynamic Systems of the German Research Foundation. OJ additionally acknowledges support through a travel grant by DAAD.

References

1. Anta, A., Tabuada, P.: To sample or not to sample: self-triggered control for nonlinear systems. IEEE Trans. Autom. Control **55**(9), 2030–2042 (2010)
2. Bellman, R.: Dynamic Programming. Princeton University Press, Princeton (1957)

3. Dellnitz, M., Froyland, G., Junge, O.: The algorithms behind GAIO-set oriented numerical methods for dynamical systems. In: Ergodic Theory, Analysis, and Efficient Simulation of Dynamical Systems, pp. 145–174, 805–807. Springer, Berlin (2001)
4. Dellnitz, M., Hohmann, A.: A subdivision algorithm for the computation of unstable manifolds and global attractors. Numerische Mathematik **75**(3), 293–317 (1997)
5. Dijkstra, E.W.: A note on two problems in connexion with graphs. Numer. Math. **1**, 269–271 (1959)
6. Fleming, W.H.: The convergence problem for differential games. J. Math. Anal. Appl. **3**, 102–116 (1961)
7. Grüne, L., Junge, O.: A set oriented approach to optimal feedback stabilization. Syst. Control Lett. **54**(2), 169–180 (2005)
8. Grüne, L., Junge, O.: Approximately optimal nonlinear stabilization with preservation of the Lyapunov function property. In: Proceedings of the 46th IEEE Conference on Decision and Control, pp. 702–707 (2007)
9. Grüne, L., Junge, O.: Global optimal control of perturbed systems. J. Optim. Theory Appl. **136**(3), 411–429 (2008)
10. Grüne, L., Müller, F.: An algorithm for event-based optimal feedback control. In: Proceedings of the 48th IEEE Conference on Decision and Control, Shanghai, China, pp. 5311–5316 (2009)
11. Grüne, L., Müller, F.: Global optimal control of quantized systems. In: Proceedings of the 18th International Symposium on Mathematical Theory of Networks and Systems — MTNS2010, Budapest, Hungary, pp. 1231–1237 (2010)
12. Grüne, L., Nešić, D.: Optimization-based stabilization of sampled-data nonlinear systems via their approximate discrete-time models. SIAM J. Control Optim. **42**(1), 98–122 (2003)
13. Junge, O.: Rigorous discretization of subdivision techniques. In: International Conference on Differential Equations, vol. 1, 2 (Berlin, 1999), pp. 916–918. World Scientific Publishing, River Edge (2000)
14. Junge, O., Osinga, H.M.: A set oriented approach to global optimal control. ESAIM Control Optim. Calc. Var. **10**(2), 259–270 (2004)
15. Lincoln, B., Rantzer, A.: Relaxing dynamic programming. IEEE Trans. Autom. Control **51**(8), 1249–1260 (2006)
16. Sethian, J.A.: A fast marching level set method for monotonically advancing fronts. Proc. Natl. Acad. Sci. U.S.A. **93**(4), 1591–1595 (1996)
17. Sethian, J.A., Vladimirsky, A.: Ordered upwind methods for static Hamilton-Jacobi equations. Proc. Natl. Acad. Sci. U.S.A. **98**(20), 11069–11074 (2001)
18. Tsitsiklis, J.N.: Efficient algorithms for globally optimal trajectories. IEEE Trans. Autom. Control **40**(9), 1528–1538 (1995)
19. Tucker, W.: Validated Numerics: A Short Introduction to Rigorous Computations. Princeton University Press, Princeton (2011)
20. von Lossow, M.: A min-man version of Dijkstra's algorithm with application to perturbed optimal control problems. In: Proceedings of the GAMM Annual Meeting, Zürich, Switzerland (2007)

An Optimal Control Derivation of Nonlinear Smoothing Equations

Jin Won Kim and Prashant G. Mehta(✉)

Coordinated Science Laboratory and Department of Mechanical Science and Engineering,
University of Illinois at Urbana-Champaign, Urbana, IL 61801, USA
{jkim684,mehtapg}@illinois.edu

Abstract. The purpose of this paper is to review and highlight some connections between the problem of nonlinear smoothing and optimal control of the Liouville equation. The latter has been an active area of recent research interest owing to work in mean-field games and optimal transportation theory. The nonlinear smoothing problem is considered here for continuous-time Markov processes. The observation process is modeled as a nonlinear function of a hidden state with an additive Gaussian measurement noise. A variational formulation is described based upon the relative entropy formula introduced by Newton and Mitter. The resulting optimal control problem is formulated on the space of probability distributions. The Hamilton's equation of the optimal control are related to the Zakai equation of nonlinear smoothing via the log transformation. The overall procedure is shown to generalize the classical Mortensen's minimum energy estimator for the linear Gaussian problem.

Keywords: Markov processes · Bayesian inference · Stochastic smoothing · Nonlinear filtering · Duality · Optimal control

1 Introduction

There is a fundamental dual relationship between estimation and control. The most basic of these relationships is the well known duality between controllability and observability of a linear system [8, Ch. 15]. The relationship suggests that the problem of filter (estimator) design can be re-formulated as a variational problem of optimal control. Such variational formulations are referred to as the duality principle of optimal filtering. The first duality principle appears in the seminal (1961) paper of Kalman-Bucy, where the problem of minimum variance estimation is shown to be dual to a linear quadratic optimal control problem. In these classical settings, the dual variational formulations are of the following two types [1, Sec. 7.3]: (i) minimum variance estimator and (ii) minimum energy estimator.

To Michael Dellnitz on the occasion of his 60th birthday.

O. Junge et al. (Eds.): SON 2020, SSDC 304, pp. 295–311, 2020.
https://doi.org/10.1007/978-3-030-51264-4_12

The classical minimum energy estimator represents a solution of the smoothing problem. The estimator is modeled as a controlled version of the state process in which the process noise term is replaced by a control input. The optimal control input is obtained by maximizing the log of the conditional (smoothed) distribution. For this reason, the estimator is also referred to as the maximum a posteriori (MAP) estimator. The MAP solution coincides with the optimal smoother in the linear-Gaussian case. The earliest construction of the minimum energy estimator is due to Mortensen [11].

A variational formulation of the nonlinear smoothing problem – the focus of this paper – leading to the conditional distribution appears in [10]. The formulation is based upon the variational Kallianpur-Striebel formula [17, Lemma 2.2.1]. The divergence is expressed as an optimal control objective function which turns out to be identical to the objective function considered in the MAP estimator [11]. The difference is that the constraint now is a controlled stochastic process, in contrast to a single trajectory in the MAP estimator. With the optimal control input, the law of the stochastic process is the conditional distribution.

The purpose of this paper is to review and highlight some connections between nonlinear smoothing and optimal control problems involving control of probability densities. In recent years, there has been a lot of interest in mean-field-type optimal control problems where the constraint is a controlled Liouville or a Fokker-Plank equation describing the evolution of the probability density [2–4]. In this paper, it is shown that the variational formulation proposed in [10] is easily described and solved in these terms. The formulation as a mean-field-type optimal control problem is more natural compared to a stochastic optimal control formulation considered in [10]. In particular, the solution with the density constraint directly leads to the forward-backward equation of pathwise smoothing. This also makes explicit the connection to the log transformation which is known to transform the Bellman equation of optimal control into the Zakai equation of filtering [7,9]. Apart from the case of the Itô-diffusion, the continuous-time Markov chain is also described. The overall procedure is shown to generalize the classical Mortensen's minimum energy estimator for the linear Gaussian problem.

The outline of the remainder of this chapter is as follows: the smoothing problem and its solution in terms of the forward-backward Zakai equation and their pathwise representation is reviewed in Sect. 2. The variational formulation leading to a mean-field optimal control problem and its solution appears in Sect. 3. The relationship to the log transformation and to the minimum energy estimator is described. The conclusions appear in Sect. 4. All the proofs are contained in the Appendix.

Notation. We denote the ith element of a vector by $[\,\cdot\,]_i$, and similarly, (i,j) element of a matrix is denoted by $[\,\cdot\,]_{ij}$. $C^k(\mathbb{R}^d;S)$ is the space of functions with continuous k-th order derivative. For a function $f \in C^2(\mathbb{R}^d;\mathbb{R})$, ∇f is the gradient vector and $D^2 f$ is the Hessian matrix. For a vector field $F \in C^1(\mathbb{R}^d;\mathbb{R}^d)$, $\text{div}(F)$ denotes the divergence of F. For a vector $v \in \mathbb{R}^d$, $\text{diag}(v)$ denotes a diagonal matrix with diagonal entries given by the vector; e^v and v^2 are defined in an element-wise manner, that is, $[e^v]_i = e^{[v]_i}$ and $[v^2]_i = ([v]_i)^2$ for $i = 1, \ldots, d$. For a matrix, $\mathrm{tr}(\cdot)$ denotes the trace.

2 Preliminaries and Background

2.1 The Smoothing Problem

Consider a pair of continuous-time stochastic processes (X, Z). The state $X = \{X_t : t \in [0,T]\}$ is a Markov process taking values in the state space $\mathbb{S}$. The observation process $Z = \{Z_t : t \in [0,T]\}$ is defined according to the model:

$$Z_t = \int_0^t h(X_s)\,\mathrm{d}s + W_t \tag{1}$$

where $h : \mathbb{S} \to \mathbb{R}$ is the observation function and $W = \{W_t : t \geq 0\}$ is a standard Wiener process.

The smoothing problem is to compute the posterior distribution $\mathsf{P}(X_t \in \cdot \,|\mathcal{Z}_T)$ for arbitrary $t \in [0,T]$, where $\mathcal{Z}_T := \sigma(Z_s : 0 \leq s \leq T)$ is the sigma-field generated by the observation up to the terminal time T.

2.2 Solution of the Smoothing Problem

The smoothing problem requires a model of the Markov process X. In applications involving nonlinear smoothing, a common model is the Itô-diffusion in Euclidean settings:

Euclidean State Space. The state space $\mathbb{S} = \mathbb{R}^d$. The state process X is modeled as an Itô diffusion:

$$\mathrm{d}X_t = a(X_t)\,\mathrm{d}t + \sigma(X_t)\,\mathrm{d}B_t, \quad X_0 \sim \nu_0$$

where $a \in C^1(\mathbb{R}^d;\mathbb{R}^d)$, $\sigma \in C^2(\mathbb{R}^d;\mathbb{R}^{d\times p})$ and $B = \{B_t : t \geq 0\}$ is a standard Wiener process. The initial distribution of X_0 is denoted as $\nu_0(x)\,\mathrm{d}x$ where $\nu_0(x)$ is the probability density with respect to the Lebesgue measure. For (1), the observation function $h \in C^2(\mathbb{R}^d;\mathbb{R})$. It is assumed that X_0, B, W are mutually independent.

The infinitesimal generator of X, denoted as $\mathcal{A}$, acts on C^2 functions in its domain according to

$$(\mathcal{A}f)(x) := a^\top(x)\nabla f(x) + \frac{1}{2}\mathrm{tr}\big(\sigma\sigma^\top(x)(D^2 f)(x)\big).$$

The adjoint operator is denoted by $\mathcal{A}^\dagger$. It acts on C^2 functions in its domain according to

$$(\mathcal{A}^\dagger f)(x) = -\,\mathrm{div}(af)(x) + \frac{1}{2}\sum_{i,j=1}^{d} \frac{\partial^2}{\partial x_i \partial x_j}\big([\sigma\sigma^\top]_{ij} f\big)(x).$$

The solution of the smoothing problem is described by a forward-backward system of stochastic partial differential equations (SPDE) (see [12, Thm. 3.8]):

$$\begin{aligned}
&\text{(forward):} && \mathrm{d}p_t(x) = (\mathcal{A}^\dagger p_t)(x)\,\mathrm{d}t + h(x)p_t(x)\,\mathrm{d}Z_t \\
& && p_0(x) = \nu_0(x), \quad \forall x \in \mathbb{R}^d && \text{(2a)} \\
&\text{(backward):} && -\mathrm{d}q_t(x) = (\mathcal{A}q_t)(x)\,\mathrm{d}t + h(x)q_t(x)\,\overleftarrow{\mathrm{d}Z_t} \\
& && q_T(x) \equiv 1 && \text{(2b)}
\end{aligned}$$

where $\overleftarrow{\mathrm{d}Z_t}$ denotes a backward Itô integral (see [12, Remark 3.3]). The smoothed distribution is then obtained as follows:

$$\mathsf{P}(X_t \in \mathrm{d}x \,|\mathcal{Z}_T) \propto p_t(x)q_t(x)\,\mathrm{d}x.$$

Each of (2) is referred to as the Zakai equation of nonlinear filtering.

2.3 Path-Wise Representation of the Zakai Equations

There is a representation of the forward-backward SPDEs where the only appearance of randomness is in the coefficients. This is referred to as the pathwise (or robust) form of the filter [14, Sec. VI.11].

Using Itô's formula for $\log p_t$,

$$\mathrm{d}(\log p_t)(x) = \frac{1}{p_t(x)}(\mathcal{A}^\dagger p_t)(x)\,\mathrm{d}t + h(x)\,\mathrm{d}Z_t - \frac{1}{2}h^2(x)\,\mathrm{d}t.$$

Therefore, upon defining $\mu_t(x) := \log p_t(x) - h(x)Z_t$, the forward Zakai Eq. (2a) is transformed into a parabolic partial differential equation (pde):

$$\begin{aligned}
&\frac{\partial \mu_t}{\partial t}(x) = e^{-(\mu_t(x)+Z_t h(x))}\big(\mathcal{A}^\dagger e^{(\mu_t(\cdot)+Z_t h(\cdot))}\big)(x) - \frac{1}{2}h^2(x) \\
&\mu_0(x) = \log \nu_0(x), \quad \forall x \in \mathbb{R}^d.
\end{aligned} \tag{3}$$

Similarly, upon defining $\lambda_t(x) = \log q_t(x) + h(x)Z_t$, the backward Zakai Eq. (2b) is transformed into the parabolic pde:

$$\begin{aligned}
&-\frac{\partial \lambda_t}{\partial t}(x) = e^{-(\lambda_t(x)-Z_t h(x))}\big(\mathcal{A} e^{\lambda_t(\cdot)-Z_t h(\cdot)}\big)(x) - \frac{1}{2}h^2(x) \\
&\lambda_T(x) = Z_T h(x), \quad \forall x \in \mathbb{R}^d.
\end{aligned} \tag{4}$$

The pde (3)–(4) are referred to as pathwise equations of nonlinear smoothing.

2.4 The Finite State-Space Case

Apart from Itô-diffusion, another common model is a Markov chain in finite state-space settings:

Finite State Space. Let the state-space be $\mathbb{S} = \{e_1, e_2, \ldots, e_d\}$, the canonical basis in $\mathbb{R}^d$. For (1), the linear observation model is chosen without loss of generality: for any function $h : \mathbb{S} \to \mathbb{R}$, we have $h(x) = \tilde{h}^\top x$ where $\tilde{h} \in \mathbb{R}^d$ is defined by $\tilde{h}_i = h(e_i)$. Thus, the function space on $\mathbb{S}$ is identified with $\mathbb{R}^d$. With a slight abuse of notation, we will drop the tilde and simply write $h(x) = h^\top x$.

The state process X is a continuous-time Markov chain evolving in $\mathbb{S}$. The initial distribution for X_0 is denoted as ν_0. It is an element of the probability simplex in $\mathbb{R}^d$. The generator of the chain is denoted as A. It is a $d \times d$ row-stochastic matrix. It acts on a function $f \in \mathbb{R}^d$ through right multiplication: $f \mapsto Af$. The adjoint operator is the matrix transpose $A^\top$. It is assumed that X and W are mutually independent.

The solution of the smoothing problem for the finite state-space settings is entirely analogous: Simply replace the generator $\mathcal{A}$ in (2) by the matrix A, and the probability density by the probability mass function. The Zakai pde is now the Zakai sde. The formula for the pathwise representation are also entirely analogous:

$$\left[\frac{\mathrm{d}\mu_t}{\mathrm{d}t}\right]_i = [e^{-(\mu_t + Z_t h)}]_i [A^\top e^{\mu_t + Z_t h}]_i - \frac{1}{2}[h^2]_i \tag{5}$$

$$-\left[\frac{\mathrm{d}\lambda_t}{\mathrm{d}t}\right]_i = [e^{-(\lambda_t - Z_t h)}]_i [A e^{\lambda_t - Z_t h}]_i - \frac{1}{2}[h^2]_i \tag{6}$$

with boundary condition $[\mu_0]_i = \log[\nu_0]_i$ and $[\lambda_0]_i = Z_T[h]_i$, for $i = 1, \ldots, d$.

3 Optimal Control Problem

3.1 Variational Formulation

For the smoothing problem, an optimal control formulation is derived in the following two steps:

Step 1. A control-modified version of the Markov process X is introduced. The controlled process is denoted as $\tilde{X} := \{\tilde{X}_t : 0 \le t \le T\}$. The control problem is to pick (i) the initial distribution $\pi_0 \in \mathcal{P}(\mathbb{S})$ and (ii) the state transition, such that the distribution of $\tilde{X}$ equals the conditional distribution. For this purpose, an optimization problem is formulated in the next step.

Step 2. The optimization problem is formulated on the space of probability laws. Let P denote the law for X, $\tilde{P}$ denote the law for $\tilde{X}$, and Q^z denote the law for X given an observation path $z = \{z_t : 0 \le t \le T\}$. Assuming these are equivalent, the objective function is the relative entropy between $\tilde{P}$ and Q^z:

$$\min_{\tilde{P}} \quad \mathsf{E}_{\tilde{P}}\Big(\log \frac{\mathrm{d}\tilde{P}}{\mathrm{d}P}\Big) - \mathsf{E}_{\tilde{P}}\Big(\log \frac{\mathrm{d}Q^z}{\mathrm{d}P}\Big).$$

Upon using the Kallianpur-Striebel formula (see [17, Lemma 1.1.5 and Prop. 1.4.2]), the optimization problem is equivalently expressed as follows:

$$\min_{\tilde{P}} \quad \mathsf{D}(\tilde{P} \| P) + \mathsf{E}\Big(\int_0^T z_t \,\mathrm{d}h(\tilde{X}_t) + \frac{1}{2}|h(\tilde{X}_t)|^2 \,\mathrm{d}t - z_T h(\tilde{X}_T)\Big). \tag{7}$$

The first of these terms depends upon the details of the model used to parametrize the controlled Markov process $\tilde{X}$. For the two types of Markov processes, this is discussed in the following sections.

Remark 1. The Schrödinger bridge problem is a closely related problem of recent research interest where one picks $\tilde{P}$ to minimize $\mathsf{D}(\tilde{P}\|P)$ subject to the constraints on marginals at time $t=0$ and T; cf., [5] where connections to stochastic optimal control theory are also described. Applications of such models to the filtering and smoothing problems is discussed in [13]. There are two differences between the Schrödinger bridge problem and the smoothing problem considered here:

1. The objective function for the smoothing problem also includes an additional integral term in (7) to account for conditioning due to observations z made over time $t \in [0,T]$;
2. The constraints on the marginals at time $t=0$ and $t=T$ are not present in the smoothing problem. Rather, one is allowed to pick the initial distribution π_0 for the controlled process and there is no constraint present on the distribution at the terminal time $t=T$.

3.2 Optimal Control: Euclidean State-Space

The modified process $\tilde{X}$ evolves on the state space $\mathbb{R}^d$. It is modeled as a controlled Itô-diffusion

$$\mathrm{d}\tilde{X}_t = a(\tilde{X}_t)\,\mathrm{d}t + \sigma(\tilde{X}_t)\big(u_t(\tilde{X}_t)\,\mathrm{d}t + \mathrm{d}\tilde{B}_t\big), \quad \tilde{X}_0 \sim \pi_0$$

where $\tilde{B} = \{\tilde{B}_t : 0 \le t \le T\}$ is a copy of the process noise B. The controlled process is parametrized by:

1. The initial density $\pi_0(x)$.
2. The control function $u \in C^1([0,T] \times \mathbb{R}^d; \mathbb{R}^p)$. The function of two arguments is denoted as $u_t(x)$.

The parameter π_0 and the function u are chosen as a solution of an optimal control problem.

For a given function $v \in C^1(\mathbb{R}^d; \mathbb{R}^p)$, the generator of the controlled Markov process is denoted by $\tilde{\mathcal{A}}(v)$. It acts on a C^2 function f in its domain according to

$$(\tilde{\mathcal{A}}(v)f)(x) = (\mathcal{A}f)(x) + (\sigma v)^{\top}(x)\nabla f(x).$$

The adjoint operator is denoted by $\mathcal{A}^{\dagger}(v)$. It acts on C^2 functions in its domain according to

$$(\tilde{\mathcal{A}}^{\dagger}(v)f)(x) = (\mathcal{A}^{\dagger}f)(x) - \mathrm{div}(\sigma v f)(x).$$

For a density ρ and a function g, define $\langle \rho, g \rangle := \int_{\mathbb{R}^d} g(x)\rho(x)\,\mathrm{d}x$.

With this notation, define the controlled Lagrangian $\mathcal{L} : C^2(\mathbb{R}^d;\mathbb{R}^+) \times C^1(\mathbb{R}^d;\mathbb{R}^p) \times \mathbb{R} \to \mathbb{R}$ as follows:

$$\mathcal{L}(\rho, v\,; y) := \frac{1}{2}\langle \rho, |\,v\,|^2 + h^2\rangle + y\,\langle \rho, \tilde{\mathcal{A}}(v)h\rangle.$$

The justification of this form of the Lagrangian starting from the relative entropy cost appears in Appendix A.1.

For a given fixed observation path $z = \{z_t : 0 \le t \le T\}$, the optimal control problem is as follows:

$$\min_{\pi_0, u} : \mathsf{J}(\pi_0, u\,; z) = \mathsf{D}(\pi_0 \| \nu_0) - z_T\langle \pi_T, h\rangle + \int_0^T \mathcal{L}(\pi_t, u_t; z_t)\,\mathrm{d}t \tag{8a}$$

$$\text{Subj.}: \frac{\partial \pi_t}{\partial t}(x) = (\tilde{\mathcal{A}}^\dagger(u_t)\pi_t)(x). \tag{8b}$$

Remark 2. This optimal control problem is a mean-field-type problem on account of the presence of the entropy term $\mathsf{D}(\pi_0\|\nu_0)$ in the objective function. The Lagrangian is in a standard stochastic control form and the problem can be solved as a stochastic control problem as well [10]. In this paper, the mean-field-type optimal control formulation is stressed as a straightforward way to derive the equations of the nonlinear smoothing.

The solution to this problem is given in the following proposition, whose proof appears in the Appendix A.3.

Proposition 1. *Consider the optimal control problem (8). For this problem, the Hamilton's equations are as follows:*

$$\textit{(forward)} \qquad \frac{\partial \pi_t}{\partial t}(x) = (\tilde{\mathcal{A}}^\dagger(u_t)\pi_t)(x) \tag{9a}$$

$$\textit{(backward)} \qquad -\frac{\partial \lambda_t}{\partial t}(x) = e^{-(\lambda_t(x) - z_t h(x))}(\mathcal{A}e^{\lambda_t(\cdot) - z_t h(\cdot)})(x) - \frac{1}{2}h^2(x) \tag{9b}$$

$$\textit{(boundary)} \qquad \lambda_T(x) = z_T h(x).$$

The optimal choice of the other boundary condition is as follows:

$$\pi_0(x) = \frac{1}{C}\nu_0(x)e^{\lambda_0(x)}$$

where $C = \int_{\mathbb{R}^d} \nu_0(x)e^{\lambda_0(x)}\,\mathrm{d}x$ *is the normalization factor. The optimal control is as follows:*

$$u_t(x) = \sigma^\top(x)\,\nabla(\lambda_t - z_t h)(x).$$

3.3 Optimal Control: Finite State-Space

The modified process $\tilde{X}$ is a Markov chain that also evolves in $\mathbb{S} = \{e_1, e_2, \ldots, e_d\}$. The control problem is parametrized by the following:

1. The initial distribution denoted as $\pi_0 \in \mathbb{R}^d$.
2. The state transition matrix denoted as $\tilde{A}(v)$ where $v \in (\mathbb{R}^+)^{d\times d}$ is the control input. After [17, Sec. 2.1.1.], it is defined as follows:

$$[\tilde{A}(v)]_{ij} = \begin{cases} [A]_{ij}[v]_{ij} & i \neq j \\ -\sum_{j\neq i}[\tilde{A}(v)]_{ij} & i = j \end{cases}$$

and we set $[v]_{ij} = 1$ if $i = j$ or if $[A]_{ij} = 0$.

To set up the optimal control problem, define a function $C:(\mathbb{R}^+)^{d\times d} \to \mathbb{R}^d$ as follows

$$[C(v)]_i = \sum_{j=1}^{d}[A]_{ij}[v]_{ij}(\log[v]_{ij} - 1), \quad i = 1,\ldots,d.$$

The Lagrangian for the optimal control problem is as follows:

$$\mathcal{L}(\rho, v; y) := \rho^\top (C(v) + \frac{1}{2}h^2) + y\,\rho^\top(\tilde{A}(v)h).$$

The justification of this form of the Lagrangian starting from the relative entropy cost appears in Appendix A.2.

For given observation path $z = \{z_t : 0 \le t \le T\}$, the optimal control problem is as follows:

$$\underset{\pi_0,u}{\text{Min}} : \mathsf{J}(\pi_0, u\,; z) = \mathsf{D}(\pi_0\|\nu_0) - z_T\pi_T^\top h + \int_0^T \mathcal{L}(\pi_t, u_t; z_t)\,\mathrm{d}t \tag{10a}$$

$$\text{Subj.}: \frac{\mathrm{d}\pi_t}{\mathrm{d}t} = \tilde{A}^\top(u_t)\pi_t. \tag{10b}$$

The solution to this problem is given in the following proposition, whose proof appears in the Appendix.

Proposition 2. *Consider the optimal control problem (10). For this problem, the Hamilton's equations are as follows:*

(forward) $$\frac{\mathrm{d}\pi_t}{\mathrm{d}t} = \tilde{A}^\top(u_t)\pi_t \tag{11a}$$

(backward) $$-\frac{\mathrm{d}\lambda_t}{\mathrm{d}t} = \text{diag}(e^{-(\lambda_t - z_t h)})\,A\,e^{\lambda_t - z_t h} - \frac{1}{2}h^2 \tag{11b}$$

(boundary) $$\lambda_T = z_T h.$$

The optimal boundary condition for π_0 *is given by:*

$$[\pi_0]_i = \frac{1}{C}[\nu_0]_i[e^{\lambda_0}]_i, \quad i = 1,\ldots,d$$

where $C = \nu_0^\top e^{\lambda_0}$. *The optimal control is*

$$[u_t]_{ij} = e^{([\lambda_t - z_t h]_j - [\lambda_t - z_t h]_i)}.$$

3.4 Derivation of the Smoothing Equations

The pathwise equations of nonlinear filtering are obtained through a coordinate transformation. The proof for the following proposition is contained in the Appendix A.5.

Proposition 3. *Suppose $(\pi_t(x), \lambda_t(x))$ is the solution to the Hamilton's Eq. (9). Consider the following transformation:*

$$\mu_t(x) = \log(\pi_t(x)) - \lambda_t(x) + \log(C).$$

The pair $(\mu_t(x), \lambda_t(x))$ satisfy path-wise smoothing Eqs. (3)–(4). *Also,*

$$\mathsf{P}(X_t \in \mathrm{d}x \,|\mathcal{Z}_T) = \pi_t(x)\,\mathrm{d}x \quad \forall t \in [0, T].$$

For the finite state-space case (11), the analogous formulae are as follows:

$$[\mu_t]_i = \log([\pi_t]_i) - [\lambda_t]_i + \log(C)$$

and

$$\mathsf{P}(X_t = e_i \,|\mathcal{Z}_T) = [\pi_t]_i \quad \forall t \in [0, T]$$

for $i = 1, \dots, d$.

3.5 Relationship to the Log Transformation

In this paper, we have stressed the density control viewpoint. Alternatively, one can express the problem as a stochastic control problem for the $\tilde{X}$ process. For this purpose, define the cost function $l : \mathbb{R}^d \times \mathbb{R}^p \times \mathbb{R} \to \mathbb{R}$ as follows:

$$l(x, v\,; y) := \frac{1}{2}|v|^2 + h^2(x) + y(\tilde{\mathcal{A}}(v)h)(x).$$

The stochastic optimal control problem for the Euclidean case then is as follows:

$$\begin{aligned}
\operatorname*{Min}_{\pi_0, U_t} &: \mathsf{J}(\pi_0, U_t\,; z) \\
&= \mathsf{E}\Big(\log \frac{\mathrm{d}\pi_0}{\mathrm{d}\nu_0}(\tilde{X}_0) - z_T h(\tilde{X}_T) + \int_0^T l(\tilde{X}_t, U_t\,; z_t)\,\mathrm{d}t\Big) && \text{(12a)} \\
\text{Subj.} &: \mathrm{d}\tilde{X}_t = a(\tilde{X}_t)\,\mathrm{d}t + \sigma(\tilde{X}_t)(U_t\,\mathrm{d}t + \mathrm{d}\tilde{B}_t). && \text{(12b)}
\end{aligned}$$

Its solution is given in the following proposition whose proof appears in the Appendix A.6.

Proposition 4. *Consider the optimal control problem (12). For this problem, the HJB equation for the value function V is as follows:*

$$\begin{aligned}
-\frac{\partial V_t}{\partial t}(x) &= \big(\mathcal{A}(V_t + z_t h)\big)(x) + h^2(x) - \frac{1}{2}|\sigma^\top \nabla(V_t + z_t h)(x)|^2 \\
V_T(x) &= -z_T h(x).
\end{aligned}$$

The optimal control is of the state feedback form as follows:

$$U_t = u_t(\tilde{X}_t)$$

where $u_t(x) = -\sigma^\top \nabla (V_t + z_t h)(x)$.

The HJB equation thus is exactly the Hamilton's Eq. (9b) and

$$V_t(x) = -\lambda_t(x), \quad \forall x \in \mathbb{R}^d, \ \forall t \in [0,T].$$

Noting $\lambda_t(x) = \log q_t(x) + h(x) z_t$, the HJB equation for the value function $V_t(x)$ is related to the backward Zakai equation for $q_t(x)$ through the log transformation (see also [7, Eqn. 1.4]):

$$V_t(x) = -\log\left(q_t(x) e^{z_t h(x)}\right).$$

3.6 Linear Gaussian Case

The linear-Gaussian case is a special case in the Euclidean setting with the following assumptions on the model:

1. The drift is linear in x. That is,

$$a(x) = A^\top x \ \text{ and } \ h(x) = H^\top x$$

   where $A \in \mathbb{R}^{d \times d}$ and $H \in \mathbb{R}^d$.
2. The coefficient of the process noise

$$\sigma(x) = \sigma$$

   is a constant matrix. We denote $Q := \sigma\sigma^\top \in \mathbb{R}^{d \times d}$.
3. The prior ν_0 is a Gaussian distribution with mean $\bar{m}_0 \in \mathbb{R}^d$ and variance $\Sigma_0 \succ 0$.

For this problem, we make the following restriction: The control input $u_t(x)$ is restricted to be constant over $\mathbb{R}^d$. That is, the control input is allowed to depend only upon time. With such a restriction, the controlled state evolves according to the sde:

$$\mathrm{d}\tilde{X}_t = A^\top \tilde{X}_t \,\mathrm{d}t + \sigma u_t \,\mathrm{d}t + \sigma \,\mathrm{d}\tilde{B}_t, \quad \tilde{X}_0 \sim \mathcal{N}(m_0, V_0).$$

With a Gaussian prior, the distribution π_t is also Gaussian whose mean m_t and variance V_t evolve as follow:

$$\frac{\mathrm{d}m_t}{\mathrm{d}t} = A^\top m_t + \sigma u_t$$
$$\frac{\mathrm{d}V_t}{\mathrm{d}t} = A^\top V_t + V_t A + \sigma\sigma^\top.$$

Since the variance is not affected by control, the only constraint for the optimal control problem is due to the equation for the mean.

It is an easy calculation to see that for the linear model,

$$(\tilde{A}(v)h)(x) = H^\top (A^\top x + \sigma v).$$

Therefore, the Lagrangian becomes

$$\mathcal{L}(\rho, v; y) = |v|^2 + |H^\top m|^2 + \text{tr}(HH^\top V) + yH^\top (A^\top m + \sigma v)$$

provided that $\rho \sim \mathcal{N}(m, V)$.

For Gaussian distributions $\pi_0 = \mathcal{N}(m_0, V_0)$ and $\nu_0 = \mathcal{N}(\bar{m}_0, \Sigma_0)$, the divergence is given by the well known formula

$$\mathsf{D}(\pi_0 \| \nu_0) = \frac{1}{2}\log\frac{|V_0|}{|\Sigma_0|} - \frac{d}{2} + \frac{1}{2}\text{tr}(V_0\Sigma_0^{-1}) + \frac{1}{2}(m_0 - \bar{m}_0)^\top \Sigma_0^{-1}(m_0 - \bar{m}_0)$$

and the term due to the terminal condition is easily evaluated as

$$\langle \pi_T, h\rangle = H^\top m_T.$$

Because the control input does not affect the variance process, we retain only the terms with mean and the control and express the optimal control problem as follows:

$$\underset{m_0,u}{\text{Minimize}} : \mathsf{J}(m_0, u; z) = \frac{1}{2}(m_0 - \bar{m}_0)^\top \Sigma_0^{-1}(m_0 - \bar{m}_0) \qquad (13a)$$

$$+ \int_0^T \frac{1}{2}|u_t|^2 + \frac{1}{2}|H^\top m_t|^2 + z_t^\top H^\top \dot{m}_t \, \mathrm{d}t - z_T^\top H^\top m_T$$

$$\text{Subject to} : \frac{\mathrm{d}m_t}{\mathrm{d}t} = A^\top m_t + \sigma u_t. \qquad (13b)$$

By a formal integration by parts,

$$\mathsf{J}(m_0, u; z) = \frac{1}{2}(m_0 - \bar{m}_0)^\top \bar{\Sigma}_0^{-1}(m_0 - \bar{m}_0)$$

$$+ \int_0^T \frac{1}{2}|u_t|^2 + \frac{1}{2}|\dot{z} - H^\top m_t|^2 \, \mathrm{d}t - \int_0^T \frac{1}{2}|\dot{z}_t|^2 \, \mathrm{d}t.$$

This form appears in the construction of the minimum energy estimator [1, Ch. 7.3].

4 Conclusions

In this paper, we provide a self-contained exposition of the equations of nonlinear smoothing as well as connections and interpretations to some of the more recent developments in mean-field-type optimal control theory. These connections suggest that the numerical approaches for mean-field type optimal control problems can also be applied to obtain approximate filters. Development of numerical techniques, e.g., particle filters to empirically approximate the conditional distribution, has been an area of intense research interest; cf., [13] and references therein. Approximate particle filters based upon approximation of dual optimal control-type problems have appeared in [6,9,13,15,16].

A Appendix

A.1 Derivation of Lagrangian: Euclidean Case

By Girsanov's theorem, the Radon-Nikodym derivative is obtained (see [13, Eqn. 35]) as follows:

$$\frac{\mathrm{d}\tilde{P}}{\mathrm{d}P}(\tilde{X}) = \frac{\mathrm{d}\pi_0}{\mathrm{d}\nu_0}(\tilde{X}_0)\ \exp\Big(\int_0^T \frac{1}{2}|u_t(\tilde{X}_t)|^2\,\mathrm{d}t + u_t(\tilde{X}_t)\,\mathrm{d}\tilde{B}_t\Big).$$

Thus, we obtain the relative entropy formula:

$$\begin{aligned}\mathsf{D}(\tilde{P}\|P) &= \mathsf{E}\Big(\log\frac{\mathrm{d}\pi_0}{\mathrm{d}\nu_0}(\tilde{X}_0) + \int_0^T \frac{1}{2}|u_t(\tilde{X}_t)|^2\,\mathrm{d}t + u_t(\tilde{X}_t)\,\mathrm{d}\tilde{B}_t\Big)\\ &= \mathsf{D}(\pi_0\|\nu_0) + \int_0^T \frac{1}{2}\langle\pi_t, |u_t|^2\rangle\,\mathrm{d}t.\end{aligned}$$

A.2 Derivation of Lagrangian: Finite State-Space Case

The derivation of the Lagrangian is entirely analogous to the Euclidean case except the R-N derivative is given according to [17, Prop. 2.1.1]:

$$\begin{aligned}\frac{\mathrm{d}\tilde{P}}{\mathrm{d}P}(\tilde{X}) = \frac{\mathrm{d}\pi_0}{\mathrm{d}\nu_0}(\tilde{X}_0)\exp\Big(-\sum_{i,j}\int_0^T [A]_{ij}[u_t]_{ij}1_{\tilde{X}_t=e_i}\Big)\\ \prod_{0<t\le T}\sum_{i\ne j}[u_{t-}]_{ij}1_{\tilde{X}_{t-}=e_i}1_{\tilde{X}_t=e_j}.\end{aligned}$$

Upon taking log and expectation of both sides, we arrive at the relative entropy formula:

$$\begin{aligned}\mathsf{D}(\tilde{P}\|P) &= \mathsf{E}\Big(\log\frac{\mathrm{d}\pi_0}{\mathrm{d}\nu_0}(\tilde{X}_0) + \int_0^T -\sum_{i,j}[A]_{ij}[u]_{ij}1_{\tilde{X}_t=e_i}\Big)\\ &\quad + \mathsf{E}\Big(\sum_{0<t\le T}\sum_{i\ne j}\log[u_{t-}]_{ij}1_{\tilde{X}_{t-}=e_i}1_{\tilde{X}_t=e_j}\Big)\\ &= \mathsf{D}(\pi_0\|\nu_0) + \int_0^T \pi_t^{\intercal}C(u_t)\,\mathrm{d}t.\end{aligned}$$

A.3 Proof of Proposition 1

The standard approach is to incorporate the constraint into the objective function by introducing the Lagrange multiplier $\lambda = \{\lambda_t : 0 \le t \le T\}$ as follows:

$$\begin{aligned}&\tilde{J}(u,\lambda;\pi_0,z)\\ &= \mathsf{D}(\pi_0\|\nu_0) + \int_0^T \frac{1}{2}\langle\pi_t, |u_t|^2 + h^2\rangle + z_t\langle\pi_t, \tilde{\mathcal{A}}(u_t)h\rangle\,\mathrm{d}t\\ &\quad + \int_0^T \langle\lambda_t, \frac{\partial\pi_t}{\partial t} - \tilde{\mathcal{A}}^{\dagger}(u_t)\pi_t\rangle\,\mathrm{d}t - z_T\langle\pi_T, h\rangle.\end{aligned}$$

Upon using integration by parts and the definition of the adjoint operator, after some manipulation involving completion of squares, we arrive at

$$\begin{aligned}\tilde{\mathsf{J}}(u,\lambda;\pi_0,z) &= \mathsf{D}(\pi_0\|\nu_0) + \int_0^T \frac{1}{2}\langle \pi_t, |u_t - \sigma^\top \nabla(\lambda_t - z_t h)|^2\rangle\,\mathrm{d}t \\ &- \int_0^T \langle \pi_t, \frac{\partial}{\partial t}\lambda_t + \mathcal{A}(\lambda_t - z_t h) - \frac{1}{2}h^2 + \frac{1}{2}|\sigma^\top\nabla(\lambda_t - z_t h)|^2\rangle\,\mathrm{d}t \\ &+ \langle \pi_T, \lambda_T - z_T h\rangle - \langle \pi_0, \lambda_0\rangle.\end{aligned}$$

Therefore, it is natural to pick λ to satisfy the following partial differential equation:

$$\begin{aligned}-\frac{\partial \lambda_t}{\partial t}(x) &= \big(\mathcal{A}(\lambda_t(\cdot) - z_t h(\cdot))\big) - \frac{1}{2}h^2(x) + \frac{1}{2}\left|\sigma^\top \nabla(\lambda_t - z_t h)(x)\right|^2 \\ &= e^{-(\lambda_t(x) - z_t h(x))}\big(\mathcal{A}e^{\lambda_t(\cdot) - z_t h(\cdot)}\big)(x) - \frac{1}{2}h^2(x)\end{aligned} \tag{14}$$

with the boundary condition $\lambda_T(x) = z_T h(x)$. With this choice, the objective function becomes

$$\begin{aligned}\tilde{\mathsf{J}}(u;\lambda,\pi_0,z) &= \mathsf{D}(\pi_0\|\nu_0) - \langle \pi_0, \lambda_0\rangle \\ &+ \int_0^T \frac{1}{2}\pi_t\left(\left|u_t - \sigma^\top\nabla(\lambda_t - z_t h)\right|^2\right)\mathrm{d}t\end{aligned}$$

which suggest the optimal choice of control is:

$$u_t(x) = \sigma^\top(x)\nabla(\lambda_t - z_t h)(x).$$

With this choice, the objective function becomes

$$\begin{aligned}\mathsf{D}(\pi_0\|\nu_0) - \langle \pi_0, \lambda_0\rangle &= \int_{\mathbb{S}} \pi_0(x)\log\frac{\pi_0(x)}{\nu_0(x)} - \lambda_0(x)\pi_0(x)\,\mathrm{d}x \\ &= \int_{\mathbb{S}} \pi_0(x)\log\frac{\pi_0(x)}{\nu_0\exp(\lambda_0(x))}\,\mathrm{d}x\end{aligned}$$

which is minimized by choosing

$$\pi_0(x) = \frac{1}{C}\nu_0(x)\exp(\lambda_0(x))$$

where C is the normalization constant.

A.4 Proof of Proposition 2

The proof for the finite state-space case is entirely analogous to the proof for the Euclidean case. The Lagrange multiplier $\lambda = \{\lambda_t \in \mathbb{R}^d : 0 \le t \le T\}$ is introduced to transform the optimization problem into an unconstrained problem:

$$\begin{aligned}\tilde{\mathsf{J}}(u,\lambda;\pi_0,z) &= \mathsf{D}(\pi_0\|\nu_0) + \int_0^T \pi_t^\top\big(C(u_t) + \frac{1}{2}h^2 + z_t\tilde{A}(u_t)h\big)\,\mathrm{d}t \\ &+ \int_0^T \lambda_t^\top\big(\frac{\mathrm{d}\pi_t}{\mathrm{d}t} - \tilde{A}^\top(u_t)\pi_t\big)\,\mathrm{d}t - z_T h^\top \pi_T.\end{aligned}$$

Upon using integral by parts,

$$\tilde{\mathsf{J}}(u,\lambda;\pi_0,z) = \mathsf{D}(\pi_0\|\nu_0) + \int_0^T \pi_t^\top\big(C(u_t) - \tilde{A}(u_t)(\lambda_t - z_t h)\big)\,\mathrm{d}t$$
$$+ \int_0^T \pi_t^\top(-\dot{\lambda}_t + \frac{1}{2}h^2)\,\mathrm{d}t + \pi_T^\top(\lambda_T - z_T h) - \pi_0^\top\lambda_0.$$

The first integrand is

$$[C(u_t) - \tilde{A}(u_t)(\lambda_t - Z_t h)]_i$$
$$= \sum_{j\neq i} A_{ij}\big([u]_{ij}(\log[u_t]_{ij} - 1)$$
$$- [u_t]_{ij}([\lambda_t - Z_t h]_j - [\lambda_t - Z_t h]_i)\big) - A_{ii}.$$

The minimizer is obtained, element by element, as

$$[u_t]_{ij} = e^{([\lambda_t - z_t h]_j - [\lambda_t - z_t h]_i)}$$

and the corresponding minimum value is obtained by:

$$[C(u_t^*) - \tilde{A}_t(\lambda_t - Z_t h)]_i = -[Ae^{\lambda_t - z_t h}]_i[e^{-(\lambda_t - z_t h)}]_i.$$

Therefore with the minimum choice of u_t above,

$$\tilde{\mathsf{J}}(u,\lambda;\pi_0,z) = \mathsf{D}(\pi_0\|\nu_0) + \int_0^T \pi_t^\top\big(-(Ae^{\lambda_t - z_t h})\cdot e^{-(\lambda_t - z_t h)}\big)\,\mathrm{d}t$$
$$+ \int_0^T \pi_t^\top(-\dot{\lambda}_t + \frac{1}{2}h^2)\,\mathrm{d}t + \pi_T^\top(\lambda_T - z_T h) - \pi_0^\top\lambda_0.$$

Upon choosing λ according to:

$$-[\dot{\lambda}_t]_i = [Ae^{\lambda_t - z_t h}]_i[e^{-(\lambda_t - z_t h)}]_i - \frac{1}{2}h_i^2, \quad \lambda_T = z_T h.$$

The objective function simplifies to

$$\mathsf{D}(\pi_0\|\nu_0) - \pi_0^\top\lambda_0 = \sum_{i=1}^d [\pi_0]_i \log\frac{[\pi_0]_i}{[\nu_0]_i e^{[\lambda_0]_i}}$$

where the minimum value is obtained by choosing

$$[\pi_0]_i = \frac{1}{C}[\nu_0]_i e^{[\lambda_0]_i}$$

where C is the normalization constant.

A.5 Proof of Proposition 3

Euclidean Case. Equation (9b) is identical to the backward path-wise Eq. (4). So, we need to only derive the equation for μ_t. Using the regular form of the product formula,

$$\begin{aligned}\frac{\partial \mu_t}{\partial t} &= \frac{1}{\pi_t}\frac{\partial \pi_t}{\partial t} - \frac{\partial \lambda_t}{\partial t} \\ &= \frac{1}{\pi_t}(\tilde{\mathcal{A}}^\dagger(u_t)\pi_t) + e^{-(\lambda_t - z_t h)}(\mathcal{A}e^{\lambda_t(\cdot)-z_t h(\cdot)}) - \frac{1}{2}h^2.\end{aligned}$$

With optimal control $u_t = \sigma^\top \nabla(\lambda_t - z_t h)$,

$$\begin{aligned}(\tilde{\mathcal{A}}^\dagger(u_t)\pi_t) = (\mathcal{A}^\dagger \pi_t) &- \operatorname{div}\left(\sigma\sigma^\top \nabla \pi_t\right) \\ &+ \pi_t \operatorname{div}\left(\sigma\sigma^\top \nabla(\mu_t + z_t h)\right) \\ &+ (\nabla \pi_t)^\top (\sigma\sigma^\top \nabla(\mu_t + z_t h))\end{aligned}$$

and

$$\begin{aligned}&e^{-(\lambda_t - z_t h)}(\mathcal{A}e^{\lambda_t(\cdot)-z_t h(\cdot)}) \\ &= \frac{1}{\pi_t}(\mathcal{A}\pi_t) - \frac{1}{2}|\sigma^\top \nabla \log \pi_t|^2 - (\mathcal{A}(\mu_t + z_t h)) \\ &\quad + \frac{1}{2}\left|\sigma^\top \nabla \log(\pi_t) - \sigma^\top \nabla(\mu_t + z_t h)\right|^2.\end{aligned}$$

Therefore,

$$\begin{aligned}\frac{\partial \mu_t}{\partial t} &= \frac{1}{\pi_t}\left((\mathcal{A}^\dagger \pi_t) + (\mathcal{A}\pi_t) - \operatorname{div}(\sigma\sigma^\top \nabla \pi_t)\right) \\ &\quad - (\mathcal{A}(\mu_t + z_t h)) + \operatorname{div}\left(\sigma\sigma^\top \nabla(\mu_t + z_t h)\right) \\ &\quad + \frac{1}{2}\left|\sigma^\top \nabla(\mu_t + z_t h)\right|^2 - \frac{1}{2}h^2 \\ &= e^{-(\mu_t(x) + z_t h(x))}\left(\mathcal{A}^\dagger e^{(\mu_t(\cdot) + z_t h(\cdot))}\right)(x) - \frac{1}{2}h^2(x)\end{aligned}$$

with the boundary condition $\mu_0 = \log \nu_0$.

Finite State-Space Case. Equation (11b) is identical to the backward path-wise Eq. (6). To derive the equation for μ_t, use the product formula

$$\begin{aligned}\left[\frac{\mathrm{d}\mu_t}{\mathrm{d}t}\right]_i &= \frac{1}{[\pi_t]_i}\left[\frac{\mathrm{d}\pi_t}{\mathrm{d}t}\right]_i - \left[\frac{\mathrm{d}\lambda_t}{\mathrm{d}t}\right]_i \\ &= \frac{1}{[\pi_t]_i}\left[\tilde{A}^\top(u_t)\pi_t\right]_i + [e^{-(\lambda_t - z_t h)}]_i[Ae^{\lambda_t + z_t h}]_i - \frac{1}{2}[h^2]_i.\end{aligned}$$

The first term is:

$$\left[\tilde{A}^\top(u_t)\pi_t\right]_i = \sum_{j=1}^{d}\left([A]_{ji}[u_t]_{ji}[\pi_t]_j - [A]_{ij}[u_t]_{ij}[\pi_t]_i\right)$$

and the second term is:

$$
\begin{aligned}
&[e^{-(\lambda_t - z_t h)}]_i [A e^{\lambda_t + z_t h}]_i \\
&\qquad = \frac{1}{[\pi_t]_i} [e^{\mu_t + z_t h}]_i \sum_{j=1}^{d} [A]_{ij} [\pi_t]_j [e^{-(\mu_t + z_t h)}]_j .
\end{aligned}
$$

The formula for the optimal control gives

$$
[u_t]_{ij} = \frac{[\pi_t]_j}{[\pi_t]_i} [e^{-(\mu_t + z_t h)}]_j [e^{\mu_t + z_t h}]_i .
$$

Combining these expressions,

$$
\begin{aligned}
\left[\frac{\mathrm{d}\mu_t}{\mathrm{d}t}\right]_i &= \sum_{j=1}^{d} [A]_{ji} [e^{-(\mu_t + z_t h)}]_i [e^{\mu_t + z_t h}]_j - \frac{1}{2}[h^2]_i \\
&= [e^{-(\mu_t + z_t h)}]_i [A^\top e^{\mu_t + z_t h}]_i - \frac{1}{2}[h^2]_i
\end{aligned}
$$

which is precisely the path-wise form of the Eq. (5). At time t=0, μ_0=log$(C[\pi_0]_i)-[\lambda_0]_i = \log[\nu_0]_i$.

Smoothing Distribution. Since (λ_t, μ_t) is the solution to the path-wise form of the Zakai equations, the optimal trajectory

$$
\pi_t = \frac{1}{C} e^{\mu_t + \lambda_t}
$$

represents the smoothing distribution.

A.6 Proof of Proposition 4

The dynamic programming equation for the optimal control problem is given by (see [1, Ch. 11.2]):

$$
\min_{u \in \mathbb{R}^p} \left\{ \frac{\partial V_t}{\partial t}(x) + (\tilde{\mathcal{A}}(u) V_t)(x) + l(x, u\,; z_t) \right\} = 0. \tag{15}
$$

Therefore,

$$
\begin{aligned}
-\frac{\partial V_t}{\partial t}(x) &= (\mathcal{A} V_t)(x) + h^2(x) + z_t (\mathcal{A} h)(x) \\
&\quad + \min_u \left\{ \frac{1}{2}|u|^2 + u^\top \big(\sigma^\top \nabla V_t(x) + z_t \sigma^\top \nabla h(x)\big) \right\}.
\end{aligned}
$$

Upon using the completion-of-square trick, the minimum is attained by a feedback form:

$$
u^* = -\sigma^\top \nabla (V_t + z_t h)(x).
$$

The resulting HJB equation is given by

$$-\frac{\partial V_t}{\partial t}(x) = \big(\mathcal{A}(V_t + z_t h)\big)(x) + h^2(x) - \frac{1}{2}|\sigma^\top \nabla (V_t + z_t h)|^2$$

with boundary condition $V_T(x) = -z_T h(x)$. Compare the HJB equation with the Eq. (14) for λ, and it follows

$$V_t(x) = -\lambda_t(x).$$

References

1. Bensoussan, A.: Estimation and Control of Dynamical Systems, vol. 48. Springer, Heidelberg (2018)
2. Bensoussan, A., Frehse, J., Yam, P., et al.: Mean Field Games and Mean Field Type Control Theory, vol. 101. Springer, Heidelberg (2013)
3. Brockett, R.W.: Optimal control of the liouville equation. AMS IP Stud. Adv. Math. **39**, 23 (2007)
4. Carmona, R., Delarue, F., et al.: Probabilistic Theory of Mean Field Games with Applications I-II. Springer, Heidelberg (2018)
5. Chen, Y., Georgiou, T.T., Pavon, M.: On the relation between optimal transport and Schrödinger bridges: a stochastic control viewpoint. J. Optim. Theory Appl. **169**(2), 671–691 (2016)
6. Chetrite, R., Touchette, H.: Variational and optimal control representations of conditioned and driven processes. J. Stat. Mech.: Theory Exp. **2015**(12), P12001 (2015)
7. Fleming, W., Mitter, S.: Optimal control and nonlinear filtering for nondegenerate diffusion processes. Stochastics **8**, 63–77 (1982)
8. Kailath, T., Sayed, A.H., Hassibi, B.: Linear Estimation. Prentice Hall, Upper Saddle River (2000)
9. Kappen, H.J., Ruiz, H.C.: Adaptive importance sampling for control and inference. J. Stat. Phys. **162**(5), 1244–1266 (2016)
10. Mitter, S.K., Newton, N.J.: A variational approach to nonlinear estimation. SIAM J. Control Optim. **42**(5), 1813–1833 (2003)
11. Mortensen, R.E.: Maximum-likelihood recursive nonlinear filtering. J. Optim. Theory Appl. **2**(6), 386–394 (1968)
12. Pardoux, E.: Non-linear filtering, prediction and smoothing. In: Stochastic Systems: the Mathematics of Filtering and Identification and Applications, pp. 529–557. Springer (1981)
13. Reich, S.: Data assimilation: the Schrödinger perspective. Acta Numerica **28**, 635–711 (2019)
14. Rogers, L.C.G., Williams, D.: Diffusions, Markov Processes and Martingales: Volume 2, Itô Calculus, vol. 2. Cambridge University Press, Cambridge (2000)
15. Ruiz, H., Kappen, H.J.: Particle smoothing for hidden diffusion processes: adaptive path integral smoother. IEEE Trans. Signal Process. **65**(12), 3191–3203 (2017)
16. Sutter, T., Ganguly, A., Koeppl, H.: A variational approach to path estimation and parameter inference of hidden diffusion processes. J. Mach. Learn. Res. **17**, 6544–80 (2016)
17. Van Handel, R.: Filtering, stability, and robustness. Ph.D. thesis, California Institute of Technology (2006)

Optimization

Structural Properties of Pareto Fronts: The Occurrence of Dents in Classical and Parametric Multiobjective Optimization Problems

Katrin Witting[1(✉)], Mirko Hessel-von Molo[2], and Michael Dellnitz[3]

[1] dSPACE GmbH, Paderborn, Germany
kwitting@dspace.de
[2] Faculty of Computer Science Fachhochschule Dortmund – University of Applied Sciences and Arts,
Dortmund, Germany
mirko.hessel-vonmolo@fh-dortmund.de
[3] Chair of Applied Mathematics, Paderborn University, Paderborn, Germany

Abstract. This contribution deals with the occurrence of "dents" in Pareto fronts of continuous and adequately smooth multiobjective optimization problems. After giving a formal definition of this notion, a system of equations is derived that characterizes points on the boundary of the dent. This can be used to obtain information about the structure of the Pareto front without computing the entire Pareto set. Furthermore, the evolution of dents in parametric multiobjective optimization problems is studied using results from bifurcation theory. Theory and computations are illustrated by several examples, whose construction is described as well.

Keywords: Multiobjective optimization · Parametric multiobjective optimization · Dents in Pareto fronts

1 Introduction

In many fields of research and industrial applications optimization plays an important role. In a variety of these not only one but several objectives are required to be optimal at the same time. For instance, in manufacturing *cost* has to be minimized, but at the same time also *quality* is desired to be maximized – at least to a certain degree. The development of theory and algorithms for the determination of solutions that are as good as possible with respect to all objectives is the task of *multiobjective optimization*. Mathematically, a continuous

M. Dellnitz—This contribution is dedicated to Michael on the occasion of his 60th birthday. He really is one of its authors, although at the time of publication he does not know it has been written.

O. Junge et al. (Eds.): SON 2020, SSDC 304, pp. 315–336, 2020.
https://doi.org/10.1007/978-3-030-51264-4_13

multiobjective optimization problem is given as

$$\min_x \{F(x) : x \in S \subseteq \mathbb{R}^n\},$$

where F is defined as the vector of the objective functions $f_1, \ldots, f_k$, $k \geq 2$, which each map from $\mathbb{R}^n$ to $\mathbb{R}$, and S denotes the feasible region. The example mentioned above already illustrates that the several objectives typically contradict each other and thus do not have identical optima. Consequently, the solution of a multiobjective optimization problem is given by the set of optimal compromises of the objectives, the so-called *Pareto set*. In the case of minimization problems the Pareto set is given by the set of solutions in which the value of any objective function can only be decreased at the cost of increasing another one.

To obtain solutions that lie within the Pareto set many algorithms have been developed. There are essentially two different types: algorithms that allow for the computation of only one or a few Pareto points, and algorithms that approximate the entire Pareto set. In the first case often a priori information, such as a specific weighting or some kind of ordering of the objectives, is required. Examples for those methods are the 'weighted sums method', the 'ε-constraint-method' and the 'lexicographic ordering' (see [8,20]). Over the past years algorithms that are able to approximate the entire Pareto set have been developed (see e. g. [2,4,5,11,19,22,23,30]). For the computations of Pareto sets in the examples given in this work set-oriented, numerical methods which are implemented in the software package GAIO have been used (see [24]).

Motivated by the fact that in the case of nonconvex objective functions it is not possible to compute all Pareto optimal solutions by the weighted sums method, in this chapter the occurrence of *dents* in Pareto fronts is studied (restricted to continuous and adequately smooth multiobjective optimization problems). In [3], Das and Dennis give a trigonometric argument why – in the case of two objectives – the weighted sums method cannot be used to compute points in the *nonconvex part*, as they call the subset of the Pareto front that contains no global optima of the weighted sum of the objectives for every weight vector. It is an interesting task to find out if a Pareto front contains nonconvex parts. If one assumes that the Pareto front is connected, then every nonconvex part contains a region in which the Pareto front 'bends inside the feasible region'. This part of the Pareto front will be called a *dent*. In Sect. 3 a formal definition of a dent is given (see also [28]).

It will be shown that at the border of a dent (seen as a subset of the Pareto front) typically the Hessian of the weighted sum of the objectives is singular. These points will be called *dent border points* and the corresponding preimages on the Pareto set will be called *dent border preimages*.

In the case of parametric multiobjective optimization problems naturally the question comes up, how dents evolve under the variation of the external parameter. This question is addressed in Sect. 4. Making use of results from bifurcation theory, it is proven that under certain assumptions dent border preimages are turning points of the Kuhn-Tucker equations

$$H_{\mathrm{KT}}^{\alpha^\star}(x,\lambda) = \sum_{i=1}^{k} \alpha_i^\star \nabla_x f_i(x,\lambda) = 0,$$

where $\alpha^\star$ ist the weight vector corresponding to the dent border preimage. Several examples of parametric multiobjective optimization problems in which the Pareto front contains dents will be given at the end of Sect. 4.

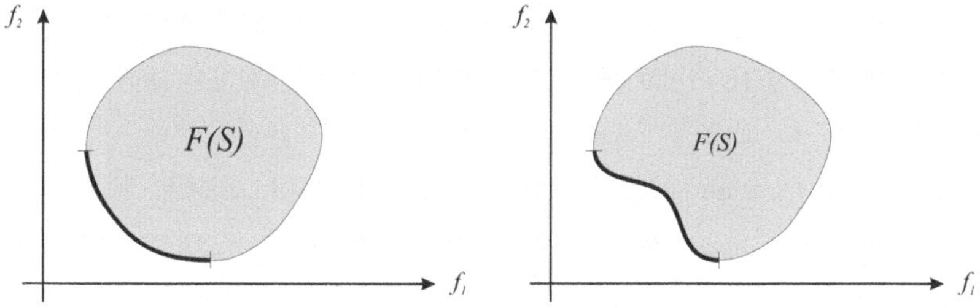

Fig. 1. Typical shape of a Pareto front for a convex problem (left figure) and possible shape in a nonconvex problem (right figure). The shaded regions represent the image of the feasible set.

2 Theoretical Background

In this section the theoretical background from multiobjective optimization, parametric multiobjective optimization and bifurcation theory needed within the context of this chapter is summarized.

2.1 Multiobjective Optimization

A continuous (constrained) *multiobjective optimization problem* (MOP) is given by

$$\min_x \{F(x) : x \in S \subseteq \mathbb{R}^n\}, \tag{MOP}$$

where F is defined as the vector of the objective functions $f_1, \dots, f_k$, $k \geq 2$, which each map from $\mathbb{R}^n$ to $\mathbb{R}$, i. e.

$$F : \mathbb{R}^n \to \mathbb{R}^k,\ F(x) = (f_1(x), \dots, f_k(x)).$$

The feasible set S is given as

$$S = \{x \in \mathbb{R}^n : h(x) = 0,\ g(x) \leq 0\}$$

with equality constraints $h : \mathbb{R}^n \to \mathbb{R}^m$, $m \leq n$ and inequality constraints $g : \mathbb{R}^n \to \mathbb{R}^q$. The MOP is called *unconstrained MOP*, if $S = \mathbb{R}^n$. In all the following considerations it is assumed that $F = (f_1, \ldots, f_k)$ consists of at least continuous objective functions.

It has to be explained what is meant by 'min' in the problem (MOP), as a vector-valued function has to be minimized. The following definition which introduces an appropriate partial order on $\mathbb{R}^k$ allows comparisons of vectors (cf. [6]).

Definition 1. Let u, v be two vectors in $\mathbb{R}^k$. Then the vector u is *less than* v (denoted by $u <_p v$) if

$$u_i < v_i \quad \text{for all } i \in \{1, \ldots, k\}.$$

In an analogous way, the relation $\leq_p$ is defined. The vector u is said to *dominate* the vector v if

$$u \leq_p v \text{ and } u_i < v_i \text{ for at least one } i \in \{1, \ldots, k\}.$$

Using the relation $\leq_p$ we define what a solution of (MOP) is.

Definition 2. A point $x^\star \in \mathbb{R}^n$ is called *globally Pareto optimal* for (MOP) (or a *global Pareto point* of (MOP)) if there exists no $x \in S \subseteq \mathbb{R}^n$ with

$$F(x) \leq_p F(x^\star) \text{ and } f_j(x) < f_j(x^\star) \text{ for at least one } j \in \{1, \ldots, k\}.$$

If this property is only valid inside a neighborhood $U(x^\star) \subset S \subseteq \mathbb{R}^n$, then $x^\star$ is called *locally Pareto optimal* (or a *local Pareto point*).

The set of all Pareto points is the *Pareto set* . Following [10], the set of the function values of all Pareto points is called the *Pareto front* .

In the literature, one can find several different names for Pareto optimal solutions. Examples are 'efficient solutions' [9,25], 'noninferior solutions' [10], 'nondominated points' [17], 'vector minimum points' [1], and 'admissible points' [15]. Especially the image of a Pareto optimal solution often is denoted as an *efficient point* .

The following classical result which goes back to Kuhn and Tucker [18] provides a necessary condition for Pareto optimality. The version of the theorem written down here can be found in [16], which itself is a reformulated version of the one given in [14].

Theorem 1 (Kuhn and Tucker, 1951 [18]**)**

Let $x^\star$ be a Pareto optimal solution of (MOP). It is assumed that $\nabla h_i(x^\star), i = 1, \ldots, m$ and $\nabla g_j(x^\star)$ for $j \in \{J : g_J(x^\star) = 0\}$ (the active constraints) are linearly independent. Then there exist vectors $\alpha \in \mathbb{R}^k$ with $\alpha_i \geq 0$ for $i = 1, \ldots, k$ and $\sum_{i=1}^k \alpha_i = 1$, $\gamma \in \mathbb{R}^m$ and $\delta \in \mathbb{R}^q$ with $\delta_j \geq 0$ for $j = 1, \ldots, q$ such that

$$\sum_{i=1}^{k} \alpha_i \nabla f_i(x^\star) + \sum_{j=1}^{m} \gamma_j \nabla h_j(x^\star) + \sum_{l=1}^{q} \delta_l \nabla g_l(x^\star) = 0 \qquad (1)$$
$$\delta_j \cdot g_j(x^\star) = 0, \; \forall j = 1, \ldots, q.$$

Following [20], points $x^\star \in \mathbb{R}^n$ that satisfy the Kuhn-Tucker condition (1) are called *substationary points* . Given a Pareto point $x^\star$ the vector of multipliers α is called the *weight vector corresponding to* $x^\star$.

Obviously the condition in the Kuhn-Tucker theorem is not sufficient for Pareto optimality in general. In the case of convex[1] objective functions, convex inequality constraints and affine[2] equality constraints, it is proven that for $\alpha > 0$ the Kuhn-Tucker conditions are also sufficient [14]. However, numerical methods often make use of this criterion.

An intuitive, classical approach to solve a multiobjective optimization problem is the weighted sums method, also called the 'weighting method', which goes back to Gass and Saaty [12] and Zadeh [29]. It is a very popular approach which makes use of the intuitive idea of converting the multiobjective optimization problem into a single objective one. For this, the objective functions are summed up, each multiplied with an individual weight. More precisely, k weights α_i are chosen such that $\alpha_i \geq 0$ for $i = 1, \dots, k$ and $\sum_{i=1}^k \alpha_i = 1$ and the problem

$$\begin{aligned} &\min_x g_\alpha(x) \\ &\text{s. t. } x \in S \subseteq \mathbb{R}^n, \end{aligned} \tag{2}$$

with $g_\alpha(x) = \sum_{i=1}^k \alpha_i f_i(x)$ is considered.

Varying the weights α_i, different Pareto points can be obtained by solving (2) – in the case of convex objective functions even all Pareto points can be computed in this way. The reason for this is that the shape of the Pareto front is also convex in such a situation. Moreover, the optimization of the weighted sums results in different points on the Pareto front for different weight vectors.

In contrast to this, for nonconvex objective functions the Pareto front can contain nonconvex parts as illustrated in Fig. 1 on the right. Here, the nonconvex part is defined to be a subset of the Pareto front that contains no global optima of the weighted sum of the objectives for every weight vector. Pareto points, which are mapped into the nonconvex part of the Pareto front, can have the same weight vector as other Pareto points whose weighted sum has a smaller value (cf. Fig. 2), as they are only local minima or saddle points of $g_\alpha(x)$.

In [3], Das and Dennis give a trigonometric argument, why – in the case of two objectives – the weighted sums method cannot be used to compute points in the nonconvex part.

It is an interesting question to find out if a Pareto front contains nonconvex parts. If one assumes that the Pareto front is connected, then any nonconvex part contains a region in which the Pareto front 'curves inside the image of the feasible region'. This part of the Pareto front will be called a 'dent'. In Sect. 3 the formal definition of a dent is given and an approach which allows the numerical computation of dents in Pareto fronts is presented.

---

[1] A function $f : \mathbb{R}^n \to \mathbb{R}$ is convex, if $f(\lambda x + (1 - \lambda)y) \leq \lambda f(x) + (1 - \lambda)f(y)$ for all $x, y \in \mathbb{R}^n$, $0 \leq \lambda \leq 1$, see e. g. [27].

[2] A function $f : \mathbb{R}^n \to \mathbb{R}$ is affine, if $f(\lambda x + (1 - \lambda)y) = \lambda f(x) + (1 - \lambda)f(y)$ for all $x, y \in \mathbb{R}^n$, $0 \leq \lambda \leq 1$, see e. g. [27].

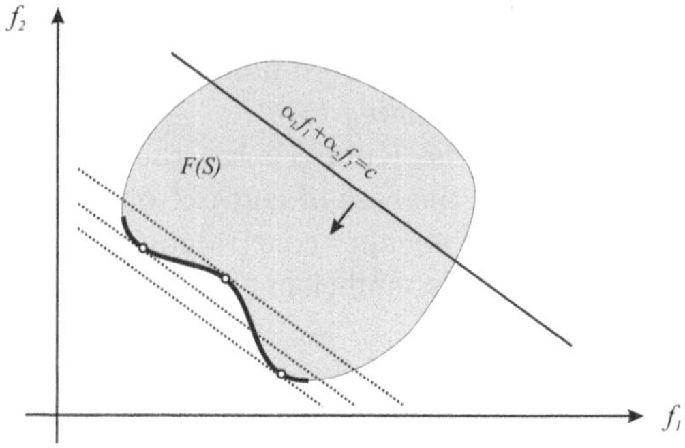

Fig. 2. Schematic illustration of the weighted sums approach

2.2 Parametric Multiobjective Optimization Problems

An unconstrained (one-) parametric multiobjective optimization problem is given as

$$\min_{x}\{F(x,\lambda) : x \in \mathbb{R}^n, \lambda \in [\lambda_{\text{start}}, \lambda_{\text{end}}] \subseteq \mathbb{R}\}, \qquad \text{(ParMOP)}$$

where F is defined as the vector of the objective functions, i. e.

$$F : \mathbb{R}^n \times [\lambda_{\text{start}}, \lambda_{\text{end}}] \to \mathbb{R}^k,\ F(x,\lambda) = (f_1(x,\lambda), \ldots, f_k(x,\lambda)).$$

The solution set of (ParMOP) is a λ-dependent family of Pareto sets.

Every point in this family satisfies the necessary condition of Kuhn and Tucker with respect to the x-variables. As (ParMOP) is an unconstrained multiobjective optimization problem, Theorem 1 reduces to the fact that there exist multipliers $\alpha_1, \ldots, \alpha_k \in \mathbb{R}_{+,0}$ such that

$$H_{\text{KT}}(x,\alpha,\lambda) = \begin{pmatrix} \sum_{i=1}^k \alpha_i \nabla_x f_i(x,\lambda) \\ \sum_{i=1}^k \alpha_i - 1 \end{pmatrix} = 0, \qquad (3)$$

where (x,λ) is a solution of (ParMOP).

Definition 3. If $x \in \mathbb{R}^n$ satisfies the Kuhn-Tucker condition (3) for a specific value of λ, then it is called – as in the non-parametric case – a *substationary point*. The set of all substationary points for the respective value of λ is denoted by S_λ.

2.3 Bifurcation Theory

Bifurcation theory analyzes the behavior of solutions of parameter-dependent systems of equations when they become singular under variation of the parameter. In the context of this work it becomes applicable when considering solutions of the Kuhn–Tucker–Eq. (3) for the parametric, unconstrained case.

It will be shown in Sect. 3 that in a dent border point a zero eigenvalue of the Hessian of g_α occurs, where $g_\alpha(x,\lambda) = \sum_{i=1}^{k} \alpha_i f_i(x,\lambda)$. The Hessian of g_α equals $\frac{\partial}{\partial x} H_{\mathrm{KT}}^{\alpha}$, as $H_{\mathrm{KT}}^{\alpha}(x,\lambda) = \nabla_x g_\alpha(x,\lambda)$. Thus, the implicit function theorem is not applicable to the Kuhn-Tucker equations (with respect to x) in a dent border point. Whenever the Jacobian with respect to x of a parametric system of equations is singular, the structure of the solution set may change. One possibility is that the system of equations has no solution before the singularity occurs, and two solutions afterwards (here, "before" and "afterwards" have to be understood in terms of the values of λ). In this case, the solution curve "turns" at the point $(x^\star, \lambda^\star)$, where the Jacobian with respect to x is singular. More formally, such a *turning point* – which sometimes is also called *saddle-node bifurcation* or *fold* in the literature – is defined as follows:

Definition 4 (Turning point (see [21]))
Consider the solutions of a nonlinear system of equations $H(x,\lambda) = 0$, where $H : \mathbb{R}^N \times \mathbb{R} \to \mathbb{R}^N$. Assume that $(x^\star, \lambda^\star)$ is such a solution which satisfies

(i) there exists $\phi^\star \in \mathbb{R}^N \backslash \{0\}$ with $\ker\left(\frac{\partial}{\partial x} H(x^\star,\lambda^\star)\right) = \mathrm{span}\{\phi^\star\}$,
(ii) $\frac{\partial}{\partial \lambda} H(x^\star,\lambda^\star) \notin \mathrm{im} \frac{\partial}{\partial x} H(x^\star,\lambda^\star)$.

Then, $(x^\star, \lambda^\star)$ is called a *turning point* .

Let $\psi^\star$ be a left eigenvector of the zero eigenvalue of $\frac{\partial}{\partial x} H(x^\star,\lambda^\star)$, i. e.

$$\psi^\star \frac{\partial}{\partial x} H(x^\star,\lambda^\star) = 0.$$

If in addition to (i) and (ii) $\psi^\star \left(\frac{\partial^2}{\partial x^2} H(x^\star,\lambda^\star)\phi^\star\phi^\star\right) \neq 0$, then the point $(x^\star, \lambda^\star)$ is called a *simple turning point* .

In Fig. 3 an example of an equation whose solution curve includes a turning point is sketched. As one can observe, the solution curve turns in the point $(0,0)$.

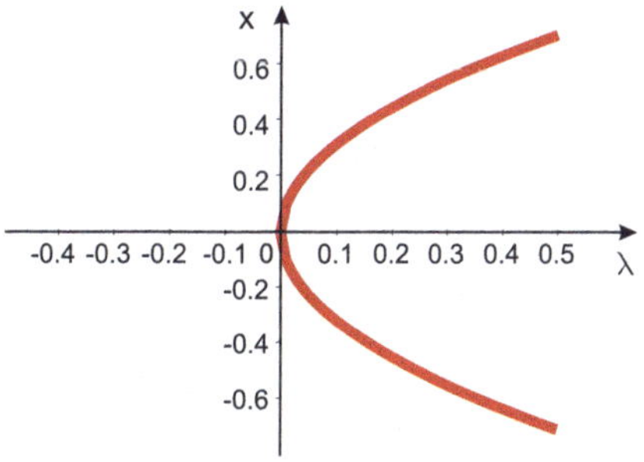

Fig. 3. For the equation $H(x,\lambda) = x^2 - \lambda = 0$, $H : \mathbb{R} \times \mathbb{R} \to \mathbb{R}$, a turning point occurs in the point $(0,0)$

3 Dents in Non-parametric Pareto Fronts

In Fig. 1 it has already been illustrated that the Pareto front may curve inside the image of the feasible region in the case of nonconvex objective functions. As already mentioned in Sect. 2 Pareto points whose images lie in such a dent cannot be computed by using the weighted sums method. The reason is that two or more Pareto points satisfy the Kuhn-Tucker equations with the same weight vector α while the weighted sum $\sum_{i=1}^{k} \alpha_i f_i(x)$ cannot be minimal for all these solutions. The following definition gives a mathematical description of a dent.

Definition 5 (Dent point, Dent preimage)
Let $P \subseteq S$ be the Pareto set of a multiobjective optimization problem $\min_{x\in S} F(x)$ with $F : \mathbb{R}^n \to \mathbb{R}^k$, $F(x) = (f_1(x), \dots, f_k(x))^T$ and f_i at least twice continuously differentiable $\forall\, i = 1, \dots, k$. For $\alpha_i \in [0, 1]$ with $\sum_{i=0}^{k} \alpha_i = 1$ define $g_\alpha : \mathbb{R}^n \to \mathbb{R}$ by

$$g_\alpha(x) = \sum_{i=1}^{k} \alpha_i f_i(x).$$

A point $x^\star \in P$ is called a *dent preimage* if it is a saddle point of g_α. The corresponding point $y^\star = F(x^\star)$ on the Pareto front is called a *dent point* .

Definition 6 (Dent, dent border, complete dent)
Let $P \subseteq S$ be the Pareto set of an at least twice continuously differentiable multiobjective optimization problem $\min_{x\in S} F : \mathbb{R}^n \to \mathbb{R}^k$ and let $PF = F(P)$ be the Pareto front. Let $y^\star \in PF$ be a dent point. Then, the connected component of dent points which includes $y^\star$ is called a *dent corresponding to* $y^\star$, denoted by $D_{y^\star}$:

$$D_{y^\star} = \{y \in PF \mid \exists \delta \geq 0 \text{ and } \exists c : [0, \delta] \to PF, c \text{ continuous, with } c(0) = y^\star, \\ c(\delta) = y, \text{ and } c(s) \text{ is a dent point } \forall\, s \in [0, \delta]\}.$$

A dent $D_{y^\star}$ is called *complete* , if

$$\partial PF \cap \overline{D_{y^\star}} = \emptyset,$$

where ∂PF is the boundary of the Pareto front PF as a subset of $\partial F(S)$ (with the induced topology from $\mathbb{R}^k$).

The boundary $\partial D_{y^\star}$ of a complete dent $D_{y^\star}$ (seen as a subset of PF) is called *dent border* and a boundary point $y_b \in \partial D_{y^\star}$ is called a *dent border point* . A point $x_b \in P$ with $F(x_b) = y_b$ is called a *dent border preimage* of y_b.

Remark 1. In [16] dents have been studied from a differential geometric point of view. In this book it has been shown that – under certain geometrical assumptions on the multiobjective optimization problem – saddle points of g_α occur if and only if the corresponding point on the Pareto front has at least one negative principal curvature.

In [16] it has already been considered what happens during the transition from a minimizer x_1 of g_{α_1} to a saddle point x_2 of g_{α_2} on a connected Pareto front i. e. during the transition of non-dent preimages to dent preimages (α_1 and α_2 denote the weight vectors corresponding to x_1 and x_2, respectively):
Assume that the non-dent preimage x_1 can be connected to the dent preimage x_2 by a continuous curve $\gamma : [0,1] \to P \subseteq S$ with $\gamma(\tau) = (x(\tau), \alpha(\tau))$, $\gamma(0) = (x_1, \alpha_1)$ and $\gamma(1) = (x_2, \alpha_2)$. To each curve point $\gamma(\tau)$ the n-tuple of eigenvalues of $\frac{\partial^2}{\partial x^2} g_\alpha(x)$, denoted by $(e_1(\tau), \dots, e_n(\tau))^T$, is assigned, where α again is the corresponding weight vector to x. This leads to another continuous curve $\tilde{\gamma} : \tau \mapsto (e_1(\tau), \dots, e_n(\tau))^T$ corresponding to γ. As x_1 is a minimizer of g_{α_1}, $\tilde{\gamma}(0) > 0$. In the saddle point x_2 of g_{α_2}, there exists an index $i \in \{1, \dots, n\}$ such that $e_i(1) < 0$. Because of the continuity of $\tilde{\gamma}$ there must exist $\bar{\tau} \in [0,1]$ with $e_i(\bar{\tau}) = 0$.

To sum up, in dent border points a zero-eigenvalue of the Hessian of g_α occurs.

Definition 7 (Simple dent border point/preimage)
Let $P \subseteq S$ be the Pareto set of an at least twice continuously differentiable multiobjective optimization problem $F : \mathbb{R}^n \to \mathbb{R}^k$ and let $PF = F(P)$ be the Pareto front. Let $y_b^\star \in PF$ be a dent border point and let $x_b^\star \in P$ be a dent border preimage of $y_b^\star$.

Then, $y_b^\star$ is called a *simple dent border point* if the zero eigenvalue of the Hessian $g_\alpha''(x_b^\star)$ is simple. In this case, $x_b^\star$ is called a *simple dent border preimage.*

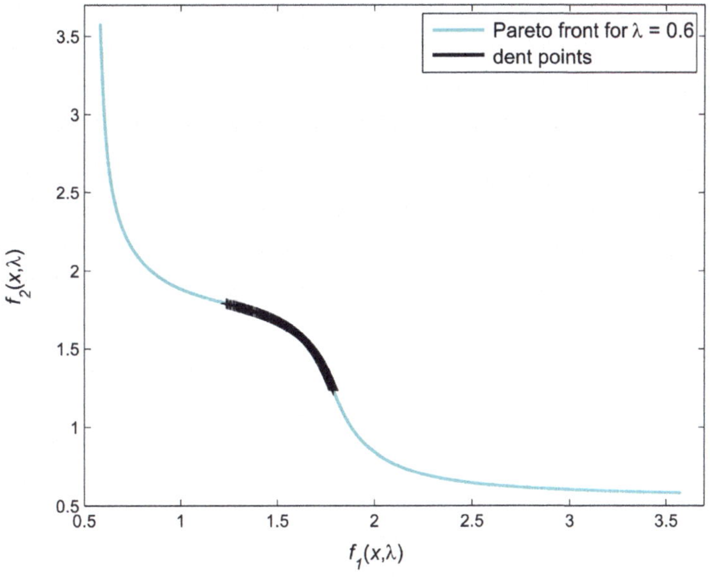

Fig. 4. Pareto front and dent points (black) for Example 1

Example 1 (Computation of dents)
Consider the bi-objective optimization problem defined by the two objectives

$$f_1(x,\lambda) = \frac{1}{2}(\sqrt{1+(x_1+x_2)^2} + \sqrt{1+(x_1-x_2)^2} + x_1 - x_2) + \lambda \cdot e^{-(x_1-x_2)^2}$$
$$f_2(x,\lambda) = \frac{1}{2}(\sqrt{1+(x_1+x_2)^2} + \sqrt{1+(x_1-x_2)^2} - x_1 + x_2) + \lambda \cdot e^{-(x_1-x_2)^2}$$

with a fixed value $\lambda = 0.6$ and $x = (x_1, x_2)$. Then, the Pareto set can be approximated e. g. by use of the set-oriented techniques.

The algorithm returns a set of boxes that covers the Pareto set. Within these boxes, a number of test points is evaluated, the best of which are in the following considered as the Pareto set. By solving the Kuhn-Tucker equations of $F = (f_1, f_2)$ for each of these points, the corresponding weight vectors $\alpha \in \mathbb{R}^k$ can be computed. Then, the eigenvalues of the Hessians of the weighted sums of the objectives are determined. All points x, in which both eigenvalues > 0 and eigenvalues < 0 exist, are dent preimages. The Pareto front and the resulting dent points for this example are visualized in Fig. 4.

4 Evolution of Dents in Parameter-Dependent Pareto Fronts

In the previous section dents in Pareto fronts have been defined motivated by the fact that these points cannot be computed by the weighted sums method . Also, dent border points have been defined. When considering parametric multiobjective optimization problems, naturally the question arises, how dents and especially dent border points evolve. The Kuhn-Tucker Eq. (3) provide a necessary condition for Pareto optimality. Within this section this parametric system of equations will be analyzed in order to obtain results about the local behavior of parameter-dependent Pareto fronts.

Solutions of parametric systems of equations have already been widely studied in bifurcation theory. The first part of this section deals with the study of properties of dent border points. It will be proven that under certain assumptions dent border preimages are turning points of the Kuhn-Tucker equations. In the second part of this section, several numerical examples for parametric multiobjective optimization problems in which dents occur are given.

Within this section it is assumed that the objective functions are at least twice continuously differentiable. Only points $x \in P_\lambda$ are considered for which the corresponding weight vector α is an element of $(0,1)^k$. Define $H^\alpha_{\mathrm{KT}} : \mathbb{R}^n \times \mathbb{R} \to \mathbb{R}^n$ by

$$H^\alpha_{\mathrm{KT}}(x,\lambda) = \sum_{i=1}^{k} \alpha_i \nabla_x f_i(x,\lambda).$$

4.1 Properties of Dent Border Points

First, it will be shown that dent border preimages can be characterized as certain turning points of the Kuhn-Tucker equations

Proposition 1. *Let $P_\lambda \subseteq S_\lambda$ be the Pareto set of a parametric multi-objective optimization problem* $\min F : \mathbb{R}^n \times \mathbb{R} \to \mathbb{R}^k$ *with* $F(x,\lambda) = (f_1(x,\lambda),\dots,f_k(x,\lambda))^T$. *Let $x^\star \in P_{\lambda^\star}$ be a simple dent border preimage. Let $\alpha^\star$ denote the weight vector corresponding to $x^\star$ and assume that the Jacobian $H_{\mathrm{KT}}^{\alpha^\star\prime}(x,\lambda)$ has full rank.*

Then, $(x^\star,\lambda^\star)$ is a turning point of $H_{\mathrm{KT}}^{\alpha^\star}(x,\lambda)$ with respect to λ.

Proof. It has been shown in [16] (see also Sect. 3) that dent border preimages $x^\star$ are solutions of the Kuhn-Tucker equations $H_{\mathrm{KT}}^{\alpha^\star}(x^\star,\lambda^\star) = 0$ in which $\frac{\partial^2}{\partial x^2} g_{\alpha^\star}(x^\star,\lambda^\star)$ is singular. Thus, there exists an eigenvector $\phi^\star$ of $\frac{\partial^2}{\partial x^2} g_{\alpha^\star}(x^\star,\lambda^\star)$ with

$$\left(\frac{\partial^2}{\partial x^2} g_{\alpha^\star}(x^\star,\lambda^\star)\right)\phi^\star = 0. \tag{4}$$

From the assumption that the dent border preimage is simple, i. e. exactly one eigenvalue of $\frac{\partial^2}{\partial x^2} g_{\alpha^\star}(x^\star,\lambda^\star)$ equals zero (cf. Definition 7), it directly follows that

$$\dim\ker\left(\frac{\partial^2}{\partial x^2} g_{\alpha^\star}(x^\star,\lambda^\star)\right) = 1. \tag{5}$$

As $H_{\mathrm{KT}}^{\alpha}(x,\lambda) = \nabla_x g_\alpha(x,\lambda)$, and thus

$$\frac{\partial}{\partial x} H_{\mathrm{KT}}^{\alpha}(x,\lambda) = \frac{\partial}{\partial x}\left(\nabla_x g_\alpha(x,\lambda)\right) = \frac{\partial^2}{\partial x^2} g_\alpha(x,\lambda),$$

(4) is equivalent to

$$\frac{\partial}{\partial x} H_{\mathrm{KT}}^{\alpha^\star}(x^\star,\lambda^\star)\varphi^\star = 0$$

and (5) is the same as

$$\dim\ker\left(\frac{\partial}{\partial x} H_{\mathrm{KT}}^{\alpha^\star}(x^\star,\lambda^\star)\right) = 1.$$

Thus, property (i) of Definition 4 is proven for $H(x,\lambda) = H_{\mathrm{KT}}^{\alpha^\star}(x,\lambda)$.

Property (ii) of Definition 4 directly follows from the assumption that the Jacobian $H_{\mathrm{KT}}^{\alpha^\star\prime}(x,\lambda)$ has full rank, i. e. rank n: if the vector $\frac{\partial}{\partial\lambda} HKT^{\alpha^\star}(x^\star,\lambda^\star)$ were in the image of the $n\times n$-matrix $\frac{\partial}{\partial x} H_{\mathrm{KT}}^{\alpha^\star}(x^\star,\lambda^\star)$, then the rank of the Jacobian $H_{\mathrm{KT}}^{\alpha^\star\prime}(x,\lambda)$ were $n-1$, in contradiction to the assumption. Thus, it has to be true that $\left(\frac{\partial}{\partial\lambda} H_{\mathrm{KT}}^{\alpha^\star}(x^\star,\lambda^\star)\right) \notin \mathrm{im}\frac{\partial}{\partial x} H_{\mathrm{KT}}^{\alpha^\star}(x^\star,\lambda^\star)$.

To sum up, both properties given in Definition 4 are satisfied, and thus $(x^\star,\lambda^\star)$ is a turning point of $H_{\mathrm{KT}}^{\alpha^\star}(x,\lambda)$ with respect to λ. □

Remark 2. Using the notation of the proof of Proposition 1 one observes that the matrix $\frac{\partial}{\partial x} H_{\mathrm{KT}}^{\alpha^\star}$ is symmetric, as it is the Hessian of the weighted sums function g_α. Thus, $\psi^\star = (\phi^\star)^T$ is a left eigenvector of $\frac{\partial}{\partial x} H_{\mathrm{KT}}^{\alpha^\star}(x^\star, \lambda^\star)$. It follows that, if additionally to (i) and (ii) of Definition 4

$$(\phi^\star)^T \left(\frac{\partial^2}{\partial x^2} H_{\mathrm{KT}}^{\alpha^\star}(x^\star, \lambda^\star) \right) \phi^\star \phi^\star \neq 0,$$

then $(x^\star, \lambda^\star)$ is a simple turning point of $H_{\mathrm{KT}}^{\alpha^\star}(x, \lambda)$ with respect to λ.

To sum up, a dent border preimage can be obtained by solving the system of equations

$$\begin{aligned} H_{\mathrm{KT}}^{\alpha^\star}(x, \lambda) &= 0 \\ \frac{\partial}{\partial x} H_{\mathrm{KT}}^{\alpha^\star}(x, \lambda) \cdot \phi &= 0 \\ l^T \phi - 1 &= 0 \end{aligned} \tag{6}$$

with an arbitrary but fixed vector $l \in \mathbb{R}^n$ which satisfies $l^T \phi^\star \neq 0$ and has non-zero entries, and $x, \phi \in \mathbb{R}^n$, $\lambda \in \mathbb{R}$. In the literature, this system of equations is also called the *extended system* of $H_{\mathrm{KT}}^{\alpha^\star}(x, \lambda)$ (cf. [21]).

Remark 3. A family of dent border preimages can be obtained by solving

$$\begin{aligned} \nabla_x g_\alpha(x, \lambda) &= 0 \\ \sum_{i=1}^{k} \alpha_i - 1 &= 0 \\ \frac{\partial^2}{\partial x^2} g_\alpha(x, \lambda) \cdot \phi &= 0 \\ l^T \phi - 1 &= 0 \end{aligned} \tag{7}$$

with an arbitrary fixed vector $l \in \mathbb{R}^n$ which satisfies $l^T \phi^\star \neq 0$ and has non-zero entries, $x, \phi \in \mathbb{R}^n$, $\lambda \in \mathbb{R}$ and $\alpha \in \mathbb{R}^k$ with $\alpha_i > 0 \; \forall i = 1, \ldots, k$.

4.2 Numerical Examples

In the following, several new examples for parametric multiobjective optimization problems are presented and its construction is motivated. The examples all have in common that the corresponding Pareto fronts contain dents for specific values of the external parameter λ. Moreover, under the variation of λ, dents originate or vanish (cf. Examples 2 and 4), or dents double or merge (cf. Example 3).

Example 2. We again consider the bi-objective optimization problem defined by the two objectives

$$f_1(x_1, x_2, \lambda) = \frac{1}{2}(\sqrt{1 + (x_1 + x_2)^2} + \sqrt{1 + (x_1 - x_2)^2} + x_1 - x_2) + \lambda \cdot e^{-(x_1 - x_2)^2}$$
$$f_2(x_1, x_2, \lambda) = \frac{1}{2}(\sqrt{1 + (x_1 + x_2)^2} + \sqrt{1 + (x_1 - x_2)^2} - x_1 + x_2) + \lambda \cdot e^{-(x_1 - x_2)^2}$$

which we have already seen in Example 1.

Before we are going to examine this example numerically, it is worthwhile to note that it can be understood geometrically and/or analytically. To see the picture, we can use the vectors $q = (1, 1)^T$ and $q^\perp = (1, -1)^T$ as a basis of $\mathbb{R}^2$ and new coordinates $u_1 = x_1 + x_2$ and $u_2 = x_1 - x_2$, so that for $x = (x_1, x_2)^T$ we have $x = 1/2 \cdot (u_1 \cdot q + u_2 \cdot q^\perp)$. Then we can write the objective as

$$\begin{aligned} F(u_1, u_2, \lambda) &= \left(\frac{1}{2} \cdot \left(\sqrt{1 + u_1^2} + \sqrt{1 + u_2^2}\right) + \lambda \cdot e^{-u_2^2}\right) \cdot q + u_2 \cdot q^\perp \\ &= \underbrace{\frac{1}{2} \cdot \sqrt{1 + u_1^2} \cdot q}_{=:F^1(u_1)} + \underbrace{\left(\frac{1}{2} \cdot \sqrt{1 + u_2^2} + \lambda \cdot e^{-u_2^2}\right) \cdot q + u_2 \cdot q^\perp}_{=:F^2(u_2,\lambda)} \end{aligned}$$

This means the $q^\perp$-component of $F(u_1, u_2, \lambda)$ is simply $u_2 \cdot q^\perp$, while the q-component is a sum of three terms, two of which are functions of u_2 only, and only one of which depends on λ. Now an easy computation shows that the q-component of F^2 is a convex function for $\lambda < 1/4$ and that it has a non-convex part around $u_2 = 0$ for $\lambda > 1/4$. This is the reason for the generation of a dent in the Pareto front. To see this, observe that the image of F^2 determines the form of the boundary of the image of F, as the F_1 term adds a component in positive q-direction only, i.e. a component that moves the image "further inside" the positive quadrant of $\mathbb{R}^2$, and has its minimum for $u_1 = 0$. Thus by adjusting the value of λ, we can control whether the boundary of the image of F is given by a convex function, that is, whether the Pareto front has a dent or not.

In Fig. 5 the Pareto sets of these two objectives for $(x_1, x_2) \in [-1.3, 1.3]^2$ are plotted for different values of the external parameter λ: for $\lambda = 0$ (green), $\lambda = 0.2$ (gray), $\lambda = 0.4$ (light blue), $\lambda = 0.6$ (cyan), and for $\lambda = 0.8$ (magenta). We see that they are all part of the line given by $u_1 = 0$.

In the same figure, an example for a λ-dependent solution path of the Kuhn-Tucker equations with a fixed weight vector $\alpha^\star \approx (0.25, 0.75)$ which corresponds to the dent border preimage $(x_1^\star, x_2^\star) \approx (-0.28, 0.28)$ of the dent border point $(y_1^\star, y_2^\star) \approx (1.23, 1.79)$, located on the Pareto front for $\lambda^\star = 0.6$, is visualized for $\lambda \in [0, 1]$ (black paths). The paths have been computed with the help of the software package AUTO2000 [7]. As one can observe, $y^\star$ is indeed a simple turning point of $H_{\mathrm{KT}}^{\alpha^\star}(x, \lambda)$ with respect to λ. Figure 6 shows the same results in objective space. Here one clearly observes that the Pareto front does not have a dent for $\lambda < 0.25$ and does have a dent for $\lambda > 0.25$.

In Fig. 7 the entire curve of dent border points in objective space is plotted. To compute this λ-dependent solution curve (red), again the software package AUTO2000 has been used. One can observe that the point $(y_1^\star, y_2^\star, \lambda^\star) =$

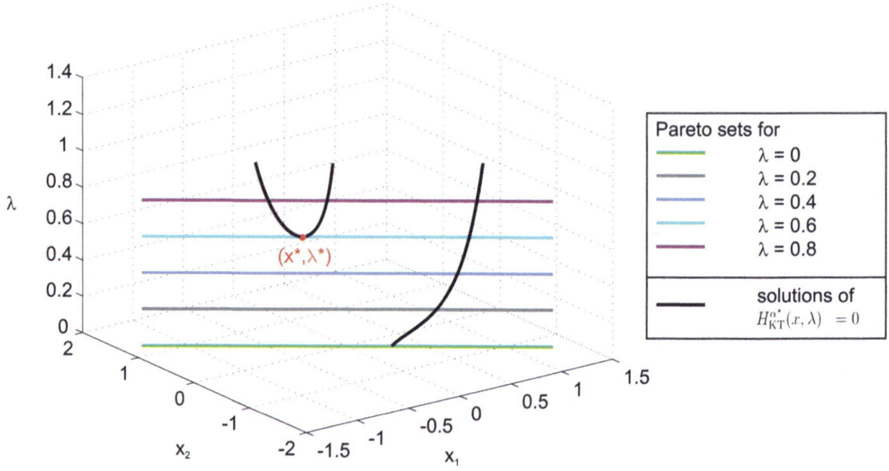

Fig. 5. Visualization of some Pareto sets and the solution curve of $H^{\alpha^\star}_{\mathrm{KT}}(x, \lambda) = 0$ (black) for the dent border preimage $x^\star$ (red) for Example 2

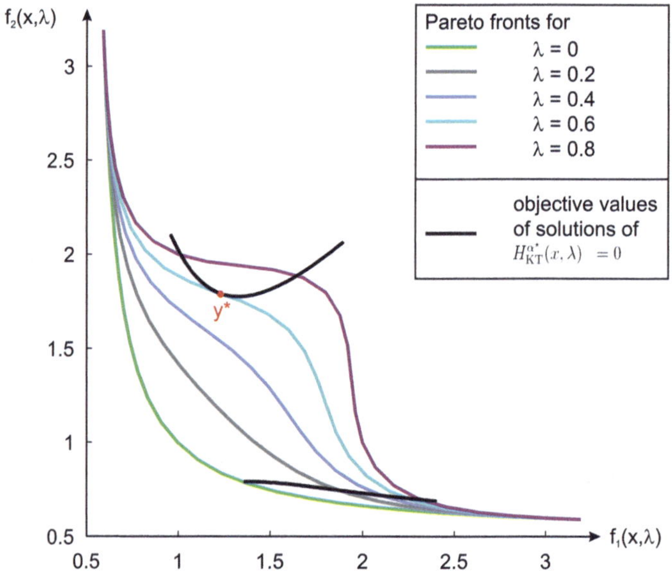

Fig. 6. Visualization of some Pareto fronts and the image of the solution curve of $H^{\alpha^\star}_{\mathrm{KT}}(x, \lambda) = 0$ (black) for a dent border point $y^\star$ (red) for Example 2

$(1.25, 1.25, 0.25)$ (with the corresponding weight vector $\alpha^\star = (0.5, 0.5)$), marked by a black dot, is specific: in this point a dent originates, i. e. for $\lambda < \lambda^\star$ the Pareto front contains no dent whereas for $\lambda > \lambda^\star$ a dent is contained in the Pareto front.

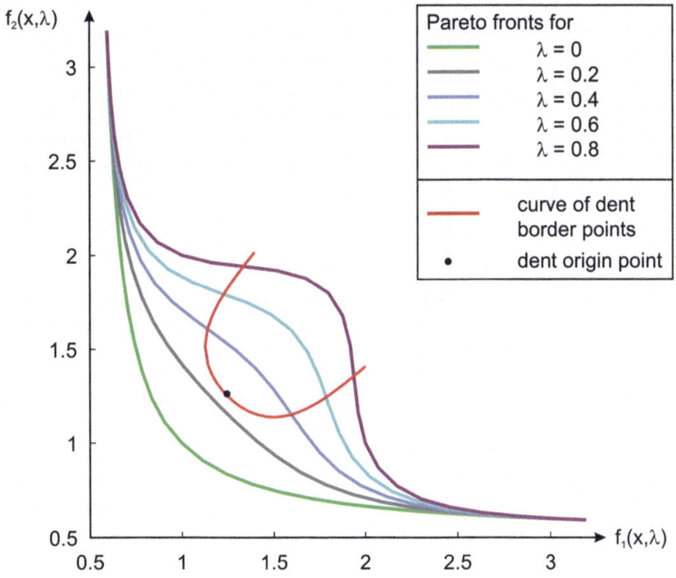

Fig. 7. Some Pareto fronts, the curve of dent border points (red) and the point in which the dent originates (black dot) for Example 2

In Fig. 8 the solutions of $H_{\mathrm{KT}}^{\alpha^\star}(x,\lambda) = 0$ with $\alpha^\star = (0.5, 0.5)$, i. e. the λ-dependent path containing the specific point in which a non-dent point changes into a dent point, are visualized for $\lambda \in [0,1]$. One can observe that a pitchfork bifurcation[3] occurs in this point. Figure 9 shows the same results in objective space.

Example 3. Consider the bi-objective optimization problem defined by the two objectives

$$f_1(x_1, x_2, \lambda) = \sqrt{1+x_1^2} + \sqrt{1+x_2^2} + e^{-(x_2-\lambda)^2} + e^{-(x_2+\lambda)^2} - x_2$$
$$f_2(x_1, x_2, \lambda) = \sqrt{1+x_1^2} + \sqrt{1+x_2^2} + e^{-(x_2-\lambda)^2} + e^{-(x_2+\lambda)^2} + x_2.$$

Using the same notation as in Example 2, we can write the objective as

$$F(x_1, x_2, \lambda) = \sqrt{1+x_1^2} \cdot q + \underbrace{\left(\sqrt{1+x_2^2} + e^{-(x_2-\lambda)^2} + e^{-(x_2+\lambda)^2}\right) \cdot q + x_2 \cdot q^\perp}_{=:F^2(x_2,\lambda)}$$

and apply a similar analysis. Here the q-component of F^2 is non-convex for every λ (in fact, for $\lambda = 0$ we have the same situation as for $\lambda = 1$ in Example 2),

[3] The definition and statements about properties of pitchfork bifurcations can be found in [13] and [26], for example.

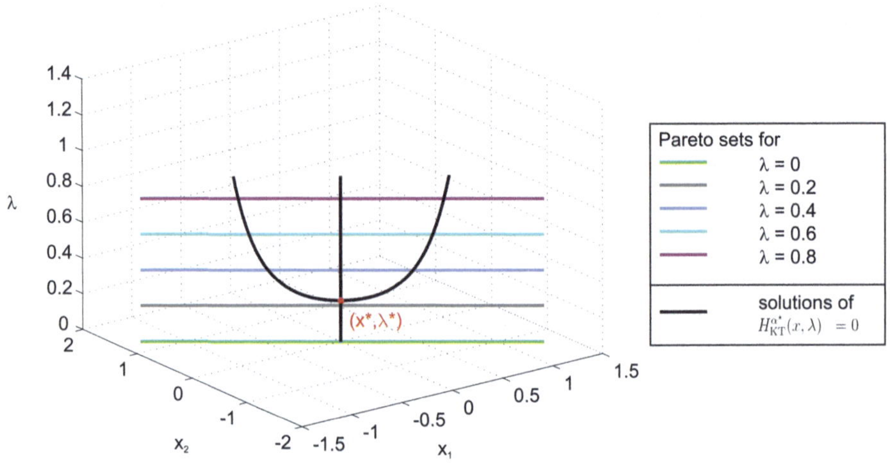

Fig. 8. Visualization of some Pareto sets and the λ-dependent solution curve of $H_{\mathrm{KT}}^{\alpha^\star}(x, \lambda) = 0$ with $\alpha^\star = (0.5, 0.5)$ (black) for Example 2

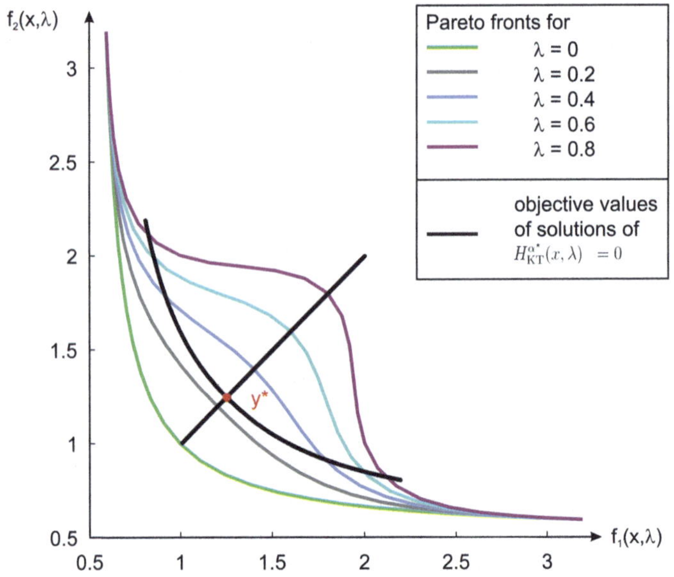

Fig. 9. Visualization of some Pareto fronts and the image of the solution curve of $H_{\mathrm{KT}}^{\alpha^\star}(x, \lambda) = 0$ with $\alpha^\star = (0.5, 0.5)$ (black) for Example 2

and a variation of λ results in the movement of the "peaks" of the exponential terms. Thus we obtain, for $|\lambda|$ sufficiently large, two separate non-convex parts of the q-component of F^2, while there is only one non-convex part for $\lambda = 0$. Correspondingly, one dent in the Pareto front for $\lambda = 0$ will split into two dents for a larger value of λ.

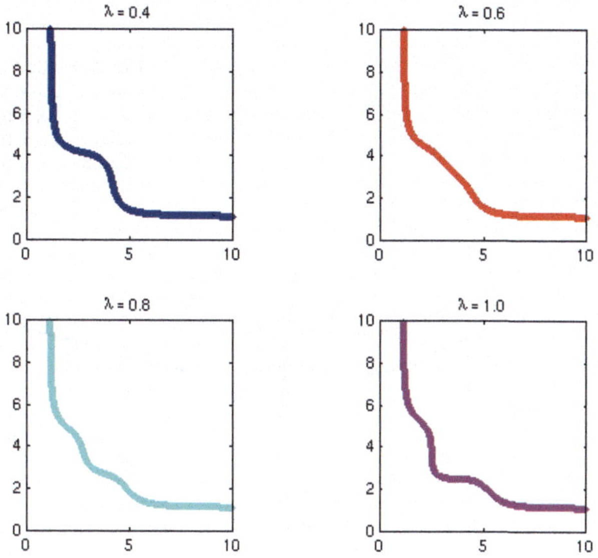

Fig. 10. Pareto fronts for the multiobjective optimization problem given in Example 3 for different values of λ

In Fig. 10 the Pareto fronts which result from the minimization of these two objectives are plotted for different values of λ. As one can observe the Pareto front contains one dent for $\lambda = 0.4$, for example. Under the variation of λ it changes into two dents (cp. for instance $\lambda = 0.8$). In between, there is a specific point in which the dent splits up into two dents, which is given by $(x_1, x_2, \lambda) \approx (0, 0, 0.5716)$ with the corresponding weight vector $\alpha^\star = (0.5, 0.5)$. In Fig. 11 the solutions of the Kuhn-Tucker equations for the fixed weight vector $\alpha^\star = (0.5, 0.5)$ are sketched. One can observe that in the point where the dent splits up into two dents a pitchfork bifurcation occurs.

Example 4. Consider the three-objective optimization problem defined by the following three objectives

$$f_1(x_1, x_2, x_3, \lambda) = \sqrt{1 + x_1^2} + \sqrt{1 + x_2^2} + \sqrt{1 + x_3^2} + \lambda \cdot e^{-(x_2^2 + x_3^2)} + \sqrt{2} x_2$$

$$f_2(x_1, x_2, x_3, \lambda) = \sqrt{1 + x_1^2} + \sqrt{1 + x_2^2} + \sqrt{1 + x_3^2} + \lambda \cdot e^{-(x_2^2 + x_3^2)} - \frac{\sqrt{2}}{2} x_2 + \sqrt{\frac{3}{2}} x_3$$

$$f_3(x_1, x_2, x_3, \lambda) = \sqrt{1 + x_1^2} + \sqrt{1 + x_2^2} + \sqrt{1 + x_3^2} + \lambda \cdot e^{-(x_2^2 + x_3^2)} - \frac{\sqrt{2}}{2} x_2 - \sqrt{\frac{3}{2}} x_3.$$

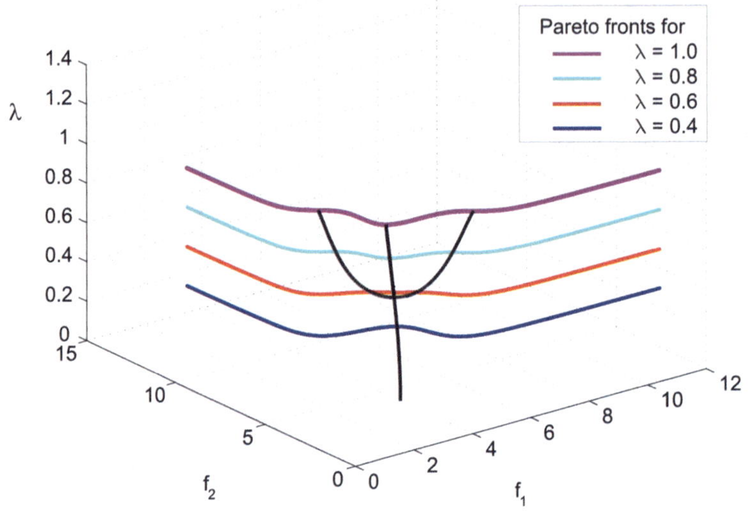

Fig. 11. Pareto fronts and solutions of the Kuhn-Tucker equations for $\alpha^\star = (0.5, 0.5)$ in (f_1, f_2, λ)-space for Example 3

In this case, the analysis is somewhat more complicated. Using the vector $q = (1,1,1)^T$, we can write

$$F(x_1,x_2,x_3,\lambda) = \left(\sqrt{1+x_1^2} + \sqrt{1+x_2^2} + \sqrt{1+x_3^2} + \lambda \cdot e^{-(x_2^2+x_3^2)} - \frac{\sqrt{2}}{3}\cdot x_2\right)\cdot q$$
$$+ \underbrace{\begin{pmatrix} \frac{8}{3\sqrt{2}} & 0 \\ -\frac{4}{3\sqrt{2}} & \sqrt{\frac{3}{2}} \\ -\frac{4}{3\sqrt{2}} & -\sqrt{\frac{3}{2}} \end{pmatrix}}_{=:q^\perp} \cdot \begin{pmatrix} x_2 \\ x_3 \end{pmatrix}$$

where, similarly to Examples 2 and 3, the matrix $q^\perp$ spans the orthogonal complement to q. Thus we see that again the q-component of F consists of a convex function independent of λ and a λ-dependent non-convex term that introduces a dent into the Pareto front for sufficiently large values of λ.

Figure 12 shows the Pareto fronts which result when minimizing these three objectives for $x \in [-2,2]^3$ for different values of λ. One can observe that under the variation of λ a dent originates.

Remark 4. It has been observed in Examples 2 and 3 that pitchfork bifurcations occur in those points where – under the variation of λ – a dent originates in the Pareto front PF_λ, or a dent splits up into two dents, respectively. Pitchfork bifurcations typically occur if the system of equations $H(x,\lambda) = 0$, in our case $H_{\mathrm{KT}}^{\alpha^\star}(x,\lambda) = 0$, includes a symmetry of the form

$$H(Sx,\lambda) = SH(x,\lambda), \qquad \text{(Z2)}$$

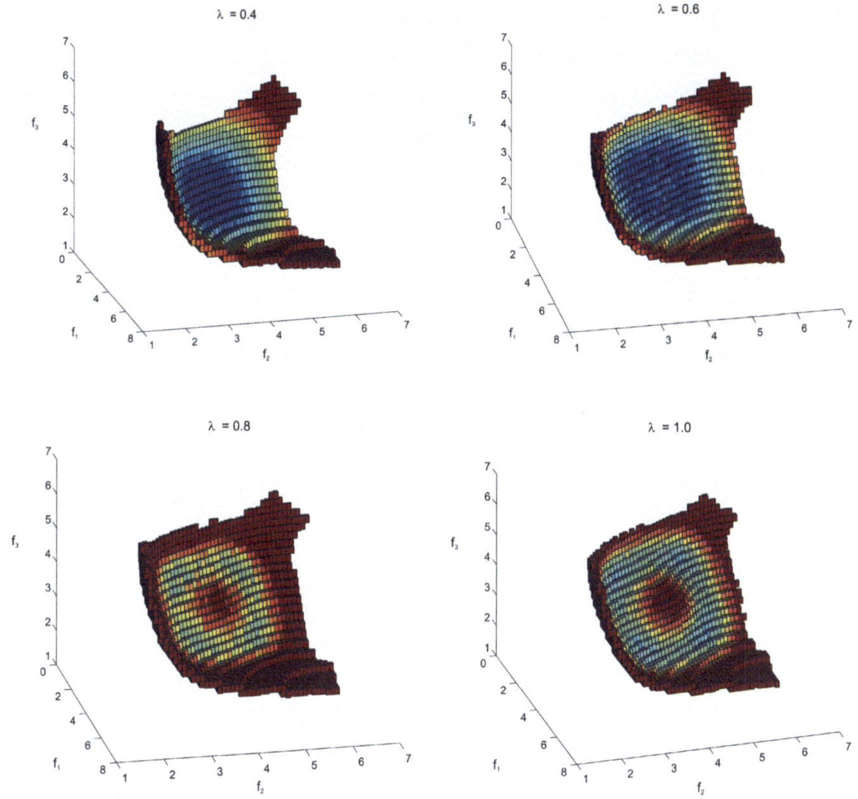

Fig. 12. Pareto fronts for the multiobjective optimization problem given in Example 4 for different values of the parameter λ

where S is a suitable symmetry matrix with $S \neq \mathbb{1}$ and $S^2 = \mathbb{1}$ (cf. [26]).

In the examples mentioned above, indeed symmetries occur. The Kuhn-Tucker equations of the objective functions given in Example 2 have a $\mathbb{Z}_2$-symmetry for $\alpha^\star = (0.5, 0.5)$. In this case, possible symmetry matrices are given as

$$S_1 = \begin{pmatrix} 0 & 1 \\ 1 & 0 \end{pmatrix} \text{ and } S_2 = \begin{pmatrix} -1 & 0 \\ 0 & -1 \end{pmatrix}.$$

The Kuhn-Tucker equations of the objective functions given in Example 3 satisfy the symmetry condition (Z2) with

$$S_1 = \begin{pmatrix} 1 & 0 \\ 0 & -1 \end{pmatrix} \text{ if } \alpha^\star = (0.5, 0.5), \text{ and}$$

$$S_2 = \begin{pmatrix} -1 & 0 \\ 0 & 1 \end{pmatrix} \text{ independent of the weight vector } \alpha.$$

The Kuhn-Tucker equations for the objective functions given in Example 4 also satisfy the symmetry condition (Z2) with

$$S = \begin{pmatrix} -1 & 0 & 0 \\ 0 & 1 & 0 \\ 0 & 0 & 1 \end{pmatrix}$$

for arbitrary weight vectors $\alpha^\star$.

5 Conclusion and Outlook

In this work the occurrence of *dents* in Pareto fronts has been studied. A formal definition of a dent has been introduced. Points at the border of a (complete) dent have a significant property. In these points a zero eigenvalue of the Hessian of the weighted sum of the objectives occurs. Thus, dent border points are solutions of a certain system of equations. Given a sufficiently smooth multiobjective optimization problem it is possible to find out if dent border points, and thus also possibly dents, occur in the Pareto front by solving this system of equations. Consequently, information about the geometry of the Pareto front can be obtained without computing the entire Pareto set. This information can for example serve as a criterion for the choice of the algorithm one wants to use for solving the multiobjective optimization problem.

Based on theoretical results from bifurcation theory, parameter-dependencies in multiobjective optimization problems have been studied in this chapter. It has been proven that dent border points are turning points of the Kuhn-Tucker equations with a fixed weight vector corresponding to the dent border point.

Several examples for parametric multiobjective optimization problems have been constructed in which dents occur. It is still an open question what happens if a dent originates or vanishes under the variation of the external parameter. The examples given at the end of Sect. 4 lead to the conjecture that in this case pitchfork bifurcations of the Kuhn-Tucker equations occur. However, the theoretical analysis of this statement has to be addressed in future work.

Acknowledgements. This work was partly developed in the course of the Collaborative Research Center 614 – Self-Optimizing Concepts and Structures in Mechanical Engineering – University of Paderborn, and was partly funded by the Deutsche Forschungsgemeinschaft.

References

1. Bigi, G., Castellani, M.: Uniqueness of KKT multipliers in multiobjective optimization. Appl. Math. Lett. **17**(11), 1285–1290 (2004)
2. Coello Coello, C.A., Lamont, G., Veldhuizen, D.V.: Evolutionary Algorithms for Solving Multi-objective Optimization Problems, 2nd edn. Springer, Berlin (2007)
3. Das, I., Dennis, J.: A closer look at drawbacks of minimizing weighted sums of objectives for Pareto set generation in multicriteria optimization problems. Struct. Optim. **14**(1), 63–69 (1997)

4. Das, I., Dennis, J.E.: Normal boundary intersection: a new method for generating the Pareto surface in nonlinear multicriteria optimization problems. SIAM J. Optim. **8**, 631–657 (1998)
5. Deb, K.: Multi-objective Optimization using Evolutionary Algorithms. Wiley-Interscience Series in Systems and Optimization. John Wiley, Chichester (2001)
6. Dellnitz, M., Schütze, O., Hestermeyer, T.: Covering Pareto sets by multilevel subdivision techniques. J. Optim. Theory Appl. **124**(1), 113–136 (2005)
7. Doedel, E.J., Champneys, A.R., Paffenroth, R.C., Fairgrieve, T.F., Kuznetsov, Y.A., Oldeman, B.E., Sandstede, B., Wang, X.J.: AUTO2000: Continuation and bifurcation software for ordinary differential equations (with homcont). Technical report California Institute of Technology, Pasadena, California USA (2000)
8. Ehrgott, M.: Multicriteria Optimization. Lecture Notes in Economics and Mathematical Systems, vol. 491. Springer, Berlin (2000)
9. Ehrgott, M.: Multicriteria Optimization, second edn. Springer, Berlin (2005)
10. Ehrgott, M., Gandibleaux, X. (eds.): Multiple Criteria Optimization, International Series in Operations Research & Management Science, vol. 52. Kluwer Academic Publishers, Boston (2002)
11. Fonseca, C.M., Fleming, P.J., Zitzler, E., Deb, K., Thiele, L.: Evolutionary multi-criterion optimization. In: Second International Conference EMO 2003. Springer (2003)
12. Gass, S., Saaty, T.: The computational algorithm for the parametric objective function. Naval Res. Logist. Q. **2**, 39–45 (1955)
13. Golubitsky, M., Schaeffer, D.G.: Singularities and groups in bifurcation theory. Vol. I, Applied Mathematical Sciences, vol. 51. Springer, New York (1985)
14. Göpfert, A., Nehse, R.: Vektoroptimierung. Teubner Verlagsgesellschaft Leipzig (1990)
15. Guddat, J., Guerra Vasquez, F., Tammer, K., Wendler, K.: Multiobjective and Stochastic Optimization Based on Parametric Optimization. Akademie-Verlag, Berlin (1985)
16. Hillermeier, C.: Nonlinear Multiobjective Optimization: A Generalized Homotopy Approach. Birkhäuser (2001)
17. Kim, B., Gel, E., Fowler, J., Carlyle, W., Wallenius, J.: Evaluation of nondominated solution sets for k-objective optimization problems: an exact method and approximations. Eur. J. Oper. Res. **173**(2), 565–582 (2006)
18. Kuhn, H., Tucker, A.: Nonlinear programming. In: Neumann, J. (ed.) Proceedings of 2nd Berkeley Symposium of Mathematical Statistics and Probability, pp. 481–492 (1951)
19. Martín, A., Schütze, O.: Pareto tracer: a predictor-corrector method for multi-objective optimization problems. Eng. Optim. **50**(3), 516–536 (2018). https://doi.org/10.1080/0305215X.2017.1327579
20. Miettinen, K.: Nonlinear Multiobjective Optimization. Kluwer Academic Publishers, Berlin (1999)
21. Moore, G., Spence, A.: The calculation of turning points of nonlinear equations. SIAM J. Numer. Anal. **17**(4), 567–576 (1980)
22. Schäffler, S., Schultz, R., Weinzierl, K.: A stochastic method for the solution of unconstrained vector optimization problems. J. Optim. Theory Appl. **114**(1), 209–222 (2002)
23. Schütze, O., Cuate, O., Martín, A., Peitz, S., Dellnitz, M.: Pareto explorer: a global/local exploration tool for many-objective optimization problems. Eng. Optim. **52**, 1–24 (2019). https://doi.org/10.1080/0305215X.2019.1617286

24. Schütze, O., Witting, K., Ober-Blöbaum, S., Dellnitz, M.: Set oriented methods for the numerical treatment of multiobjective optimization problems. In: Tantar, E., Tantar, A.A., Bouvry, P., Moral, P.D., Legrand, P., Coello Coello, C.A., Schütze, O. (eds.) EVOLVE – A Bridge Between Probability, Set Oriented Numerics, and Evolutionary Computation, pp. 187–219. Springer (2013)
25. Steuer, R.E.: Multiple Criteria Optimization: Theory, Computation, and Application. Wiley Series in Probability & Mathematical Statistics, John Wiley Inc. (1986)
26. Werner, B., Spence, A.: The computation of symmetry-breaking bifurcation points. SIAM J. Numer. Anal. **21**(2), 388–399 (1984)
27. Werner, D.: Funktionalanalysis. Springer, Heidelberg (2005)
28. Witting, K.: Numerical algorithms for the treatment of parametric multiobjective optimization problems and applications. Dissertation, Universität Paderborn (2012). http://digital.ub.uni-paderborn.de/urn/urn:nbn:de:hbz:466:2-8617
29. Zadeh, L.: Optimality and non-scalar-valued performance criteria. IEEE Trans. Autom. Control **8**, 59–60 (1963)
30. Zitzler, E., Laumanns, M., Thiele, L.: SPEA2: Improving the strength of pareto evolutionary algorithms for multiobjective optimization. In: Evolutionary Methods for Design, Optimisation and Control with Application to Industrial Problems (EUROGEN 2001), pp. 95–100 (2002)

An Image Set-Oriented Method for the Numerical Treatment of Bi-Level Multi-objective Optimization Problems

Alessandro Dell'Aere(✉)

Chair of Applied Mathematics, Institute for Mathematics University of Paderborn, Paderborn, Germany
sante_@web.de

Abstract. In this chapter, we consider equality constrained bi-level multi-objective optimization problems, where the lower level problem is convex. Based on a suitable reformulation of the Kuhn-Tucker equations, we present an image set-oriented algorithm of reference point type for the approximation of the solution set, the Pareto set respectively its image, the Pareto front, of such a problem. The algorithm is designed such that the generated representation of the Pareto front is well-distributed with respect to the higher level image space. We first prove convergence for this algorithm and further on indicate its efficiency on two academic test problems.

Keywords: Bi-Level multi-objective optimization · Bi-level optimization · Multi-objective optimization · Hierarchical optimization · Image set-oriented methods · Reference point methods

1 Introduction

Both, *multi-objective optimization problems* as well as *bi-level optimization problems* have been considered thoroughly during the last decades. The relatively new class of optimization problems considered in this article can be understood as a combination of the two above mentioned problems in the sense that both the higher and the lower level problem of such a bi-level optimization problem are given by multi-objective optimization problems. In other words, we are concerned with a multi-objective optimization problem (the *higher level* of the bi-level optimization problem), where the feasible set itself is restricted by the solution set of another (parametrized) multi-objective optimization problem (the *lower level* of such a bi-level optimization problem). Therefore, we call these problems *bi-level multi-objective optimization problems* (BLMOP). To demonstrate the relevance of such problems from a practical point of view, consider the following example. For the design of a perfect passenger car, two important goals are the fuel consumption (to be minimized) and the power of the engine (to be maximized) leading to a bi-objective optimization problem in the higher level. However, due

O. Junge et al. (Eds.): SON 2020, SSDC 304, pp. 337–354, 2020.
https://doi.org/10.1007/978-3-030-51264-4_14

to safety reasons there is the restriction that in the first place optimality concerning the mechanical guidance of the undercarriage in both horizontal and vertical direction have to be optimized leading to another bi-objective problem in the lower level.

In this chapter, we will concentrate on problems with equality constraints[1] for both the higher and lower level problem. Moreover, we assume that the lower level problem is convex, that is, the lower level objectives are assumed to be convex and the lower level constraints are assumed to be affine-linear.

The outline of this article is as follows. In Sect. 2 we review the basic definitions and concepts of multi-objective optimization and bi-level optimization needed to understand the contents of the subsequent sections. The proposed algorithm for the solution of a BLMOP is presented in Sect. 3. In Sect. 4, we prove convergence of the algorithm. Then, in Sect. 5, we indicate the efficiency of the algorithm on two academic examples. Finally, we draw our conclusions in Sect. 6.

2 Background and Related Work

In the following we briefly review the relevant definitions and concepts of multi-objective optimization and bi-level optimization. Next, we describe in detail the bi-level multi-objective optimization problem (BLMOP) that we will consider in this article. We also state a Kuhn-Tucker based reformulation of the given BLMOP, which is used for the construction of the subproblem to be solved repeatedly in order to compute the individual points of the solution set as our new BL-Recovering-IS algorithm presented in Sect. 3 proceeds.

In a multi-objective optimization problem (MOP) one is faced with the problem that several objectives have to be optimized at the same time. Mathematically, a continuous MOP can be expressed as

$$\min_{x \in S} F(x). \tag{MOP}$$

Hereby, the map F is defined by the individual objective functions F_i, i.e.,

$$F : S \to \mathbb{R}^k, \qquad F(x) = (F_1(x), \ldots, F_k(x))^T, \tag{1}$$

where we assume all functions $F_i : S \to \mathbb{R}$, $i = 1, \ldots, k$, to be continuous. Problems with $k = 2$ objectives are termed bi-objective optimization problems (BOPs).

The domain or feasible set $S \subset \mathbb{R}^n$ of F can in general be expressed by equality and inequality constraints,

$$S = \{x \in \mathbb{R}^n \mid G_i(x) \leq 0,\ i = 1, \ldots, l, \text{ and } H_j(x) = 0,\ j = 1, \ldots, p\}. \tag{2}$$

[1] Hereby, we assume that involved inequality constraints can be transformed into equality constraints, e.g., via the use of slack variables.

If $S = \mathbb{R}^n$, we call the MOP unconstrained.

Optimality of a MOP is based on the concept of dominance.

Definition 1. *(a) Let $v, w \in \mathbb{R}^k$. Then the vector v is* less than *w (in short: $v <_p w$), if $v_i < w_i$ for all $i \in \{1, \dots, k\}$. The relation $\leq_p$ is defined analogously.*

(b) A vector $y \in S$ is called strictly dominated *(or simply* dominated*) by a vector $x \in S$ ($x \prec y$) with respect to (MOP) if*

$$F(x) \leq_p F(y) \qquad \text{and} \qquad F(x) \neq F(y),$$

else y is called non-dominated *by x.*

If a feasible point x dominates another feasible point y, then we can consider x to be 'better' than y with respect to the given MOP. The definition of optimality (i.e., the definition of the 'best' solutions) in multi-objective optimization is now straightforward.

Definition 2. *(a) A point $x \in S$ is called* (Pareto) optimal *or a* Pareto point *of (MOP) if there exists no $y \in S$ that dominates x.*

(b) The set of all Pareto optimal solutions is called the Pareto set, *i.e.,*

$$P_S := \{x \in S \;:\; x \; is \, a \, Pareto \, point \, of \; (MOP)\}. \tag{3}$$

(c) The image $F(P_S)$ of P_S is called the Pareto front.

If all the objectives and constraint functions of the MOP are differentiable one can state a necessary condition for optimality which is analog to 'classical' scalar objective optimization problems (SOPs, i.e., MOPs with $k = 1$).

Theorem 1 ([25]). *Let x^* be a Pareto point of (MOP), where S is as in (2), and all objectives and constraint functions are differentiable in x^*. Further, let the vectors $\nabla H_i(x^*)$, $i = 1, \dots, p$, be linearly independent. Then there exist vectors $\alpha^* \in \mathbb{R}^k$, $\lambda^* \in \mathbb{R}^l$, and $\mu^* \in \mathbb{R}^p$ such that the tuple $(x^*, \alpha^*, \lambda^*, \mu^*)$ satisfies*

$$\begin{aligned} \sum_{i=1}^{k} \alpha_i^* \nabla F_i(\mathbf{x}^*) + \sum_{i=1}^{l} \lambda_i^* \nabla G_i(\mathbf{x}) + \sum_{i=1}^{p} \mu_i^* \nabla H_i(\mathbf{x}^*) &= 0 \\ \alpha_i^* \geq 0, \quad i &= 1, \dots, k \\ \sum_{i=1}^{k} \alpha_i^* &= 1 \\ \lambda_i^* \geq 0, \quad i &= 1, \dots, l \\ \lambda_i^* G_i(x^*) = 0, \quad i &= 1, \dots, l. \end{aligned} \tag{4}$$

Moreover, it is known that these conditions are already sufficient under the assumptions used in this article.

Theorem 2 ([25]). *Assume that the objectives $F_i, i = 1, \ldots, k$, are convex. Further, let the problem contain no inequality constraints and let all equality constraints $H_i, i = 1, \ldots, p$, be affine-linear. Then the conditions stated in Theorem 1 are sufficient for a solution of (MOP).*

A *reference point* $t \in \mathbb{R}^k$ can be regarded as a vector that consists of desirable objective values called *aspiration levels* or *targets*, $t_i, i = 1, \ldots, k$.

In the following we will focus on *distance function based approaches*, which are relevant for our new algorithm presented in Sect. 3. As indicated by its notation, distance function based approaches use a *distance function*, which is typically based on a norm, to measure the distance between a reference point and a given point in image space. To state the auxiliary problem corresponding to a target vector $t \in \mathbb{R}^k$, let $\delta : \mathbb{R}^k \times \mathbb{R}^k \to \mathbb{R}_+$ be a distance function derived from a norm, i.e., $\delta(a, b) = ||a - b||$ for some norm $|| \cdot || : \mathbb{R}^k \to \mathbb{R}_+$. Then the auxiliary problem to be solved is

$$\min_{x \in S} \delta(F(x), t). \tag{RPP}$$

If we have $\delta(F(x^\star), t) > 0$, where $x^\star$ is a solution to RPP, then we know that $F(x^\star)$ is on the boundary of the image $F(S) = \{F(x) : x \in S \subset \mathbb{R}^n\}$. Moreover, if in addition $t <_p F(x^\star)$ we can expect that $x^\star$ is (at least a local) Pareto point. Thus, local Pareto points can be found by first choosing suitable targets and then solving RPP. Indeed, Theorem 3, which was taken from [13], guarantees that, under certain assumptions, $x^\star$ is a Pareto point. For this, recall that a norm $|| \cdot || : \mathbb{R}^k \to \mathbb{R}_+$ is called *strictly monotonically increasing*, if $||y^1|| < ||y^2||$ for all $y^1, y^2 \in \mathbb{R}^k$ with $|y_j^1| \leq |y_j^2|, j = 1, 2, \ldots, k$, and $|y_j^1| \neq |y_j^2|$ for some j.

Theorem 3 ([13]). *Let $|| \cdot ||$ be a strictly monotonically increasing norm and assume $t_i = \min\{F_i(x) : x \in S\}$ for $i = 1, 2, \ldots, k$. If $x^\star$ is an optimal solution of RPP, then $x^\star$ is a solution of MOP.*

We stress that throughout this article it will be $|| \cdot || = || \cdot ||_2$ unless stated differently.

The analytic expression of the entire (exact) Pareto set/front is except for some academic test problems in general not possible. In literature, a huge variety of different methods can be found for the effective numerical treatment of MOPs. There are, for instance, mathematical programming (MP) techniques, point-wise iterative search techniques that generate a sequence of solutions that can converge toward *one* optimal solution (e.g., [16,30] and references therein.). The most widely used sub-class of the MP techniques is given by scalarization methods that replace the given MOP into a suitable auxiliary SOP (e.g., [19,20,32,38,39]). Via identifying a clever sequence of such SOPs a suitable approximation of the entire Pareto set/front can be obtained in certain cases (e.g., [5,14,17,18,24,30,33]). *Reference point methods* use feasible or infeasible reference points for the construction of scalar valued auxiliary functions. For an overview on different types of reference point methods the reader is referred to [16].

Another class of methods are given by continuation-like methods that take advantage of the fact that the solution set forms at least locally a manifold. Such methods start from a given solution and perform a search along the solution manifold ([22,27–29,34,36,37,43]). However, one potential drawback of all the above mentioned methods is that they are of local nature, i.e, that they may get stuck in local Pareto optimal solutions of the given MOP depending on the chosen starting point and the chosen method to solve the auxiliary SOP.

Next to these point-wise iterative methods there exist specialized set oriented methods such as multi-objective evolutionary strategies (MOEAs, e.g., [2,3,6]), subdivision techniques [10,23,41,42] or cell mapping techniques [21,31,35,45,46, 48,49]. These methods have in common that they use entire sets in an iterative manner and are thus able to deliver an approximation of the solution set in one run of the algorithm. Further, the set based approach allows a more global view on the problem leading to a reduced probability to get stuck in local optimal solutions. Cell mapping techniques are particularly advantageous over other methods if a thorough investigation of the entire (low or moderate dimensional) system is of interest as they deliver next to Pareto set/front approximations also approximations of the set of nearly optimal solutions as well as the set of local solutions, as we will discuss in the following.

A **bi-level optimization problem** can be understood as an optimization problem *(the higher level problem)*, where the feasible set is restricted by the solution set of another (parametrized) optimization problem *(the lower level problem)*.

Many different approaches for solving (classical) bi-level optimization problems have been proposed in the past, as there are for example descent algorithms, bundle algorithms, penalty methods, trust region methods, smoothing methods, and branch-and-bound methods. Many of these approaches are based on the conversion of the bi-level problem to an ordinary (or classical) optimization problem (a one-level problem). One possibility is to replace the lower level objective f by an additional non-differentiable equation $f(x,y) = \varphi(y)$, where $\varphi(y) = \min_x\{f(x,y) : g(x,y) \leq 0, h(x,y) = 0\}$. Other approaches use the implicit function theorem to derive a local description of the function $x(y) : \mathbb{R}^m \to \mathbb{R}^n$, which is then inserted into the higher level problem. Another concept is to replace the lower level problem by its Kuhn-Tucker conditions. In general, the resulting one-level problem, which is a *mathematical program with equilibrium constraints* or MPEC, see [26], is not equivalent to the original problem, but the desired equivalence is ensured in the particular case where the lower level problem is a convex one. For an overview on bi-level optimization the reader is referred to [1,4,7,11,12,15,44,47].

In this article, we are concerned with the case where both the higher and lower level problem are given by multi-objective optimization problems. Such problems are called **bi-level multi-objective optimization problem** (BLMOP), see [9,14,15].

The higher level problem of a BLMOP can be written as

$$\min_y \min_x \{F(x,y) : x \in \psi(y)\} \qquad \text{(BLMOP-H)}$$
$$\text{s.t. } H(x,y) = 0,$$

where $\psi(y)$ denotes for every fixed $y \in \mathbb{R}^m$ the solution, that is, the Pareto set of the lower level problem given by

$$\min_x f(x,y) \qquad \text{(BLMOP-L)}$$
$$\text{s.t. } h(x,y) = 0,$$

It should be mentioned that in the notions of [11], BLMOP-H and BLMOP-L correspond to an *optimistic formulation* of the general Bi-Level Optimization Problem. Since in this article we concentrate on the case with a convex lower level problem, the lower level problem can be replaced by the corresponding Kuhn-Tucker conditions stated in Theorem 1 to obtain an expression which is equivalent to BLMOP-H. For this, we assume that the higher level problem includes k objective functions $F_i : \mathbb{R}^n \times \mathbb{R}^m \to \mathbb{R}$, which are collected in the vector valued function $F : \mathbb{R}^n \times \mathbb{R}^m \to \mathbb{R}^k$, $F(x,y) = (F_1(x,y), \ldots, F_k(x,y))^t$, and r equality constraints $H_i : \mathbb{R}^n \times \mathbb{R}^m \to \mathbb{R}$, which are collected in the vector valued function $H : \mathbb{R}^n \times \mathbb{R}^m \to \mathbb{R}^r$, $H(x,y) = (H_1(x,y), \ldots, H_r(x,y))^t$. Analogously, we assume that the lower level problem includes l objectives $f_i : \mathbb{R}^n \times \mathbb{R}^m \to \mathbb{R}$, which are collected in the vector valued function $f : \mathbb{R}^n \times \mathbb{R}^m \to \mathbb{R}^l$, $f(x,y) = (f_1(x,y), \ldots, f_l(x,y))^t$ and p equality constraints $h_i : \mathbb{R}^n \times \mathbb{R}^m \to \mathbb{R}$, which are collected in the vector valued function $h : \mathbb{R}^n \times \mathbb{R}^m \to \mathbb{R}^p$, $h(x,y) = (h_1(x,y), \ldots, h_p(x,y))^t$. We denote by

$$\mathcal{L}(x,y,\alpha,\lambda) := \sum_{i=1}^{l} \alpha_i f_i(x,y) + \sum_{i=1}^{p} \lambda_i h_i(x,y)$$

the lower level Lagrangian and by $\nabla_x \mathcal{L}$ the gradient of $\mathcal{L}$ with respect to x. According to Theorem 2, $x \in \psi(y)$ if and only if

$$\begin{gathered} h(x,y) = 0, \\ \nabla_x \mathcal{L}(x,y,\alpha,\lambda) = 0, \\ \sum_{i=1}^{l} \alpha_i = 1, \\ \alpha_i \geq 0, \quad i = 1, \ldots, l \end{gathered}$$

for some $\alpha \in \mathbb{R}^l$ and $\lambda \in \mathbb{R}^p$. Let

$$\hat{F}(x,y,\alpha,\lambda,s) := \begin{pmatrix} h(x,y) \\ H(x,y) \\ \nabla_x \mathcal{L}(x,y,\alpha,\lambda) \\ \sum_{i=1}^{l} \alpha_i - 1 \\ \alpha - (s \circ s) \end{pmatrix},$$

where $s \in \mathbb{R}^l$ is a vector of l slack variables and $a \circ b$ denotes the component-wise product of two vectors a, b. Moreover, let $z := (x, y, \alpha, \lambda, s)$, $\hat{S} := \{z : \hat{F}(z) = 0\}$, and denote by $\pi(z)$ the projection of z to the (x, y)-space $\mathbb{R}^{n+m}$. Observe that $\{\pi(z) : z \in \hat{S}\}$ is the feasible set of BLMOP-H and therefore the desired reformulation for the given Problem can be written as follows:

$$\min_{z \in \hat{S}} F(\pi(z)), \qquad \text{(BLMOP-R)}$$

where again minimization has to be understood in the sense of Definition 1. In order to handle BLMOP-R by the use of reference point methods, we define the following variant of RPP:

$$\min_{z \in \hat{S}} \delta(F(\pi(z)), t) \qquad \text{(RPP-R)}$$

Note that RPP-R will be the method used for the computation of the individual Pareto points of the given BLMOP while our BL-Recovering-IS algorithm presented in Sect. 3 proceeds.

3 Algorithm and Realization

We present the *BL-Recovering-IS algorithm* for the solution of equality constrained BLMOPs with a convex lower level problem. This algorithm can be understood as an extension of our algorithm for the solution of unconstrained MOPs described in [8]. In addition, we state some theoretical results which apply both to the algorithm presented here and to the algorithm presented in [8].

The aim of the BL-Recovering-IS algorithm is to generate both a box covering and a discrete representation of the entire Pareto front of the given BLMOP (see also Fig. 1).

We assume that this representation is required to be well-distributed in higher level image space in the following sense: Denote by $Q \subset \mathbb{R}^k$ the region of interest in image space. For formal reasons denote by $\mathcal{P}_d$ a *complete* partition[2] of the set Q into boxes of subdivision size – or *depth* – d, which are generated by successive bisection of Q. These boxes are understood to be half-open, that is, they can be written as cartesian products $[a_1, b_1) \times, \ldots, \times [a_k, b_k)$ of half-open intervals $[a_i, b_i), i = 1, \ldots, k$. Then there exists for every point $\bar{F} \in Q$ exactly one box $B(\bar{F}, d) \in \mathcal{P}_d$ such that $\bar{F} \in B(\bar{F}, d)$. The algorithm computes both a covering

$$\mathcal{B} = \bigcup_{B \in \mathcal{P}_d \cap F(P)}$$

and a discrete representation of the Pareto set P. The discrete representation is well-distributed in the sense that for every $B \in \mathcal{P}_d \cap F(P)$ there is at least one computed point $(x, y) \in \mathbb{R}^{n+m}$ such that $F(x, y) \in B$.

[2] $\mathcal{P}_d$ has *not* to be explicitly computed by our algorithm.

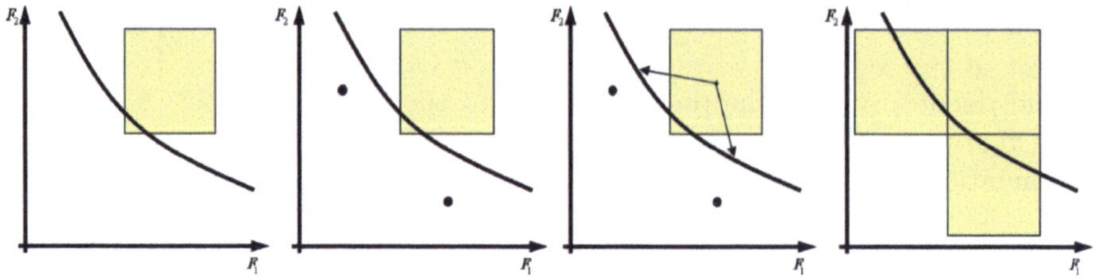

Fig. 1. Idea fo the BL-Recovering-IS algorithm: use boxes to obtain a uniform spread of solutions around the Pareto front.

In order to compute these points, our new image set-oriented algorithm presented in the following repeatedly solves a variant of RPP-R while the targets are varying. To state a corollary which guarantees that the corresponding solutions are at least locally Pareto optimal we denote $T = (T_1, \ldots, T_k)$, where

$$T_i = \min\{F_i(\pi(z)) : z \in \hat{S}\} \qquad \text{for } i = 1, 2, \ldots, k,$$

and define for a given target vector $t \in \mathbb{R}^k$ the modified feasible set

$$\hat{S}_t = \{z \in \hat{S} : F_i(\pi(z)) \geq t_i,\ i = 1, \ldots, k\}.$$

Furthermore, we define variants of BLMOP-R and RPP-R, respectively, by replacing $\hat{S}$ by $\hat{S}_t$:

$$\min_{z \in \hat{S}_t} F(\pi(z)), \qquad \text{(BLMOP – R')}$$

$$\min_{z \in \hat{S}_t} \delta(F(\pi(z)), t). \qquad \text{(RPP – R')}$$

Now, with these notations we can state the following result.

Corollary 1. *Let F be continuous on the compact domain $\hat{S}$. Moreover, let $||\cdot||$ be a strictly monotonically increasing norm and assume that $T <_p t <_p F(\pi(z^\star))$, where $z^\star$ is an optimal solution of RPP-R'. Then $\pi(z^\star)$ is a local solution of the given BLMOP.*

Proof. Since F_i is continuous and since there are $\bar{z}^i, z^\star \in \hat{S}$ with $F_i(\pi(\bar{z}^i)) = T_i < t_i < F_i(\pi(z^\star))$, there exist $z^i \in \hat{S}$ such that $F_i(\pi(z^i)) = t_i$ for all $i = 1, 2, \ldots, k$. ¿From construction of $\hat{S}_t$ it is obvious, that $z^\star \in \hat{S}_t$ and $t_i = \min\{F_i(\pi(z)) : z \in \hat{S}_t\}$. Thus, Theorem 3 guarantees that $z^\star$ solves BLMOP-R'. Since $\hat{S}_t$ is constructed from $\hat{S}$ just by constraining the image of F, such that $F(\pi(\hat{S}_t))$ contains a part of a local Pareto optimal set in image space, $z^\star$ is a local solution of BLMOP-R, that is, $\pi(z^\star)$ is a local solution of the given BLMOP. □

In practice, a randomly chosen point $t \in \mathbb{R}^k$ does not necessarily belong to the image $F(\pi(\hat{S})) = \{F(\pi(z)) : z \in \hat{S}\}$, that is, we do not know a priori whether there is any $z \in \hat{S}$ such that $F(\pi(z)) = t$. Moreover, if $F(\pi(z)) = t$ for some $z \in \hat{S}$, we do not know whether $\pi(z)$ is Pareto optimal. To get an answer to these questions, we solve the auxiliary problem RPP-R'. If $t <_p F(\pi(z^\star))$ for a solution $z^\star$ of RPP-R', then we know that – under suitable assumptions – $\pi(z^\star)$ is at least locally Pareto optimal. Otherwise, if $t = F(\pi(z^\star))$, then we repeatedly have to vary t and solve RPP-R' until $t <_p F(\pi(z^\star))$. A strategy for the choice and variation of the targets t can be found later on in this section. In the algorithm described below and for the remainder of this article the distance function δ is based on the norm $||\cdot||_2$ that is, $\delta(a,b) = ||a-b||_2$ for all $a, b \in \mathbb{R}^k$. Our algorithm belongs to the family of continuation methods ([8,22,40]), that is, the aim of every step is to compute Pareto points in the neighborhood of Pareto points already found in a previous step. Accordingly, we assume that at least one box B along with a point $z^\star$ with $F(\pi(z^\star)) \in F(P)$ has been computed previously, e.g., by the solution of RPP-R' for the target $t = (t_1, \ldots, t_k)$, $t_i = \min\{F_i(\pi(z)) : z \in \hat{S}\}$ for $i = 1, 2, \ldots, k$.

Then, for a given box collection $\mathcal{B}_j \subset \mathbb{R}^k$ (in image space) of subdivision depth d and denoting by z_B and F_B the previously generated solution (in parameter and image space, respectively) associated with a box $B \in \mathcal{B}_j$, a step of the BL-Recovering-IS algorithm can be written as shown in Algorithm 1.

It remains to answer the question of how to choose the target vectors t_i, $i = 1, 2, \ldots, n_t$, near a current box B in order to compute Pareto points which are well-distributed in the sense mentioned above. Efficient strategies for the choice of target vectors can be defined, particularly by using local information on the Pareto set, e.g. orientation or curvature, which can be calculated via objective derivatives (or numerical approximations of the derivatives). In the following we will focus on a particular strategy for the choice of the targets which was originally designed for problems with smooth objectives, but is also applicable and works satisfactorily in the case of more general objectives. Let us assume that the higher level image $F(P) \subset \mathbb{R}^k$ of the Pareto set P is smooth and forms a $(k-1)$-dimensional manifold in a neighborhood $N_\varepsilon(F^\star)$ of a given Pareto optimal point $F^\star \in F(P)$ in higher level image space. Since an approximation of $F(P)$ at $F^\star$ is given by the tangent space $T_{F^\star}F(P)$, there are certainly further Pareto points near $T_{F^\star}F(P) \cap N_\varepsilon(F^\star)$. Consequently, we can expect that there are $\lambda \in \mathbb{R}$ and $p \in T_{F^\star}F(P) \cap N_\varepsilon(F^\star)$, such that suitable targets needed for the computation of further Pareto points can be expressed by $p + \lambda d$, where $d \leq_p 0$ denotes a basis vector of the 1-dimensional space $(T_{F^\star}F(P))^\perp$. Thus, to apply this idea in practice, we first have to construct d and a basis $V := \{b_1, b_2, \ldots, b_{k-1}\}$ of $T_{F^\star}F(P)$ and then to specify targets

$$t_i = F_i^\star + \sum_{j=1}^{k-1} \alpha_{i,j}\, b_j + \lambda_i\, d, \quad i = 1, 2, \ldots, n_t$$

by determining the coefficients $\alpha_{i,j}$ and λ_i. Fortunately, as stated in the following lemma, if $F(P)$ forms a smooth manifold in a neighborhood of $F^\star$ and if $F^\star$ was

Algorithm 1. Algorithm BL-Recovering-IS

Require: current box collection $\mathcal{B}_j$
Ensure: new box collection $\mathcal{B}_{j+1}$
for all $B \in \mathcal{B}_j$ **do**
    $B.active := TRUE$
end for
for $k = 1, \ldots, MaxStep$ **do**
    $\hat{\mathcal{B}} := \mathcal{B}_j$
    for all $B \in \{B \in \mathcal{B}_j \ : \ B.active == TRUE\}$ **do**
        choose target vectors $\{t_i\}_{i=1,\ldots,n_t}$ near B with $t_i <_p F_B$
        find $z_i^\star = \arg\min_{z \in \hat{S}_{t_i}} ||F(\pi(z)) - t_i||, i = 1, \ldots, n_t \ i = 1, \ldots, n_t$
        $F_i^\star := F(\pi(z_i^\star)), i = 1, \ldots, n_t$
        $B.active := FALSE$
        for all $i = 1, \ldots, n_t$ **do**
            if $B(F_i^\star, d) \notin \hat{\mathcal{B}}$ **then**
                $\breve{B} := B(F_i^\star, d), \ z_{\breve{B}} := z_i^\star, \ F_{\breve{B}} := F_i^\star$
                $\breve{B}.active := TRUE$
                $\hat{\mathcal{B}} := \hat{\mathcal{B}} \cup \breve{B}$
            end if
        end for
    end for
    if $\hat{\mathcal{B}} == \mathcal{B}_j$ **then**
        BREAK
    end if
end for
$\mathcal{B}_{j+1} := \hat{\mathcal{B}}$
Return $\mathcal{B}_{j+1}$

found in a previous step by solving RPP-R' for a given target $t^\star, t^\star <_p F^\star$, then d can be obtained without any additional effort by $d := t^\star - F^\star$.

Lemma 1. *Let $F_i \in C^1(\mathbb{R}^n, \mathbb{R})$ for $i = 1, \ldots, k$, and consider the multi-objective optimization problem*

$$\min_{z \in \hat{S}} F(\pi(z)).$$

Denote by P the corresponding Pareto set and let $F^\star := F(\pi(z^\star))$, where $z^\star$ is the unique solution of RPP-R' associated with the target $t^\star <_p F^\star$. Moreover, assume that $F(P)$ makes up a $(k-1)$-dimensional smooth manifold in a neighborhood of $F^\star$. Then

$$F^\star - t^\star \in T_{F^\star} F(P)^\perp.$$

Proof. Let ∂F denote the boundary of $\{F(\pi(z)) \, : \, z \in \hat{S}\}$. Since $F(P)$ forms a differentiable manifold in a neighborhood of $F^\star$, there exists a differentiable curve $\alpha : [-1, 1] \to \partial F$ with $\alpha(0) = F^\star$, $\alpha'(0) \in T_{F^\star} F(P)$ and $\alpha(\lambda) \in F(P)$ for all $\lambda \in [0, 1]$. Then, since $z^\star$ is a solution of RPP', $\lambda = 0$ is a solution of

$$\min_{\lambda \in [-1,1]} ||\alpha(\lambda) - t^\star||^2$$

and therefore

$$\frac{d}{d\lambda}||\alpha(\lambda) - t^\star||^2 = \frac{d}{d\lambda}\langle\alpha(\lambda) - t^\star, \alpha(\lambda) - t^\star\rangle = 2\langle\alpha(\lambda) - t^\star, \alpha'(\lambda)\rangle = 0$$

for $\lambda = 0$. With $\alpha(0) = F^\star$ we obtain

$$\langle F^\star - t^\star, \alpha'(0)\rangle = 0,$$

that is, $F^\star - t^\star \in T_{F^\star}F(P)^\perp$. □

Once d is available, any standard method for the construction of an orthogonal basis, e.g. the Grahm-Schmidt method can be used to obtain the required basis V. For all $i = 1, 2, \ldots, n_t$, the coefficients $\alpha_{i,j}$ are chosen such that $p_i := \sum_{j=1}^{k-1} \alpha_{i,j}\, b_j$ is located inside a neighbor box of the current box. Moreover, the p_i should be well-distributed around $F^\star$. With this heuristic, it is very likely to find new boxes containing the image of Pareto points. For the choice of λ_i an adaptive concept has to be applied, because a computed solution z of RPP-R' can only be accepted, if $t_i <_p F(\pi(z))$ is satisfied. Such an adaptive concept should be guided by the fact that $t_i <_p F(\pi(z))$ certainly holds if λ_i is sufficiently large, but it should also be considered that RPP-R' is ill-conditioned if λ_i is too large.

4 Convergence

Since the described BL-Recovering-IS algorithm is realized by minimizing a reformulation of the BLMOP, which can be understood as an equality constrained MOP, in the following we prove convergence for the more general class of image set-oriented recovering algorithms for the solution of MOP as defined in Sect. 2. This includes in particular the *Recovering-IS* algorithm presented in [8].

The proof is carried out in two steps: first, Theorem 4 states that for every subset $B \subset \mathbb{R}^k$ containing a part of the Pareto optimal solution in image space, there is a minimal set of targets, such that for at least one of these targets the corresponding distance minimization subproblem leads to a Pareto point $x^\star$ with $F(x^\star) \in B$. Then, this result is used in Corollary 2 to complete the proof from the global point of view. In the following, let

$$\text{dist}(y, \mathcal{X}) = \min_{t \in \mathcal{X}} ||y, t||$$

be the distance between a point $y \in \mathbb{R}^k$ and a subset $\mathcal{X} \subset \mathbb{R}^k$.

Theorem 4. *Let $F : \mathbb{R}^n \to \mathbb{R}^k, S \subset \mathbb{R}^n$ and denote by P the Pareto set of the constrained MOP:*

$$\min_{x \in S} F(x).$$

Assume that the norm $||\cdot||$ is strictly monotonically increasing. Let $B \subset \mathbb{R}^k$ be an open subset such that $B \cap F(P) \neq \emptyset$. Then there is $d > 0$, such that for any set $\mathcal{X} \subset B$ of targets with $\text{dist}(y, \mathcal{X}) < d$ for all $y \in B \cap F(P)$ there exists a target $t \in \mathcal{X}$ with $F(x^\star) \in F(P) \cap B$, where $x^\star := \arg\min_{x \in S_t} ||F(x) - t||$.

Proof. There are $\bar{y} \in F(P)$ and $\varepsilon > 0$, such that $U_\varepsilon(\bar{y}) \subset B$. Let $d := \frac{\varepsilon}{8\sqrt{k}}$ and $c := \bar{y} - 2\,d\sum_{i=1}^{k} e_i$, where e_i denotes the i-th standard basis vector in $\mathbb{R}^k$. Then, for every $y \in U_d(c)$, we have

$$||y - \bar{y}|| \leq ||y - c|| + ||c - \bar{y}|| \leq d + 2d\sqrt{k} = \frac{\varepsilon}{8\sqrt{k}} + \frac{\varepsilon}{4} < \frac{\varepsilon}{2},$$

that is, $U_d(c) \subset U_\varepsilon(\bar{y})$. Consequently, there is a target $t = c + v \in \mathcal{X}, ||v|| \leq d$, such that

$$\min_{x \in S_t} ||F(x) - t|| \leq ||t - \bar{y}|| < \frac{\varepsilon}{2}$$

and

$$t_i = c_i + v_i = \bar{y}_i - 2d + v_i < \bar{y}_i \quad \text{for} \quad i = 1, \ldots, k.$$

With $x^\star = \arg\min_{x \in S_t} ||F(x) - t||$, it follows that

$$||F(x^\star) - \bar{y}|| \leq ||F(x^\star) - t|| + ||t - \bar{y}|| < \frac{\varepsilon}{2} + \frac{\varepsilon}{2} = \varepsilon$$

and therefore

$$F(x^\star) \in U_\varepsilon(\bar{y}) \subset B.$$

Now we have to show that $F(x^\star)$ is not dominated by any $\hat{y} \in F(P) \cap S_t$. For $F(x^\star) = \hat{y}$ this nondominance is obvious. For the case $F(x^\star) \neq \hat{y}$ we have to show that $F_i(x^\star) < \hat{y}_i$ for at least one $i = 1, \ldots, k$. To see this, assume that the opposite is true. Then $F_i(x^\star) \geq \hat{y}_i > t_i$ for all $i = 1, \ldots, k$, where, since $F(x^\star) \neq \hat{y}$, strict inequality holds for at least one $i \in \{1, \ldots, k\}$. Consequently, since $|| \cdot ||$ is strictly monotonically increasing,

$$||F(x^\star) - t|| > ||\hat{y} - t||,$$

which is a contradiction to $||F(x^\star) - t|| = \min_{x \in S_t} ||F(x) - t||$. Finally, since $F(x^\star)$ is not dominated by any $\hat{y} \in F(P) \cap S_t$, we have $F(x^\star) \in F(P)$, which completes the proof. □

To guarantee that an image set-oriented recovering method converges towards the union of those connected components of $F(P)$ which correspond to the initial box collection $\mathcal{B}_0$, in every step of the algorithm the set of targets t_i has to be chosen properly, such that all desired boxes are found, that is, boxes which are both neighbors of the boxes generated in the respective previous step and contain a part of the respective connected component of $F(P)$. To this end, we denote by $\bar{B}$ the closure of a box B and we state the following

Corollary 2. *Using the notations of Theorem 4 and denoting by $\mathcal{B}_0$ a box collection of (fixed) subdivision depth d covering a part of $F(P)$, assume that every step of the Recovering-IS or BL-Recovering-IS algorithm, respectively, is realized in a way such that for every $B \in \mathcal{B}_j \backslash \mathcal{B}_{j-1}$ targets are chosen according to Theorem 4 within all boxes $C \in \{C : \bar{C} \cap \bar{B} \neq \emptyset, C \notin \mathcal{B}_j\}$. Moreover, assume that $F(P)$ is bounded. Then, the algorithm terminates after a finite number of steps such that the final box collection covers those connected components of $F(P)$, which correspond to at least one $B \in \mathcal{B}_0$.*

5 Numerical Results

In the following we demonstrate the working principle and strength of the proposed algorithm on two academic benchmark problems.

5.1 Example 1

In our first example we consider a classical (i.e., one-level) bi-objective optimization problem in order to demonstrate the working principle of the IS recovering techniques. For this, let the BOP be given by

$$
\begin{aligned}
F &= (F_1, F_2)^t : \mathbb{R}^3 \to \mathbb{R}^2 \\
F(x_1, x_2, x_3) &= \begin{pmatrix} (x_1 - 1)^2 + (x_2 - 1)^2 + (x_3 - 1)^4 \\ (x_1 + 1)^4 + (x_2 + 1)^2 + (x_3 + 1)^2 \end{pmatrix}
\end{aligned}
$$

We assume that the decision maker is only interested in solutions for which both objective values are located within the interval $I := [0, 20]$, and therefore define

$$S := \{x \in \mathbb{R}^3 : F_i(x) \in I, i = 1, 2\}.$$

Figure 2 shows the solutions generated by the Recovering-IS algorithm using different box sizes (depths). Here, the reader can get an impression of how the density of the computed representation can be controlled by choosing the box size.

5.2 Example 2

Next, we consider the following equality constrained bi-level multi-objective optimization problem with a convex lower level problem:

$$\min_{x \in \mathbb{R}^3,\, y \in \mathbb{R}} F(x, y) = \begin{pmatrix} 4((x_1 + 1)^2 + (x_2 - 1 - y)^4 + x_3^2) \\ (x_1 - 1)^2 + (x_2 + 1 - y)^2 + (x_3 - 0.5)^4 \end{pmatrix},$$

such that $H(x, y) = x_1^2 + x_3 - y^2 = 0,$
and x solves:

$$\min_{x \in \mathbb{R}^3} f(x, y) = \begin{pmatrix} (x_1 - 1)^2 + 0.5(x_2 + y)^2 + (x_3 - 0.5)^4 \\ (x_1 + 1)^2 + 0.5(x_2 + y)^2 + (x_3 + 1)^2 \\ x_1^2 + x_2^2 + (x_3 + 1)^2 \end{pmatrix}$$

such that $h_1 = x_1 - x_2 y = 0.$

The solution of this problem was computed by the presented BL-Recovering-IS algorithm. For this, we have chosen $Q = [0, 10]^2$ for the domain of interest in higher level image space. The partition $\mathcal{P}_d$ was chosen corresponding to 5 virtual subdivisions in each coordinate, such that all boxes $B \in \mathcal{P}_d$ are of the size 0.3125^2. The computed solution in higher level image space along with the generated boxes is shown on top of Fig. 3. As expected, the solution is well-distributed in the sense that there is at least one computed point in every box of the box collection covering the Pareto set. The projection of the corresponding Pareto set to the x-space is shown on the bottom of Fig. 3.

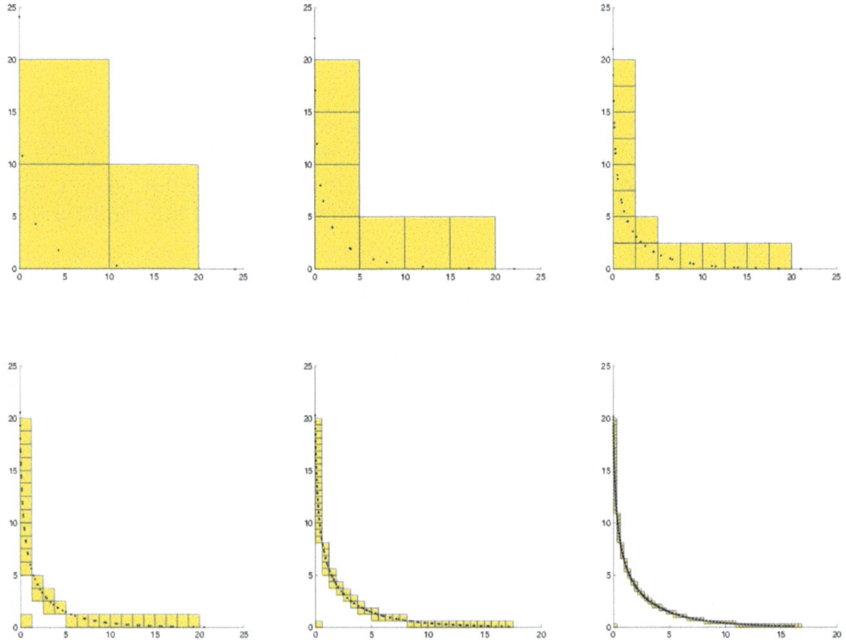

Fig. 2. Numerical results on Example 1 computed by the image set-oriented recovering algorithm using different box sizes in image space.

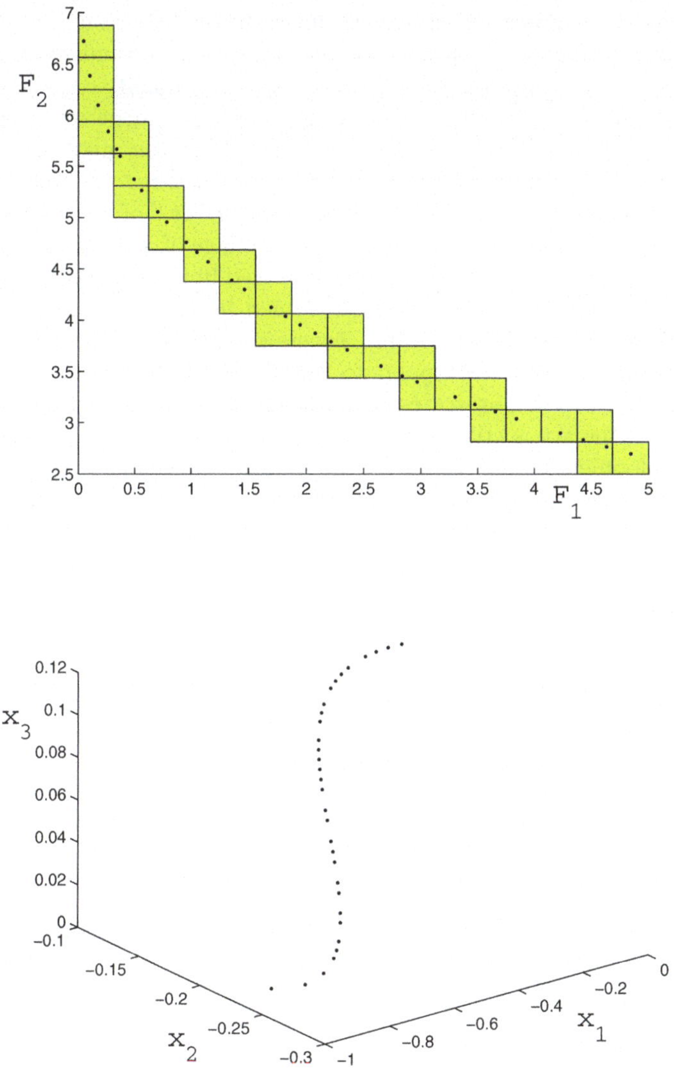

Fig. 3. The Pareto set of the example problem computed by our algorithm in higher level image space (top) and in parameter space (projection to the x-space)(bottom).

6 Conclusions

In this chapter, we have considered the class of bi-level multi-objective optimization problems (BLMOP) with equality constraints for both the higher and lower level problem. The lower level problem was assumed to be convex, that is, the lower level objectives are convex and the lower level equality constraints are affine-linear. Due to the concentration to this particular subclass, we have been able to write down an equivalent formulation based on the well-known Kuhn-Tucker optimality conditions for multi-objective optimization problems. The resulting reformulation has the form of a general equality constrained multi-objective optimization problem. We have presented an image set-oriented algorithm for the approximation of the Pareto set P of the given BLMOP. The representation of P computed by this algorithm turns out to be well-distributed in the sense that in every box $B \subset \mathbb{R}^k$ with $B \cap F(P) \neq \emptyset$ of a given partition $\mathcal{P}_d$ of the higher level image space $\mathbb{R}^k$, there is the image of at least one of the computed Pareto points. Convergence has been proved in the sense that after a finite number of iterations, the box collection formed by those boxes containing the images of the computed Pareto points, covers the image of the entire connected components of P, which correspond to the given initial points. The efficiency of the algorithm was demonstrated on an academic example, where comparison to the state-of-the-art is still missing, which we leave for future work. Finally, also the development of algorithms for the solution of the more general BLMOP, which includes non-convex lower level problems, shall be investigated in the future.

References

1. Bard, J.F.: Practical Bilevel Optimization: Algorithms and Applications. Kluwer Academic Publishers, Berlin (1998)
2. Bogoya, J.M., Vargas, A., Schütze, O.: The averaged Hausdorff distances in multi-objective optimization: a review. Mathematics **7**(10), 894 (2019)
3. Coello Coello, C.A., Lamont, G.B., Van Veldhuizen, D.A.: Evolutionary Algorithms for Solving Multi-Objective Problems, 2nd edn. Springer, New York (2007). ISBN 978-0-387-33254-3
4. Colson, B., Marcotte, P., Savard, G.: Bilevel programming: a survey. Q. J. Oper. Res. **3**, 87–107 (2005)
5. Das, I., Dennis, J.E.: Normal-boundary intersection: a new method for generating the Pareto surface in nonlinear multicriteria optimization problems. SIAM J. Optim. **8**(3), 631–657 (1998)
6. Deb, K.: Multi-Objective Optimization Using Evolutionary Algorithms. Wiley, Hoboken (2001)
7. Deb, K., Sinha, A.: Solving bilevel multi-objective optimization problems using evolutionary algorithms. In: Ehrgott, M., et al. (eds.) Evolutionary Multi-Criterion Optimization (2009)
8. Dell'Aere, A.: Multi-objective optimization in self-optimizing systems. In: Proceedings of the 32nd Annual Conference of the IEEE Industrial Electronics Society (2006)
9. Dell'Aere, A.: Numerical methods for the solution of bi-level multi-objective optimization problems. Ph.D. thesis, University of Paderborn (2008)
10. Dellnitz, M., Schütze, O., Hestermeyer, T.: Covering Pareto sets by multilevel subdivision techniques. J. Optim. Theory Appl. **124**(1), 113–155 (2005)
11. Dempe, S.: Foundations of Bilevel Programming. Kluwer Academic Publishers, Berlin (2002)
12. Dempe, S.: Annotated bibliography on bilevel programming and mathematical programs with equilibrium constraints. Optimization **52**, 333–359 (2003)

13. Ehrgott, M.: Multicriteria Optimization. Lecture Notes in Economics and Mathematical Systems (2000)
14. Eichfelder, G.: Adaptive Scalarization Methods in Multiobjective Optimization. Springer, Heidelberg (2008)
15. Eichfelder, G.: Multiobjective bilevel optimization. Math. Program. **123**, 419–449 (2010)
16. Figueira, J., Greco, S., Ehrgott, M.: Multiple Criteria Decision Analysis. Lecture Notes in Economics and Mathematical Systems. Springer (2005)
17. Fliege, J.: Gap-free computation of Pareto-points by quadratic scalarizations. Math. Methods Oper. Res. **59**, 69–89 (2004)
18. Fliege, J., Fux Svaiter, B.: Steepest descent methods for multicriteria optimization. Math. Methods Oper. Res. **51**(3), 479–494 (2000)
19. Gass, S., Saaty, T.: The computational algorithm for the parametric objective function. Naval Res. Logist. Q. **2**(1), 39–45 (1955)
20. Gembicki, F.W., Haimes, Y.Y.: Approach to performance and multiobjective sensivity optimization: the goal attainment method. IEEE Trans. Autom. Control **20**, 769–771 (1975)
21. Hernández, C., Naranjani, Y., Sardahi, Y., Liang, W., Schütze, O., Sun, J.-Q.: Simple cell mapping method for multi-objective optimal feedback control design. Int. J. Dyn. Control **1**(3), 231–238 (2013)
22. Hillermeier, C.: Nonlinear Multiobjective Optimization - A Generalized Homotopy Approach. Birkhäuser, Basel (2001)
23. Jahn, J.: Multiobjective search algorithm with subdivision technique. Comput. Optim. Appl. **35**(2), 161–175 (2006)
24. Klamroth, K., Tind, J., Wiecek, M.: Unbiased approximation in multicriteria optimization. Math. Methods Oper. Res. **56**, 413–437 (2002)
25. Kuhn, H.W., Tucker, A.W.: Nonlinear programming. In: Proceedings of the Second Berkeley Symposium on Mathematical Statistics and Probability, 1950, Berkeley and Los Angeles, pp. 481–492. University of California Press (1951)
26. Luo, Z.Q., Pang, J.S., Ralph, D.: Mathematical Programs with Equilibrium Constraints. Cambridge University Press, Cambridge (1996)
27. Martín, A., Schütze, O.: Pareto tracer: a predictor corrector method for multi-objective optimization problems. Eng. Optim. **50**(3), 516–536 (2018)
28. Martin, B., Goldsztejn, A., Granvilliers, L., Jermann, C.: Certified parallelotope continuation for one-manifolds. SIAM J. Numer. Anal. **51**(6), 3373–3401 (2013)
29. Martin, B., Goldsztejn, A., Granvilliers, L., Jermann, C.: On continuation methods for non-linear bi-objective optimization: towards a certified interval-based approach. J. Global Optim., 1–14 (2014)
30. Miettinen, K.: Nonlinear Multiobjective Optimization. Kluwer Academic Publishers, Berlin (1999)
31. Naranjani, Y., Hernández, C., Xiong, F.-R., Schütze, O., Sun, J.-Q.: A hybrid algorithm for the simple cell mapping method in multi-objective optimization. In: Emmerich, M., et al. (eds.) EVOLVE–A Bridge between Probability, Set Oriented Numerics, and Evolutionary Computation IV. Advances in Intelligent Systems and Computing, pp. 207–223. Springer, Heidelberg (2013)
32. Pascoletti, A., Serafini, P.: Scalarizing vector optimization problems. J. Optim. Theory Appl. **42**, 499–524 (1984)
33. Peitz, S., Dellnitz, M.: A survey of recent trends in multiobjective optimal control surrogate models, feedback control and objective reduction. Math. Comput. Appl. **23**(2) (2018)

34. Pereyra, V., Saunders, M., Castillo, J.: Equispaced Pareto front construction for constrained bi-objective optimization. Math. Comput. Model. **57**(9–10), 2122–2131 (2013)
35. Qin, Z.-C., Xiong, F.-R., Ding, Q., Hernandez, C., Fernandez, J., Schütze, O., Sun, J.-Q.: Multi-objective optimal design of sliding mode control with parallel simple cell mapping method. J. Vib. Control (2015)
36. Recchioni, M.C.: A path following method for box-constrained multiobjective optimization with applications to goal programming problems. Math. Methods Oper. Res. **58**, 69–85 (2003)
37. Ringkamp, M., Ober-Blöbaum, S., Dellnitz, M., Schütze, O.: Handling high dimensional problems with multi-objective continuation methods via successive approximation of the tangent space. Eng. Optim. **44**(6) (2012)
38. Roy, B.: Problems and methods with multiple objective functions. Math. Program. **1**, 239–266 (1971)
39. Schandl, B., Klamroth, K., Wiecek, M.M.: Introducing oblique norms into multiple criteria programming. J. Global Optim. **23**, 925–942 (2002)
40. Schütze, O., Dell'Aere, A., Dellnitz, M.: On continuation methods for the numerical treatment of multi-objective optimization problems. In: Branke, J., Deb, K., Miettinen, K., Steuer, R.E. (eds.) Practical Approaches to Multi-Objective Optimization, number 04461 in Dagstuhl Seminar Proceedings. Internationales Begegnungs- und Forschungszentrum (IBFI), Schloss Dagstuhl, Germany (2005). http://drops.dagstuhl.de/opus/volltexte/2005/349
41. Schütze, O., Vasile, M., Junge, O., Dellnitz, M., Izzo, D.: Designing optimal low thrust gravity assist trajectories using space pruning and a multi-objective approach. Eng. Optim. **41**(2), 155–181 (2009)
42. Schütze, O., Witting, K., Ober-Blöbaum, S., Dellnitz, M.: Set oriented methods for the numerical treatment of multiobjective optimization problems. In: Tantar, E., et al. (eds.) EVOLVE - A Bridge between Probability, Set Oriented Numerics and Evolutionary Computation, pp. 187–219. Springer, Heidelberg (2013)
43. Schütze, O., Cuate, O., Martín, A., Peitz, S., Dellnitz, M.: Pareto explorer: a global/local exploration tool for many-objective optimization problems. Eng. Optim. (2019)
44. Sinha, A., Deb, K.: Bilevel multi-objective optimization and decision making. In: Talbi, E.G. (ed.) Metaheuristics for Bi-Level Optimization. Springer, Heidelberg (2013)
45. Sun, J.-Q., Jia, T., Xiong, F.-R., Qin, Z.-C., Wu, W., Ding, Q.: Aircraft landing gear control with multi-objective optimization using generalized cell mapping. Trans. Tianjin Univ. **21**(2), 140–146 (2015)
46. Sun, J.-Q., Xiong, F.-R., Schütze, O., Hernández, C.: Cell Mapping Methods - Algorithmic Approaches and Applications. Springer, Heidelberg (2007)
47. Vicente, L.N., Calamai, P.H.: Bilevel and multilevel programming: a bibliography review. J. Global Optim. **5**, 291–306 (1994)
48. Xiong, F.-R., Qin, Z.-C., Hernández, C., Sardahi, Y., Narajani, Y., Liang, W., Xue, Y., Schütze, O., Sun, J.-Q.: A multi-objective optimal pid control for a nonlinear system with time delay. Theoret. Appl. Mech. Lett **3**(6), 140–146 (2013)
49. Xiong, F.-R., Qin, Z.-C., Xue, Y., Schütze, O., Ding, Q., Sun, J.-Q.: Multi-objective optimal design of feedback controls for dynamical systems with hybrid simple cell mapping algorithm. Commun. Nonlinear Sci. Numer. Simul. **19**(5), 1465–1473 (2014)

The Gradient Subspace Approximation and Its Application to Bi-objective Optimization Problems

Oliver Schütze[1(✉)], Lourdes Uribe[2], and Adriana Lara[2]

[1] Computer Science Department, Cinvestav-IPN, Mexico City, Mexico
schuetze@cs.cinvestav.mx
[2] ESFM del Instituto Politécnico Nacional, Mexico City, Mexico
{lourdesur,adriana}@esfm.ipn.mx

Abstract. Evolutionary algorithms are very popular and are frequently applied to many different optimization problems. Reasons for this success include that methods of this kind are of global nature, very robust, and only require minimal assumptions on the optimization problem. It is also known that such methods need quite a few resources to generate accurate approximations of the solution sets. As a remedy, researchers have used hybrid (or memetic) algorithms, i.e., evolutionary algorithms coupled with local search for which mainly techniques from mathematical programming are utilized. Such hybrids typically yield satisfying results, the problem, however, remains that the algorithms are relatively expensive since the gradients have to be computed or approximated at each given candidate solution that is designated for local search.

In this chapter, we review the Gradient Subspace Approximation (GSA) which allows to compute a descent direction in a best fit manner from given neighborhood information that is e.g. already given in evolutionary algorithms. The computation of such directions comes hence for free in terms of additional function evaluations of the given problem which opens the door for the realization of low-cost local search engines within evolutionary algorithms. In a next step, we show how GSA can be applied to the context of bi-objective optimization. Finally, to demonstrate the benefit of the method we present some results on a hybrid that is based on the evolutionary algorithm NSGA-II.

Keywords: Gradient Subspace Approximation · Gradient free optimization · Bi-objective optimization · Descent directions

1 Introduction

In many problems in engineering and finance the problem arises that several objectives have to be optimized concurrently [3,7,8,11,12,14,17,22,27,28,30, 37]. One main challenge of such *multi-objective optimization problems* (MOPs)

O. Junge et al. (Eds.): SON 2020, SSDC 304, pp. 355–390, 2020.
https://doi.org/10.1007/978-3-030-51264-4_15

is that their solution set—the so-called Pareto set respectively its image, the Pareto front—typically forms a $(k-1)$-dimensional object, where k is the number of objectives involved in the problem. This is in contrast to classical scalar optimization problems (SOPs), where one expects that the optimum is taken at one single solution. Modern heuristics such as Multiobjective Evolutionary Algorithms (MOEAs, e.g., [2,9,10,23]) are able to provide an approximation of the entire Pareto set/front of a given MOP in one single run of the algorithm due to their set oriented approach. These methods are very robust e.g. to initial conditions and only require minimal assumptions from the model (e.g., no derivative information). Evolutionary algorithms as well as other related heuristics are hence very popular for the numerical treatment of MOPs as well as other optimization problems. One drawback, however, that most of them suffer is that they need quite a few function evalutions in order to obtain accuate approximations of the Pareto sets/fronts. As a remedy, many researchers have in the past hybridized the (global) evolutionary algorithms with local search techniques mainly coming from Mathematical Programming that utilize derivative information from the objectives and the constraint functions. Such methods are termed hybrid evolutionary algorithms or memetic algorithms. While such methods yield in almost all cases satisfying results, they are still relatively expensive since the derivative information is required for every point that is designated for local search.

In this chapter, we review several recently developed tools that allow to realize a local search within a population based optimization algorithm with low computational cost. The basic idea of the Gradient Subspace Approximation (GSA, [33]) is to utilize existing neighborhood information to estimate the most greedy direction within the search space that is spanned by the samples. One advantage of this approach is that it can easily be extended to the context of constrained problems. The focus of this chapter is on bi-objective optimization problems (BOPs, i.e., MOPs with $k = 2$ objective functions). For this, we will consider the descent direction for BOPs proposed in [25] and present recent adaptations for constrained problems ([38]). Next, we will show how GSA can be used to build a low-cost local search engine that can be used within a population based algorithm such as a MOEA. In order to show the efficiency of the method, we will show some numerical results from a hybrid evolutionary algorithm that coupled the GSA based local search engine with the famous evolutionary algorithm NSGSA-II ([10]) which is state-of-the-art for MOPs with two and three objectives.

The remainder of this chapter is organized as follows: in Sect. 2, we review the Gradient Subspace Approximation for the approximation of the most greedy search direction out of given neighborhood information. In Sect. 3, we review a particular descent direction for bi-objective optimization problems for unconstrained, equality and inequality contrained problems. We further combine these two concepts in order to build a low-cost local searcher within a chosen set-oriented optimization method (such as an evolutionary algorithm). In Sect. 4

we present some numerical examples on a hybrid evolutionary algorithm that is based on the famous NSGA-II. Finally, we draw our conclusions Sect. 5.

2 Gradient Subspace Approximation

In this section, we will review the Gradient Subspace Approximation that aims to compute a descent direction at a given point x_0 and a given scalar optimization problem using existing neighborhood information. For more details the reader is referred to [33].

2.1 Background and Related Work

In this section, we will consider continuous scalar optimization problems (SOP) of the following form

$$\begin{aligned} &\min_x f(x) \\ &\text{s.t. } g_i(x) \leq 0, \; i = 1, \ldots, p \\ &\qquad h_j(x) = 0, \; j = 1, \ldots, m. \end{aligned} \tag{1}$$

Hereby, $f : \mathbb{R}^n \to \mathbb{R}$ is called the objective function, and the functions $g_i : \mathbb{R}^n \to \mathbb{R}$ and $h_j : \mathbb{R}^n \to \mathbb{R}$ are called the inequality and equality constraints, respectively. We assume that all functions f, g_i and h_j are differentiable.

A point x is called feasible if it satisfies all constraints. A point x^* is called a solution to (1) if it is feasible and if there exists no other feasible point y that has a lower objective value.

The object of interest in this section is the gradient of a function at a given point x_0. Formally, the gradient of the objective at x_0 is defined by

$$\nabla f(x) = \left(\frac{\partial f}{\partial x_1}(x), \ldots, \frac{\partial f}{\partial x_n}(x) \right)^T \in \mathbb{R}^n. \tag{2}$$

A vector $\nu \in \mathbb{R}^n$ is called a *descent direction* for f at x_0 if

$$\langle \nabla f(x_0), \nu \rangle < 0;$$

in that case, it holds for all sufficiently small step sizes $t > 0$ that $f(x_0 + t\nu) < f(x_0)$.

When the gradient of a function is not available in analytic form, there are several ways to obtain either $\nabla f(x_0)$ at a given point x_0 or approximations of this vector. The most prominent and widely used technique is to use finite differences (e.g., [29]). The method presented in this section, GSA is also using a finite difference approach. The variance, however, is that GSA can gather the sampling points from all directions whereas the classical finite difference method utilizes samples in coordinate directions. GSA is hence more suited to the use within population based algorithms since in this case the neighboring samples are already given (and typically not aligned in coordinate directions). In [6], a very similar method is proposed to approximate the Jacobians of the objective map of a given unconstrained multi-objective optimization problem. This work, however, does not discuss how to address constrained problems. The Hill Climber with Sidestep [26] and the Directed Search Method [35,36] both use neighborhood samples in order to determine promising search directions for the local search within hybrid evolutionary algorithms. The difference to the GSA is that these works do not directly aim to approximate the gradient. Automatic Differentiation (AD, [18]) can be used to evaluate the exact gradient at a given point x_0 if the function is specified by a computer program. One drawback of AD is that it can not be applied if this computer program is provided in form of binary code.

Further, there are several methods that replace the original objective by easier models. One example is response surface methodology (RSM), where the objective function f is replaced by low-order polynomials $\tilde{f}$ (mainly of degree one and two) those gradients are approximated using least squares techniques [24]. If a first-order model is chosen, the match of the gradients $\nabla\tilde{f}(x_0)$ and $\nabla f(x_0)$ is typically quite good for a nonlinear function f if the chosen point x_0 is sufficiently far away from the optimum. For second-order models, the match is in general much better, however, this accuracy comes with an additional cost since n^2 parameters have to be fitted at every point x_0. Further works that can utilize scattered samples can be found in [13,20]. In [20], a least squares regression is performed while in [13] statistical expectation is used. In both works, the authors restrict themselves to unconstrained problems.

2.2 The Basic Idea

The task is to compute a cost-free good approximation of the normalized gradient

$$n(x_0) := -\frac{\nabla f(x_0)}{\|\nabla f(x_0)\|_2}, \tag{3}$$

evaluated at a given point x_0 within a particular subspace of the $\mathbb{R}^n$. For this, we make use of the fact that the gradient is the direction of the steepest ascent, and hence $n(x_0)$ can be seen as the solution of the following optimization problem:

$$\begin{aligned} &\min_{\nu\in\mathbb{R}^n} \langle \nabla f(x_0), \nu\rangle \\ &\text{s.t. } \|\nu\|_2^2 = 1. \end{aligned} \tag{4}$$

To avoid to directly compute the gradient we can make use of neighboring information as follows: assume that we are given the points $x_1, \ldots, x_r$ in the vicinity of x_0 (e.g., these samples may be contained within the population or archive of a set based optimization method such as an evolutionary algorithm) as well as their function values $f(x_i)$. In that case, we can use this information to approximate the directional derivatives in the directions

$$\nu_i := \frac{x_i - x_0}{\|x_i - x_0\|_2}, \qquad i = 1, \ldots, r. \tag{5}$$

Since it holds

$$f'_{\nu_i}(x_0) = \langle \nabla f(x_0), \nu_i \rangle = \frac{f(x_i) - f(x_0)}{\|x_i - x_0\|_2} + O(\|x_i - x_0\|_2), \tag{6}$$

where O denotes the Landau symbol, and since we are given the points x_0 and x_i with their respective objective values $f(x_0)$ and $f(x_i)$, we can hence use the approximations

$$f'_{\nu_i}(x_0) \approx \frac{f(x_i) - f(x_0)}{\|x_i - x_0\|_2}, \qquad i = 1, \ldots, r. \tag{7}$$

Given the samples $x_1, \ldots, x_r$ and ν_i as above we define the subspace S as

$$S = span\{\nu_1, \ldots, \nu_r\} \tag{8}$$

and are interested in a best approximation of $n(x_0)$ within S. Since every $\nu \in S$ can be written as

$$v = \sum_{i=1}^{r} \lambda_i \nu_i \tag{9}$$

for some $\lambda = (\lambda_1, \ldots, \lambda_r) \in \mathbb{R}^r$, and

$$\langle \nabla f(x_0), \nu \rangle = \sum_{i=1}^{r} \lambda_i \langle \nabla f(x_0), \nu_i \rangle \tag{10}$$

we can state problem (4), where we restrict the search to S, as follows:

$$\begin{aligned} \min_{\lambda \in \mathbb{R}^r} \; & \sum_{i=1}^{r} \lambda_i \langle \nabla f(x_0), \nu_i \rangle \\ \text{s.t.} \; & \left\| \sum_{i=1}^{r} \lambda_i \nu_i \right\|_2^2 = 1. \end{aligned} \tag{11}$$

Hence, when using an approximation of the directional derivatives as in (7) via using neighboring samples, we can avoid to directly compute the gradient. One advantage of using problem (11) is that constraint information can directly be incorporated.

In the following, we will analyze the best fit approximations of $n(x_0)$ within the subspace S both for unconstrained and constrained SOPs. For this, we will first consider the ideal scenario where we assume that we are given all directional derivatives, and later on we will discuss the gradient-free realizations.

2.3 Gradient Subspace Approximation

We will in the following discuss how to approximate the most greedy search direction out of the given data, separately for unconstrained, equality constrained, and inequality constrained problems.

2.3.1 Unconstrained Problems

We are given the problem

$$\min_{x \in \mathbb{R}^n} f(x), \tag{12}$$

where $f : \mathbb{R}^n \to \mathbb{R}$. Further, we are given a point $x_0 \in \mathbb{R}^n$ and the samples $x_1, \ldots, x_r \in \mathbb{R}^n$ in the vicinity of x_0, together with their objective values $f(x_i)$. Further, for purpose of a better analysis of the problem we also assume that we are given the directional derivatives

$$\langle \nabla f(x_0), \nu_i \rangle, \qquad i = 1, \ldots, r. \tag{13}$$

Define the matrix V by

$$V = (\nu_1, \ldots, \nu_r) \in \mathbb{R}^{n \times r}. \tag{14}$$

Then, the most greedy search direction within the subset

$$S = span\{\nu_1, \ldots, \nu_r\} \tag{15}$$

is given by the solution of the following problem

$$\begin{aligned} \min_{\lambda \in \mathbb{R}^r} \; & \sum_{i=1}^{r} \lambda_i \langle \nabla f(x_0), \nu_i \rangle \\ \text{s.t. } & \lambda^T V^T V \lambda - 1 = 0. \end{aligned} \tag{16}$$

The following result shows that the solution of (16) can be computed in closed form.

Proposition 1. *Let $\nu_1, \ldots, \nu_r \in \mathbb{R}^n$, $r \leq n$, be linearly independent and*

$$\tilde{\lambda}^* := -(V^T V)^{-1} V^T \nabla f(x_0). \tag{17}$$

Then

$$\lambda^* := \frac{\tilde{\lambda}^*}{\|V\lambda^*\|_2^2} \tag{18}$$

is the unique solution of (16) and

$$\nu^* = \frac{-1}{\|V\lambda^*\|_2^2} V (V^T V)^{-1} V^T \nabla f(x_0) \tag{19}$$

is the most greedy search direction in S.

Proof. The Karush-Kuhn Tucker (KKT) system of (16) reads as

$$\nabla_\lambda L(\lambda, \mu) = V^T \nabla f(x_0) + 2\mu V^T V \lambda = 0 \tag{20}$$

$$h(\lambda) = \lambda^T V^T V \lambda - 1 = 0. \tag{21}$$

Apparently, Equation (21) is only used for normalization. If we omit this equation and the factor 2μ in (20) we can rewrite (20) as the following normal equation system

$$V^T V\lambda = -V^T \nabla f(x_0). \tag{22}$$

To solve the entire KKT system we have to choose $2\mu = \|V\lambda^*\|_2^2$. Finally, the claim follows since the Hessian of the Lagrangian

$$\nabla^2_{\lambda\lambda} L(\lambda, \mu) = V^T V \tag{23}$$

is positive definite since the directions ν_i are linearly independent. □

Next, we discuss how to approximate the most greedy solution ν^* without explicitly computing or approximating the gradients. Since

$$V^T \nabla(x_0) = \begin{pmatrix} \langle \nabla f(x_0), \nu_1 \rangle \\ \vdots \\ \langle \nabla f(x_0), \nu_r \rangle \end{pmatrix} \tag{24}$$

we can do the approximation as follows: let $d = (d_1, \ldots, d_r) \in \mathbb{R}^r$, where

$$d_i := \frac{f(x_i) - f(x_0)}{\|x_i - x_0\|_2}, \qquad i = 1, \ldots, r, \tag{25}$$

and $\tilde{\lambda}$ be the vector that solves the system of linear equations

$$V^T V \tilde{\lambda} = -d, \tag{26}$$

then the most greedy search direction can be approximated as

$$\tilde{\nu}^* = \frac{-1}{\|V\tilde{\lambda}\|_2^2} V(V^T V)^{-1} d. \tag{27}$$

Remark 1. (a) To compute ν^* one has to solve system (22). It is hence advisable to avoid to choose directions ν_i that nearly point into the same directions. The linear equation system yields the best condition number if the directions are chosen orthogonal to each other. In this case, we obtain

$$\nu^* = \frac{-1}{\|\lambda^*\|_2^2} V V^T \nabla f(x_0), \tag{28}$$

i.e., the orthogonal projection of $\nabla f(x_0)$ onto S. That is, ν^* is the best approximation of $n(x_0)$ in S.

(b) In the special case that the coordinate directions are chosen, i.e., if we choose

$$x_i = x_0 + t_i e_{j_i}, \;\; i = 1, \ldots, r, \tag{29}$$

for the samples, where e_j denotes the j-th unit vector, we obtain for the j_i-th entry of $\tilde{\nu}^*$ (without normalization)

$$\tilde{\nu}^*_{j_i} = \frac{f(x_0 + t_i e_{j_i}) - f(x_0)}{|t_i|}. \tag{30}$$

That is, if we for instance choose $x_i = x_0 + t_i e_i$, $i = 1, \dots, n$ (i.e., all coordinate directions), the search direction ν^* coincides with the forward difference quotient.

(c) The idea of GSA is to utilize existing data whenever possible. However, it may be the case that for a given point x_0 that the existing data is not sufficient (e.g., not enough individuals of the current population are close enough to x_0). A possible remedy may be to sample further points in order to compute a search direction. In that case it makes sense to choose the points so that the resulting directions ν_i are orthogonal to each other as well as to all existing directions. See [1] for a possible realization.

Example 1. We consider the objective $f : \mathbb{R}^6 \to \mathbb{R}$, where

$$f(x) = \sum_{i=1}^{6} x_i^2. \tag{31}$$

Let $x_0 = [1,1,1,1,1,1]^T$, then $\nabla f(x_0) = [2,2,2,2,2,2]^T$ and

$$g = \frac{-1}{\sqrt{24}} [2,2,2,2,2,2]^T \tag{32}$$

for which $\langle \nabla f(x_0), g \rangle = -4.8990$.

First, we choose three orthogonal search directions that form the matrix V^T as follows:

$$V^T = \begin{pmatrix} 2\,0\,1\,0\,0\,0 \\ 0\,1\,0\,3\,0\,0 \\ 0\,0\,0\,0\,3\,1 \end{pmatrix}. \tag{33}$$

Doing so, we obtain $\langle \nabla f(x_0), \frac{v_1}{\|v_1\|} \rangle = 2.6833$, $\langle \nabla f(x_0), \frac{v_2}{\|v_2\|} \rangle = 2.5298$ and $\langle \nabla f(x_0), \frac{v_3}{\|v_3\|} \rangle = 2.5298$. If we solve problem (16) we get

$$v^* = [-0.5367, -0.1789, -0.2683, -0.5367, -0.5367, -0.1789]^T, \tag{34}$$

for which $\langle \nabla f(x_0), v^* \rangle = -4.4721$.

Next, we use the samples

$$x_i = x_0 + 0.1 v_i, \quad \text{for } i = 1, 2, 3, \tag{35}$$

and via formula (27) we obtain

$$\tilde{v}^* = [-0.5237, -0.1813, -0.2618, -0.5438, -0.5438, -0.1813]^T$$

which leads to $\langle \nabla f(x_0), \tilde{v}^* \rangle = -4.4714$.

If choosing the non-orthogonal search directions

$$V^T = \begin{pmatrix} 1\,1\,0\,0\,0\,0 \\ 0\,2\,1\,2\,0\,0 \\ 1\,0\,2\,0\,1\,2 \end{pmatrix}, \tag{36}$$

we obtain $\langle \nabla f(x_0), \frac{v_1}{\|v_1\|}\rangle = 2.8284$, $\langle \nabla f(x_0), \frac{v_2}{\|v_2\|}\rangle = 3.3333$ and $\langle \nabla f(x_0), \frac{v_3}{\|v_3\|}\rangle = 3.7947$.

Solving (16) leads to

$$v^* = [-0.3769, -0.4744, -0.5686, -0.3054, -0.2080, -0.4159]^T \tag{37}$$

with $\langle \nabla f(x_0), v^*\rangle = -4.6985$. For the discretized problem we obtain

$$\tilde{v}^* = [-0.3595, -0.4674, -0.5770, -0.3171, -0.2092, -0.4184]^T \tag{38}$$

with $\langle \nabla f(x_0), \tilde{v}^*\rangle = -4.6972$.

2.3.2 Equality Constrained Problems

Next, we assume that the SOP contains some equality constraints, i.e., that we are given the following problem

$$\begin{aligned} &\min_{x\in\mathbb{R}^n} f(x) \\ &\text{s.t. } h_i(x) = 0,\ i = 1,\dots,p, \end{aligned} \tag{39}$$

where we assume that each $h_i : \mathbb{R}^n \to \mathbb{R}$ is differentiable.

Analogously to the unconstrained case discussed above the most greedy search direction at x_0 in the entire space $\mathbb{R}^n$ is given by

$$\begin{aligned} &\min_{\nu\in\mathbb{R}^n} \langle \nabla f(x_0), \nu\rangle \\ &\text{s.t. } \|\nu\|_2^2 = 1 \\ &\qquad \langle \nabla h_i(x), \nu\rangle = 0,\ i = 1,\dots,p, \end{aligned} \tag{40}$$

and the most greedy direction at x_0 within the subspace S is given by

$$\begin{aligned} &\min_{\lambda\in\mathbb{R}^r} \sum_{i=1}^r \lambda_i \langle \nabla f(x_0), \nu_i\rangle \\ &\text{s.t. } \lambda^T V^T V \lambda - 1 = 0 \\ &\qquad \sum_{i=1}^r \lambda_i \langle \nabla h_j(x_0), \nu_i\rangle = 0,\ j = 1,\dots,p. \end{aligned} \tag{41}$$

Denote the matrix H by

$$H = \begin{pmatrix} \nabla h_1(x_0)^T \\ \vdots \\ \nabla h_p(x_0)^T \end{pmatrix} \in \mathbb{R}^{p\times n}. \tag{42}$$

As for the unconstrained case, we can also express the most greedy solution for an equality constrained problem analytically.

Proposition 2. *Let $\nu_1, \ldots, \nu_r \in \mathbb{R}^n$ be linearly independent where $p \leq r \leq n$, let $rank(H) = p$, and*

$$\begin{pmatrix} \tilde{\lambda}^* \\ \tilde{\mu}^* \end{pmatrix} = \begin{pmatrix} V^T V & (HV)^T \\ HV & 0 \end{pmatrix}^{-1} \begin{pmatrix} -V^T \nabla f(x_0) \\ 0 \end{pmatrix}, \tag{43}$$

then

$$\lambda^* := \frac{\tilde{\lambda}^*}{\|V\lambda^*\|_2^2} \tag{44}$$

is the unique solution of (41) and thus

$$\nu^* = \frac{-1}{\|V\lambda^*\|_2^2} V(V^T V)^{-1} V^T \nabla f(x_0) \tag{45}$$

is the most greedy search direction in $span\{\nu_i, \ldots, \nu_r\}$.

Proof. The KKT system of (41) is given by

$$V^T \nabla f(x_0) + 2\mu_0 V^T V \lambda + (HV)^T \mu = 0 \tag{46}$$

$$HV\lambda = 0 \tag{47}$$

$$\lambda^T V^T V \lambda - 1 = 0, \tag{48}$$

and via applying the same "normalization trick" as above we can transform the KKT equations into

$$\begin{pmatrix} V^T V & (HV)^T \\ HV & 0 \end{pmatrix} \begin{pmatrix} \lambda \\ \mu \end{pmatrix} = \begin{pmatrix} -V^T \nabla f(x_0) \\ 0 \end{pmatrix}. \tag{49}$$

To show that the matrix is regular, let $y \in \mathbb{R}^r$ and $z \in \mathbb{R}^p$ such that

$$\begin{pmatrix} V^T V & (HV)^T \\ HV & 0 \end{pmatrix} \begin{pmatrix} y \\ z \end{pmatrix} = 0. \tag{50}$$

It follows that $HVy = 0$ and hence that

$$0 = \begin{pmatrix} y \\ z \end{pmatrix}^T \begin{pmatrix} V^T V & (HV)^T \\ HV & 0 \end{pmatrix} \begin{pmatrix} y \\ z \end{pmatrix} = yV^T V y. \tag{51}$$

Thus, it is $y = 0$ since $V^T V$ is positive definite. Further, by (50) it follows that $V^T H^T z = 0$. Since $V^T \in \mathbb{R}^{n \times r}$ has rank $r \geq p$, it follows that $V^T H^T$ has rank p. This implies that also $z = 0$, and thus, that the matrix in (43) is regular.

The rest follows by the discussion above setting $2\mu_0 = \| \sum_{i=1}^r \tilde{\lambda}_i^* \nu_i \|_2^2$ and since the Hessian of the Lagrangian $\nabla^2_{\lambda\lambda} L(\lambda, \mu) = V^T V$ is positive definite. □

The key for a gradient-free approximation of the search direction is the matrix HV. Since

$$(HV)_{ij} = \nabla h_i(x_0)^T \nu_j,$$

we compute an approximation $M = (m_{ij})$ of HV via

$$m_{ij} := \frac{h_i(x_j) - h_i(x_0)}{\|x_j - x_i\|_2}, \quad i = 1, \ldots, p, \; j = 1, \ldots, r. \tag{52}$$

Doing so, we can now solve the system

$$\begin{pmatrix} V^T V & M^T \\ M & 0 \end{pmatrix} \begin{pmatrix} \lambda \\ \mu \end{pmatrix} = \begin{pmatrix} -d \\ 0 \end{pmatrix} \tag{53}$$

which leads to λ^*. To obtain ν^* we proceed as for the unconstrained case using the approximation $V^T \nabla f(x_0) \approx d$.

Example 2. We consider the SOP from Example 1 and impose the constraint

$$h(x) = x_1 + x_2 + x_3 = 0. \tag{54}$$

For $x_0 = [-1, 0, 1, 1, 1, 1]^T$ we have $\nabla f(x_0) = [2, 2, 2, 2, 2, 2]^T$, $g = \frac{-1}{\sqrt{24}}[2, 2, 2, 2, 2, 2]^T$, $\nabla_h(x_0) = [1, 1, 1, 0, 0, 0]^T$ and $\langle \nabla f(x_0), g \rangle = -4.8990$.

First, we choose again the three orthogonal search directions

$$V^T = \begin{pmatrix} 2\,0\,1\,0\,0\,0 \\ 0\,1\,0\,3\,0\,0 \\ 0\,0\,0\,0\,3\,1 \end{pmatrix}; \tag{55}$$

and, we obtain $\langle \nabla f(x_0), \frac{v_1}{\|v_1\|} \rangle = 2.6833$, $\langle \nabla f(x_0), \frac{v_2}{\|v_2\|} \rangle = 2.5298$, $\langle \nabla f(x_0), \frac{v_3}{\|v_3\|} \rangle = 2.5298$, and

$$v^* = [0.1210, -0.1815, 0.0605, -0.5444, -0.7662, -0.2554]^T \tag{56}$$

with $\langle \nabla f(x_0), v^* \rangle = -3.1322$.

Using the above setting and the samples

$$x_i = x_0 + 0.1 v_i, \quad \text{for } i = 1, 2, 3 \tag{57}$$

we obtain the search direction

$$\tilde{v}^* = [0.1293, -0.1939, 0.0646, -0.5817, -0.7368, -0.2456]^T$$

which leads to $\langle \nabla f(x_0), \tilde{v}^* \rangle = -3.1281$.

In a next step, we consider the non-orthogonal search directions

$$V^T = \begin{pmatrix} 1\,1\,0\,0\,0\,0 \\ 0\,2\,1\,2\,0\,0 \\ 1\,0\,2\,0\,1\,2 \end{pmatrix} \tag{58}$$

leading to $\langle \nabla f(x_0), \frac{v_1}{\|v_1\|}\rangle = 2.8284$, $\langle \nabla f(x_0), \frac{v_2}{\|v_2\|}\rangle = 3.3333$ and $\langle \nabla f(x_0), \frac{v_3}{\|v_3\|}\rangle = 3.7947$.

When solving (41) we obtain

$$v^* = [0.4050, 0.2244, -0.6294, -0.3963, -0.2156, -0.4312]^T$$

with $\langle \nabla f(x_0), v^*\rangle = -2.0863$ for the idealized problem and for the discretized problem we obtain

$$\tilde{v}^* = [0.4490, 0.1617, -0.6107, -0.4742, -0.1868, -0.3736]^T$$

with $\langle \nabla f(x_0), \tilde{v}^*\rangle = -2.0691$.

2.3.3 Inequality Constrained Problems

Finally, we assume we are given an inequality constrained SOP of the form

$$\begin{aligned} &\min_{x\in\mathbb{R}^n} f(x) \\ &\text{s.t. } g_i(x) \leq 0,\ i = 1, \ldots, m, \end{aligned} \tag{59}$$

where we for simplicity assume that all inequalities are active at a given point x_0. The most greedy search direction at x_0 is given by

$$\begin{aligned} &\min_{\nu\in\mathbb{R}^n} \langle \nabla f(x_0), \nu\rangle \\ &\text{s.t. } \|\nu\|_2^2 = 1 \\ &\qquad \langle \nabla g_i(x), \nu\rangle \leq 0,\ i = 1, \ldots, m, \end{aligned} \tag{60}$$

and the related subspace optimization problem reads as

$$\begin{aligned} &\min_{\lambda\in\mathbb{R}^r} \sum_{i=1}^r \lambda_i \langle \nabla f(x_0), \nu_i\rangle \\ &\text{s.t. } \lambda^T V^T V \lambda - 1 = 0 \\ &\qquad \sum_{i=1}^r \lambda_i \langle \nabla g_j(x_0), \nu_i\rangle \leq 0,\ j = 1, \ldots, m. \end{aligned} \tag{61}$$

One way to find a solution to (61) is to use gradient projection which is advantageous in particular if m is small and $r \gg m$. In the following, we first discuss the special case $m = 1$ (i.e., one active inequality constraint) and will later on discuss the general case.

The classical gradient projection approach is to take the solution ν^* of the underlying unconstrained problem (16) and to project it to the space $\nabla g(x_0)^\perp$ that is orthogonal to $\nabla g(x_0)$ (see Fig. 1): given a QR decomposition of $\nabla g(x_0)$, i.e.,

$$\nabla g(x_0) = QR = (q_1, \ldots, q_n)R, \tag{62}$$

then the vectors $q_2, \ldots, q_n$ build an orthonormal basis (ONB) of $\nabla g(x_0)^\perp$. Using $Q_g = (q_2, \ldots, q_n)$, the projection is hence given by

$$\nu_{new} = Q_g Q_g^T \nu^*. \tag{63}$$

It is of course not advisable to follow this approach directly since $\nabla g(x_0)$ is neither given, nor do we want to approximate it. Alternatively, we propose to proceed as follows: let

$$M := \nabla g(x_0)^T V = (\langle \nabla g(x_0), \nu_1 \rangle, \ldots, \langle \nabla g(x_0), \nu_r \rangle) \in \mathbb{R}^{1 \times r}. \tag{64}$$

Note that if w is a kernel vector of M then Vw is perpendicular to $\nabla g(x_0)$ and vice versa. Thus, we can compute the matrix

$$K = (k_1, \ldots, k_{r-1}) \in \mathbb{R}^{r \times (r-1)} \tag{65}$$

those column vectors build an ONB of the kernel of M. If the search directions ν_i are orthogonal, then also the vectors $Vk_1, \ldots, Vk_{r-1}$ are orthogonal to each other. The latter are the column vectors of $VK \in \mathbb{R}^{n \times (r-1)}$ (if the ν_i's are not orthogonal to each other, VK has to be orthogonalized via another QR decomposition). Doing so, the projected vector to the kernel of M is given by

$$\tilde{\nu}_{new} = VK(VK)^T \nu^* = VKK^T V^T \nu^*. \tag{66}$$

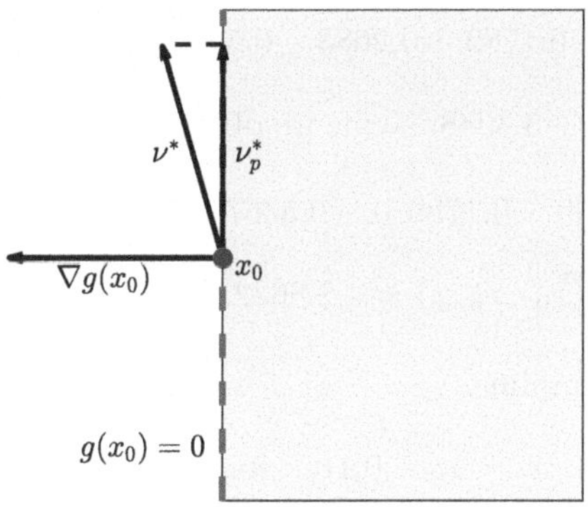

Fig. 1. Handling inequality constraints using gradient projection.

For a general number m of inequality constraints we can extend the method as follows: let

$$M = GV = (\langle \nabla g_i(x_0), \nu_j \rangle)_{\substack{i=1,\ldots,m \\ j=1,\ldots,r}}, \tag{67}$$

and perform the following steps

(1) compute an orthonormal basis $K \in \mathbb{R}^{r\times(r-m)}$ of the kernel of M
(2) compute $VK = QR = (q_1, \ldots, q_{r-m}, \ldots, q_n)R$ and set $O := (q_1, \ldots, q_{r-m}) \in \mathbb{R}^{n\times(r-m)}$
(3) $\tilde{\nu}_{new} = \tilde{Q}\tilde{Q}^T\nu^*$

Example 3. We consider again the SOP from Example 1 but this time we impose the inequality

$$g(x) = 1 - x_1 \leq 0. \tag{68}$$

For $x_0 = [1,1,1,1,1,1]^T$ we have $\nabla f(x_0) = [2,2,2,2,2,2]^T$, $g = \frac{-1}{\sqrt{24}}[2,2,2,2,2,2]^T$ and $\nabla_c(x_0) = [-1,0,0,0,0,0]^T$. Thus $\langle\nabla f(x_0), g\rangle = -4.8990$.

First, we again choose the three orthogonal search directions

$$V^T = \begin{pmatrix} 2\,0\,1\,0\,0\,0 \\ 0\,1\,0\,3\,0\,0 \\ 0\,0\,0\,0\,3\,1 \end{pmatrix}, \tag{69}$$

and obtain $\langle\nabla f(x_0), \frac{v_1}{\|v_1\|}\rangle = 2.6833$, $\langle\nabla f(x_0), \frac{v_2}{\|v_2\|}\rangle = 2.5298$ and $\langle\nabla f(x_0), \frac{v_3}{\|v_3\|}\rangle = 2.5298$.

For the search direction we obtain

$$v^* = [-0.5367, -0.1789, -0.2683, -0.5367, -0.5367, -0.1789]^T$$

with $\langle\nabla f(x_0), v^*\rangle = -4.4721$. Then by the gradient projection approach and v^*, we obtain the projected vector

$$v_{new} = [0, -0.1789, -0.2683, -0.5367, -0.5367, -0.1789]^T$$

with $\langle\nabla f(x_0), v_{new}\rangle = -3.3988$. Next, we obtain

$$\tilde{v}_{new} = [0, -0.1789, 0, -0.5367, -0.5367, -0.1789]^T$$

via Eq. (54) with $\langle\nabla f(x_0), \tilde{v}_{new}\rangle = -2.8622$.

Next we use the sampling

$$x_i = x_0 + 0.1v_i, \quad \text{for} \quad i = 1, 2, 3 \tag{70}$$

which leads to the search direction

$$\tilde{v}^* = [-0.5237, -0.1813, -0.2618, -0.5438, -0.5438, -0.1813]^T$$

with $\langle\nabla f(x_0), \tilde{v}^*\rangle = -4.4714$. Then by the gradient projection approach and $\tilde{v}^*$, we obtain $\tilde{v}_{new} = [0, -0.1813, 0, -0.5438, -0.5438, -0.1813]^T$ and $\langle\nabla f(x_0), \tilde{v}_{new}\rangle = -2.9004$.

3 Bi-objective Optimization

In this section, we will review a descent direction for bi-objective optimization problems, and will show how GSA can be used approximate these directions gradient-free. For details the reader is referred to [38].

3.1 Background and Related Work

In many applications one is faced with the problem that several objectives have to be optimized concurrently leading to a multi-objective optimization problem (MOP, e.g., [3,8,12,14,17,27,28,30,37]).

A continuous MOP can be expressed mathematically as

$$\begin{aligned} \min_x \; & (f_1(x), \dots, f_k(x))^T \\ \text{s.t. } & g_i(x) \leq 0, \; i = 1, \dots, p \\ & h_j(x) = 0, \; j = 1, \dots, m, \end{aligned} \tag{71}$$

where the f_i, $i = 1, \dots, k$ are the objectives to be minimized, and the g_i's and h_j's are the inequalities and equalities, respectively. Denote by Q the feasible set. We assume that all objectives and all constraint functions are differentiable. To define optimality of a MOP the concept of Pareto dominance is used: let $v, w \in \mathbb{R}^k$, then we say that the vector v is *less than* the vector w ($v <_p w$), if $v_i < w_i$ for all $i \in \{1, \dots, k\}$; the relation $\leq_p$ is defined analogously. A vector $y \in Q$ is *dominated* by a vector $x \in Q$ ($x \prec y$) with respect to (71) if $F(x) \leq_p F(y)$ and $F(x) \neq F(y)$, else y is called non-dominated by x. A point $x^* \in \mathbb{R}^n$ is Pareto optimal to (71) if there is no $y \in Q$ which dominates x. The set of all the Pareto optimal points is called the Pareto set and its image is the Pareto front. Both Pareto set and front form under certain (mild) smoothness assumptions a $(k-1)$-dimensional object ([21]). We will in this section focus on bi-objective problems (BOPs), i.e., on MOPs where $k = 2$.

A Multi-objective Descent Direction (MODD) ν at a point x_0 and a given MOP is a direction in which a sufficiently small movement yields dominating solutions, i.e.,

$$x_0 + t\nu \prec x_0 \qquad \forall t \in (0, \bar{t}) \tag{72}$$

for a certain $\bar{t} > 0$. In [25], a descent direction has been proposed for unconstrained BOPs.

Proposition 3 ([25]). *Let $x \in \mathbb{R}^n$ and $f_1, f_2 : \mathbb{R}^n \to \mathbb{R}$ define an unconstrained BOP. If $\nabla f_i(x) \neq 0$ for $i \in \{1, 2\}$ then*

$$\nu_L := -\left[\frac{\nabla f_1(x)}{\|\nabla f_1(x)\|} + \frac{\nabla f_2(x)}{\|\nabla f_2(x)\|} \right], \tag{73}$$

is a descent direction of x at F.

In the sequel we will show how to adapt this to the context of constrained BOPs, and how to make a gradient-free realization via utilizing GSA.

It is worth mentioning that there exist some proposals to compute MODDs in general; they make use of first or second-order information related to the objective functions. Among these proposals is what is called the Steepest Descent Direction [15], which requires the solution to a quadratic programming problem involving the Jacobian of F. This method is valid for both convex or non-convex Pareto fronts. In [32], the authors introduced a mathematical formula that uses the solution of a stochastic differential equation related to the Karush-Kuhn-Tucker conditions in order to generate the solution set. This technique requires all the involved functions to be continuously differentiable, and it only works for unconstrained MOPs. One particular work to mention is [5] where appears the idea of the descent cone. The descent cone gets determined by the intersection of the negative half-spaces generated by the objective function gradients. In [4] the authors notice that computing a MODD yields again a multi-objective optimization problem since the negative gradient of every objective function is in conflict with the remaining objectives' gradients. In this same work, the authors propose a way to calculate the entire set of MODDs to take one of them at random finally. Another proposal [19], called the Pareto Descent Method, computes a set of possible Pareto descent directions by solving a linear programming problem with the information from the descent cones. This method applies just for unconstrained and inequality CMOP. There exist also approaches based on the Newton method as [14] where at each iteration of a line search, one minimization subproblem has to be solved to obtain the direction to follow. In this method, the Hessians of all functions need to be available. In this direction, also, [12,16,34] proposed an extension of the multi-objective Newton method for equality constrained problems. In this chapter, we decided to work with expression (73) since it is the easiest and no-cost proposal since no linear or quadratic programming solvers are necessary for its computation. This feature also makes it suitable for the application presented in Sect. 4 when this direction will be introduced into a population-based heuristic.

3.2 A Descent Direction for Constrained BOPs

3.3 Equality Constrained BOPs

Here we consider equality constrained BOPs of the form

$$\begin{aligned} \min_{x\in\mathbb{R}^n} \; & F(x) := [f_1(x), f_2(x)]^T, \\ \text{s.t} \quad & h_j(x) = 0, \qquad j = 1, \ldots, m. \end{aligned} \tag{74}$$

We will in the following discuss separately the cases where x_0 is feasible and infeasible.

Feasible Case

We first assume that x_0 is feasible, i.e., that $h_j(x_0) = 0$ for all $j = 1, \ldots, m$. One way to obtain a MODD for the given BOP is to project the MODD of the related unconstrained BOP to the tangent space $T_{h^{-1}(0)}(x_0)$ of the feasible set $h^{-1}(0)$ at x_0: let ν be the MODD for the unconstrained BOP, i.e.,

$$\nu := -\left[\frac{\nabla f_1(x_0)}{\|\nabla f_1(x_0)\|} + \frac{\nabla f_2(x_0)}{\|\nabla f_2(x_0)\|}\right], \tag{75}$$

and let

$$H^T = (\nabla h_1(x_0) \; \ldots, \; \nabla h_m(x_0)) = QR = (q_1, \ldots, q_m, q_{m+1}, \ldots, q_n)\, R \tag{76}$$

be a QR-factorization of H^T. Then, the vectors $q_1, \ldots, q_m$ build an orthonormal basis of the image of H^T. Further, since the image of H^T is the orthogonal complement of the kernel of H, we have for $i = m+1, \ldots, n$:

$$Hq_i = 0 \qquad \Leftrightarrow \qquad (\nabla h_j(x_0)^T q_i = 0, \quad j = 1, \ldots, m). \tag{77}$$

That is, the column vectors of

$$\tilde{Q} := (q_{m+1}, \ldots, q_n) \tag{78}$$

form an orthonormal basis of the tangent space $T_{h^{-1}(0)}(x_0)$ of $h^{-1}(0)$ at x_0. The orthogonal projection of ν onto $T_{h^{-1}(0)}(x_0)$ is hence given by (see also Algorithm 1)

$$\nu_p := \tilde{Q}\tilde{Q}^T \nu. \tag{79}$$

The following result establishes criteria under which ν_p is a MODD.

Algorithm 1. Computation of the search direction ν_p for equality constrained BOPs, feasible case

---

Require: BOP of form (), x_0 with $\nabla f_i(x_0) \neq 0$, $i = 1, 2$
Ensure: search direction ν_p

1: $\nu := -\left[\frac{\nabla f_1(x_0)}{\|\nabla f_1(x_0)\|} + \frac{\nabla f_2(x_0)}{\|\nabla f_2(x_0)\|}\right]$

2: $H := \begin{pmatrix} \nabla h_1(x_0)^T \\ \vdots \\ \nabla h_m(x_0)^T \end{pmatrix}$

3: compute Q and R s.t. $H^T = QR = (q_1, \ldots, q_m, q_{m+1}, \ldots, q_n)R$
4: $\tilde{Q} := (q_{m+1}, \ldots, q_n)$
5: $\nu_p := \tilde{Q}\tilde{Q}^T \nu$
6: **return** ν_p

Proposition 4. *Let a BOP of the form (74) be given and x_0 with $\nabla f_i(x_0) \neq 0$ for $i \in \{1,2\}$ and $h_j(x_0) = 0$, $j = 1,\dots,m$. Further, let ν_p by given as in Eq. (79) such that $\langle \nu_p, \nabla h_j(x_0)\rangle = 0$ for $j \in \{1,\dots,m\}$. Then the following holds:*

(a) If $\nabla f_1(x_0)^T\tilde{Q}\tilde{Q}^T\nabla f_2(x_0) > 0$, then ν_p is a MODD of F at x_0.
(b) If $\nabla f_1(x_0)^T\tilde{Q}\tilde{Q}^T\nabla f_2(x_0) = 0$ and $\tilde{Q}^T\nabla f_i(x_0) \neq 0$ for an index $i \in \{1,2\}$, then ν_p is a MODD of F at x_0.
(c) If $\nabla f_1(x_0)^T\tilde{Q}\tilde{Q}^T\nabla f_2(x_0) < 0$, then ν_p is not a descent direction of F at x_0.

Proof. Note that for the first objective we obtained

$$\begin{aligned}\nabla f_1(x_0)^T\nu_p &= \nabla f_1(x_0)^T\tilde{Q}\tilde{Q}^T\nu_L\\ &= -\left[\frac{\nabla f_1(x_0)^T\tilde{Q}\tilde{Q}^T\nabla f_1(x_0)}{\|\nabla f_1(x_0)\|} + \frac{\nabla f_1(x_0)^T\tilde{Q}\tilde{Q}^T\nabla f_2(x_0)}{\|\nabla f_2(x_0)\|}\right], \qquad (80)\end{aligned}$$

and for the second objective

$$\begin{aligned}\nabla f_2(x_0)^T\nu_p &= \nabla f_2(x_0)^T\tilde{Q}\tilde{Q}^T\nu_L\\ &= -\left[\frac{\nabla f_2(x_0)^T\tilde{Q}\tilde{Q}^T\nabla f_1(x_0)}{\|\nabla f_1(x_0)\|} + \frac{\nabla f_2(x_0)^T\tilde{Q}\tilde{Q}^T\nabla f_2(x_0)}{\|\nabla f_2(x_0)\|}\right]. \qquad (81)\end{aligned}$$

Further, since $0 < \|\nabla f_1(x_0)\|$ and $0 < \|\nabla f_2(x_0)\|$, and $\tilde{Q}\tilde{Q}^T$ is symmetric we obtain

$$\nabla f_1(x_0)^T\tilde{Q}\tilde{Q}^T\nabla f_2(x_0) = \nabla f_2(x_0)^T\tilde{Q}\tilde{Q}^T\nabla f_1(x_0) \qquad (82)$$

and

$$\begin{aligned}\nabla f_i(x_0)^T\tilde{Q}\tilde{Q}^T\nabla f_i(x_0) &= \left\langle \tilde{Q}^T\nabla f_i(x_0), \tilde{Q}^T\nabla f_i(x_0)\right\rangle\\ &= \|\tilde{Q}^T\nabla f_i(x_0)\|^2 \geq 0 \text{ for } i \in \{1,2\}. \qquad (83)\end{aligned}$$

Then, three cases arise:

Case 1. If $\nabla f_1(x_0)^T\tilde{Q}\tilde{Q}^T\nabla f_2(x_0) > 0$ holds, then by (80), (81) and (83) we obtain

$$\nabla f_1(x_0)^T\nu_p < 0 \text{ and } \nabla f_2(x_0)^T\nu_p < 0;$$

which means that ν_p is a descent direction of F at x_0.

Case 2. If $\nabla f_1(x_0)^T\tilde{Q}\tilde{Q}^T\nabla f_2(x_0) = 0$, then
(i) If $\|\tilde{Q}^T\nabla f_i(x_0)\| > 0$ for $i \in \{1,2\}$, then by (80), (81) and (83) $\nabla f_i(x_0)^T\nu_p < 0$ for $i \in \{1,2\}$, i.e., ν_p is a descent direction for F at x_0.
(ii) If $\|\tilde{Q}^T\nabla f_1(x_0)\| > 0$ and $\|\tilde{Q}\nabla f_2(x_0)\| = 0$, then $\nabla f_1(x_0)^T\nu_p < 0$ and $\nabla f_2(x_0)^T\nu_p = 0$, i.e., ν_p is a descent direction for F at x_0.

(iii) If $\|\tilde{Q}^T \nabla f_1(x_0)\| = 0$ and $\|\tilde{Q} \nabla f_2(x_0)\| > 0$, then $\nabla f_1(x_0)^T \nu_p = 0$ and $\nabla f_2(x_0)^T \nu_p < 0$, i.e., ν_p is a descent direction for F at x_0.

Therefore, if $\nabla f_1(x_0)^T \tilde{Q}\tilde{Q}^T \nabla f_2(x_0) = 0$ and $\tilde{Q}^T \nabla f_i(x_0) \neq 0$ for an index $i \in \{1, 2\}$, then ν_p is a descent direction of F at x_0.

Case 3. If $\nabla f_1(x_0)^T \tilde{Q}\tilde{Q}^T \nabla f_2(x_0) < 0$ assume, for the sake of contradiction, that ν_p is a descent direction for F at x_0.

Then, if $\nabla f_1(x_0)^T \nu_p < 0$ we have by Eq. (80) the following:

$$\begin{aligned} \frac{\nabla f_1(x_0)^T \tilde{Q}\tilde{Q}^T \nabla f_2(x_0)}{\|\nabla f_2(x_0)\|} &< \frac{\nabla f_1(x_0)^T \tilde{Q}\tilde{Q}^T \nabla f_1(x_0)}{\|\nabla f_1(x_0)\|} \\ \Leftrightarrow \frac{\langle \tilde{Q}^T \nabla f_1(x_0), \tilde{Q}^T \nabla f_2(x_0)\rangle}{\|\nabla f_2(x_0)\|} &< \frac{\langle \tilde{Q}^T \nabla f_1(x_0), \tilde{Q}^T \nabla f_1(x_0)\rangle}{\|\nabla f_1(x_0)\|} \\ \Leftrightarrow \frac{\|\tilde{Q}^T \nabla f_1(x_0)\| \|\tilde{Q}^T \nabla f_2(x_0)\| \cos\theta}{\|\nabla f_2(x_0)\|} &< \frac{\|\tilde{Q}^T \nabla f_1(x_0)\|^2}{\|\nabla f_1(x_0)\|} \\ \Leftrightarrow \frac{\|\tilde{Q}^T \nabla f_2(x_0)\| \cos\theta}{\|\nabla f_2(x_0)\|} &< \frac{\|\tilde{Q}^T \nabla f_1(x_0)\|}{\|\nabla f_1(x_0)\|} \\ \Leftrightarrow \left\|\tilde{Q}^T \frac{\nabla f_2(x_0)}{\|\nabla f_2(x_0)\|}\right\| \cos\theta &< \left\|\tilde{Q}^T \frac{\nabla f_1(x_0)}{\|\nabla f_1(x_0)\|}\right\| \\ \Leftrightarrow \cos\theta &< \frac{\left\|\tilde{Q}^T \frac{\nabla f_1(x_0)}{\|\nabla f_1(x_0)\|}\right\|}{\left\|\tilde{Q}^T \frac{\nabla f_2(x_0)}{\|\nabla f_2(x_0)\|}\right\|}. \end{aligned} \tag{84}$$

Analogously, if $\nabla f_2(x_0)^T \nu_p < 0$ then by Eq. (81) we have

$$\cos\theta < \frac{\left\|\tilde{Q}^T \frac{\nabla f_2(x_0)}{\|\nabla f_2(x_0)\|}\right\|}{\left\|\tilde{Q}^T \frac{\nabla f_1(x_0)}{\|\nabla f_1(x_0)\|}\right\|}.$$

When considering $-1 < \cos\theta < 1$,

$$\frac{\left\|\tilde{Q}^T \frac{\nabla f_1(x_0)}{\|\nabla f_1(x_0)\|}\right\|}{\left\|\tilde{Q}^T \frac{\nabla f_2(x_0)}{\|\nabla f_2(x_0)\|}\right\|} < 1 \text{ and } \frac{\left\|\tilde{Q}^T \frac{\nabla f_2(x_0)}{\|\nabla f_2(x_0)\|}\right\|}{\left\|\tilde{Q}^T \frac{\nabla f_1(x_0)}{\|\nabla f_1(x_0)\|}\right\|} < 1$$

leads to

$$\left\|\tilde{Q}^T \frac{\nabla f_1(x_0)}{\|\nabla f_1(x_0)\|}\right\| < \left\|\tilde{Q}^T \frac{\nabla f_2(x_0)}{\|\nabla f_2(x_0)\|}\right\| \text{ and } \left\|\tilde{Q}^T \frac{\nabla f_2(x_0)}{\|\nabla f_2(x_0)\|}\right\| < \left\|\tilde{Q}^T \frac{\nabla f_1(x_0)}{\|\nabla f_1(x_0)\|}\right\|,$$

which is not possible. Thus, we conclude that if $\nabla f_1(x_0)^T \tilde{Q}\tilde{Q}^T \nabla f_2(x_0) < 0$, then ν_p is not a descent direction of F at x_0, and we are done. □

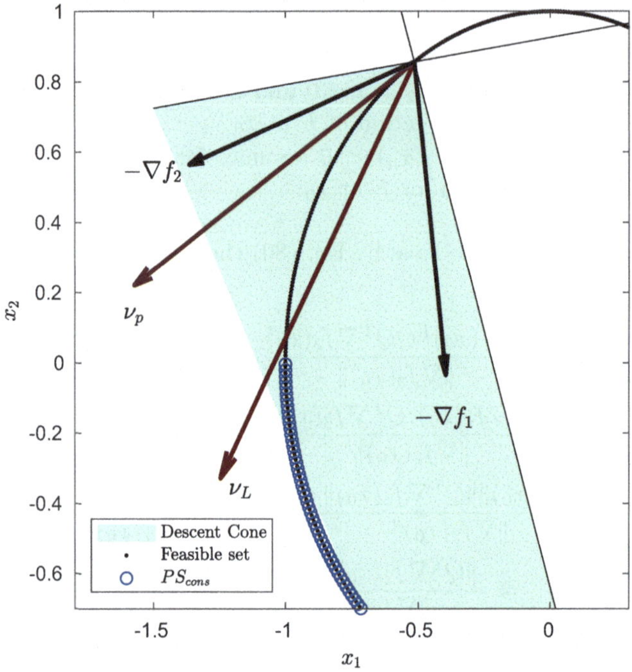

Fig. 2. Considering a feasible starting point when computing ν_p using Eq. (73). Here ν_p is a MODD of BOP (85). This figure illustrates the case presented in Example 4.

Example 4. Consider the following BOP:

Minimize

$$\begin{aligned} f_1(x_1, x_2) &= x_1^2 + (x_2 + 3)^2 \\ f_2(x_1, x_2) &= (x_1 + 3)^2 + x_2^2 \end{aligned} \tag{85}$$

subject to

$$h(x_1, x_2) = x_1^2 + x_2^2 - 1 = 0.$$

For this example, consider the computation of direction ν_p starting from a feasible initial point. Figure 2 illustrates the case when ν_p lays over the descent cone, hence ν_p is a descent direction.

Infeasible Case

Next, we consider that the initial point x_0 is infeasible, i.e., that for at least one $j \in \{1, \ldots, m\}$ it holds $h_j(x) \neq 0$. Further, we assume that all equalities are linear, i.e., we are given the problem

$$\min_{x\in\mathbb{R}^n} F(x) := [f_1(x), f_2(x)]^T, \quad (86)$$
$$\text{s.t} \quad h(x) = Ax - b = 0,$$

where $A \in \mathbb{R}^{m\times n}$. In this case we can consider the Newton method applied to the residual $r : \mathbb{R}^{n+m+k} \to \mathbb{R}^{n+m+k}$ which is given by

$$r(x, \alpha, \nu) = \begin{pmatrix} J(x)^T\alpha + A^T\nu \\ Ax - b \\ \bar{e}^T\alpha - 1 \end{pmatrix}, \quad (87)$$

where J denotes the Jacobian of F and $\bar{e} = [1, \ldots, 1]^T \in \mathbb{R}^k$. The first order Taylor approximation of r near an estimate $y = (x, \alpha, \nu) \in \mathbb{R}^{n+k+m}$ is given by

$$r(y + z) \approx r(y) + Dr(y)z, \quad (88)$$

where $Dr(y)$ is the Jacobian of r at y. The Newton step Δy for the Newton method applied to r solves the following linear system of equations:

$$Dr(y)\Delta y = -r(y). \quad (89)$$

Denote

$$W_\alpha := \sum_{j=1}^{2} \alpha_j \nabla^2 f_j(x), \quad (90)$$

then

$$Dr(x, \alpha, \nu) = \begin{pmatrix} W_\alpha & J(x)^T & A^T \\ A & 0 & 0 \\ 0 & \bar{e}^T & 0 \end{pmatrix}, \quad (91)$$

and the Newton step is given by the vector

$$\Delta y = (\Delta x, \Delta\alpha, \Delta\nu)$$

that solves

$$\begin{pmatrix} W_\alpha & J(x)^T & A^T \\ A & 0 & 0 \\ 0 & \bar{e}^T & 0 \end{pmatrix} \begin{pmatrix} \Delta x \\ \Delta\alpha \\ \Delta\nu \end{pmatrix} = - \begin{pmatrix} J(x)^T\alpha + A^T\nu \\ Ax - b \\ \bar{e}^T\alpha - 1 \end{pmatrix}. \quad (92)$$

If we set $\nu^+ := \nu + \Delta\nu$, we can rewrite the above system as

$$\begin{pmatrix} W_\alpha & J(x)^T & A^T \\ A & 0 & 0 \\ 0 & \bar{e}^T & 0 \end{pmatrix} \begin{pmatrix} \Delta x \\ \Delta\alpha \\ \nu^+ \end{pmatrix} = - \begin{pmatrix} J(x)^T\alpha \\ Ax - b \\ \bar{e}^T\alpha - 1 \end{pmatrix}. \quad (93)$$

The following discussion shows that the norm of r decreases for sufficiently small step sizes in direction Δy: it is

$$\frac{d}{dt}\|r(y + t\Delta y)\|^2\Big|_{t=0} = -2r(y)^T Dr(y)\Delta y$$
$$= -2r(y)^T r(y). \quad (94)$$

Taking out the square leads to

$$\frac{d}{dt}\|r(y+t\Delta y)\|\Big|_{t=0} = -r(y)^T r(y) = -\|r(y)\| \tag{95}$$

which is negative at y with $r(y) \neq 0$.

In the following we summarize this result.

Proposition 5. *Let a BOP be of the form* (86) *and suppose x_0 is given such that $h_j(x_0) \neq 0$ for at least one $j \in \{1, \ldots, m\}$. The Newton step on the residual r as defined in* (87) *is given by the vector that solves equation system* (93)*, and $\|r\|$ decreases for sufficiently small steps in direction of the Newton step.*

Proof. It follows by the above discussion. □

Example 5. Consider

$$F : \mathbb{R}^5 \to \mathbb{R}^2$$

subject to one linear equality constraint as

$$\begin{aligned} f_j(x) &= \textstyle\sum_{i=1, i\neq j}^{5} \left(x_i - a_i^j\right)^2 + \left(x_j - a_j^j\right)^4, j = 1, 2 \\ \text{s.t. } \tfrac{1}{2}x_1 &= x_2. \end{aligned} \tag{96}$$

Here, $a^1 = [1, ..., 1]^T \in \mathbb{R}^5$ and $a^2 = -a^1$. We apply Newton's method for the initial infeasible point $p_0 = a^1$ with $h(p_o) = -0.5$. Figure 3 shows the obtained solutions in each Newton step in the *(a)* variable and *(b)* objective space. In the fourth step we obtain the final solution

$$p_4 = [0.2668, 0.1334, 0.3750, 0.3750, 0.3750]^T$$

with $h(p_4) = -2.7756e{-}17$, which can be considered to be feasible.

3.4 Inequality Constrained BOPs

Next we consider inequality constrained BOPs of the form

$$\begin{aligned} \min_{x\in\mathbb{R}^n} \; & F(x) := [f_1(x), f_2(x)]^T, \\ \text{s.t} \quad & g_j(x) \leq 0, \qquad j = 1, \ldots, p. \end{aligned} \tag{97}$$

Let x_0 be given and denote by

$$I(x_0) := \{g_{i1}(x_0), g_{i2}(x_0), \ldots, g_{is}(x_0)\} \tag{98}$$

the set of active inequalities at x_0. Assume that $I(x_0)$ is not empty, i.e., that $s \geq 1$. Denote by

$$G := \begin{pmatrix} \nabla g_{i1}(x_0)^T \\ \vdots \\ \nabla g_{is}(x_0)^T \end{pmatrix} \in \mathbb{R}^{s\times n} \tag{99}$$

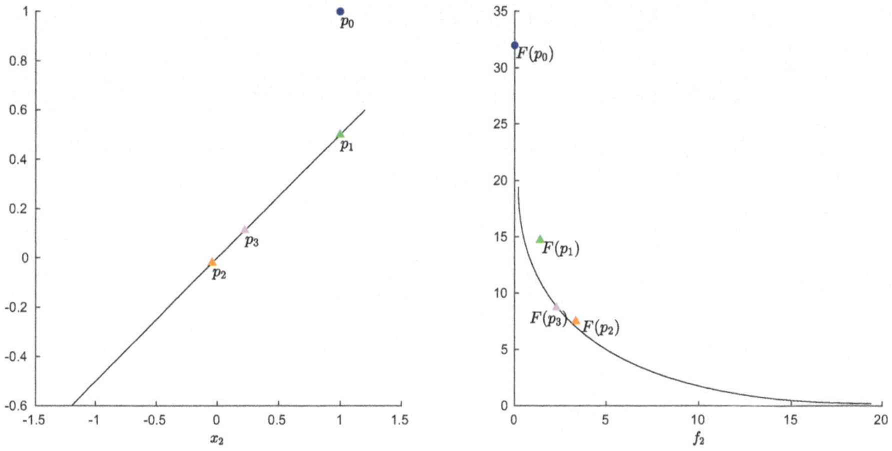

(a) Variable space **(b) Objective space**

Fig. 3. Newton steps starting from p_0 which is an infeasible point for Problem (96). This figure illustrates Example 5.

the matrix formed by the gradients of the active inequality constraints.

Similarly to the feasible case for equality constrained BOPs, we can also in this case generate descent directions via projection as follows: suppose that $rank(G) = s$ (i.e., maximal), then we can compute the factorization

$$G^T = QR = (q_1, \ldots, q_s, q_{s+1}, \ldots, q_n)\, R, \tag{100}$$

where $Q \in \mathbb{R}^{n \times n}$ is orthogonal and $R \in \mathbb{R}^{n \times s}$ right upper triangular. Doing so, the last $n - s$ column vectors of Q form an orthonormal basis of the tangent space of the feasible set $g_i^{-1}(0)$ at x_0 with $\nabla f_i(x_0) \neq 0$ and $g_i(x_0) = 0$. The orthogonal projection ν^p onto $T_{g_i^{-1}(0)}(x)$ is hence given by

$$\nu_p := \tilde{Q}\tilde{Q}^T \nu_L, \tag{101}$$

where ν_L is as in (73) and

$$\tilde{Q} := (q_{s+1}, \ldots, q_n)\,. \tag{102}$$

Similarly to the equality constrained case, ν_p is a descent direction under certain conditions (see also Algorithm 2).

Proposition 6. *For a BOP of the form (97), suppose $\nabla f_i(x_0) \neq 0$ for $i \in \{1, 2\}$. Let ν_p be as in Eq. (101) such that $\langle \nu_p, \nabla g_i(x_0)\rangle = 0$. Let $x \in \mathbb{R}^n$ such that $g_i(x_0) = 0$ for every $i \in I(x_0)$. Then*

(a) If $\nabla f_1(x_0)^T \tilde{Q}\tilde{Q}^T \nabla f_2(x_0) > 0$, then ν_p is a MOPP of F at x_0.

(b) If $\nabla f_1(x_0)^T \tilde{Q}\tilde{Q}^T \nabla f_2(x_0) = 0$ *and* $\tilde{Q}^T \nabla f_i(x_0) \neq 0$ *for an index* $i \in \{1, 2\}$*, then* ν_p *is a MODD of* F *at* x_0*.*
(c) If $\nabla f_1(x_0)^T \tilde{Q}\tilde{Q}^T \nabla f_2(x_0) < 0$*, then* ν_p *is not a descent direction of* F *at* x_0*.*

Proof. Note that by construction $\nabla g_i(x_0)^T \nu_p = 0$ for all active $i \in I(x_0)$; thus, the proof is analog to the one from Proposition 4. □

Algorithm 2. Computation of the search direction ν_p for inequality constrained BOPs

Require: BOP of form (97), x_0 with $\nabla f_i(x_0) \neq 0$, $i = 1, 2$
Ensure: search direction ν_p
1: $\nu := -\left[\frac{\nabla f_1(x_0)}{\|\nabla f_1(x_0)\|} + \frac{\nabla f_2(x_0)}{\|\nabla f_2(x_0)\|}\right]$
2: $G := \begin{pmatrix} \nabla g_{i1}(x_0)^T \\ \vdots \\ \nabla g_{is}(x_0)^T \end{pmatrix} \in \mathbb{R}^{s \times n}$
3: compute Q and R s.t. $G^T = QR = (q_1, \ldots, q_s, q_{s+1}, \ldots, q_n)R$
4: $\tilde{Q} := (q_{s+1}, \ldots, q_n)$
5: $\nu_p := \tilde{Q}\tilde{Q}^T \nu$
6: **return** ν_p

Example 6. Consider the following BOP proposed in [8]:

Minimize

$$f_1(x) = x_1 \tag{103}$$
$$f_2(x) = g(x) * \left(1 - \sqrt{\frac{x_1}{g(x)}}\right),$$
$$\text{with} \quad g(x) = 1 + x_2.$$

Subject to

$$\cos\theta\,(f_2(x_1, x_2) - e) - \sin\theta f_1(x_1, x_2) \geq \tag{104}$$
$$a|\sin\left(b\pi\left(\sin\theta\,(f_2(x_1, x_2) - e) + \cos\theta f_1(x_1, x_2)\right)^c\right)|^d, \tag{105}$$

where $0 \leq x_i \leq 1$ for $i \in \{1, 2\}$. And $a = 0.1, b = 10, c = 2, d = 0.5, e = 1$ and $\theta = -0.2 * \pi$. Fig. 4 shows an example of the proposed MODD. Note that ν_p lays over the descent cone and fulfills the above criterion; thus ν_p is a descent direction.

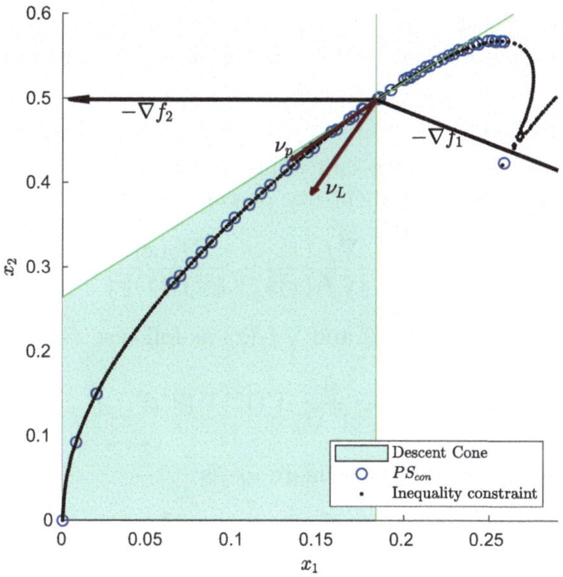

Fig. 4. Considering a starting point for an active constraint when computing ν_p using Eq. (73), it can be a descent direction or not. This figure illustrates the behavior in decision variable space for Example 6.

3.5 A Gradient Free Approximation of a MODD for CBOPs

In the following we use the GSA method to compute the above discussed descent directions for (unconstrained or constrained) bi-objective optimization problems.

We assume that we are given a candidate solution x_0 that is designated for local search, and that we are given sample points $x_1, x_2, \ldots, x_r$ in the vicinity of x_0. We further assume that their objective functions values $f(x_i)$ are already known, which is indeed the case if the x_i's are chosen from a given population within a MOEA. Recall from GSA that we can use

$$\nu_i := \frac{x_i - x_0}{\|x_i - x_0\|_2}, \quad d_i := \frac{f(x_i) - f(x_0)}{\|x_i - x_0\|_2}, \; i \in \{1, \ldots, r\}, \tag{106}$$

Thus, (16) turns into

$$\begin{aligned} \min \; & \lambda^T d \\ \text{s.t.} \; & \lambda^T V^T V \lambda - 1 = 0, \end{aligned} \tag{107}$$

which is the problem we have to solve. Then, consider the matrix

$$\tilde{V} := (\nu_1, \ldots, \nu_r), \tag{108}$$

to finally obtain

$$\tilde{\nu^*} = -\frac{1}{\|\tilde{V}\tilde{\lambda^*}\|} \tilde{V}(\tilde{V}^T\tilde{V})^{-1} d. \tag{109}$$

Then, $\tilde{\nu^*}$ is the most greedy direction for a single-objective. It is worth to notice that this derivation can be done for all objective functions in a MOP.

Unconstrained Case

For the unconstrained case, we can apply the previous idea to approximate Eq. (73), hence obtaining a gradient-free descent direction. Recall that

$$\nu_L := -\left[\frac{\nabla f_1(x)}{\|\nabla f_1(x)\|} + \frac{\nabla f_2(x)}{\|\nabla f_2(x)\|}\right], \tag{110}$$

then we can approximate $\nabla f_1(x)$ and $\nabla f_2(x)$ as follows:

$$\tilde{\nu_j^*} = -\frac{1}{\|\tilde{V}\lambda^*\|}\tilde{V}(\tilde{V}^T\tilde{V})^{-1}d^j, \tag{111}$$

for $j = \{1, 2\}$. Then, we can approximate ν_L as

$$\tilde{\nu}_L := -\left[\frac{\tilde{\nu_1^*}}{\|\tilde{\nu_1^*}\|} + \frac{\tilde{\nu_2^*}}{\|\tilde{\nu_2^*}\|}\right]. \tag{112}$$

Equality Constrained Case

For this scenario assume that x_0 is a feasible solution, and that we are given $x_1, x_2, \ldots, x_r$ which are sample points in the neighborhood of x_0; also, that their objective functions values $f(x_i)$ are already known. From the constrained case of GSA recall that the Matrix $M := (m_{ji}) \in \mathbb{R}^{m \times r}$ is given by

$$m_{j,i} := \frac{h_j(x_i) - h_j(x_0)}{\|x_i - x_0\|_2},\ i \in \{1, \ldots, r\},\ j \in \{1, \ldots, m\}. \tag{113}$$

Via Eqs. (108) and (113) we can compute an approximation $\tilde{H}^T$ of the Jacobian matrix H^T of the constraint functions given by

$$\tilde{H}^T = \tilde{V}(\tilde{V}^T\tilde{V})^{-1}M^T \tag{114}$$

that will be used to compute the projection of ν_p. Then, considering a feasible solution x_0, we proceed analogously to the unconstrained case, and first compute $\tilde{\nu}_L$ as in Eq. (112). Next, in order to compute $\tilde{\nu}_p$, we compute a QR decomposition of $\tilde{H}^T$, and define

$$\tilde{Q} := (\tilde{q}_{m+1}, \ldots, \tilde{q}_n),$$

where $\tilde{q}_i$, $i \in \{m+1, \ldots, n\}$, are the last $n-m$ column vectors of the orthogonal matrix Q obtained by the QR-decomposition of $\tilde{H}^T$. Then

$$\tilde{\nu}_p := \tilde{Q}\tilde{Q}^T\tilde{\nu}_L, \tag{115}$$

is the orthogonal projection of $\tilde{\nu}_p$ onto the set of feasible directions. Assuming the notation for $\tilde{V}$ and d_i^j as in Eq. (108) and (106), the following criteria manages the application of our gradient-free proposal:

For a BOP of the form (74), with $\nabla f_i(x) \neq 0$ for $i \in \{1,2\}$ and $h_j(x) = 0$. Compute $\tilde{\nu}_p$, as in Eq. (115) and

$$C_g := d^{1T}(\tilde{V}^T\tilde{V})^{-1}\tilde{V}^T\tilde{Q}\tilde{Q}^T\tilde{V}(\tilde{V}^T\tilde{V})^{-1}d^2. \tag{116}$$

Then we proceed as follows:

1. If $C_g > 0$, then perform a line search over $\tilde{\nu}_p$.
2. If $C_g = 0$ and $d^{iT}(\tilde{V}^T\tilde{V})^{-1}\tilde{V}^T\tilde{Q}\tilde{Q}^T\tilde{V}(\tilde{V}^T\tilde{V})^{-1}d^i \neq 0$ for an index $i \in \{1,2\}$, then perform a line search over $\tilde{\nu}_p$.
3. If $C_g < 0$, then the line search is not applied.

Note that the above criteria allow us to decide, during the algorithm's running time, when the information available is likely enough to have an approximation of a MODD. After deciding to approximate such direction, we compute the new iterative point x_i as follows:

$$x_i := x_0 + t\tilde{\nu}_p, \tag{117}$$

where t is a suitable step length. In this work, we computed t by a backtracking procedure based on the Armijo's condition [29]. The description in Algorithm 3 corresponds to the standalone gradient-free algorithm for equality constrained MOPs.

Inequality Constrained Case

In the case that inequality constraints are present, the consideration is made over $I(x)$, that is the set of active inequality constraints at x. Thus we obtain the new approximation of $\tilde{\nu}_L^p$ as follows:

$$\tilde{\nu}_L^p := \tilde{Q}\tilde{Q}^T\tilde{\nu}_L. \tag{118}$$

Assuming the notation for $\tilde{V}$ and d_i^j as in Eq. (108) and (113), the following result states the criteria for the application of the gradient-free proposal:

For a BOP of the form (97), with $\nabla f_i(x) \neq 0$ for $i \in \{1,2\}$ and $g_i(x) = 0$ for every $i \in I(x)$. Compute $\tilde{\nu}_L^p$ as in Eq. (118) and

$$C_g := d^{1T}(\tilde{V}^T\tilde{V})^{-1}\tilde{V}^T\tilde{Q}\tilde{Q}^T\tilde{V}(\tilde{V}^T\tilde{V})^{-1}d^2. \tag{119}$$

Then we proceed as follows:

1. If $C_g > 0$, then perform a line search over $\tilde{\nu}_p$.
2. If $C_g = 0$ and $d^{iT}(\tilde{V}^T\tilde{V})^{-1}\tilde{V}^T\tilde{Q}\tilde{Q}^T\tilde{V}(\tilde{V}^T\tilde{V})^{-1}d^i \neq 0$ for an index $i \in \{1,2\}$, then perform a line search over $\tilde{\nu}_p$.
3. If $C_g < 0$, then the line search is not applied.

Algorithm 3. Standalone gradient-free MODD for equality constrained BOPs.

Require: x_0 :initial solution, r :number of neighbors, ε : threshold for C_g
Ensure: x_f :final solution.

```
while Stopping Criterion does not fulfill do
  $i \leftarrow 1$;
  $x_i \leftarrow x_0$;
  Compute $x_1, \ldots, x_r$ neighbor points for $x_0$;
  Compute $\nu_i$ and $d$ as in Eq. (113);
  Compute $\tilde{V}$ as in Eq. (108);
  if $x_i$ is a feasible solution then
    Compute $C_g$ as in Eq. (116).
    $c_j \leftarrow d^{jT}(\tilde{V}^T\tilde{V})^{-1}\tilde{V}^T\tilde{Q}\tilde{Q}^T\tilde{V}(\tilde{V}^T\tilde{V})^{-1}d^j \quad j = 1, 2$;
    if $C_g > 0$ then
      Compute $\tilde{\nu_p}$ as in Eq. (115);
      Compute $t \in \mathbb{R}^+$;
      $x_{i+1} \leftarrow x_i + t\tilde{\nu_L}$;
      $i \leftarrow i + 1$;
    else if $|C_g| < \varepsilon$ then
      if $c_1 > 0$ or $c_2 > 0$ then
        Compute $\tilde{\nu_p}$ as in Eq. (115);
        Compute $t \in \mathbb{R}^+$, a suitable step size
        $x_{i+1} \leftarrow x_i + t\tilde{\nu_L}$;
        $i \leftarrow i + 1$;
      else
        It is not a descent direction.
      end if
    else
      It is not a descent direction.
    end if
  end if
end while

return $x_f \leftarrow x_i$;
```

After deciding to approximate such direction, we compute the new iterative point x_i as follows:

$$x_i := x_0 + t\tilde{\nu}_L^p, \tag{120}$$

where t is a suitable step length. In this work, we compute t via a backtracking procedure based again on Armijo's condition. The standalone algorithm for inequality CMOPs is described in Algorithm 4.

We have presented in this section some criteria to decide when a solution is a MODD. The potential of these criteria will be displayed in the next section when used in combination with population-based heuristics.

Algorithm 4. Standalone gradient-free MODD for inequality CBOPs.

Require: x_0 :initial solution, r :number of neighbors, ε : threshold for C_g, $I(x)$:active set.

Ensure: x_f :final solution.

while Stopping Criterion does not fulfill **do**
  $i \leftarrow 1$;
  $x_i \leftarrow x_0$; Compute $x_1, \ldots, x_r$ neighbor points for x_0;
  Compute ν_i and d as in Eq. (113);
  Compute $\tilde{V}$ as in Eq. (108);
  Compute C_g as in Eq. (119);
  $c_j \leftarrow d^{jT}(\tilde{V}^T\tilde{V})^{-1}\tilde{V}^T\tilde{Q}\tilde{Q}^T\tilde{V}(\tilde{V}^T\tilde{V})^{-1}d^j \quad j = 1, 2$;
  if $C_g > 0$ **then**
    Compute $\tilde{\nu_p}$ as in Eq. (115);
    Compute $t \in \mathbb{R}^+$, a suitable step size
    $x_{i+i} \leftarrow x_i + t\tilde{\nu_L}$;
    $i \leftarrow i + 1$;
  else if $|C_g| < \varepsilon$ **then**
    if $c_1 > 0$ or $c_2 > 0$ **then**
      Compute $\tilde{\nu_p}$ as in Eq. (115);
      Compute $t \in \mathbb{R}^+$;
      $x_{i+1} \leftarrow x_i + t\tilde{\nu_L}$;
      $i \leftarrow i + 1$;
    else
      It is not a descent direction;
    end if
  else
    It is not a descent direction;
  end if
end while

return $x_f \leftarrow x_i$;

4 Application: Use of GFDD Within NSGA-II

In order to apply the developed ideas, we will in the following show some examples of the integration the above ideas to perform a multi-objective local search within the execution of the well-known algorithm NSGA-II [10] as demonstrator. This is a state-of-the-art algorithm for bi- and three-objective optimization problems that makes use of an archiving strategy based on the *crowding distance.* This strategy will play an important role in the hybridization as it will help us to decide which individual is a suitable starting point to perform the local search.

Algorithm 5 shows a pseudo code of a hybrid of GFDD and NSGA-II which also shows that such a coupling can be done relatively easily with in principle

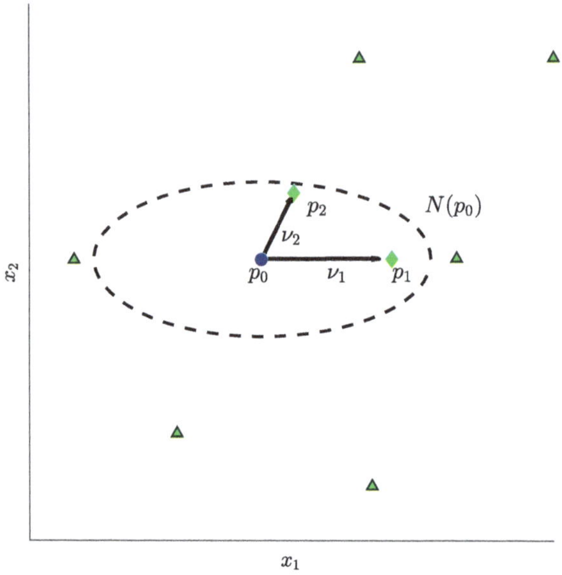

Fig. 5. Points p_1 and p_2 are considered to be neighbors of p_0. The green triangles outside the circle are population elements which are not used for the GSA computation.

Algorithm 5. Pseudocode of $NSGA - II + Localsearch$

Require: P_s := Population size, $size$:= Problem Size, P_c := crossover probability, P_m := mutation probability
Ensure: Set of approximated solutions
Population ← `InitializePopulation`(P_s, $size$)
`FitnessEvaluation`(Population)
`FastNondominatedSort(Population)`
Selected ← `SelectParentsByRank`(Population, P_s)
Offspring ← `CrossoverAndMutation`(Selected, P_c, P_m)
while ¬StopCondition() **do**
  `FitnessEvaluation`(Offspring)
  Union ← Merge(Population, Offspring)
  Fronts ← `FastNondominatedSort`(Union)
  `CrowdingDistanceAssignment`($Fronts$)
  Selected ← `SelectParentsByRankAndDistance`(Parents, P_s)
  Population ← Offspring
  Selected individual x_s
  `ApplyLocalSearch`(x_s) if suitable
  Offspring ← `CrossoverAndMutation`(Selected, P_c, P_m)
end while
return Offspring

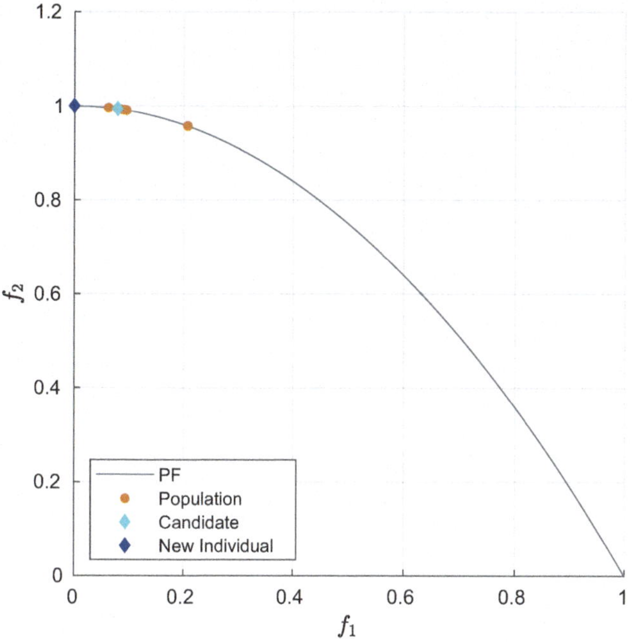

Fig. 6. This figure illustrates a particular instant of a certain iteration in the population-based algorithm. The population is represented by circles, while the initial and final solutions involved in the local search procedure are marked by light and dark diamonds correspondingly. The tested function corresponds to the CZDT2 from [31].

any other MOEA.The first part of the algorithm (lines 1 to 12) coincides with the evolutionary process of the NSGA-II. Then, the interleaving of our proposal starts; in lines 13 and 14 we select an individual x_s related to the biggest crowding value. If x_s is feasible we decide based on the propositions presented above if the local search is suitable. For the case of inequality constraints we just consider the set of active constraints. Once the proposed low-cost MODD is successfully computed, we apply a regular line search through it with a suitable step size control provided with a traditional backtracking tuning.

At each generation, we applied the local search only to one selected individual mainly because we do not want to make big changes through the entire population of the evolutionary algorithm. If we apply the local search from many individual the possibility of diversity losses or premature convergence increases. Also, the computational cost (in terms of function evaluations) will increase due to step size computation. When selecting the starting point for the local search, the straight decision is to choose the individual with the most significant crowding distance value in order to assure the existence of close neighbors. These

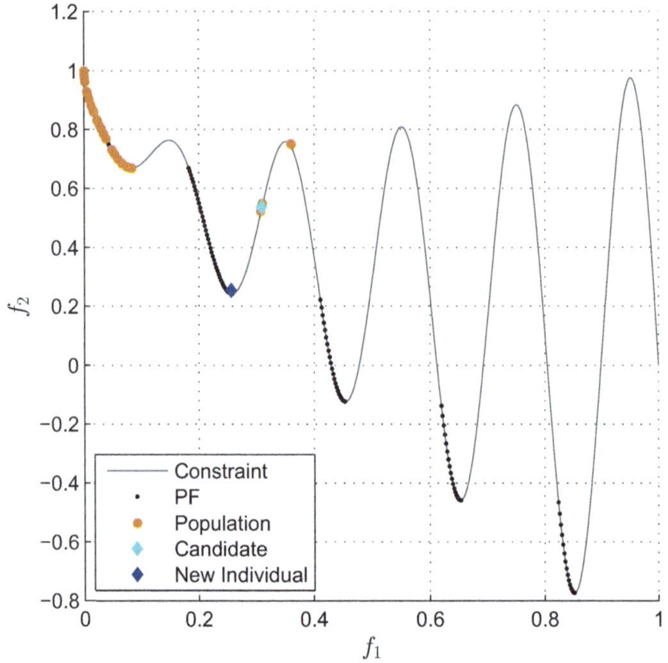

Fig. 7. This figure illustrates a particular instant of a certain iteration in the population-based algorithm. The population is represented by circles, while the initial and final solutions involved in the local search procedure are marked by light and dark diamonds correspondingly. The image of the constraint function is also marked in the figure. The tested function corresponds to the CZDT3 from [31].

neighbors are used to approximate the gradient information required by the proposed operator (see Fig. 5). By doing this, there is a chance to generate a new individual such that: (i) it is not that far from x_s; but (ii) it can be deleted by the crowding process itself. Therefore, there is a compromise between choosing a candidate that has enough neighbors to approximate the gradient-free MODD and the chances of losing the new candidate due to crowding. A better idea could be to choose an individual x_s with an average crowding value and at least r close neighbors.

Numerical results that support the advantages of the use of this proposal can be found in [38]. Next, we present some examples of the performance of a population-based algorithm when applying this proposed low-cost local search. Figures 6 and 7 illustrates the application of NSGA-II applied to the CZDT2 and CZDT3 benchmark functions given in [31]. These functions are bi-objective optimization problems with equality constraints. The figures show a certain

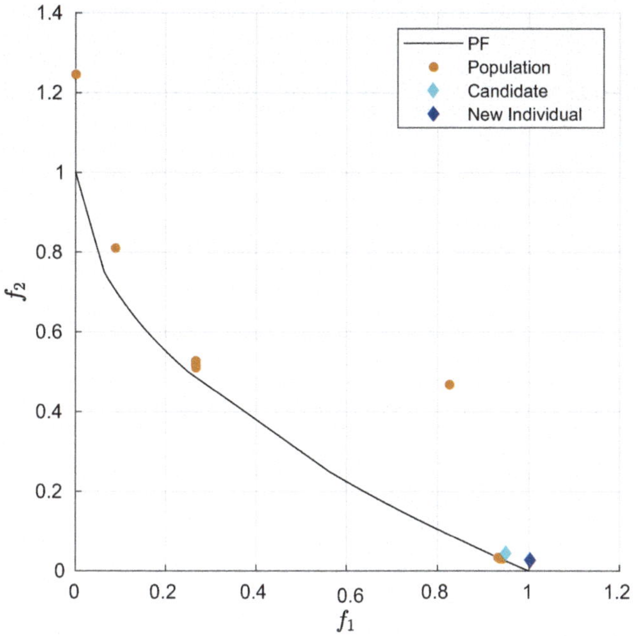

Fig. 8. This figure illustrates a particular instant of a certain iteration in the population-based algorithm. The population is represented by circles, while the initial and final solutions involved in the local search procedure are marked by light and dark diamonds correspondingly. The tested function corresponds to the CF2 from [39].

generation of the Multi-objective Evolutionary Algorithm (MOEA) when the local search is applied to generate a new individual. Figures 8 and 9 illustrates the application of NSGA-II applied to the CF2 and CF4 benchmark functions [39]. These functions are bi-objective optimization problems with inequality constraints. The figures show a certain generation of the MOEA when the local search is applied to generate a new individual.

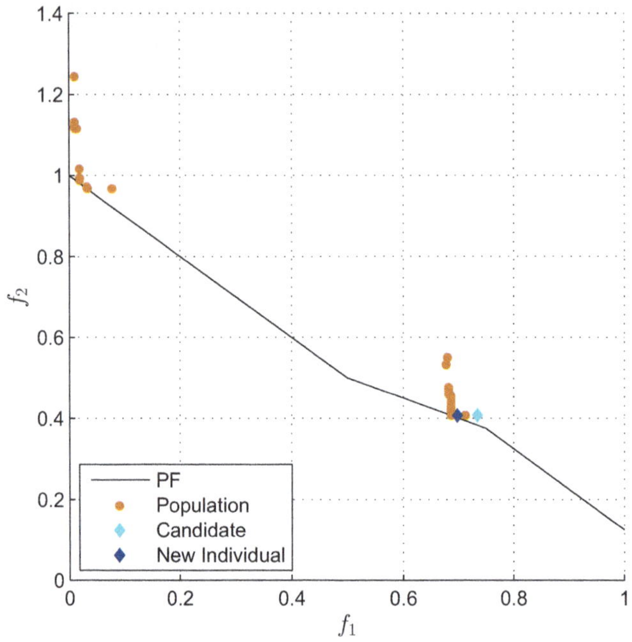

Fig. 9. This figure illustrates a particular instant of a certain iteration in the population-based algorithm. The population is represented by circles, while the initial and final solutions involved in the local search procedure are marked by light and dark diamonds correspondingly. The tested function corresponds to the CF4 from [39].

5 Conclusions

In this chapter we have reviewed some tools that allow to realize a local search engine for consrained bi-objective optimization problems with a low cost when using within set based optimization strategies such as evolutionary algorithms. The basic idea of the Gradient Subspace Approximation is to compute the most greedy search direction at a given point out of given neighborhood information. Next, we have presented a way of how to compute descent directions for bi-objective optimization problems. The method can be applied both to unconstrained as well as to constrained problems. Next, we have shown how to utilize GSA in order to obtain a "gradient free" approximation of these search directions. Finally, we have demonstrated the possible benefit of such a resulting low cost searcher on a hybrid evolutionary algorithm, which couples the proposed search technique with the widely used evolutionary algorithm NSGA-II. We stress, however, that the gradient free local search engine can in principle be integrated into any other evolutionary algorithm or any other set-oriented search heuristic.

Acknowledgements. The authors acknowledge support from Conacyt Basic Science project no. 285599, SEP-Cinvestav project no. 231, and IPN project no. SIP20201381.

References

1. Alvarado, S., Segura, C., Schütze, O., Zapotecas, S.: The gradient subspaceapproximation as local search engine within evolutionary multi-objectiveoptimization algorithms. Computación y Sistemas **22**(2) (2018)
2. Beume, N., Naujoks, B., Emmerich, M.: SMS-EMOA: multiobjective selection based on dominated hypervolume. Eur. J. Oper. Res. **181**(3), 1653–1669 (2007)
3. Bogoya, J., Vargas, A., Cuate, O., Schütze, O.: A (p, q)-averaged Hausdorff distance for arbitrary measurable sets. Math. Comput. Appl. **23**(3), 51 (2018)
4. Bosman, P.A.: On gradients and hybrid evolutionary algorithms for real-valued multiobjective optimization. IEEE Trans. Evol. Comput. **16**(1), 51–69 (2011)
5. Brown, M., Smith, R.E.: Effective use of directional information in multi-objective evolutionary computation. In: Genetic and Evolutionary Computation Conference, pp. 778–789. Springer (2003)
6. Brown, M., Smith, R.E.: Directed multi-objective optimization. Int. J. Comput. Syst. Signals **6**(1), 3–17 (2005)
7. Coello, C.C.A., Lamont, G.B., Van Veldhuizen, D.A.: Evolutionary Algorithms for Solving Multi-Objective Problems, 2nd edn. Springer, New York (2007). ISBN 978-0-387-33254-3
8. Deb, K.: Multi-Objective Optimization Using Evolutionary Algorithms. Wiley, New York (2001)
9. Deb, K., Jain, H.: An evolutionary many-objective optimization algorithm using reference-point-based nondominated sorting approach, part I: solving problems with box constraints. IEEE Trans. Evol. Comput. **18**(4), 577–601 (2014)
10. Deb, K., Pratap, A., Agarwal, S., Meyarivan, T.: A fast and elitist multiobjective genetic algorithm: NSGA-II. IEEE Trans. Evol. Comput. **6**(2), 182–197 (2002)
11. Dellnitz, M., Schütze, O., Hestermeyer, T.: Covering Pareto sets by multilevel subdivision techniques. J. Optim. Theory Appl. **124**(1), 113–155 (2005)
12. Dilettoso, E., Rizzo, S.A., Salerno, N.: A weakly Pareto compliant qualityindicator. Math. Comput. Appl. **22**(1), 25 (2017)
13. Domínguez, I.S., Aguirre, A.H., Valdez, S.I.: A new EDA by a gradient-driven density. In: International Conference on Parallel Problem Solving from Nature, pp. 352–361. Springer (2014)
14. Fliege, J., Drummond, L.G., Svaiter, B.F.: Newton's method for multiobjective optimization. SIAM J. Optim. **20**(2), 602–626 (2009)
15. Fliege, J., Svaiter, B.F.: Steepest descent methods for multicriteria optimization. Math. Methods Oper. Res. **51**(3), 479–494 (2000)
16. Gebken, B., Peitz, S., Dellnitz, M.: A descent method for equality and inequality constrained multiobjective optimization problems. In: Numerical and Evolutionary Optimization, pp. 29–61. Springer (2017)
17. Gebken, B., Peitz, S., Dellnitz, M.: On the hierarchical structure of Pareto critical sets. In: AIP Conference Proceedings, vol. 2070, p. 020041. AIP Publishing (2019)
18. Griewank, A., Walther, A.: Evaluating Derivatives: Principles and Techniques of Algorithmic Differentiation, vol. 105. Siam, Philadelphia (2008)
19. Harada, K., Sakuma, J., Kobayashi, S.: Local search for multiobjective function optimization: Pareto descent method. In: Proceedings of the 8th annual conference on Genetic and evolutionary computation, pp. 659–666. ACM (2006)
20. Hazen, M., Gupta, M.R.: A multiresolutional estimated gradient architecture for global optimization. In: 2006 IEEE International Conference on Evolutionary Computation, pp. 3013–3020. IEEE (2006)

21. Hillermeier, C.: Nonlinear Multiobjective Optimization: A Generalized Homotopy Approach, vol. 135. Springer, Heidelberg (2001)
22. Jahn, J.: Multiobjective search algorithm with subdivision technique. Comput. Optim. Appl. **35**(2), 161–175 (2006)
23. Jain, H., Deb, K.: An evolutionary many-objective optimization algorithm using reference-point based nondominated sorting approach, part II: handling constraints and extending to an adaptive approach. IEEE Trans. Evolu. Comput. **18**(4), 602–622 (2014)
24. Kleijnen, J.P.: Response surface methodology. In: Handbook of simulation optimization, pp. 81–104. Springer (2015)
25. Lara, A.: Using gradient based information to build hybrid multi-objective evolutionary algorithms. Ph.D. thesis, Computer Science Department, CINVESTAV-IPN (2012)
26. Lara, A., Sanchez, G., Coello, C.A.C., Schutze, O.: HCS: a new local search strategy for memetic multiobjective evolutionary algorithms. IEEE Trans. Evol. Comput. **14**(1), 112–132 (2009)
27. Martín, A., Schütze, O.: Pareto tracer: a predictor-corrector method for multi-objective optimization problems. Eng. Optim. **50**(3), 516–536 (2018)
28. Miettinen, K.: Nonlinear Multiobjective Optimization, vol. 12. Springer, Heidelberg (2012)
29. Nocedal, J., Wright, S.: Numerical Optimization. Springer, Heidelberg (2006)
30. Peitz, S., Dellnitz, M.: A survey of recent trends in multiobjective optimal control-surrogate models, feedback control and objective reduction. Math. Comput. Appl. **23**(2), 30 (2018)
31. Saha, A., Ray, T.: Equality constrained multi-objective optimization. In: 2012 IEEE Congress on Evolutionary Computation, pp. 1–7 (2012)
32. Schäffler, S., Schultz, R., Weinzierl, K.: Stochastic method for the solution of unconstrained vector optimization problems. J. Optim. Theory Appl. **114**(1), 209–222 (2002)
33. Schütze, O., Alvarado, S., Segura, C., Landa, R.: Gradient subspace approximation: a direct search method for memetic computing. Soft Comput. **21**, 6331–6350 (2016)
34. Schütze, O., Esquivel, X., Lara, A., Coello, C.C.A.: Using the averaged Hausdorff distance as a performance measure in evolutionary multiobjective optimization. IEEE Trans. Evol. Comput. **16**(4), 504–522 (2012)
35. Schütze, O., Lara, A., Coello, C.C.: The directed search method for unconstrained multi-objective optimization problems. In: Proceedings of the EVOLVE–A Bridge Between Probability, Set Oriented Numerics, and Evolutionary Computation, pp. 1–4 (2011)
36. Schütze, O., Martín, A., Lara, A., Alvarado, S., Salinas, E., Coello, C.A.C.: The directed search method for multi-objective memetic algorithms. Comput. Optim. Appl. **63**(2), 305–332 (2016)
37. Sun, J.Q., Xiong, F.R., Schütze, O., Hernández, C.: Cell mapping methods - algorithmic approaches and applications. Springer (2018)
38. Uribe, L., Lara, A., Schütze, O.: On the efficient computation and use of multi-objective descent directions within constrained MOEAs. Swarm Evol. Comput. **52**, 100617 (2020)
39. Zhang, Q., Zhou, A., Zhao, S., Suganthan, P.N., Liu, W., Tiwari, S.: Multiobjective optimization test instances for the CEC 2009 special session and competition. Technical report special session on performance assessment of multi-objective optimization algorithms, University of Essex, Colchester, UK and Nanyang technological University, Singapore **264** (2008)

Author Index

O. Junge et al. (Eds.): SON 2020, SSDC 304, pp. 391–392, 2020.
https://doi.org/10.1007/978-3-030-51264-4

The manufacturer's authorised representative in the EU is Springer
Nature Customer Service Centre GmbH, Europaplatz 3, 69115 Heidelberg,
Germany. If you have any concerns regarding our products, please
contact ProductSafety@springernature.com

Printed and bound by CPI Group (UK) Ltd, Croydon, CR0 4YY
30/09/2026
02225557-0012